Much ado about nothing

Theories of space and vacuum from the Middle Ages to the Scientific Revolution

MUCH ADO
ABOUT NOTHING

*Theories of space and vacuum
from the Middle Ages to the Scientific
Revolution*

EDWARD GRANT

*Professor of History and Philosophy of Science and History
Indiana University*

Cambridge University Press

*Cambridge
London   New York   New Rochelle
Melbourne   Sydney*

CAMBRIDGE UNIVERSITY PRESS
Cambridge, New York, Melbourne, Madrid, Cape Town, Singapore, São Paulo

Cambridge University Press
The Edinburgh Building, Cambridge CB2 8RU, UK

Published in the United States of America by Cambridge University Press, New York

www.cambridge.org
Information on this title: www.cambridge.org/9780521229838

First published 1981
This digitally printed version 2008

A catalogue record for this publication is available from the British Library

Library of Congress Cataloguing in Publication data
Grant, Edward, 1926—
Much ado about nothing.
Bibliography: p.
Includes index.
1. Space and time – History.
2. Vacuum – History.
3. Science – History.
4. Science – Philosophy.
I. Title.
QC173.59.S65G72   530.1′1   80—13876

ISBN 978-0-521-22983-8 hardback
ISBN 978-0-521-06192-6 paperback

To
MARSHALL CLAGETT
PIERRE DUHEM
ALEXANDRE KOYRÉ
ANNELIESE MAIER

who taught me the meaning of conceptual history

Contents

Preface   *page xi*

Part I   Intracosmic space

Scope of study   3

1   Aristotle on void space   5

2   Medieval conceptions of the nature and
     properties of void space   9
     1. What is vacuum?   9
     2. Two types of void space: external and internal   14
        a. Internal space   14
        b. External space   17
     3. External void space always occupied by body   19
        a. The distinction between material
           and immaterial dimensions   19
     4. External void space unoccupied by body and the
        application to it of the distinction between
        material and immaterial dimensions   21

3   The possibility of motion in void space   24
     1. The problem of instantaneous motion   24
     2. Resistant media and finite, natural motion in a vacuum   25
     3. The *distantia terminorum* argument as justification
        of finite, natural motion in a void   27
        a. The *distantia terminorum* identified as resistance
           to motion in a void *ex parte medii*   28
        b. On the possible origins of the
           distantia terminorum argument   29
     4. Criticisms of the *distantia terminorum* argument   31
     5. Resistance in the void *ex parte mobilis*   38
        a. Averroes' explication of the relationship of motive
           and resistive forces in one and the same body   40
     6. Violent motion in the void *ex parte mobilis*   41
     7. Natural motion in the void *ex parte mobilis*   44
        a. The natural motion of elemental bodies   45

b. The natural motion of mixed bodies   49

c. How the concept of mixed body destroyed
the homogeneity of the vacuum   55

d. On the equality of velocities of homogeneous,
mixed bodies falling in void space   57

8. Motion in a vacuum in the sixteenth and seventeenth
centuries: Galileo and the medieval tradition   60

4 Nature's abhorrence of a vacuum   67

1. Formulation of the concept   67

2. Universal and particular natures   69

3. The evidence for and against nature's abhorrence
of an interstitial vacuum (*vacuum imbibitum*)   70

a. The problem of condensation and rarefaction   71

b. Nicholas of Autrecourt's defense
of the interstitial vacuum   74

4. The evidence for nature's abhorrence of a
separate vacuum (*vacuum separatum*)   77

a. The burning candle in an enclosed vessel   77

b. Liquids in reeds, siphons, and sealed vessels   80

c. The bellows   82

d. The clepsydra   83

e. The separation of two surfaces in direct contact   86

5. The significance and brief subsequent history
of the experiments and experiences demonstrating
nature's abhorrence of a vacuum   95

Part II Infinite void space beyond the world

Largely a theological problem   103

5 The historical roots of the medieval concept
of an infinite, extracosmic void space   105

1. Aristotle's rejection of extracosmic
void and the reaction in Greek antiquity   105

2. The Condemnation of 1277   108

3. Did a void space exist before the
creation of the world?   110

4. Where is God?   112

6 Late medieval conceptions of extracosmic
("imaginary") void space   116

1. The meanings of the term "imaginary"
in the expression "imaginary space"   117

2. The God-created, independent, separate,
extracosmic void space   121

a. Is extracosmic, created void dimensional and
are measurements possible within it? 123
b. Can God create an actual infinite vacuum? 127
3. The God-filled, dependent, extracosmic void space 135
a. Bradwardine's first corollary 135
b. Bradwardine's second corollary 136
c. Bradwardine's third corollary and the medieval
dictum that "God is an infinite sphere whose
center is everywhere and circumference nowhere" 138
d. Bradwardine's fourth corollary 141
e. Bradwardine's fifth corollary 141
f. Bradwardine's legacy 142
4. Medieval opposition to a God-filled, infinite void 144

7 Extracosmic, infinite void space in sixteenth- and
seventeenth-century scholastic thought 148
1. John Major as a possible link between medieval
and early modern scholastics 148
2. In defense of God's omnipresence in
extracosmic, infinite space 152
a. Francisco Suarez 153
b. Pedro Fonseca and the Coimbra Jesuits 157
c. God's immensity and space as privation or negation 163
d. Bartholomeus Amicus 165
e. Emanuel Maignan: imaginary space
as virtual extension 174
f. Franciscus Bona Spes: imaginary space as
true space 178

8 Infinite space in nonscholastic thought during the
sixteenth and seventeenth centuries 182
1. The impact of the new Greek treatises 182
2. The infinite universe of Lucretius and
the Greek atomists: Giordano Bruno 183
a. The relations of God and space in Bruno's thought 190
3. Finite void space and the influence of John Philoponus 192
a. Bernardino Telesio 192
b. Tommaso Campanella 194
4. Infinite space in the Stoic tradition 199
a. Francesco Patrizi 199
b. Pierre Gassendi 206
c. Gassendi, Otto von Guericke, and the
scholastic tradition on space 213
d. Extended space as God's attribute: Henry More 221

e. Beyond Henry More: Benedict Spinoza and the
   conflation of God, space, and matter   228
f. The direct influence of Henry More:
   Joseph Raphson   230
g. Was space "geometrized" in the seventeenth century?   232
h. "Parts" in an indivisible space   234
i. Space as pure potentiality: the interpretation
   of Isaac Barrow   236
j. The indirect influence of Henry More:
   John Locke and Isaac Newton   238
k. John Locke   238
l. Isaac Newton: fruition of a long tradition   240
m. The tumultuous climax: Newton and the
   Leibniz–Clarke correspondence   247
n. Aftermath   254

Part III   Summary and reflections

9   Summary and reflections   259

Notes   265
Bibliography   419
Index   439

Preface

As the place within which bodies could move and rest and the potential frame of the universe, empty space was a central theme in physical thought from Greek antiquity to the Scientific Revolution. Despite a great awakening interest in the history of science in this century, only a few significant studies have appeared to shape our understanding of earlier spatial conceptions. However much these works may differ from each other, they share one common feature: Medieval and early modern scholastic contributions are excluded from serious consideration, and in most cases mentioned not at all. Given the almost universally held assumption that Aristotelian scholastics had nothing to contribute, their omission caused no comment. Serious discussions on space were thought to have begun in the Renaissance with the likes of Nicholas of Cusa or Bernardino Telesio.

The historical realities, however, are quite otherwise. Scholastics did indeed play a significant role in the development of a concept of infinite space that would be acceptable to the majority of nonscholastics for whom the relationship between God and space was still a vital matter. Demonstration of this claim is a major objective of my study, which is divided in two major parts (the third being a conclusion). The first concerns the existence of intracosmic void, which was always conceived as if it were surrounded by the innermost surface of one or more material bodies. In this part, which by the nature of the conditions was discussed largely in Aristotelian terms, the scholastic contribution is not immediately apparent because virtually all scholastics were agreed with Aristotle that no such separate extended void could exist. Here the interesting medieval arguments are hypothetical and quite problematic as to their subsequent influence. Driven by theological considerations, however, some scholastics devised the concept of a real, infinite, extracosmic void space that they identified with God's own immensity. It was this view, with its accompanying arguments, that helped shape nonscholastic spatial interpretations in the seventeenth century and that constitutes the major theme of the second part of this volume. Although Newton's conception of space was significantly different from the traditional scholastic interpretation, the two are intimately linked by a profound concern for the nature of the relationship between God and space: How was God's omnipresence and operation to be conceived spatially?

As this study should make clear, the history of spatial theory to Newton cannot be properly understood without inclusion of the scholastic tradition, a tradition that has thus far been ignored. Once the scholastic role
became evident and undeniable, it seemed essential to conclude the substantive, historical parts of the volume, which was originally envisioned
as exclusively medieval in scope, with a reexamination of the great nonscholastic thinkers of the sixteenth and seventeenth centuries. The eighth
chapter, by far the longest, represents an attempt to interpret nonscholastic
developments within the context of the entire tradition and complex of
spatial ideas then available, thus including not only the newly received
inheritance from Greek antiquity but also a considerable stock of ideas
and arguments from the hitherto invisible scholastic tradition. The virtual
exclusion of Galileo from the eighth chapter and the absence from it of a
section devoted exclusively to Descartes may occasion surprise, but it is
fully justified. Galileo fails even to discuss the subject, and Descartes's interpretations are akin to Aristotle's and played only a negative role, especially in relation to Henry More. In truth, neither contributed to the
development of the major themes of the second part of this study, namely,
the conception of an infinite separate space distinct from matter and associated in some manner with an omnipresent divine immensity.

The concepts and arguments that were instrumental in the historical
development of theories of space and vacuum fall naturally into conceptual or intellectual history. They form a remarkably cohesive and independent tradition that was virtually immune to social, political, economic,
educational, and religious change. The enormous transformations that
altered European society between the late Middle Ages and the early eighteenth century appear to have left the problem of space unaffected. Interpretations of space were formulated within the context of a purely bookish
tradition. They were the product of a fusion of ancient pagan Greek speculations with Christian theology and its conception of the divine nature.
As regards the doctrine of space, it is no exaggeration to claim that, despite
radically different solutions, Isaac Newton operated within the same intellectual framework in the seventeenth century as did Thomas Bradwardine in the fourteenth.

It is now a pleasurable duty to acknowledge my debts. For their patience
and forbearance, I am especially grateful to my wife (Sydelle) and chil-
dren (Robyn, Marshall, and Jonathan), who early on realized that my
frequently distracted gaze signified nothing more than a temporary in-
vasion of my thoughts by one or more of the cast of characters that fill this
book. Without the timely and generous assistance of the National Science
Foundation (NSF), the American Council of Learned Societies (ACLS),
and the Institute for Advanced Study, Princeton, New Jersey, the research

that produced this volume would not yet have materialized. To the Division of Social Sciences of the NSF, I am indebted for three successive research grants between 1966 and 1974. Within this period, which included a crucial year abroad (1969–1970, spent largely at the British Museum), most of the basic research was completed. A fellowship from the ACLS during 1975–1976 enabled me to begin the actual composition of the volume. In March 1976, I was also privileged to receive a visiting membership from the School of Historical Studies of the Institute for Advanced Study, where I absorbed much of what would eventually become the seventh chapter of this work. It is also with gratitude and pleasure that I now acknowledge the contributions of those splendid departmental secretaries, Kim Rumple, Donna Hazel, and Karen Blaisdell, who labored mightily to convert my thoughts into a proper typescript.

Of the four historians of science to whom I have dedicated this volume and who have profoundly shaped my approach, a brief explanation about Alexandre Koyré is in order. As will be evident in the eighth chapter, my interpretation of the conceptualization of infinite space in the seventeenth century is radically different from his. That difference, however, has in no way diminished my admiration for and heavy dependence on his brilliant contributions. The true measure of my respect for Alexandre Koyré should be gauged by the inclusion of his name on the dedicatory page of this volume.

E. G.

PART I

Intracosmic space

Scope of study

The concept of space that will be of primary concern in this volume is one in which space was regarded as something separate from material body. With a few rare exceptions, that separate space was associated with a vacuum that was assumed to exist either within the confines of a finite, spherical universe, as described by Aristotle, or beyond the bounds of that universe, extending infinitely in every direction.

Although our study focuses on the period between the thirteenth and seventeenth centuries in Western Europe and, except for the last two chapters, is largely confined to the Aristotelian scholastic tradition, the central role of vacuum in the history of spatial theories was firmly established in the fifth century B.C., when Greek atomists designated void as a "nonbeing" that was as real as the hard, impenetrable, and eternal atoms that randomly moved and collided through its infinite extent.[1] The atomist identification of real, albeit empty, space with "nothing" guaranteed that the history of spatial concepts would, from its inception, be rooted in paradox and enigma. Described and defined as nothing by the terms that came to represent it – *kenon* in Greek; *inane, vacuum,* and *nihil* in Latin – the void was from the outset, and almost inevitably, subjected to a double entendre. Was it an unintelligible, total privation incapable of existence – a true "nothing"? Or was it a nothing conceived of as a something, a something with definite properties that could range from a pure dimensionless emptiness to a three-dimensional magnitude, and even be conceived of as God's infinite and omnipresent immensity? Such a nothing was capable of evoking powerful imagery, as it did in the seventeenth century, when Otto von Guericke described it as something that "contains all things" and is "more precious than gold, . . . more joyous than the perception of bountiful light" and "comparable to the heavens."[2]

For all the ambiguities and paradoxes, the history of spatial theories represents, at the very least, an attempt to describe and define the properties and characteristics of a real, though for the most part only hypothetically real, separate nothing. It is thus the nothingness of space conceived of ontologically as a something, however minimal, that defines the scope of our study. Had there been unanimous agreement that a separate empty space was conceptually unintelligible and utterly incapable of ex-

3

istence, or even description, the history of space would have been largely confined to an analysis of worlds such as Aristotle's, in which spatial modes are restricted to differentiations of a plenum filled everywhere with extended material bodies. It was Greek atomism that prevented this turn of events and compelled even its mortal enemies to debate, if only to refute, the merits of void space as a means of explaining physical phenomena. By virtue of its association with atomism, however, the doctrine of separate void space was destined to fare badly. A natural philosophy, in which the random and mechanical motion of an infinitude of uncreated atoms in an infinite empty space gave rise to an infinity of worlds that came into being and passed away just as randomly and mechanically, was committed to a denial of providence and purposeful creation. It is hardly surprising that Christianity viewed atomism with alarm and abhorrence. Not until Lucretius' *De rerum natura* was published in 1473, after having been "rediscovered" earlier in the fifteenth century by Gian Francesco Poggio,[3] did natural philosophers once again pay serious attention to atoms and the void.

During the approximately fifteen centuries of neglect, the atomist conception of space was known largely through Aristotle's hostile critique in the fourth book of the *Physics*. To lay the foundation for what is to follow, it will be necessary to summarize briefly Aristotle's discussion.

I

Aristotle on void space

In arriving at his definition of the place of a body as "the boundary [or inner surface] of the containing body at which it is in contact with the contained body,"[1] Aristotle had occasion to reject three other possibilities, namely that place is (1) shape or form; (2) the matter of a body; and (3) the extension between the bounding surfaces of a containing body.[2] The last of these represented the concept that place is a three-dimensional void space. By defining void as "place bereft of body,"[3] it was easy to infer that a void place is something distinct from the bodies that occupy it. Indeed because most people believed "that everything is somewhere and in place," it would seem that, like Hesiod's chaos, place, or space, would necessarily precede the things that must move and rest in it. "If this is its nature," Aristotle declared, "the potency of place must be a marvellous thing, and take precedence of all other things. For that without which nothing else can exist, while it can exist without the others, must needs be first; for place does not pass out of existence when the things in it are annihilated."[4]

If Hesiod and the atomists could find justification for the existence of such a separate, empty space, Aristotle found it an unintelligible conception from which only absurdities and impossibilities followed. For if place or space were three-dimensional, it would be a body. "But the place [of a body] cannot *be* body," Aristotle insisted, "for if it were there would be two bodies in the same place,"[5] a patent absurdity if one accepts the impossibility of the interpenetration of bodies. As a subsequent illustration of this absurdity, Aristotle compared the immersion of a wooden cube in a material medium, such as water or air, with its "immersion" in a dimensional empty space. In water or air, the cube would, of course, displace an equal volume of water or air, respectively. When the cube was placed in empty space, however, the void space would not be displaced but, on the contrary, would "penetrate" the cube throughout its extent. As Aristotle explained, it would be "just as if the water or air had not been displaced by the wooden cube, but had penetrated right through it."[6] Because the interpenetration of material medium and material cube is impossible, so also is it impossible for the dimensions of an alleged void space to interpenetrate with the dimensions of a material cube. The simultaneous interpenetration and coexistence of separate dimensional entities is thus

5

impossible; or, to express the idea in its most concise form, two bodies cannot occupy the same place simultaneously. Thus was Aristotle convinced that void space could not possibly exist. His explicit identification of three-dimensional void with three-dimensional material body was destined to play a significant role in the controversy over the possible existence of a separate space in the period from the Middle Ages to the seventeenth century. The equation of dimensional void with body would not only be used to reject void space, as Aristotle had intended, but, as we shall see, would eventually be turned against Aristotle and used to justify motion in an extended vacuum.

But if the idea of extended void space was deemed ridiculous because it implied acceptance of the simultaneous occupation of one and the same place by two extended dimensions, or "bodies," it was also to be rejected on grounds of superfluousness.[7] To demonstrate this notion, Aristotle asked that we imagine the impossible and conceive the dimensions, or volume, of that same wooden cube as if they existed independently of all their other inseparable properties, such as hot or cold, heavy or light, and so forth. Under these circumstances, the abstracted volume of the cube would "occupy an equal amount of void, and fill the same place, as the part of place or of void equal to itself." "How then," Aristotle asked, "will the body of the cube differ from the void or place that is equal to it? And if there can be two such things, why cannot there be any number coinciding?"[8] Thus if a void space differs in no way from the abstracted dimensions of the body that occupies it, is it not plausible to assume the superfluity of separate void space? And if one void space is superfluous because indistinguishable from the dimensions of the body that occupies it, then any number of such spaces could coincide simultaneously, which is patently absurd. If the dimensions of a body are one of its fundamental attributes, an attribute that it retains wherever it may be, why assume that it requires yet another equal void dimension in which to be located?

Although Aristotle's most effective arguments against a separate void space were those just described, it was by no means the limit of his efforts. So great was the danger of void space to his world system that he also formulated a variety of arguments[9] to show that motion in a separate void space was either impossible or that the consequences of such motion were absurd and contrary to nature. That motion in a void would be impossible followed for Aristotle from the homogeneous nature of empty space. Because every part of a void is identical to every other part, bodies would necessarily remain at rest. Lack of differentiation would offer no good reason for a body to move in one direction rather than another.[10] Natural and violent motions would also prove impossible. Natural motion would fail because up and down are indistinguishable in void. "For in so far as it is infinite, there will be no up or down or middle, and in so far as it is a void, up differs no whit from down."[11] Bodies in a void would thus be immobilized and incapable of determining the region of the world in which they could fully actualize their potentialities. Violent motion would also be prevented because in a void, the external medium, or propellant, deemed essential for the continued motion of a body away from its natural place would be lacking.[12]

But even if motion is assumed, impossibilities would unavoidably follow. A body set in motion in a void, undifferentiated space would not come to rest, "for why should it stop *here* rather than *here*? So that a thing will either be at rest or must be moved *ad infinitum*, unless something more powerful get in its way."[13] Since a perpetual motion without apparent cause was considered unintelligible, Aristotle's inertial consequence was probably an effective argument.[14]

Two further arguments are noteworthy because they trade on the absence of resistance in void space. Without resistance, void would yield equally in every direction,[15] so that a body would tend to move in any direction, presumably in whatever direction it was pushed by some motive force; or perhaps it would move in all directions simultaneously.[16] The second argument, which will play a significant role in this study, derives instantaneous motion as a consequence of motion in a void. Here Aristotle argued that because "there is no ratio in which the void is exceeded by body, as there is no ratio of o to a number," it follows that "the void can bear no ratio to the full, and therefore neither can movement through the one to movement through the other, but if a thing moves through the thickest medium such and such a distance in such and such a time, it moves through the void with a speed beyond any ratio."[17]

And last among the arguments on motion that will be mentioned here is one in which Aristotle insisted that even if motion in a void were finite, rather than instantaneous, bodies of dfferent weights would fall in a void with equal velocities, when they ought to fall with speeds that are directly proportional to their respective weights. For in a plenum one body moves faster than another by virtue of its greater weight, which enables it to cleave through the resisting medium more easily. Because there is no medium to cleave in a void, no plausible reason or cause can be offered to explain why one body should move with a greater speed than another.[18]

The assumption of void was beset with yet other difficulties and puzzles. If a void dimension is the place of a body, will it not also be the place of a surface? And if it is the place of a surface, will it not also serve as the place of a point? In its latter capacity, however, void cannot be distinguished from the point itself. But "if the place of a point is not different from the point, no more will that of any of the others be different, and place will not be something different from each of them."[19] With perhaps the atomists in mind, Aristotle in this argument appeared to assume continuous divisibility terminating at the point. As continua, both body and void are continuously divisible. Thus if we can divide a body into surfaces, then each surface will occupy a void space as its place. When our continuous division reaches the point as ultimate entity,[20] it would obviously be indistinguishable from the void space it occupies. But if surfaces and bodies are ultimately constituted of points, the void places of those surfaces and bodies will also be indistinguishable from their void spaces. In effect, the void will have no separate existence and will be utterly superfluous. Moreover, because a void place has size but no body,[21] and because it fails to qualify as one of the four basic causes,[22] what can it be? But if it is in some sense a real entity, it must be somewhere and consequently, as Zeno argued, it must also have a place. Therefore place will have a place, and so on, ad infinitum.[23] For Aristotle, then, void dimensional space was incapable of existence whether understood as a separate dimension or as an entity perpetually occupied by body.[24]

In his carefully structured sequence of arguments against the possibility of the existence of void space, Aristotle sought to close all avenues of escape to his opponents by showing that motion in a void is impossible; that even if it could occur, it would be unending or instantaneous; and that if it were finite, all bodies, whatever their weights, would fall with equal velocities.[25] The absurdities described above were derived by Aristotle from his own physics and cosmology, which presupposed a highly structured, finite, spherical world filled everywhere with matter, a world in which the resistance of media, natural places, and natural and violent motions were fundamental concepts. Because none of these vital principles and ideas were functional and viable in void space, Aristotle sought not only to demonstrate the nonexistence of void by every means available but also to reveal how useless was the concept of empty space for the explanation of physical phenomena.

2

Medieval conceptions of the nature and properties of void space

Throughout the Middle Ages, the three-dimensional void space rejected by Aristotle as a candidate for place was regularly described as the "common" or "vulgar" opinion.[1] Not only did Aristotle himself suggest this in the *Physics*,[2] but ironically, the characterization of three-dimensional void space as common or vulgar was invoked in the Middle Ages in order to save Aristotle from the charge that he himself had advocated the rejected opinion. For it was generally acknowledged that in the *Categories*,[3] or *Predicamenta*, as it was known in the Middle Ages, Aristotle had actually assigned the properties of tridimensionality and divisibility to place and body. According to one scholastic interpretation, Aristotle was believed to have declared in the *Predicamenta* that "as are the dimensions of a body so are the dimensions of its place and conversely; and that the parts of a body are joined (*copulantur*) to the parts of its place."[4] No exegesis, however subtle, could explicate the passage in the *Predicamenta* as suggestive of place as a two-dimensional surface, the position upheld by Aristotle in the *Physics* and known to be his true opinion. The dilemma was resolved by appeal to Averroes, who, in his *Commentary on the Metaphysics*, explained that in the *Predicamenta* Aristotle had frequently described common, or vulgar, opinions, whereas he sought to determine the truth in other parts of philosophy.[5] From this it followed that Aristotle's genuine opinion on place was to be found in the *Physics* and not the *Predicamenta*.

1. WHAT IS VACUUM?

In the Middle Ages, it was commonly assumed that if vacuum existed, its definition would be that given by Aristotle in *De caelo*, where it is described as "that in which the presence of body, though not actual, is possible."[6] Elsewhere Aristotle explained that "the void is thought to be a place with nothing in it" because "people take what exists to be body, and

9

hold that while every body is in place, void is place in which there is no body, so that where there is no body, there must be void."[7] But we have already seen that for Aristotle void was nonexistent, or a privation of being. He had argued that if void existed, it would be something – that is, a place without body but capable of receiving body. Convinced, however, for a variety of reasons that it did not exist, he described it as nonexistent, or a privation of being. Despite the denial of existence, Aristotle had, in medieval terms, formulated the accepted *quid nominis*, or definition, of vacuum.[8]

Thus did Aristotle define vacuum in positive terms but conclude that it was a privation, or nonexistent entity. During the Middle Ages his approach was often subsumed under a twofold conception of vacuum, one privative, the other positive. Thomas Bradwardine, for example, explained[9] that vacuum could be conceived in one way privatively as pure privation, or the lack of a pure plenum; in another way, it could be taken positively in two senses: either as a corporeal dimension that was separate from all natural forms, in the manner of a mathematical body or a quantity that was independent of all other natural things;[10] or as an active power and the cause of motion.[11]

Although most scholastics agreed with Aristotle that vacuum is nonexistent, their discussions about its definition and nature led some to accord it a certain peculiar ontological status that conflicted with Aristotle. A major departure derived from the formulation of a question that Aristotle did not even raise. If a vacuum as defined by Aristotle existed, would it be the privation or negation of anything? In the context of Aristotelian physics, in which all existents were either substances or attributes of substances, it seemed natural to raise such a question. One who did was John of Jandun, who argued[12] that if we define vacuum as "a space separated from every natural body [and presently] not filled by any natural body, but capable of being filled," and if we further define privation as "a negation in a subject that is naturally apt for possessing the disposition [in question]," then if such a vacuum existed, it would not be a privation because vacuum, as a separate, empty dimension, is not the negation of any disposition or attribute naturally appropriate to any subject. Jandun allowed, however, that a separate, dimensional vacuum could be a privation if it were conceived as the negation of the state of actually having a body, or the negation of an aptitude for having or receiving a body. In the end Jandun decided for the first option, denying the status of privation to vacuum and insisting with Aristotle that outside of the mind, vacuum is an absolute nonbeing. For if it were an existent thing, its magnitude would have to be sensible and mobile.[13]

Whereas Aristotle seemed to equate privation of being with nonbeing and nonexistence, Jandun and other medieval authors distinguished between them. A vacuum conceived as a privation of a property actually possessed by a body was judged to be something rather than nothing. From this interpretation of privation or negation, a kind of existence was assignable to vacuum. The problem was essentially one of determining whether vacuum is a nothing that could be conceived as a something or a nonexistent nothing. Does a vacuum totally deprived of body[14] denote something or not? Or is it something that can be assigned meaning but has no referent? Although the problem of vacuum as a nothing is a significant one in the history of physical thought, it also falls into the larger philosophical domain that attempts to cope with the likes of "centaurs and unicorns, carnivorous cows, republican monarchs, and wife-burdened bachelors."[15] It seems no exaggeration to hold that "ever since Parmenides laid it down that it is impossible to speak of what is not, broke his own rule in the act of stating it, and deduced himself into a world where all that ever happened was nothing, the impression has persisted that the narrow path between sense and nonsense on this subject is a difficult one to tread and that the less said of it the better."[16]

For better or worse, however, the ontological status of vacuum became a primary medieval concern. By asking whether the existence of a separate, dimensional vacuum would be a privation or negation, and determining that under certain circumstances it could, John of Jandun and those who adopted his approach conferred more upon vacuum than Aristotle had. But other scholastics remained steadfastly with Aristotle by insisting that the mere assignation of a proper definition to vacuum did not entail any degree of reality or existence. Thus Albert of Saxony argued that although "vacuum" is a privative term in the sense that it is defined as a "place that is not filled with body," it surely does not follow that a place without body, or vacuum, actually exists.[17] Indeed Albert seemed to argue that the concept of vacuum is useful in a negative sense, because it is by the denial of its existence that we arrive at negative propositions about place and body. For if we say that "no vacuum exists," we come to know that "no place is without body."[18]

But the nature and concept of vacuum were such that the issue of its existence or nonexistence could not long be neglected or avoided. William of Ockham confronted it directly when he asked whether vacuum is something or nothing. He assumed that the proposition "vacuum exists" was equivalent to the proposition that "between certain bodies between which there was, or can be, a positive medium without the local motion of those bodies, there is [now] no positive medium."[19] Given this definition, is vacuum something or nothing? Ockham distinguished two senses of something and two of nothing as they pertain to the definition. As presented above, the definition does in one sense signify something, for it signifies that there is now no positive medium where previously one existed or could exist. But in another sense it does not signify something because the term "vacuum" cannot be predicated affirmatively of something that actually exists or can exist. For if *something* actually were demonstrated there, it surely would be false to say that it is a vacuum, presumably on the grounds that a vacuum is a negative concept.[20]

In a similar vein, if we say that a vacuum is nothing, two senses of nothing ought to be distinguished. The first sense is the definition itself. For to say that between certain bodies there is no positive medium where formerly there was, or could be, is to say that there is nothing between those bodies; and this sense of nothing is true. But it would be false to assume that "nothing is something, or can be something, which would be a vacuum" and then truly predicate this nothing of the vacuum. Ockham insisted that it is impossible to assign any meaningful sense to a statement such as "vacuum is nothing or nothing is a vacuum."[21]

Thus did Ockham attempt to render intelligible the difficult relationship between vacuum and nothing. The latter was in and of itself unintelligible, and to render the meaning of vacuum in terms of nothing was to make nonsense of the concept of vacuum. Vacuum could, however, be an intelligible notion if conceived in terms of the absence of something, namely the absence, or privation, of a positive material medium.[22]

But if vacuum could be described in terms of privation, or the absence of a positive medium, what was it the privation of? In coping with this problem, echoes of the problem of nothing are to be heard. Should the privation called vacuum belong to the interval between the surface (or surfaces) of the body (or bodies) surrounding a vacuum, or should it be assigned to the surrounding surface itself? Upon imagining that God has annihilated everything within the inferior world – that is, the world contained within the concave surface of the lunar sphere – John Buridan decided that it would be improper to describe that interval as a vacuum because, as he put it, "below the heaven there would be nothing." It is rather the surface that formerly contained the things that had been destroyed that is truly void. Although he did not employ the term "privation," Buridan had it clearly in mind when he assigned vacuum to the containing surface of the heaven, or celestial sphere, because "the concave surface of the heaven is now filled with body or bodies, and then it would be a place not filled with body."[23] Empty though it now is, the containing and surrounding surface formerly contained bodies and is capable of doing so again. Hence the void is a privation of the condition of fullness for that surface, and the vacuum "belongs" to the surface and not to the empty interval within that surface, which Buridan described as nothing.[24] Using

the terms "privation" and "negation," Gabriel Vasquez (1551–1604) offered much the same argument in the sixteenth century.[25] The void interval between the surface, which Vasquez said is called "imaginary space"
(*spatium imaginarium*),[26] cannot be considered a privation[27] because the
negation, or nothingness, that lies between the surface does not have a
subject and must therefore be characterized as a "pure negation" (*pura
negatio*), or absolutely nothing.[28] The negation of a body, which is the
nothingness or void under discussion, belongs rather to the surrounding
surface and is called a privation with respect to that surface. "For just as
the surface, which contains a body within itself, is said to be filled, so also
when it has no body within itself, it is said to be deprived, that is, not
filled, but yet capable of being filled; and it is this vacuum which we call
privation."[29] As we shall see later (Chapter 7), Vasquez's distinction between negation (or privation) and pure negation, which had already
been made by John of Jandun in the Middle Ages (see note 28), would
play a significant role in early modern scholastic analyses of imaginary
space.

The paradoxical character of void space is well illustrated by the third
of five types of void distinguished by Roger Bacon, who allowed that
although a void space could exist beyond the world, it would be a void in
which there is not only no body whatever but one that is naturally incapable of receiving bodies.[30] Aristotle had argued that body, place, void,
and time could not exist beyond our finite, spherical world.[31] Because we
have seen that for Aristotle the accepted definition of void is an extended
dimension devoid of body but capable of receiving it, it follows that Aristotle had denied the existence of a dimensional space beyond the world.
Indeed, apart from certain briefly mentioned unchangeable spiritual entities that are not in any place,[32] Aristotle had concluded that nothing of
a physical nature exists beyond our world. In an extraordinary interpretation of this passage, Bacon chose to identify as a vacuum what Aristotle
had described as total nonexistence. Thus did Bacon conceive extracosmic
vacuum as if it were a something that simply lacked the property to receive bodies! It seems plausible to conclude that Bacon's vacuum, unlike
Aristotle's conception of it, lacked dimension. In this incredible manner,
Aristotle's insistence that nothing existed beyond the world, not even vacuum, was interpreted by Bacon to mean that something did indeed exist
to which we can give the name "vacuum," although this vacuum is incapable of containing anything.[33] It is as if Bacon believed that the term
"nothing" in the phrase "nothing exists" must have a referent to which
we may assign the name "vacuum," even though no positive properties can
be assigned to this vacuum, of which it can only be said that it is incapable of receiving or containing bodies. Where Bacon was willing to con-

fer existence only on extracosmic vacuum, others would eventually conceive it as an incredible and magnificent something, especially when God's immensity became associated with it. Some five centuries after Bacon, Otto von Guericke would make of Nothing the most magnificent entity in the universe.[34]

The dilemma of determining whether vacuum was to be nothing or something, or perhaps a combination of the two, is plainly evident throughout the period that is studied here.[35] We shall have occasion later to consider other ideas about the ontological status of vacuum, especially in connection with extracosmic space and its alleged role as God's infinite immensity.

2. TWO TYPES OF VOID SPACE: EXTERNAL AND INTERNAL

Whatever the ontological status of vacuum, medieval scholastics learned from Averroes that the separate, external, empty space rejected by Aristotle could be distinguished in two ways: either as (1) something distinct from, but always occupied by, body or (2) a separate dimensional entity distinct from body but also capable of independent, albeit temporary, existence when unoccupied by body.[36] From the thirteenth to the seventeenth centuries, in virtually all discussions of external void space, Aristotle's hostile arguments played a fundamental role, serving either to encourage conformity or to stimulate significant departures and even radical innovations. Most frequently cited was the argument in which Aristotle assumed that void was like a body, from which it followed, on the assumption of the impenetrability of bodies, that void spaces could not receive material bodies. Aristotle's demonstration of the superfluous nature of separate void space was also employed, though to a lesser extent.

But empty space could also be conceived as an integral part, or property, of material body, as in the concept of internal space or in the identification of prime matter with void.[37] Because it avoided the Aristotelian criticisms used against external space and the fact that it was little discussed, it will be convenient to consider internal space first.

a. INTERNAL SPACE

The equation of space with material extension yielded internal space.[38] In the conflation of space with material extension, the problem of interpenetration was completely avoided by reducing to one the number of extensions involved. This position could have been derived as a consequence of Aristotle's argument described above, namely his insistence on the superfluousness of external space, because all the dimensions of such a space are already found in the extension of any material body. From this interpretation, one could easily deny the existence of external space on the grounds that all the space a body needed was already within it, in the form of its own extension or dimension. It is also possible that John Philoponus played a significant role in the development of the concept of an internal space. In his *De aeternitate mundi contra Proclum* of 529 (a work unknown in the Latin Middle Ages but translated into Arabic and then from Greek into Latin in the sixteenth century, when it was widely read) Philoponus argued that the substance of a corporeal entity is its three-dimensional extension.[39] Rejecting the traditional interpretation of Aristotle's concept of prime matter as something without definite properties, Philoponus assigned three-dimensional extension to it.[40] Although Philoponus seems not to have conceived this three-dimensionality as the internal space of bodies – [41] as we shall see below, he believed in the existence of a separate void space – those who rejected any kind of external space could easily have adapted his ideas on extended substance to the concept of internal space.

Among those who subsequently adopted the idea of internal space, we may mention John Buridan in the fourteenth century, Franciscus Toletus (1532–1596) and Francisco Suarez (1548–1617) in the sixteenth, and René Descartes (1596–1650) in the seventeenth. As a typical expression of internal space, the Jesuit commentator, Toletus, described it as the extended quantity of a body's matter. The existence of internal space was undeniable because body and space "imply each other as a mutual consequence. For if there is a body, there is a space; and if there is a true space, there is a body in it."[42] A compelling illustration of the identification of internal space and material extension was offered by Buridan and Toletus. Thus Buridan, who believed that space is nothing but the dimension of a body, asked that we imagine a man located beyond the last sphere or heaven. Should that man now raise his arm there, Buridan explained that "before you raise your arm outside this [last] sphere nothing would be there; but after your arm has been raised, a space would be there, namely the dimension of your arm."[43] Offering much the same example of internal space, Toletus observed that it matters little whether this internal space is really (*realiter*) and truly distinct from quantity, or only formally (*formaliter*) so. Despite his declared indifference, however, Toletus concluded that it is not a separate space but only a proper accident inhering in a body[44] and therefore presumably distinguishable only formally, or by reason alone.

Internal space, or the dimension of a body, was easily linked with mathematical magnitude. Duns Scotus observed that the volume of a body, or its *quantum*, as it was called in Latin, is naturally prior to its qualities and, although it does not exist apart from the body, is the primary concern of mathematics when it is considered per se. That Scotus' *quantum* is akin to, if not identical with, internal space may be seen from his further claim that it is the *quantum* that causes separation of the sides of a body.[45]

The concept of internal space, whether medieval or early modern, is the physical counterpart of the purely geometric space of Euclid's *Elements*. Just as the internal space or dimension of a material body is inseparable from that body so that, whatever its location, a body has its own space, so also does a geometric figure in Euclid's *Elements* possess its own "internal" space, namely the space of its own configuration, which accompanies it wherever it might be located or moved.[46] There is nothing in Euclid's geometry to suggest that he assumed an independent, infinite, three-dimensional, homogeneous space in which the figures of his geometry were located. In a purely geometric sense, such a space would have been superfluous because every geometric figure has its own internal space. To assume that it also lies in a separate and independent three-dimensional space would have served no purpose whatever. Moreover, if the space of the geometric figure and the independent space it is alleged to occupy are conceived as indistinguishable, an infinity of spaces could be postulated in one and the same place. Indeed a separate dimensional space might also have confronted Euclid with the Aristotelian problem of penetrability: Could the extension of a geometric figure penetrate the dimensionality of the separate space? Thus if Euclid had assumed the existence of a separate, three-dimensional, homogeneous, infinite void space in which his geometric figures are assumed to be located, he would have confronted the mathematical analogy to the problems formulated by Aristotle with respect to the relationship between material bodies and separate dimensional space. Perhaps the dilemmas posed by Aristotle were known to Euclid, who, as a consequence, found it prudent to refrain from the postulation of a separate space in his geometry, a space he in no way needed.[47] It is a mistake, then, to insist that "Euclidean geometry appears to require a three dimensional infinite space, whose subsistence . . . is wholly independent of any relation it may have to bodies."[48] Euclidean geometric space was the space of geometric figures of any size whatever and when applied to material bodies was conceived as an internal space.[49]

Perhaps it was the intrinsically mathematical nature of the concept of internal space that appealed to that great geometer René Descartes, whose name is most conspicuously associated with it. More explicitly than his predecessors, Descartes identified internal place with space[50] and assumed that "the same extension in length, breadth, and depth, which constitutes space, constitutes body."[51] It would appear, then, that like Buridan and Toletus before him, Descartes avoided the Aristotelian problem of interpenetration of separate extensions by a total identification of matter with its internal place or space, a move that implied rejection of a separate, three-dimensional space divorced from materiality.

And yet Descartes occasionally spoke as if a separate, empty, extended space might exist, as when he explained that "in body we consider extension as particular and conceive it to change just as body changes; in space, on the contrary, we attribute to extension a generic unity, so that after a body which fills a space has been changed, we do not suppose that we have also changed the extension of that space, because it appears to us that the same extension remains so long as it is of the same magnitude and figure, and preserves the same position in relation to certain other bodies, whereby we determine this space";[52] or when, upon imagining all attributes stripped from body, we find "that there is nothing remaining in the idea of body excepting that it is extended in length, breadth, and depth; and this is comprised in our idea of space, not only of that which is full of body, but also of that which is called a vacuum."[53] Thus it would appear that Descartes might have allowed for the possible existence of a three-dimensional vacuum, which could then be equated with body, a move that was indeed made in the Middle Ages, as will be seen below. Such an interpretation must not, however, be attributed to Descartes, who insisted that every extension is associated with substance, that is, with material body, "because it is absolutely inconceivable that nothing should possess extension."[54] Thus any alleged void space assumed to have extension must also be assumed to have substance, a conception that enabled Decartes, and all who adopted the doctrine of internal space, to avoid the Aristotelian dilemma wherein two separate extensions might occupy one and the same place.

b. EXTERNAL SPACE

In the Aristotelian tradition of the thirteenth to seventeenth centuries, the existence of a separate, extended void space, whether always filled or occasionally devoid of body, was usually rejected on the basis of the familiar arguments of impenetrability and superfluousness. A typical defense of Aristotle's position was offered by John of Jandun (d. 1328), who argued that if place were a space or corporeal dimension, it would follow that two bodies, or dimensions, could coincide simultaneously in the same place. And if two bodies could coexist in the same place, so could an infinite number. Indeed the whole heaven, or any quantity equal to it, could be simultaneously located in the place occupied by a single millet seed. To illustrate, Jandun imagined the heavens divided into as many parts as would result if each part equaled a millet seed. Because any two such parts could occupy the same place, and therefore any number whatever, the absurdity would follow that all the particles into which the heaven was divided could occupy the same place simultaneously. The place of a single millet seed could thus contain the heaven, or for that matter, the whole world.[55] Although scholastics would readily have conceded that God could, if He wished, create two or more bodies in the same place simultaneously,[56] they were agreed that it was naturally impossible except perhaps in an equivocal sense, which plays no role in what will be described here.[57]

Invoking the superfluous nature of a separate space, Albert of Saxony (ca. 1316–1390) argued that if a body required a separate space as its place, this would be so only because of the body's dimensions rather than because of its matter, form, or qualities, all of which were assumed to inhere in those dimensions. But if the dimensions of a natural body are located in a separate, three-dimensional space, that dimensional space would itself require a separate dimensional place, and so on, ad infinitum.[58] Extending the argument to the cosmos itself, John Buridan (ca. 1300–ca. 1358) insisted that the world needed no separate space or place because it was itself a magnitude with a dimension, as were all its parts. Thus God did not require a preexisting dimensional space in which to create the world. For if such a space did exist, God could by His absolute power destroy it and subsequently re-create it. But in re-creating it, God would surely not require yet another space in which to place it. To argue that God needed a space in which to create a space would be to limit His infinite power to create a space without something there to receive it. In the same manner, then, in which God does not require a preexisting space in which to create a possible space for the world, so also does He not require a separate dimensional space or magnitude in which to create the world itself.[59] Because what applies to the world applies to any part of it, Buridan concluded that a separate space is superfluous.[60]

Acceptance of the idea of a three-dimensional, separate void space was thus confronted with formidable obstacles. This space was conceived either as a corporeal dimension and thus incapable of receiving material bodies because of the obvious violation of the self-evident principle that two bodies cannot occupy the same place simultaneously; or it was simply superfluous. Unless plausible and acceptable solutions to these problems could be formulated, the idea of a separate void space appeared doomed. The dilemmas posed by Aristotle somehow had to be answered. Would a three-dimensional vacuum be truly incapable of receiving a three-dimensional material body because of the impossibility of the interpenetration of bodily dimensions? Were other options and interpretations available that might plausibly, and perhaps even compellingly, avoid interpenetration? The various responses to the problems posed by Aristotle's criticisms against a separate space represent a significant aspect of the history of spatial doctrines up to approximately 1700. We have already seen how the concept of internal space avoided these dilemmas, and now we must describe some of the solutions that were proposed for external space.

3. EXTERNAL VOID SPACE ALWAYS OCCUPIED BY BODY

Because the external void space that was conceived as something always occupied by body was held by relatively few in the period covered by this study, we shall consider it first. With Averroes as their source of information, scholastics in the Middle Ages were well aware that John Philoponus was the major proponent of this opinion.[61] But they were aware of little else because Averroes did not include any of Philoponus' arguments, which may explain the little attention that was paid to them in the Middle Ages. Perhaps the most interesting, though brief, reaction came in the thirteenth century from Pseudo-Siger of Brabant, who improperly assumed that Philoponus conceived void space as nothing in itself. From this assumption, Pseudo-Siger argued that Philoponus' position avoided the Aristotelian dilemma of interpenetration of extensions. For if the void space in which bodies exist and move is itself nothing at all, then the bodies in those empty spaces will not interpenetrate with anything. To this problem Pseudo-Siger responded by arguing that "if such dimensions are assumed to be a place, *and* since such dimensions are nothing, it follows that place is nothing, which is impossible."[62]

Although Philoponus' conception of a void space as the container of all bodies, and indeed of the whole material world, gained little support during the Middle Ages, it fared well in the sixteenth century, when the works of Philoponus became available in the original Greek as well as in Latin translations. The Italian natural philosophers – Pico della Mirandola,[63] Telesio, Patrizi, Bruno, and Campanella – would assume the existence of a separate void filled with some kind of matter.[64]

a. THE DISTINCTION BETWEEN MATERIAL AND IMMATERIAL DIMENSIONS

Perhaps no less significant in winning support for the idea of a separately existing space were Philoponus' arguments against Aristotle's charge of the impossibility of interpenetration between body and extended space.[65] In rejecting Aristotle's concept of place as the two-dimensional, concave surface of the containing body in contact with the contained body, Philoponus had declared in favor of the interval between the containing surfaces as the true place of the bodies that came to occupy it; that is, he assumed that the place of all bodies is a three-dimensional void space. To meet Aristotle's criticism that such a dimensional space would, by virtue of its tridimensionality, be a body and therefore unable to receive any other body, Philoponus simply denied that three-dimensional void space is a body. In effect, he distinguished between material body and immaterial three-dimensional extension; the former is a substance, the latter is not. Although two or more material bodies cannot occupy one and the same place, a material body can occupy an equal empty space with which it coincides.

Indeed Philoponus insisted that the void space that a body occupies serves as its volumetric measure. That such a volumetric measure is independent of the things that occupy it, and is therefore a separate empty space, followed for Philoponus from an analysis of the successive occupations of a pitcher by different bodies. For if a pitcher contains air, the dimensions between the pitcher's inner surfaces must not only equal the volume of air but serve as its exact measure. That the dimensions lying between the pitcher's inner surfaces are independent of the air is evident when the air is displaced by water. Now it is the three dimensions of the water that are exactly measured by the volumetric dimensions within the pitcher. Philoponus thus assumed that the fixed volume within the pitcher measures first the air and then the water that displaces it. Because the pitcher's inner volumetric dimensions constitute the measure of successive and different material and extended bodies, the pitcher's dimensions cannot themselves be corporeal, for then two corporeal entities, namely the dimensions of the pitcher and the dimensions of the occupying body, would coincide and occupy the same exact place, which is impossible. Philoponus therefore concluded that the place of successive occupants of the interior of a pitcher, for example, air and water, is a three-dimensional, incorporeal void, which is but part of an absolute, three-dimensional void space that not only contains the entire cosmos but is coterminous with it. This finite void space is thus the place and measure of all bodies in the universe. It is obvious, then, that place cannot be a two-dimensional surface, as Aristotle contended, because a two-dimensional entity cannot function as the measure of three-dimensional bodies.

In Philoponus' cosmos, bodies move in an absolutely immobile, three-dimensional void space. When a body moves, it leaves behind successive parts of that void equal to itself and occupies other parts equal to itself. Although bodies occupy and then depart from successive parts of an absolute void space, the latter remains immobile. By virtue of its absolute immobility, then, no part of void space can be transported elsewhere to occupy another part of void space. But even if one part of cosmic void were admitted into another part, this coincidence of equal voids would no more cause an increase or decrease of the original void than would the superposition of any number of equal lines or any number of equal mathematical surfaces. Thus did Philoponus reply to another of Aristotle's major criticisms against extended void space, namely the claim that an infinity of places would coincide if the space between the inner surfaces of a vessel were conceived as the place of the bodies that occupy it.[66] For Philoponus, however, it mattered not at all whether one conceives absolute void space as a single, three-dimensional volume or as a series of two or more superposed equal volumes. Because the latter conception must ultimately reduce to the former, the two modes are, in the final analysis, identical. An infinity of superposed equal void extensions differs in no way from any one of them taken singly. Thus Philoponus not only resolved the dilemma of interpenetration involving corporeal and incorporeal extensions but also explained how two or more incorporeal, dimensional extensions could coincide, thereby countering another of Aristotle's criticisms.

4. EXTERNAL VOID SPACE UNOCCUPIED BY BODY AND THE APPLICATION TO IT OF THE DISTINCTION BETWEEN MATERIAL AND IMMATERIAL DIMENSIONS

The ideas formulated by Philoponus were of the utmost importance. Not only did he distinguish between material and immaterial dimensions,[67] and thus destroy the basis of one of Aristotle's most powerful arguments against the existence of vacuum, but his conception of vacuum as a three-dimensional extension always filled with body and never existent per se was accepted in some form by the Italian natural philosophers of the sixteenth century, some of whom filled their separate, empty space with light (Patrizi) or ether (Bruno). The second of the voids described by Averroes – that is, the extended void that could receive body but could also exist independently – would emerge as a significant rival to Aristotle's plenum only in the seventeenth century.

But even before the commentaries of Averroes became available, the idea that place could be three-dimensional rather than two-dimensional, as Aristotle insisted, had already become part of the Latin tradition. By reading either or both of two Latin translations of Nemesius' Greek treatise *De natura hominis* (*On the Nature of Man*),[68] scholastics would have learned that although every body has three dimensions, not everything that has three dimensions is a body, as, for example, place and quality.[69] And despite considerable ambiguity, they could have derived a similar sense of place in that most famous of all medieval theological treatises, the *Sentences* of Peter Lombard, where, in a direct quotation from Saint Augustine, we are told that "a place in a space is what is occupied by the length, width, and depth of a body."[70] Thus did medieval scholastics possess definitions of place as a three-dimensional, incorporeal entity capable of receiving bodies. But neither of the authors mentioned bothered to identify this three-dimensional place as a vacuum and neither indicated whether the place they had defined was capable of existence without the presence of body.

The opinion that a three-dimensional vacuum could exist independently of body, though capable of receiving body, thus came to be identified primarily with the Greek atomists, whose views were transmitted to the European Middle Ages by Aristotle and his most famous commentator, Averroes. Few, however, were prepared to adopt the atomist conception. One who did was the Spanish Jew, Hasdai Crescas (ca. 1340–1412), who, in a Hebrew treatise titled *Or Adonai* (*The Light of the Lord*), believed with Philoponus (whose work he does not appear to have known directly) that our cosmos is a material plenum located in a three-dimensional vacuum. But unlike Philoponus, he assumed that the three-dimensional vacuum extended infinitely beyond our world in every direction.[71] Cognizant of Aristotle's argument about the impenetrability of body and dimensional void, Crescas formulated a clear and significant response, one that bears a striking resemblance to that of Philoponus. Crescas insisted that only *material* dimensions are mutually impenetrable but that an *immaterial* dimension, such as a vacuum, could receive and accommodate a material dimension.[72] A three-dimensional vacuum could therefore receive a material, extended body and function as its place. The bogeyman of impenetrability was thus rejected, if not destroyed. To the further Aristotelian argument that if a dimensional vacuum were a place, then, insofar as it was also like a body, it would require a three-dimensional void place, and so on, ad infinitum, Crescas argued that only material dimensions require places. Immaterial dimensions are not in anything else and have no need of separate places.[73]

Thus did Crescas, possibly following in the path of Philoponus, deny that three-dimensional, empty space was a body. The problem of the impenetrability of bodies was not applicable to the relationship between a dimensional vacuum and the material bodies it could receive. Crescas, and before him Philoponus, had thus anticipated Pierre Gassendi, John Locke, and others in the seventeenth century who made an infinite, three-dimensional void space the basis of a new cosmology. Like his predecessors, Gassendi insisted that incorporeal dimensions, such as volume or space, could exist independently of corporeal dimensions, the only kind acknowledged by Aristotle. As representative of the ancient tradition opposed to Aristotle, Gassendi quoted Nemesius, who insisted that "Every body is endowed with three dimensions. But not everything endowed with three dimensions is a body. For of this sort are Place and Quality, which are incorporeal entities."[74] For Gassendi, not only are incorporeal, extended spaces infinite and immobile but, by virtue of their incorporeality, they "have no resistance, or can be penetrated by bodies, or as it is even commonly said, can coexist with them; so that wherever there is a body either permanently or transiently, it accordingly occupies an equal part of space."[75]

Even more explicitly than Gassendi did John Locke distinguish between corporeal and incorporeal extension. Reacting against the Cartesian identification of space and body, Locke, relying on the concept of clear and distinct ideas, insisted that the idea of space was clearly distinct from that of solidity. Indeed the very controversy about the existence of vacuum was evidence of this distinction. For "it is not necessary to prove the real existence of a *Vacuum*, but the *Idea* of it; which 'tis plain Men have, when they enquire and dispute, whether there be a *Vacuum* or no? For if they had not the *Idea* of Space without Body, they could not make a question about its existence: And if their *Idea* of Body did not include in it something more than the bare *Idea* of Space, they could have no doubt about the plenitude of the World."[76] Assuming the clear difference between void space and body, – after all, no one would dare deny that God could annihilate a body at rest and thereby leave behind a total vacuum[77] – Locke lamented the exclusive application of the term "extension" to body. Those who conceive body as "pure Extension without Solidity" must then designate a vacuum as an extension "without Extension. For *Vacuum*, whether we affirm or deny its existence, signifies Space without Body."[78] Such absurdities might have been avoided, Locke observes, if only separate terms had been employed for matter and space, for example, if "the Name *Extension* were applied only to Matter, or the distance of the Extremities of particular Bodies, and the Term *Expansion* to Space in general, with or without solid Matter possessing it, so as to say *Space* is *expanded*, and *Body extended*."[79]

Despite arguments of the kind just described, Crescas, operating within a Hebrew tradition, was one of the very few, if not the only one, in the European Middle Ages to reject Aristotle's identification of void with body, an identification that, left unchallenged, allowed Aristotle's supporters to argue that a body could not occupy a void space because the former would be unable to penetrate the latter. The doctrine of the impossibility of the interpenetration of bodies was thus ready at hand to render acceptance of a separate, three-dimensional void space untenable, a situation that obtained until Crescas's approach prevailed in the sixteenth and seventeenth centuries. But if the doctrines of impenetrability and superfluity were indicative of the absurdity of the very concept of a separate void space and served to operate against its acceptance, the medieval reaction to Aristotle's arguments against such a space based on the impossibility of motion was destined to elicit a more favorable response, one that would make extended, empty space seem less absurd, even if still unacceptable as a natural reality within the cosmos.

3

The possibility of motion in void space

1. THE PROBLEM OF INSTANTANEOUS MOTION

Among Aristotle's arguments denying the possibility of motion in a vacuum (see Chapter 1), none was viewed as more fundamental than the deduction that such motions would be instantaneous. As Aristotle expressed it, the speeds of bodies moving in a void would be "beyond any ratio"; or, to put it another way, a body would occupy the termini of its motion, and all intervening points, simultaneously.[1] Before describing the subsequent history of this powerful argument, it will be well to take note of a paradoxical feature that was implicit in many discussions of it. Did the assumption of instantaneous motion in a vacuum categorize one as a proponent of motion in a vacuum, even though that motion is instantaneous or of infinite velocity? Or rather, did the assumption of instantaneous motion imply that its proponent actually denied motion in a vacuum because the very concept of instantaneous motion is absurd and impossible? One can scarcely doubt that Aristotle was of the latter opinion. For him, the consequence that motion in a vacuum would be instantaneous was equivalent to a denial of motion. Thus instantaneous motion in a vacuum is no motion at all. But already in the thirteenth century, Roger Bacon distinguished between those motions in a void that were instantaneous (and presumably nontemporal) and those that were successive (and therefore finite and temporal).[2] Not long after, Aegidius Romanus (ca. 1245–1316) conferred on instantaneous motion in a vacuum the same status as that of light moving through a medium. Both were conceived as instantaneous and successive (see Section 4). Instantaneous motion was for Aegidius as real as finite motion. Although few went as far as Aegidius, many would treat it, perhaps inadvertently, as if it were a real motion to be contrasted to finite motion. In the fourteenth century, John Buridan made this problem the subject of a special question when he inquired whether a heavy body could move in a vacuum, if the latter existed. Some would argue that because Aristotle had proved that such

24

motions would be instantaneous, it followed that a heavy body could indeed be moved in a vacuum.[3]

By the late sixteenth and early seventeenth centuries, some Aristotelian scholastic commentators would include anyone who opted for instantaneous motion in a vacuum among those who believed in the possibility of motion in a vacuum. Assuming the hypothetical existence of a vacuum, discussants of the question were sometimes divided into those who thought such motion impossible and those who thought it possible. The second group was then further subdivided into those who thought such motion would be instantaneous and those who believed it would be finite and successive. The confusions that could arise are illustrated in a seventeenth-century commentary on Aristotle's *Physics* by Bartholomeus Mastrius (1602–1673) and Bonaventura Bellutus (1600–1676). Adopting the distinctions just described, they observed that those who accepted the possibility of instantaneous motion deduced from that the impossibility of motion in a vacuum. In describing the opinions of Walter Burley and John of Jandun, Mastrius and Bellutus compounded the inherent difficulties even further. Convinced that Burley and Jandun were among those who denied the possibility of motion in a vacuum, our authors described the derivation of this position first by attributing to Burley and Jandun the belief that instantaneous motion in a void is possible, from which both are alleged to have deduced the impossibility of such motion![4] The difficulty described here undoubtedly arose from the expression "instantaneous motion" or its equivalent. Much as the term "vacuum" was conceived as a nothing that was a something (see Chapter 2, Section 1), the concept of instantaneous motion was accorded ontological status by virtue of its opposition to finite motion. It was thus easy to interpret it as some kind of motion while at the same time denying its very possibility. Fortunately, we need not detain ourselves further with this peculiar classificatory problem because it failed to obscure the real issue, which divided those who thought motion in a vacuum would be instantaneous – whether that motion was conceived as an infinite velocity or no motion at all – and those who thought it would be temporal and successive, as were all motions actually observed in nature.

2. RESISTANT MEDIA AND FINITE, NATURAL MOTION IN A VACUUM

Opposition to Aristotle's derivation of instantaneous motion in a vacuum is already evident in late antiquity, when in the sixth century A.D., John Philoponus (or John the Grammarian), in his *Commentary on Aristotle's Physics*, insisted on the possibility of finite motion. Philoponus regarded the material medium through which a body moves as a hindrance to its motion and quite unessential to it. When a body moves downward with a finite motion, the sole cause of that motion is its natural downward tendency – a tendency that was conceived as proportional to the total, not specific, weight of the body. Because the medium functions only as a retarding factor, Philoponus concluded that true motion could occur only in a void, where the time in which a given distance was traversed represented the original, or true, time of motion for that particular body. When traversing the same distance through a material medium, the same body required an additional time proportional to the density of the medium.[5]

Although Philoponus' *Commentary on Aristotle's Physics* remained unknown in the Latin West until the sixteenth century, it may have exerted an influence through Avempace (Ibn Bajja [d. 1138/1139]), a Spanish Arab who expressed a similar opinion[6] that was subsequently reported by Averroes in the latter's *Commentary on the Physics of Aristotle*, book 4, text 71.[7] We learn from Averroes that Avempace denied that the motion of a stone, for example, takes time only because it is moved in a medium.[8] For "if we make this assumption, it would imply that motion only takes time because of something resisting it – for the medium seems to impede the thing moved. And, if this were the case, then the heavenly bodies would be moved instantaneously as they have no medium resisting them."[9] But this is patently false because we observe in the heavenly motion "the greatest slowness, as in the case of the motion of the fixed stars, and also the greatest speed, as in the case of the diurnal rotation."[10] Because the celestial bodies exhibit a variety of finite speeds as they course through the material medium of the celestial ether, Avempace inferred that the temporal motion of a body is not caused solely by the material medium through which it moves. According to Averroes, Avempace would subtract the retardation caused by the resistance of the medium to arrive at the body's true, natural motion. Averroes, who disagreed with Avempace's analysis, equated motion through a resistanceless, material medium with motion in a vacuum and attributed to Avempace the opinion that "as every motion involves time, that which is moved in a void is also necessarily moved in time and with a divisible motion."[11]

In reading book 4, text 71 of Averroes' *Commentary on the Physics of Aristotle*, Latin scholastics were confronted with a fundamental choice of interpretations. They could side with Averroes and insist that a resistant, material medium was required for the successive motion of all bodies in the sublunar realm,[12] or they could adopt the opinion of Avempace that a resistant medium was in no way essential for temporal motion, so that finite rather than instantaneous speeds could occur in a hypothetical vacuum. In arriving at a decision, however, scholastics would have found nothing in the passage from Avempace, or in Averroes' discussion, that explained how finite motion in a void could occur in the absence of a material medium that both Aristotle and Averroes believed essential to prevent instantaneous motion. Avempace had only provided an obscure explanation of the differences in the speeds of celestial motions, which he attributed "to a difference in perfection (*nobilitas*) between the mover and the thing moved," so that "when the mover is of greater perfection, that which is moved by it will be swifter; and when the mover is of less perfection, it will be nearer [in perfection] to the thing moved, and the motion will be slower."[13] Although the speed of a celestial body involved a relationship between mover and moved, scholastic authors could hardly determine from this passage whether for Avempace motion in a void was determined by the weight of the body, its dimensions, or perhaps by some indwelling motive force. As we shall see, a number of causal explanations were offered, one of the earliest of which, frequently identified with Thomas Aquinas, could not possibly represent Avempace's true intent.

3. THE *DISTANTIA TERMINORUM* ARGUMENT AS JUSTIFICATION OF FINITE, NATURAL MOTION IN A VOID

The most significant and widely cited argument in defense of finite motion in a void was designated by the scholastics as *distantia terminorum* or *incompossibilitas terminorum*.[14] It was an argument that sought to explain how finite motion could occur in a vacuum, or resistanceless medium. To render such motion intelligible, vacuum was conceived, implicitly or explicitly, analogically with a material plenum, or body. Just as motion through a plenum, or body, with distinct and separate termini is successive and temporal, so also would it be successive and temporal in a vacuum if the latter is assumed to possess dimension and extension. The rationale for finite motion in a void was thus rooted in the idea that a dimension is divisible into parts that must, of necessity, be traversed in sequence. Because a dimensional vacuum is like a body, Roger Bacon, one of the first to have left an argument on this problem, stated that motion through a vacuum would be finite because "every body is divisible and has a distance [or separation] between its boundaries; therefore it has a prior part and another posterior part, and will traverse one part and then another. For this reason, before and after [or prior and posterior] occur within the parts of the magnitude and there will be a before and after in the motion and, therefore, in the time."[15] Because motion could not occur unless a body traversed a prior and then a posterior part, Bacon even declared that the cause of the motion through that void space is the "before and after" (*prius et posterius*) of the space itself. In Bacon's view, it was precisely the spatial before and after that presupposed a temporal before and after and thus guaranteed that motion in a vacuum, if such existed, would be finite.[16] From this idea it followed that if a three-dimensional vacuum existed, finite and successive motion ought necessarily to occur in it because a body could not simultaneously occupy any two of the distinct termini of an extended vacuum. Without explicit use of the later terminology, Bacon fully and precisely described the *distantia terminorum* argument and also conceived it as a cause of temporal motion, an opinion that was often repeated.[17]

The causal nature of the *distantia terminorum* was not, however, destined to become associated with the name of Roger Bacon, but rather with that of Thomas Aquinas,[18] who nowhere identified it as a cause of motion. Considering the problem of motion in a vacuum at approximately the same time as Bacon, and fully aware of the conflict between Averroes and Avempace, Aquinas sided with the latter, though without mention of his name. Aquinas flatly rejected Aristotle's contention that no ratio could obtain between motions in void and plenum. "Indeed," he argued, "any motion has a definite velocity [arising] from a ratio of motive power to mobile – even if there should be no [external] resistance. This is obvious by example and reason." As an example, Aquinas offered Avempace's illustration of celestial bodies "whose motions are not impeded by anything, and yet they have a definite speed in a definite time." For his appeal to reason, he declared that "just as in a magnitude through which there is a motion we understand that there is a before and after in the motion; [and] moreover [that] the before and after are in the motion by reason of time, it follows [from this] that motion takes place in a definite time."[19]

a. THE *DISTANTIA TERMINORUM* IDENTIFIED AS RESISTANCE TO MOTION IN A VOID *EX PARTE MEDII*

Bacon's sense of the causal nature of the *distantia terminorum* was sometimes taken as a sufficient explanation for finite motion in a void. That is, the necessity for a body to traverse the prior parts of a dimensional vacuum before the posterior parts was accepted as an adequate causal explanation for the finitude of the motion. Others, however, sought to explain the finitude of motion in a vacuum in more Aristotelian terms by identifying the *distantia terminorum* with a resistance to motion *ex parte medii*, that is, by conceiving the extension of void in a manner similar to a material medium. In medieval physics, the primary function of resistance was to prevent instantaneous motion or, to put it another way, to cause finite motion. Within the sublunar realm of the Aristotelian cosmos, the ubiquitous material plenum served this function. But in the absence of all matter, as in a vacuum, what could serve in this capacity? One possibility was the separation of the termini of motion, that is, the *distantia terminorum* itself. Already in the thirteenth century, Pseudo-Siger of Brabant, a probable contemporary of Bacon and Aquinas, identified the *repugnantia terminorum*, which was an equivalent expression, with resistance "because a motion cannot be traversed from extreme to extreme except through the middle."[20] In the fourteenth century, Albert of Saxony would include the *distantia terminorum*, or the *incompossibilitas terminorum*, as he called it, as one of seven kinds of external resistance.[21]

The *distantia terminorum* argument enunciated by Bacon and Aquinas was an attempt to explain Avempace's claim that motion in a resistanceless medium, or vacuum, would be finite rather than instantaneous. According to Averroes, Avempace eliminated resistant media as the sole causal factor in motion and implied that a true natural motion would occur under conditions in which all resistance to motion was absent. But apart from vague talk about the relationship of perfection between mover and moved in celestial bodies, Averroes, in his account of Avempace's opinion, reported no proper mechanism that might justify finitude of motion in a vacuum. It was to remedy this deficiency, and to render Avempace's position tenable and plausible, that the *distantia terminorum* argument was introduced. Essentially a kinematic argument, because it relied on space and time alone, the *distantia terminorum* was also made to function as a dynamic cause of motion and as a resistance similar to a resistant medium (*ex parte medii*). We must now inquire whether this explanation was an original contribution first formulated in the thirteenth century by Bacon or Aquinas or some other scholastic commentator; or whether the idea was derived from Greek and Arabic sources. The latter alternative seems overwhelmingly likely, at least for the kinematic aspects of the argument and, perhaps by implication, for the dynamic aspects as well.

b. ON THE POSSIBLE ORIGINS OF THE *DISTANTIA TERMINORUM* ARGUMENT

In the thirteenth century, a number of possible sources of the *distantia terminorum* argument were available to Bacon and Aquinas. Aristotle himself provided at least two passages that clearly qualify. In the fourth book of the *Physics*, in discussing the nature of time, Aristotle declared that "since 'before' and 'after' hold in magnitude, they must hold also in movement, these corresponding to those. But also in time the distinction of 'before' and 'after' must hold, for time and movement always correspond with each other."[22] Although this passage alone might have served as the basis of the *distantia terminorum* argument,[23] Aristotle expressed a similar opinion in defending his belief in the instantaneous transmission of light. Reporting the claim of Empedocles that "light from the sun arrives first in the intervening space before it comes to the eye, or reaches the earth," Aristotle, in *De sensu*, acknowledged the *prima facie* appeal of Empedocles' opinion: "For whatever is moved [in space] is moved from one place to another; hence there must be a corresponding interval of time also in which it is moved from the one place to the other."[24]

Quite similar arguments appear in at least two Arabic authors whose works were available in Latin translation. In his *Sufficientia*, or *Liber Sufficientiae*, Avicenna explained that "when something is moved in a void, it is necessary that it should cleave the void space with a motion in time, or not in time. But it is impossible that it should not be in time, since it falls part of a distance before it falls the whole distance. Therefore it is necessary that it should occur in time."[25] Although Avicenna rejected the argument presented here and even drew a contradiction from it, he has clearly described the *distantia terminorum* argument and applied it to motion in a void. That Avicenna may have played a significant, if not crucial, role in justifying motion in a void is indicated by Albertus Magnus, who mistakenly linked Avicenna with Avempace as one who believed that if motion occurred in a void it would be temporally measurable.[26]

Finally, it is possible that Alhazen also contributed to the medieval Latin development of the *distantia terminorum*. In his *Perspectiva*, or *Optics*, which had been translated from Arabic to Latin in the late twelfth or early thirteenth century,[27] Alhazen rejected the widely held Aristotelian opinion of the instantaneous transmission of light and sided with Empedocles in arguing for its finite speed.[28] But it remained for Roger Bacon to discuss the transmission of light in void space and to apply the *distantia terminorum* argument to justify its finite speed. Employing both the kinematic and causal aspects of the *distantia terminorum*, Bacon argued that "if there were a void medium without any natural coarseness, and if light or any body whatsoever could pass through it, there would be some succession caused by the medium simply as a result of the prior and posterior [character] of space itself, quite apart from any resisting coarseness of the medium."[29]

On the basis of the evidence cited above, it would appear that the kinematic aspect of the *distantia terminorum* argument was readily available from a variety of sources. The causal and resistive aspects, in which the finitude of speed is attributed to the essential divisibility of plenum or void, is, however, not readily apparent in the earlier sources. Because one or the other, or both, is clearly employed by Bacon, Pseudo-Siger of Brabant, and numerous later Latin scholastics, these aspects may have been first applied to justify finite motion in a void in the course of the thirteenth century.

4. CRITICISMS OF THE
DISTANTIA TERMINORUM ARGUMENT

In medieval terminology, the *distantia terminorum* argument justified finite motion in a hypothetical vacuum *ex parte medii*. That is, by analogy with a material plenum through which bodies moved, vacuum was conceived as a three-dimensional magnitude divisible into successive and sequential parts that could not be traversed instantaneously. The causal and resistive attributes of the vacuum derived from this conception. By treating the vacuum as if it were an extended body or a material medium, the *distantia terminorum* seemed to have provided a satisfactory mechanism to account for the possibility of finite motion in a vacuum. But if an extended vacuum is analogous to and even identified with body, how could a material body be received into a vacuum? Would this not imply that two dimensions, body and void, occupy the same place simultaneously? How could the *distantia terminorum* argument come into play if bodies could not interpenetrate with the vacuum? Without interpenetration, bodies could not enter any vacuum whatever and therefore could not move within it. The analogy, or identification, of void with body, that made the *distantia terminorum* argument acceptable also made it impossible. Aristotle's argument on the impossibility of the interpenetration of void with body (see Chapter 1) either had to be met or the *distantia terminorum* argument had to be abandoned as impossible. Confronted with this dilemma, Roger Bacon abandoned the *distantia terminorum*. Quite aware that the latter argument depended on the possibility of a body being received into an empty dimension, and unable to counter Aristotle's argument on impenetrability, Bacon finally concluded that no motion at all, whether instantaneous or successive, could occur in a vacuum.[30] The principle of the impenetrability of bodies triumphed over the *distantia terminorum*.

Although the problem detected by Bacon was often ignored by those who accepted the *distantia terminorum* argument,[31] some of those who had an awareness of it sought to avoid the dilemma of impenetrability by a different approach. Ironically, this involved a further extension of the analogy, or identification, of body and extended void. For if an extended vacuum is really like a body, or material medium, with respect to divisibility, then just as a material medium yields to a body moving through it, so also should an extended vacuum yield to a body moving through it.[32] On this assumption, as well as another wherein the matter surrounding a hypothetical dimensional vacuum would not collapse inward to fill the void that nature so violently abhorred, John Duns Scotus allowed that "the motion of a heavy body in a vacuum would be successive because a prior part of the vacuum would yield (*cederet*) first, and the whole heavy body would traverse that part of space and then this [part]."[33]

Duns Scotus' all too brief allusion to, and apparent acceptance of, the yielding of void space to body won few adherents in the Middle Ages. John of Jandun, for example, denied the possibility that the parts of a vacuum could yield to a body moving through them. For if a dimensional vacuum existed, which he denied, it would by virtue of its quantity be divisible into prior and posterior parts. But a body could move through those parts of void only if the latter yielded to its motion; otherwise two three-dimensional entities would occupy the same place simultaneously. Jandun, however, denied that parts of a vacuum could yield to body and concluded therefrom that successive motion in a vacuum was impossible because void and body could not interpenetrate.[34]

Aegidius Romanus also considered the problem when he asked whether a vacuum would yield (*cederet*) to a body that might be placed in it. He insisted that bodies yield to each other only when at least one of them is rarefiable and condensible. But a vacuum is a separate space distinct from all qualities and affections and cannot undergo rarefaction and condensation. Therefore it cannot receive the dimensions of another body because two dimensions would occupy the same place simultaneously, a natural impossibility. Those philosophers who

assume the existence of a vacuum in order to account for motion, are compelled to concede the existence of two bodies [in the same place] simultaneously, and must accept the absurdity that two bodies can exist [in the same place] simultaneously. For if a body is received into a plenum, it is not necessary that its dimensions exist simultaneously with those of the plenum because the latter yields. But since a vacuum does not yield, if a body is received into it, the dimensions of the vacuum will exist simultaneously with those of the body ....[35]

This is impossible. But there was an apparent attempt to demonstrate how the dimensions of a void might yield to an entering body and yet avoid the interpenetration of bodies. Perhaps this was a deliberate effort to pursue the implications of Duns Scotus' statement. A report of this attempt was made by Walter Burley in the course of a highly significant and lengthy discussion of motion in a vacuum. Although he would reject this novel opinion, Burley first distinguished two kinds of separate, extended vacua, one in which the principle of impenetrability is inoperative and the other in which it is assumed effective. In the first kind, which Burley attributed to certain ancients, the vacuum is assumed capable of receiving bodies and being coextensive with them, so that two quantities could occupy the same place simultaneously.[36] The second kind is assumed incapable of receiving bodies, although it could yield to them, from which it followed that two quantities could not occupy the same place simultaneously.[37] It is the second vacuum that is relevant to our discussion.

The obvious problem in this situation is how to explain the possibility of motion in a vacuum that was conceived as a body and that therefore could not receive another body. How could the *distantia terminorum* apply to an impenetrable vacuum? To imagine motion in such a vacuum, we must analyze the manner in which bodies move through any material medium. When, for example, a body moves through air, the latter simultaneously yields and resists. Because two bodies cannot occupy the same place simultaneously, a portion of air equal in volume to the body is continually displaced. Where air exists, there is no moving body; and where the moving body is located at any moment, air is absent. And of course, as the process of displacement continues, the air that surrounds the body resists its motion. A separate, extended void space must be conceived as operating in much the same manner. Wherever body is located, an equal volume of void space is imagined to have been displaced, so that where body exists, void does not; and where void exists, body is absent.[38] It follows that body can move through void but never coincide with it, so that two bodies will not occupy the same place simultaneously. Moreover, to push the analogy further, the extended void surrounding the moving body functions as a resistance by virtue of the *distantia terminorum*, that is, by reason of its extension.

That those who adopted the second kind of vacuum described by Burley had all this in mind is made apparent by their further extension of the analogy of void with body or plenum. If void space yields to material body, does this not imply that empty space is somehow capable of rarefaction and condensation, that is, of yielding and resisting by expansion and contraction, in a manner analogous to a plenum that yields to a body moving through it? Because every quantity is necessarily rare or dense,[39] and because void space is assumed to be a quantity, it would seem to follow that extended, empty space is capable of condensation and rarefaction.

But how is it possible to assign qualities such as rarity and density to an empty space that is assumed to be separated from all qualities, including rarity and density? And if extended space could not possess the qualities rare and dense, it could bear no ratio to a full material medium, from which it followed that motion made in a void space could bear no ratio to motion in a plenum. Motion in such a vacuum would be instantaneous, just as Aristotle had declared.[40]

Although Burley would eventually reject the claim that finite motion could occur in an extended void that yields to but does not receive bodies, he first attempted to offer some justification for assigning rarity and density, and therefore resistance, to void space. He distinguished two kinds of rarity and density: qualitative and quantitative.[41] If rarity and density are conceived as tangible qualities, then truly separate, dimensional space will be neither rare nor dense because void space is assumed devoid of qualities.[42] But if rarity and density are taken quantitatively, that is, associated with parts of a quantity in greater or lesser proximity to each other, then because every quantity is necessarily rare or dense, we may properly conclude that even a quantity separated from every substance or quality must also be rare and dense. The analogy, nay identification, of extended space and material body guarantees this conclusion. It follows, then, that a "ratio of medium to medium in rarity is as the ratio of motion to motion in speed."[43] These comparisons are possible because quantity as an active principle has the capacity to resist,[44] a capacity that arises from the *distantia terminorum*, that is, from the impossibility that a body can be in two places simultaneously in an extended space. Indeed supporters of this position, or Burley himself, explicitly identified the *distantia terminorum* as a privative resistance, in contrast to a positive resistance, which always involved the reaction of a material medium to the body in motion, as when a body penetrates earth or moves against flowing water or air.[45] Positive resistance was obviously inapplicable in a vacuum. But air, or any material medium, need not always offer violent, or positive, resistance to a body. We can imagine that a stone occupies a quantity of air equal to itself, so that stone and air coincide in the same place simultaneously. Under these circumstances, the air would, by assumption, offer no positive resistance to the stone. But the diverse parts of the air that surround the stone can be said to offer a privative resistance (*resistentia privativa*) to the stone in the sense that the stone cannot occupy these places simultaneously. It is obvious that privative resistance is nothing other than the *distantia terminorum*. Because a body moving through any medium, whether full or empty, cannot occupy all places simultaneously, partisans of this position concluded that privative resistance suffices for local motion.[46]

Supporters of the ideas described by Burley are unknown. They represent an interesting, if strange, attempt to make sense of motion in a void by accepting Aristotle's identification of void with body[47] and deriving from it the consequence that body cannot occupy void but only displace it. Extended void space was thus assigned the property of yielding to body. In principle, a body could move through a vacuum without occupying any part equal to itself in volume and shape. But the persistent effort to extend the analogy between void and body, or plenum, resulted in the attribution to void of properties that made its behavior incomprehensible. No sense could be made of a void that could be displaced by body and that could expand and contract, thus becoming alternately rare and dense.

Nor was anything gained by assigning a new name, "privative resistance," to the *distantia terminorum*.

In the end, Burley rejected all efforts to justify finite motion in an extended vacuum that were based on the *distantia terminorum* argument. Distance alone could not operate as a genuine resistance. But for Burley, as for Aristotle and Averroes, resistance was essential for the occurrence of finite motion in a plenum or vacuum. In its capacity as a medium (*ex parte medii*), then, the extensive magnitude of a void space could not resist the bodies moving through it, which as a consequence would move instantaneously.[48] Burley insisted, with Aristotle, that no relationship, or ratio, could obtain between void and plenum in rarity (or density), from which it followed that no ratio could obtain between motion in a void and one in a plenum. Moreover, as Aristotle had also argued, even if motion could occur in a void, another absurd consequence would follow, namely that one and the same body could traverse equal distances in plenum and void in the same time.[49] Finite motion was thus impossible in both types of void that Burley described and discussed *ex parte medii*. Thus, in the end, it made no difference whether or not a void space could receive bodies. The *distantia terminorum* in an extended vacuum could not of itself cause any motion to be finite.[50] But this did not mean that Burley repudiated the possibility of finite motion in a void. On the contrary, as we shall see in Section 7b, he sought to justify such motion *ex parte mobilis*, that is, from the standpoint of the properties of the body itself rather than from the nature of the void.

Although attempts to justify finite motion in a vacuum by the *distantia terminorum* produced some interesting developments concerning the nature and properties of void space, this concept evoked harsh criticisms. We have now seen that Bacon rejected it because of the impossibility of interpenetration of body and void, whereas Burley abandoned it because he could find no good arguments for believing that mere extension, or magnitude, could function as a resistance to the motion of bodies in empty space. The idea that, in its purely spatial characteristics, the *distantia terminorum* could act as a resistance and cause successive motion was also rejected by Nicholas Bonetus, who offered a different definition of resistance to justify motion in a void. For Bonetus, resistance to a body's motion, which gives rise to its successiveness, derives from all the forms that it acquires prior to the acquisition of its ultimate terminus. These intermediate forms (*formae*), or positions (*ubi*), resist the motive power and the mobile itself in order to prevent a finite power from causing a mobile to arrive at its *terminus ad quem* instantaneously.[51] Thus it is not the space itself that acts as a resistance by virtue of its prior and posterior parts, as in the concept of *distantia terminorum*. Bonetus redefined the resistance of a medium in terms of the series of positions, or locations, that a body must occupy before it reaches its final terminus.

Even earlier than Bonetus and Canonicus (see note 51), Aegidius Romanus had formulated a similar attack on the *distantia terminorum* but added concepts that were truly astonishing. Aegidius inquired whether it is space itself that determines the temporal nature of motion. At first glance, one might assume the affirmative because we see that, all other things being equal, it takes a greater time to traverse a greater distance than a smaller distance.[52] On closer analysis, however, Aegidius rejected this interpretation. Motion with respect to some end or form does not require time because of any quantitative entities, but rather because of a "contrary disposition" (*contraria dispositio*). Matter has a contrary disposition toward receipt of a new form when it resists the introduction of that form. But when it is actively disposed toward the acceptance of a particular form, it will receive that form instantaneously, presumably because it has no contrary disposition toward that form. Aegidius argued that just as the contrary disposition operates in subjects, so does a resistant medium function in the motion of light and heavy bodies. Where a resistant medium exists, there we shall also find a contrary disposition to the reception of bodies moving through it. Thus a body's motion is slowed in such a medium, and its movement is temporal and successive. Under these circumstances, a body acquires its form – its place, or location, in this case – successively and temporally. But when a resistant medium is absent, as in a vacuum, motion occurs in no time at all and the form or place of a body is acquired instantaneously, "just as light is multiplied in no time through the whole of a medium, since there is no contrary disposition to light in a medium."[53]

But although Aegidius insisted that motion in a vacuum would be instantaneous, he also argued that it would be successive. For despite the absence of resistance in a vacuum, the points of its extended dimension would be ordered in succession with respect to any instantaneous motion. A body moving through a void space would thus move from one point to the next in a durationless instant. Because the sum total of such instants is zero, the body will have moved with a successive motion through all the successive points in a durationless instant. In this manner did Aegidius believe that a body could move through a void instantaneously but successively.[54]

Aegidius' argument is, of course, absurd. It presupposes succession and instantaneity of motion simultaneously. If instantaneous motion is understood as the simultaneous occupation of all the points of a given distance, those points cannot also be said to have been traversed successively and, therefore, temporally. As an argument against the *distantia terminorum*, however, it was deemed a telling blow by John Buridan, one of the greatest natural philosophers of the Middle Ages. For what if God should cause a body in the heavens to reach the earth instantaneously? Would it not be the case that such a body would pass through all the intervening points between heaven and earth? Because those points are ordered into prior and posterior parts, it surely must follow that the motion would be successive and also instantaneous, for who would dare deny that God could produce such an instantaneous motion?[55] The *distantia*, or *incompossibilitas, terminorum* argument is thus inadequate because the successiveness of the parts of a space or distance is no guarantee that a motion will be finite and temporal. Instantaneous motion was apparently compatible with successiveness. The problem might also be expressed another way. Instead of a durationless instant, the distance between two distinct termini in a void space might be traversed in an infinitely small time by an infinitely great velocity.[56] Although an infinite velocity was impossible by natural means, it was possible supernaturally, in which event all the intervening points would be traversed successively, a consequence that was, ironically, guaranteed by the *distantia terminorum* itself. If successiveness was as compatible with instantaneous as with finite motion, the *distantia terminorum* argument would have lost its rationale because it was devised to demonstrate that motion in an extended vacuum with prior and posterior parts must be finite and temporal because it unavoidably traversed those prior and posterior parts successively.

One of the most devastating critiques of the *distantia terminorum* position was formulated by Marsilius of Inghen in the second half of the fourteenth century. As the first of four arguments against those who would assign to vacuum a resistive capacity that would cause motion to be successive and finite, Marsilius declared that a void space could offer no more resistance to one kind of body than to another. And yet differences of speed must occur when bodies of different power and heaviness move through the same resistant medium. But in a void this does not occur because, unlike a plenum, a vacuum does not resist a heavier body less than a lighter body; nor does it bear a greater ratio of resistance to a lighter than to a heavier body. Thus the *distantia*, or *incompositas, terminorum* cannot account for differences in speed that must necessarily arise from bodies of varying heaviness.[57]

The second argument follows as a consequence of the first. For if the void space offers the same resistance to all bodies so that a heavy body moves with the same speed as a lighter body, it follows that all bodies will move with equal velocities in a void. Without any cause to explain differences of speed in a void, we must conclude that no such differences will exist.[58] Thus the *distantia terminorum* cannot account for the generation of different speeds by bodies of varying weights and thereby fails to explain an obvious fact of experience. Moving a step further, Marsilius now argued that the equal speed with which all bodies will move in a vacuum operating under the constraints of the *distantia terminorum* will be infinite. This absurd consequence follows from the assumption that the *distantia*, or *incompositas, terminorum* is the only resistance in a vacuum. But because, as we have already seen, that resistance cannot determine any specific velocity, it follows that for any given velocity you mention, the body can be conceived as moving with yet a greater velocity. And because this process may go on ad infinitum, the velocity with which all bodies move in a vacuum is taken as infinite.[59]

From all these arguments Marsilius drew a fourth one in which he declared that velocity in a vacuum could not be measured in the usual manner as a ratio of motive power to resistive force. Because the motion of all bodies of whatever heaviness and motive force is equal, no specific quantity of resistance can be assigned to the *distantia terminorum*. Thus the universally accepted measurement of a velocity as the resultant interaction between a motive force and a resistance became meaningless on the assumption that the *distantia terminorum* functioned as a resistance.[60]

As a result of the cumulative impact of the kinds of criticisms just described, an awareness may have been generated that the *distantia terminorum* argument could not of itself justify the belief that motion in an extended vacuum would of necessity be both successive and finite. During the later Middle Ages, it was a commonplace that successive and finite motions resulted either from the resistance to motion offered by the mobile itself (*ex parte mobilis*) or from the resistance posed by the medium through which a body moved (*ex parte medii*), or both.[61] The perennial problem for those who believed in the possibility of finite motion in a vacuum was to identify something that might plausibly function as a resistance in a vacuum. We have already seen that in the *distantia terminorum* tradition, efforts to identify the resistance exclusively within the extended void medium itself, which then qualified as a resistance *ex parte medii*, left many natural philosophers unsatisfied. The attempts by Bonetus and Canonicus to redefine the resistance of a medium in terms of the successive positions, or forms, that a body acquired were of little avail. Thus it was that the next significant advance in justifying motion in a void would come from those who sought to locate motive power and resistance in the mobile itself (*ex parte mobilis*).[62]

5. RESISTANCE IN THE VOID
EX PARTE MOBILIS

Earlier in this study (see Section 3), we saw how Thomas Aquinas defended the Avempacean concept that finite motion could occur without the resistance of an external medium. Let us now examine more closely his statement that "any motion has a definite velocity [arising] from a ratio of motive power to mobile – even if there should be no [external] resistance." In support of this bold assertion, Aquinas cited the celestial motions as empirical evidence and made an appeal to reason by invoking the *distantia terminorum*. The latter, with its prior and posterior parts of motion and magnitude, guaranteed that motion in a void would occur in a measurable time. But it was nowhere identified by Aquinas as a resistance to motion. And yet Aquinas shared the common conviction of all medieval scholastics that both motive force and resistance were essential for every motion, whether in plenum or void. With the external medium, whether material or void, deemed inessential for motion, Aquinas located the primary and fundamental resistance to motion in the moved body itself. For if the form of a heavy or light body were removed, the magnitude or quantity of that body would remain, a magnitude that Aquinas called the *corpus quantum*. It was a body's *corpus quantum* that functioned as a resistance to the motive force and, together with that force, produced a velocity.[63] Thus whereas the *distantia terminorum* would guarantee the successiveness of motion in a void, it was a body's *corpus quantum* that provided the essential resistance that enabled it to be moved in the first instance.[64] It seems that the *corpus quantum* could function in a vacuum as a resistance in both violent and natural motion. In the former, the motive power would have been an external body or a power pushing the *corpus quantum* of the moved body; in the latter, the *corpus quantum* must be conceived as resisting the *generans*, or agent, namely the force or power that directly produced the body and conferred upon it all its natural properties, including a downward tendency for heavy bodies and an upward tendency for light bodies.[65]

The concept of a *corpus quantum* as a sufficient resistance for the occurrence of motion in a vacuum never gained much support.[66] The tradition of an external medium essential to motion was too strongly entrenched in thirteenth-century Aristotelian physical thought. The *distantia terminorum* had already marked a radical departure from Averroes' interpretation of Aristotle. In place of Averroes' insistence on the necessity of a material, external medium as a resistance for natural and violent motion, some thirteenth-century scholastics, such as Bacon, had invoked the *distantia terminorum* and allowed a void, external medium to function with the same resistive capacity. But scholastics were not yet prepared to eliminate completely the role of the external medium and replace it with a *corpus quantum*. Indeed many of them, for example Aegidius Romanus, not only rejected the *corpus quantum* but found the idea of natural motion of inorganic bodies without a material resistance totally unacceptable. Even if the form of a heavy or light inorganic body were abstracted, nothing would remain of the body but its matter, or quantity – Aquinas' *corpus quantum* – which, in Aegidius' opinion, could not offer resistance to a motive force. The finite, natural motion of heavy and light inorganic, terrestrial bodies could occur only in an external, material medium; in a void such motions would be instantaneous.[67]

In the preceding paragraph, attention was focused on the natural motion of inorganic bodies. The medieval problem of motion in a void was primarily concerned with natural motion, and especially the most difficult to explain of all motions, that of inorganic bodies. The *distantia terminorum* and the *corpus quantum* were essentially responses to problems about the natural motion of inorganic bodies in void space. To comprehend this issue in its larger context, it will prove helpful to describe a widely held interpretation concerning the ways in which the motions of celestial bodies, animals, and inorganic bodies were conceived.

a. AVERROES' EXPLICATION OF THE RELATIONSHIP OF MOTIVE AND RESISTIVE FORCES IN ONE AND THE SAME BODY

As part of his lengthy and widely known reply to Avempace in his comment on book 4, text 71 of Aristotle's *Physics*, Averroes argued that motion must always arise from a conjoint action of two forces or powers, one motive, the other resistive, the latter usually identified with the thing moved. It is the resistive power that ensures that motions will not be instantaneous. Averroes explained that a resistance obtains between mover and moved when the two are distinct, as happens in celestial motion and the motion of animals. In a celestial sphere, the mover is an actually existent immaterial intelligence that is distinct from the material celestial sphere *qua* moved thing. An animal similarly has its mover, the soul, distinct from its body.[68] But the situation is quite different with all terrestrial, inorganic bodies – that is, with the four elements and the compounds or mixed bodies formed from them[69] that undergo both natural and violent motion. The latter poses no problem because the mover is always a separate body, as when air is assumed to push a body away from its natural place. The difficulties arise in the natural motion of inorganic bodies.

In an inorganic body, the mover is identified as the form of the body. However, body itself as the moved thing is matter, which has only potential existence. Thus within all such bodies, the form, or "essence," as mover is not actually distinct from the body conceived as the material thing that is moved. According to Averroes, however, no motion can occur in a body unless there is an actual distinction between mover and moved, for only then can a resistance arise. Because the form is not a distinct entity within an inorganic body, it cannot be the efficient cause of natural motion. In short, an inorganic body cannot move itself because its form cannot act on its matter; and conversely, because its matter, that is, the thing moved, cannot be distinguished, it cannot presumably act as a resistance to a mover. How, then, do such bodies move naturally? Averroes explained that the form of an inorganic body moves that body *per accidens* by overcoming the resistance of an external medium. Such bodies can move only if they are located in an external, material medium. The form of the body, plus an external medium, are thus essential for the occurrence of motion. Without that medium, as in a vacuum, such bodies would move instantaneously.[70]

Averroes' distinction between mover and moved became the crucial criterion in determining whether a finite, temporal motion could occur in a vacuum. Such motion was possible whenever mover and moved were actually distinct, as in violent motion, the motion of animals, and indeed, anything moved by itself. In all such instances, resistance to motion was offered essentially by the mobile itself,[71] not by the external resistance of a medium. It was the mobile's resistance that produced the finite motion. An animal, for example, moves by means of desire that arises in its soul. The resistance to that desire is the animal's bodily heaviness, which functions as the resisting mobile. Because mover and moved thing are clearly distinguished, any animal could move itself progressively in a vacuum.[72] But if an animal fell from a height, would it fall with a finite speed? Clearly not. Under these conditions, the animal differs in no way from an inorganic body that might fall from the same height. Because its downward plunge is not caused by virtue of its own desire, mover and moved thing, or mobile, are thus not distinguished. Without an external, resistant medium to cause temporal motion, the animal will fall instantaneously, just as would any inorganic body.[73] For the latter lacks a separate mover that can be distinguished from the body itself. Inorganic bodies are thus dependent for temporal motion on the resistance of an external medium, which when absent, as in a vacuum, will yield an instantaneous motion.[74] The *corpus quantum* of a light or heavy inorganic body cannot serve as a resistance because it is nothing other than the quantitative matter of a body stripped of all qualities and form (see above and note 67). Only an external resistant medium could cause its motion to be finite.

6. VIOLENT MOTION IN THE VOID
EX PARTE MOBILIS

With this issue in mind, attempts were made in the fourteenth century to explain how finite natural motion might occur in a vacuum. Before describing them, let us consider first the manner in which violent motion, much the easier case, was explained. We have seen that for Aegidius

Romanus temporal violent motion was possible in a vacuum. Although he did not offer a mechanism for such motion, it was obvious to all that violent motion, defined as motion away from a body's natural place, could arise only when mover and moved thing were distinct. As the separate mover in violent motion, Aristotle had identified the external medium, usually air, which pushed the body away from its natural place. When a stone, for example, is hurled upward away from its natural place at the center of the earth, the initial cause of its motion would be the thrower and the continuing cause the ambient air, which is continually in contact with the stone.[75] Physical contact between mover and moved was thus essential to violent motion. Air, for example, could serve as both primary mover and secondary resistance, with the body as primary resistance. In a vacuum, however, no external material medium could function either as primary mover or as resistance. But the body itself could obviously function as a resistance to some mover. But what could be identified as an independent mover? The solution was found in the emerging concept of impressed force, or *impetus*.[76] Two of the earliest authors who applied impressed force as a causal explanation for finite, violent motion in a vacuum were Johannes Canonicus and Nicholas Bonetus (see Section 4 and note 51), who were active at Paris in the 1320s.

Violent motion in a vacuum is intelligible and plausible, according to Johannes Canonicus, only if we abandon the untenable notion that air or any material medium is required for the continuation of projectile motion. Not only is an external medium unnecessary for projectile motion, but Canonicus insisted that such motion would occur more readily in a vacuum than in a plenum, "since a projected object is moved naturally by the power [or force] of the projector and not by parts of the medium through which it is carried." When an arrow is projected through the air, the latter is obviously not the cause of the arrow's impulsion. On the contrary, air is nothing but an obstacle to its movement.[77] In this brief discussion, Canonicus identified the cause of violent motion in a vacuum as "the power [or force] of the projector," which under the circumstances can only be an impressed force, or *impetus*, as it was called some years later by John Buridan.[78]

Fortunately, Nicholas Bonetus was more expansive than Canonicus on the problem of motion in a void and even indicated the nature of the impressed force. According to Bonetus, violent motion could be produced in a void either by a real conjunction or by a virtual, or causal, conjunction. As illustrations of real conjunctions, in which the mover is actually in contact with the mobile, Bonetus proposed a light body that is "conjoined to a heavy body which draws [the light body] to the center of the world," or when a man moving through a void carries a stone in his hand. A virtual conjunction might occur when an abstract intelligence, which is far removed from a certain heavy body, causes the latter to move violently. Thus physical contact is not essential.[79]

But what if there is neither a real nor a virtual conjunction between mover and moved? For example, could a man hurl a stone upward in a void? In the absence of any apparent motive force, it would appear not, "since the air does not move – for there is no air in a void – nor, indeed, anything else. Moreover, it does not seem that anything could [project] bodies in the void because the projector would cease projecting immediately after [the initial] projection."[80] To these objections, Bonetus responded that some believe violent motion is possible without the real or virtual conjunction of a motive force with a mobile. "The reason for this is that in a violent motion some non-permanent and transient form is impressed in the mobile so that motion in a void is possible as long as this form endures; but when it disappears, the motion ceases."[81]

Thus did Nicholas Bonetus explain violent motion in a void by a self-expending impetus. That Bonetus actually accepted this possibility, and was not merely reporting the opinion of others, is borne out by his subsequent argument in favor of the claim that violent motion in a void would be slower at the end than in the beginning, because "that force, or impressed form, continually fails and diminishes in moving the mobiles, and, as a consequence, moves slower [i.e., causes the body to mover slower]. Thus violent motion made in a void has to be slower in the end than in the beginning, just as [violent motion] in a plenum."[82]

Whether or not Bonetus was one of the first, if not the first, of the Latins to justify possible violent motion in a vacuum,[83] his arguments are significant. Not only did he propose an impressed motive force for violent motion in a void, but by characterizing that impressed force as self-expending, he also provided a mechanism for terminating every such motion. For those who were interested, and few apparently were, Bonetus had provided a reply to Aristotle's inertial consequence, namely that motion in a void would be interminable unless it met some obstacle that would obstruct its movement (see Chapter 1). Self-expending impetus made every motion in a vacuum of finite duration and extent, even if the void were conceived as infinite.

The significance of Bonetus' discussion is best comprehended by contrasting his approach to that of John Buridan, who opted for a permanent impetus, which he applied so effectively to violent and natural motion in external media as well as to celestial motion.[84] But Buridan avoided consideration of violent motion in a void. This is hardly surprising when it is realized that for Buridan impetus was a "thing of permanent nature, distinct from local motion in which the projectile is moved" and corrupted only by external resistances or by the natural inclination of bodies to seek their natural places when otherwise unimpeded. As with Bonetus, a body set in motion in a void, and having acquired a quantity of impetus in the process, would have fulfilled for Buridan the basic Aristotelian requirements for motion, namely the conjoint action of a motive force (the impressed *impetus*) and resistance (the body itself). Within the context of medieval physics, however, Buridan's permanent impetus would have produced an impossible situation. If the void were assumed infinite, any inorganic body moving under the agency of a permanent impressed force would move with a uniform speed forever because no cause could operate to diminish or increase its quantity of impetus; and because an infinite void has no natural places, the body would move with an indefinite inertial motion, an impossible consequence in medieval physics.[85]

Even if the void were finite, Buridan's impetus would have posed serious problems. In the Middle Ages, finite vacua were usually imagined as spaces that would be left if God annihilated all or part of the matter between earth and moon. A body that had acquired a permanent impetus would eventually collide with the concave surface of the moon and either rebound and continue its motion until the next rebound or simply come to rest at the inner surface of the lunar sphere. It might even seek its natural place, provided that such places were still meaningful in a vacuum lying between earth and moon. The problems posed under such hypothetical conditions would have been so formidable that Buridan and other permanent impetus theorists avoided them altogether.

Bonetus' self-expending impressed force was quite another matter. It provided a basic and intelligible mechanism for the "natural" cessation of motion in a vacuum, thus eliminating the problems that would have confronted proponents of a permanent impetus. Had the problem of violent motion in a void become a regular feature of Aristotelian commentaries on the fourth book of the *Physics*, the concept of a self-expending impressed force, which played a role in Galileo's early thought, might have added significant new elements to medieval physical discussions. With self-expending impetus, the concept of violent motion in a vacuum was rendered as intelligible as violent motion in a plenum. But the problem of violent motion in a vacuum was largely ignored during the late Middle Ages. It was rather the difficulties associated with natural motion in empty space that attracted the attention of medieval natural philosophers.

7. NATURAL MOTION IN THE VOID
EX PARTE MOBILIS

Employing distinctions made by Aristotle,[86] it was customary to distinguish two kinds of bodies capable of natural motion: elemental and mixed, or compound, bodies. The former were pure, homogeneous substances identified with one of the four elements: earth, water, air, and fire; the latter were composed of at least two elements in varying proportions.[87] Although bodies ordinarily observed in the sublunar region were compounds,[88] scholastics were intensely interested in the behavior of the pure elements because these were the basic constituents of all perceptible mixed bodies. Moreover, within every mixed body, Aristotle had assumed that one of the elements was necessarily dominant and would determine its natural motion.[89] If the natural motion of a mixed body were observed, its dominant element could be directly inferred. A body that fell naturally toward the center of the earth – that is, toward the center of the universe – and came to rest naturally on the earth's surface was said to be dominated by the element earth. If it ascended to the region of fire just below the lunar sphere, fire could be assumed the predominant element. Bodies in which the intermediate elements, air or water, prevailed could rise or fall naturally, depending on their initial location. A body that came to rest naturally in air after falling from fire or rising from earth or water had air as its dominant element, whereas one that rested in water after falling naturally from fire or air or rising from the earth was dominated by the element water. Although Aristotle was the source of the distinction between elemental and mixed bodies, he made no use of it in his arguments against motion in a vacuum. It was, however, a distinction that would prove crucial in medieval discussions, and each of the two types of bodies must be considered separately.

a. THE NATURAL MOTION OF ELEMENTAL BODIES

For those who ignored or rejected the *distantia terminorum* as a resistance and who did not associate any other form of resistance with pure, elemental bodies, it seemed an inevitable consequence that any elemental body would necessarily fall through a void instantaneously. Among this group, those who further believed that instantaneous motion was impossible, as most did, it followed inevitably that elemental bodies could not possibly fall successively and temporally in extended vacua. Avempace's famous argument in favor of finite, successive motion in a resistanceless medium was thus rendered untenable for pure, elemental bodies.

Two of Aristotle's most widely received arguments were often invoked to destroy the Avempacean position. If one assumed, as some did in the fourteenth century, that the cause of the natural motion of a heavy, pure, elemental body was its heaviness, the lack of identifiable resistance would imply that an elemental body would either move instantaneously or, if its motion were assumed finite, would fall with the same speed in plenum and vacuum, which was equally absurd. Both of these arguments were invoked by John Buridan,[90] who accepted the general rule in medieval dynamics that speed varies as a ratio of mover to resistance. Because void and plenum could not be related or compared as resistances, the velocities of a body falling in one and the other could form no ratio. And yet, if a body could fall with a finite speed in a vacuum, a ratio of speeds would indeed be possible. The impossibility of such a ratio of finite times signified for Buridan that motion in a vacuum would occur in an instant.[91] But if elemental motion in a vacuum is assumed to occur in a finite time – and this is the second argument – we may conclude that the same body would fall the same distance in a plenum in a proportionately longer time. Let us assume that the time of fall in a plenum is 100 times greater than in a vacuum. Now it is possible to imagine that plenum made 100 times rarer than at present. The body should now fall through that plenum 100 times faster than before, or in 1/100 of the time, so that its speed is equal to its free fall in the vacuum. Thus one and the same body will fall with equal finite speed in vacuum and plenum, which is absurd.[92] These two arguments were typical defenses of Aristotle's rejection of the vacuum and the possibility of motion in it. To counter these arguments, a resistance of some kind had to be associated with the body itself because efforts to assign a resistive role or capacity to vacuum, as we have seen, had never gained much credibility.

The major defense of the natural motion of elemental bodies thus came to center on the body itself. To justify temporal motion of elemental bodies in void space, some scholastic assumed that every elemental body was capable of a finite maximum speed that it could not exceed and that it could attain only in the absence of all external resistance. According to Albert of Saxony, who reported this opinion, the heavier the elemental body, the greater its natural maximum velocity.[93] The natural maximum velocity of every elemental body was a direct function of the finitude of its motive power, or agent, which was probably identified with the body's weight or heaviness. Avempace's defenders now had their resistance to justify finite motion in a void. The cause of successive, temporal motion in a void for pure, elemental bodies was the limitation, or finitude, of the motive power or agent (*limitatio agentis*), which Albert of Saxony identified as an "internal resistance" (*resistentia intrinseca*). In effect, the body's own heaviness served to resist its own motion; the greater its weight, the greater its natural limitation and, therefore, the greater its natural velocity in the absence of external resistance.[94]

To all who believed with Aristotle that the illumination of a medium was instantaneous, the internal resistance, or limitation, of the agent described here was unacceptable. If all natural agents, including the natural cause of light, are of finite power, and if the finitude of power is the real cause of succession, it would follow, Albert of Saxony informs us, that light should be transmitted successively in a measurable time. Common experience, however, indicated that the speed of light was instantaneous. Because at least one finite power can cause instantaneous motion, we may not properly infer that the finitude of an agent is the cause of succession in motion.[95]

The medieval mind conceived yet other forms of internal resistance, most of which were hardly to be taken seriously.[96] One of the less frivolous proposed that one part of an elemental body might resist another, so that the body's downward motion would be impeded. Few, however, were prepared to argue that a heavy elemental body would fall naturally with one or more violent motions occurring within it. It was the whole body, not just parts of it, that moved naturally toward the center of the world. Because none of its parts tended in the opposite direction, no internal resistance could develop to obstruct its downward motion.[97]

A curious argument in favor of the finitude of motion of descent, and one that was applicable to the natural motion of elemental bodies, was reported by Walter Burley. On the basis of Aristotle's assumption that every mobile always occupies a place equal to itself, it would follow that a body that descended instantaneously in a vacuum would be simultaneously up, down, and in the middle of that particular space. Consequently, if the smallest heavy body, or minimum particle of earth, were moved in the greatest vacuum, it would occupy the whole of that vacuum simultaneously. The smallest heavy body would thus equal the greatest vacuum, which is impossible.[98] Because instantaneous motion produces the absurdity just described, one is presumably compelled to choose the more plausible alternative that the descending motion must be finite, an alternative that Burley himself rejected.

Those who repudiated the *distantia terminorum* and internal resistance, as Marsilius of Inghen did, usually believed that a material medium was essential for successive, finite motion.[99] For them, elemental motion in a vacuum was simply impossible. Under certain specially contrived conditions, however, an elemental body that lacked internal resistance could be conceived to move successively in a vacuum that lacked the resistance of an external medium. A few such illustrations were described by Albert of Saxony. One of the most interesting involved infinitely great velocities.[100] As the context of this conception, Albert imagined the creation of a vacuum by the supernatural destruction of everything within the surfaces of the sky, by which he probably meant everything within the concave surface of the lunar sphere. Within this vacuum, an elementary body, say a piece of pure earth, is located at the concave surface of the sky, or lunar sphere. Albert now argued that this heavy body would descend infinitely quickly through the vacuum "because [it will] fall more than twice as fast as any body descending with a definite [or finite] velocity; and more than four times as fast; and more than eight times as fast; and so on into infinity, because it would descend without any internal or external resistance." But if a heavy body can move infinitely quickly between two distinct points in a vacuum in the manner just described, we must assume that it has actually moved. And if it has actually moved, its motion must be temporal because every motion must occur in some time, however small. If this is conceded, a heavy simple, or elemental, body will not move in a vacuum with an instantaneous motion, as Aristotle had argued, but in time. It is quite clear that in the motion of the elemental body, "infinitely quickly" is not to be taken as motion in a durationless instant – that is, as an absolutely infinite velocity – but as a motion that would occur in an infinitely small time.

Albert's example depends on a crucial distinction between categorematic and syncategorematic infinites.[101] The infinite speed in question is not to be conceived as an actual, or categorematic, infinite beyond which no greater speed can be imagined. Were this the case, the heavy body would indeed move through the vacuum in an instant, as Aristotle would have it. Albert's infinite speed is syncategorematic, that is, a potential infinite, which signifies that the heavy, simple body will fall infinitely quickly only in the sense that however large its assigned speed, a greater speed can yet be conceived, as in Albert's example. Although Albert may not have been the first to propose this interesting interpretation, its significance lies in the attempt to conceive successive motion in a vacuum without invoking internal or external resistances. The distinction between syncategorematic and categorematic made this possible. However large the velocity assigned, a greater one could subsequently be assigned with a concomitant diminution in the time of motion. The concept of a syncategorematic infinite velocity allowed velocities to occur in infinitely small times, thus guaranteeing succession without internal or external resistances.[102]

Other situations and conditions were imagined in which elemental bodies could move in a vacuum with finite and successive motions without invoking either internal resistance or the external resistance of media. In these circumstances, however, an external resistance, usually in the form of a balance, was introduced. Among the illustrations cited by Albert of Saxony are those in which homogeneous, elemental spheres are placed on or hung from balances conceived as suspended in a vacuum between earth and sky. For example, if two unequal but homogeneous spheres are placed on a balance suspended in this sublunar vacuum, the larger will descend successively because it has an external resistance in the lesser sphere, which acts as a counterpoise to slow the larger body.[103] In another situation, two equal weights are suspended from the balance. By adding another weight, however small, to either of these equal spheres, we can cause that body to descend with a successive motion.[104]

These and other examples are obviously so contrived and arbitrary that they fail to qualify as plausible and appropriate responses to the problem that produced them, namely the possibility of identifying a resistance that might be associated with elemental bodies and cause them to fall with finite, successive motion. Without the *distantia terminorum* to function as an external resistance, there seemed no generally acceptable way to justify finite motion for pure, elemental bodies falling or rising naturally in a hypothetical vacuum. But what about mixed bodies? Because they were real bodies observed in nature, plausible mechanisms might be conceived to show how they would fall in empty space. Failure to do so with elemental bodies might be dismissed as relatively unimportant because they were, after all, hypothetical entities conceived as falling in hypothetical vacua. Perhaps real bodies might behave differently. Here, then, was the crucial problem. For if mixed bodies could not fall in empty space with a range of finite speeds, then it might well be concluded that the existence of vacuum is truly impossible, because it would be an entity in which motion of real bodies would be impossible.

b. THE NATURAL MOTION OF MIXED BODIES

The initial prospects for a solution were not promising. If, as Aristotle had maintained, one element was dominant in every mixed body, the latter ought to behave in a vacuum in much the same manner as an elemental body. A compound body in which, for example, earth was dominant would presumably behave no differently in a vacuum than would a body of pure earth. All the arguments for denying the successive and finite natural motion of elemental bodies in vacua would thus also seem applicable to mixed bodies.[105]

The solution devised by medieval natural philosophers involved the abandonment of Aristotle's opinion that a single dominant element determined the natural motion of a mixed body. This step had already been taken by the 1320s, by which time some scholastics were prepared to argue that the direction of motion of a compound body was determined by the relationship between light and heavy elements conceived as contraries. If light elements dominated, the mixed body was characterized as a "light mixed body" wherein the light elements functioned as a motive force that would cause the body to rise naturally against the internal resistance of the combined activity of the heavy elements. Should the heavy elements prevail, a "heavy mixed body" would result wherein the aggregate of heavy elements would constitute the motive force and cause a downward motion resisted internally by the natural tendency of the light elements to move upward.[106] Every mixed body thus had within itself contrary light and heavy elements, the greater or more powerful of which functioned as a motive force with the other opposing it as an internal resistance. Because the conjoint action of motive force applied against an internal resistance met the fundamental Aristotelian condition for motion, it would appear that any mixed body could move with finite speed in a vacuum. Although the principle of internal resistance in mixed bodies was destined to raise havoc with Aristotelian physics, it was nevertheless an attractive theory that won a number of adherents of the first rank. Among these was Walter Burley, whose detailed discussion in his *Physics* commentary qualifies as one of the earliest, if not the first, presentations of the theory.

Here we already find a distinction between "perfect" and "imperfect" mixed bodies, though one that differs from our earlier description. Rather than using the number of elements in the compound as the basis of the distinction (see note 87), Burley differentiated perfect from imperfect by the degree of integration of the elements in the compound. An "imperfect mixed body" (*mixtum imperfectum*) is so designated because not only are its constituent elements actually distinct from each other, but they also occupy different positions within the body. In "perfect mixed bodies" (*mixta perfecta*), by contrast, none of the constituent elements exists in an actually independent and complete state; the mixed body possesses a single, uniform form (*forma mixti uniformis*).[107]

Burley further observed that two opinions have been proposed for perfect mixed bodies. In the first, the constituent elements are assumed to have been corrupted and replaced by a certain mean, substantial form. Even the heavinesses and lightnesses of the elements are replaced by mean forms. Every perfect mixed body of this type would be dominated by a single, uniform, substantial form and behave much like a pure, homogeneous element. Like a pure element, it would lack internal resistance, and because the vacuum itself offers no resistance, Burley concluded that "such a mixed body could no more be moved in a vacuum than a pure element."[108]

In the second opinion on the nature of perfect mixed bodies, Burley reported that the elements in the compound were assumed to retain their actual identities and to be more integrated than in imperfect bodies, where, as we have seen, they were not only distinct but located in different parts of the body. Because the elements retained their identities and qualities sufficiently to allow contrary qualities to act against each other, perfect and imperfect mixed bodies could possess internal resistances[109] and be capable of finite, successive, natural motion in a void.[110]

The principle of internal resistance was of great significance in medieval physics generally and was especially crucial for the justification of finite motion in a void. Not only did it alter Aristotle's concept of mixed bodies, but it also provided a mechanism that made possible the idea of finite, natural motion in a vacuum. By conceiving the actions and relationships of the elements in terms of their contrary qualities of lightness and heaviness, Burley and others established the basis for opposing a motive force to a resistance in every mixed body.[111] Because this met the universally accepted Aristotelian condition for motion in a way that was far more satisfactory than the *distantia terminorum* concept, which it replaced, the possibility that natural motion in a vacuum would differ in no way from motion in a plenum gained plausibility. It comes as no surprise to learn that some of the greatest scholastic natural philosophers of the fourteenth century – for example, Thomas Bradwardine, Albert of Saxony, John Buridan, and Marsilius of Inghen – followed Burley in adopting the new concept of mixed bodies with their internal motive forces and resistances.

But if the new ideas seemed simple, their application to a host of specific hypothetical situations raised serious problems about the nature of mixed bodies and the void spaces through which they were assumed to fall. It also posed difficulties about the nature of place. Although attempting to support the most basic Aristotelian principles of motion, such as the necessity of a motive force acting against a resistance, medieval natural philosophers often derived consequences from the new ideas that jeopardized other important, and seemingly self-evident, Aristotelian principles.

These difficulties are already evident in Walter Burley's significant discussion. We soon discover that it is not only mixed bodies with a single, uniform, substantial form that are unable to move with finite motions in a vacuum; under certain conditions other types of mixed bodies will also fail to do so because they may lack internal resistance. To illustrate, Burley first imagined a vacuum in the natural place of fire and then located therein an imperfect mixed body that lacked only the element fire. Under these conditions, the imperfect mixed body would lack internal resistance because its constituent elements – earth, water, and air – are all heavy in the region of fire. Without internal resistance, however, the body would fall instantaneously through the void that extends over the natural place of fire.[112] In all other natural places, however, this same body would fall with finite speed.[113] For if the same mixed body were located in a vacuum imagined to exist in the natural place of earth or water, its natural motion would be finite because it would then contain an internal motive force and resistance. Thus if the hypothetical vacuum were assumed in the natural place of earth, then if the combined lightnesses of water and air in the body were greater than the earth in it, the body would rise because the lightness serving as motive force would overcome the heaviness of the earth functioning as resistance. Should the earth's natural downward inclination exceed the capacity of the water and air to ascend, the body would descend. Now earth serves as motive force and air and water combine to resist.

Analysis of these examples not only reveals the significant role played by the doctrine of natural place but also shows that void spaces were not always conceived as homogeneous and identical. Apparently they differed according to their location in the sublunar realm – that is, according to the natural place of the void space. It is also obvious that every element in a mixed body is active when the body is located in the natural place of that element. These concepts represented major departures from Aristotelian physics, which treated all vacua as homogeneous and undifferentiable and also took it as axiomatic that an element was inactive in its natural place. During the fourteenth century, Burley[114] and other scholastic natural philosophers changed this scheme completely.

The vast, extended vacuum that was left behind when God was imagined to have annihilated all matter between the concavity of the lunar surface and the center of the world was neither homogeneous nor devoid of natural places, as Aristotle declared. The new concept of a mixed body composed of elements that somehow retained their identities and powers made it virtually impossible to abandon the concept of natural places. Whether an element would rise or fall depended on its place in the sublunar domain. Fixing that element in a mixed body and annihilating all matter external to it made no fundamental difference to its directionality. The essential feature of a mixed body as it was conceived in the fourteenth century was the distinction made between its internal motive and its resistive powers, a distinction that Averroes had denied to all inorganic bodies (see Section 5a). By restricting Averroes' claim to elemental bodies and denying it to mixed bodies, Latin scholastics made a momentous departure from traditional Aristotelian physics. In order to generate a dynamic relationship between internal motive and resistive powers, it was essential to oppose the contrary qualities of lightness and heaviness. But these qualities were the partially absolute and partially relative properties of the constituent elements of a mixed body. Whereas the element fire was always light and the element earth always heavy, water and air were relatively light and heavy, depending on their location. The sum total of lightness as opposed to the sum total of heaviness, which determined whether the body would rest or move and in what direction it would move, was thus dependent not only on the number and relative proportions of the elements in the mixed body but also on the location of the body, that is, whether it was in the natural place of earth, water, air, or fire. Whether these natural places were filled with their usual matter or made void by supernatural action made no difference to the internal relationships, which were wholly isolated from external conditions. Although apparently left implicit, the concept of void space is, however, seriously affected by these concepts. If the totality of the matter in the sublunar region is assumed to have been destroyed by supernatural means, that vast empty space must still be conceived as divided into four distinctive regions or natural places, each with different properties. For despite a physical homogeneity, their effect on the elements in mixed bodies must be assumed to differ. Otherwise no good reason could be found why the element air in a mixed body should behave differently in the region of space that corresponds to the former natural place of air than it behaves in the region that corresponds to the former natural place of water. Thus the potencies of the different natural places within the sublunar realm remain even when the entire region is left void of all matter. Here was a significant consequence of the new doctrine of mixed bodies, albeit one that went undetected or at least undiscussed.

No scholastic author better exemplified these developments than Albert of Saxony. By the time Albert considered the problem in the second half of the fourteenth century, it had become customary to assign arbitrary numerical values to represent the degree of lightness or heaviness that a given element possessed. When a numerical value was assigned to each of the constituent elements of a mixed body, the total degree of heaviness could be summed and compared with the total degree of lightness. If heaviness represented the greater sum, it would function as motive force and cause the body to descend in opposition to the total lightness acting as internal resistance. These relationships would, of course, be reversed if the total lightness exceeded the aggregate of heaviness. In arriving at the totals, however, the location of the mixed body was crucial – that is, whether it was in the natural place of earth, water, air, or fire.

Let us now examine two examples concerning the behavior of mixed bodies that Albert of Saxony presented in his *Questions on the Physics*. Not only do they incorporate much that has already been described but they include startling consequences that were incompatible with certain basic features of Aristotelian physics.

In the first example,[115] Albert assumed a mixed body that consists of earth, assigned a value of 3, and air, assigned a value of 2. He then imagined that everything below the sphere of fire is annihilated, leaving behind a vast vacuum. Because fire remains in its sphere and would offer resistance to any body moving through it, Albert assigned to fire a resistive capacity of 1.[116] If the mixed body falls first through the sphere of fire, the values of earth and air will be added to constitute a motive force of 5 "because both earth and air [of the mixed body] are outside their natural places, and because both seek to descend through fire." Because the fire offers a resistance of 1, the ratio of motive force to resistance – external resistance in this case – will be $F/R = 5/1$, where F is motive force and R the external resistance of the fire.

But when our heavy mixed body reaches the natural place of air, which is now void,

the air [in the mixed body], which is as 2, begins to resist the earth,[117] which is as 3, because the air strives to remain there. But since this mixed body is dominated by earth, it descends further so that only the earth in it moves [i.e., causes motion] while the air resists, whereas previously in the full medium [of fire] both were moving. Thus the mixed body descends further into the vacuum by a ratio of 3 to 2, which is less than a quintuple ratio, as is the motion it produces. And so, the same mixed body has a much slower motion in a vacuum than in a plenum, namely fire.

Of great interest in this example is Albert's conclusion that, under certain conditions, one and the same heavy mixed body might descend with a greater velocity in a plenum than in a vacuum, a state of affairs that contradicted Aristotelian physics. Because such examples were easily fabricated, it seemed that the theory of mixed bodies was either productive of absurdities, and therefore to be rejected, or it was not unreasonable to accept Albert's consequence given the circumstances that produced it. After all, as Albert explained, "we have never experienced the existence of vacuum, and so we do not readily know what would happen if a vacuum existed."[118] To understand what might occur in a vacuum, we must devise hypothetical situations and derive consequences from them, recognizing, however, that such conclusions are only conjectural.[119]

Although numerous examples of this kind had been devised, Burley, who had himself constructed one,[120] cautioned that it was nevertheless impossible for one and the same mixed body to be moved equally quickly in void and plenum when the same internal resistance was operative in both places.[121] That is, a given mixed body could not possibly fall with the same speed in the natural place of air when the air was there and when it was absent. A fortiori, then, a mixed body could not fall more quickly in such a void than in a plenum. Generally Burley and other medieval scholastics thought it impossible for a mixed body to fall equally quickly, or more quickly, in a plenum than in a void when the comparisons were made for the same natural place. For if the same magnitude of internal resistance operated in void and plenum, then the external resistance of the plenum itself must, of necessity, further slow the mixed body. Only when the comparisons were made for different natural places, one a plenum and the other imagined as void, was it thought possible.[122] But even this was a significant departure from Aristotelian physics and truly subversive of it. That it was not actually used to undermine Aristotelian physics is also typical of numerous medieval physical ideas that were developed but never exploited.

In the second of his informative examples, Albert of Saxony assumed a mixed body composed of all four elements, which are assigned the following degrees: 1 to fire; 1 to air; 1 to water; and 4 to earth. Assuming the existence of a vacuum in the entire sublunar region, Albert demonstrated that the natural downward motion of a mixed body in a homogeneous void space can be quicker at the beginning than at the end, which was contrary to the universally accepted opinion in Aristotelian physics that natural downward motion is faster at the end than at the beginning. In support of his conclusion, Albert declared that "if the mixed body were placed where the fire was, then this mixed body would descend more quickly through the vacuum of fire (*vacuum ignis*) than through the vacuum of air (*vacuum aeris*), and so on, as can easily be deduced from this case." Although Albert did not produce the ratios of motion, it would appear that the ratio of motive force to internal resistance in the "vacuum of fire" would be 6 to 1; in the "vacuum of air" 5 to 2; in the "vacuum of water" 4 to 3, the ratio it would retain through the "vacuum of earth" until it came to rest at the center of the world.[123] Because the ratio of force to internal resistance diminishes in each successive natural place, the speed will also decrease and the body would be slower at the end of its motion than at the beginning.[124] "What should be said, therefore," asked Albert, "about the common assertion that natural motion is quicker in the end than at the beginning? One can say," he replied, "that it is universally true of the motion of heavy and light [elemental] bodies but not of the motion of heavy and light mixed [or compound] bodies."[125]

c. HOW THE CONCEPT OF MIXED BODY DESTROYED THE HOMOGENEITY OF THE VACUUM

Just as significant as the dramatically anti-Aristotelian consequences that Albert drew from the doctrine of mixed bodies falling in vacua is the insight, perhaps unintentional, that he provided into the nature of the void space through which bodies fell or rose. By using the expressions "vacuum of fire" (*vacuum ignis*) and "vacuum of air" (*vacuum aeris*), Albert gave utterance to what was apparently an implicit assumption of all who considered the problem of the natural motion of mixed bodies in vacua: When the whole of a material element is imagined as annihilated in its natural place, the void space left behind retains the fundamental properties associated with the natural place of that element. Albert's terminology was not a mere convenience for locating mixed bodies in the vast vacuum that might be imagined in the sublunar region; it was intended to signify real differences that occur in the behavior of mixed bodies as a consequence of their spatial location. Thus if God annihilated all matter in the sublunar region, the total vacuum left behind would be subdivided into four concentric vacuous regions in each of which one and the same body would behave differently. Not only would its velocity differ in each region, but in regions where no internal resistance operated, some, like Burley, believed its motion would be instantaneous. In effect, although the entire void space within the concavity of the lunar sphere is obviously homogeneous – for one part could not physically differ from another – the behavior of a single mixed body would differ in each of its four natural places, or subdivisions.

But was this variation in the behavior of a given mixed body attributable solely to the different powers of the natural places of the void space itself, or was it caused only by the internal constitution of the mixed body? Or was it a combination of the two – mixed body and void space? The proper explanation must be sought in the latter, in the relationship between mixed body and space. Each element in a mixed body that is placed in a void acts as if it were independent of the other constituent elements. It somehow "knows" its location and acts accordingly. Water in a mixed body will rise in the natural place of earth and fall in the natural places of fire and air. The ultimate direction of the natural motion of a mixed body and its actual speed are thus determined by the total number of degrees of lightness versus the total number of degrees of heaviness: When the former is in excess, the body rises; when the latter predominates, the body falls. In each case, the speed will vary as the ratio of heaviness to lightness. Because each element behaves differently in different natural places, the totals of heaviness and lightness will alter with the location of the body. Now one might argue that it is basically the mixed body itself that determines the motion of the body and that the void space, which is homogeneous throughout, has no effect on the actual behavior of the constituent elements of the mixed body. In this interpretation, the homogeneous void space is conceived as physically inert and undifferentiated, with the constituent elements somehow capable of knowing or sensing the specific natural place of the body at any given moment. The interactions of the constituent elements are thus solely responsible for the resultant direction and speed of motion.

But this account is surely unsatisfactory. More is needed. It does not explain why, if the sublunar void space is completely homogeneous with respect to all possible properties, any particular element in a mixed body should abruptly change its behavior when it crosses an invisible line of demarcation between, say, the sphere of fire and the sphere of air. Although medieval natural philosophers appear to have ignored explicit consideration of this problem, they implicitly assumed that any hypothetical void space between the lunar sphere and the center of the world was objectively distinguished into the same four natural places as would have existed if the space were in its usual plenistic state. That an element could alter its behavior because it somehow knows when it has passed from one natural place to another must have been attributed by them to actual differences in the natural places themselves and hence to distinctions in the "homogeneous" void space. The natural places of the sublunar world thus retained their capacity to affect the behavior of the elements, whether or not they were filled with matter.[126] We may then conclude that elements fixed in a mixed body were still able to sense these spatial differences, or natural places, even when all the matter was removed from them. In this explanation, the elements in the mixed body and the external void space itself are simultaneously efficacious in producing the resultant motions.

Medieval scholastics thus took seriously Aristotle's sense of the marvelous potency of place, which, as Aristotle put it, "does not pass out of existence when the things in it are annihilated" (for the complete passage, see Chapter 1). Their interpretations, however, would hardly have pleased Aristotle, who emphasized the lack of distinguishing features between the parts of an extended void space and who, for that reason, denied the possibility of natural places in such a homogeneous emptiness. Without natural places, he also denied that violent or natural motions were possible (see Chapter 1.) These ideas were dramatically altered in the fourteenth century. The potency of natural place became far greater than dreamed of by Aristotle. For medieval natural philosophers, natural place was efficacious in void as well as plenum. However absurd the implication that one part of void space could differ from another, this was almost an inevitable consequence for all in the fourteenth century who assumed that the regular Aristotelian natural places continued to exist and function even when all those places were void of matter. Apparently unaware of the full impact of their doctrine of mixed bodies, many in the Middle Ages committed themselves, however unwittingly, to a belief in the nonhomogeneity of void space.

d. ON THE EQUALITY OF VELOCITIES OF HOMOGENEOUS, MIXED BODIES FALLING IN VOID SPACE

The new conception of mixed bodies in the fourteenth century produced at least one additional major deviation from traditional Aristotelian physics. On the assumption that bodies fall in a single, uniform plenum with velocities directly proportional to their respective weights, Aristotle demonstrated that in a void, bodies of unequal weight would fall with equal velocities. Without a material medium to cleave, he believed that no plausible reason could be offered to explain why a heavier body should fall with a proportionately greater speed than a lighter body (see Chapter 1). In a void, all bodies of whatever weight or shape would descend with equal facility and therefore with equal speed. The absurd idea that in a void a very heavy body would fall with the same speed as a very light body was taken by Aristotle as further evidence that the existence of void space was impossible.

The doctrine of mixed bodies changed this idea. Although most scholastics agreed with Aristotle that pure, elemental bodies would behave as he had described, *homogeneous* mixed bodies would not. Unequal bodies of the latter kind would indeed fall with equal velocities in the void. It was a proper and reasonable consequence that followed from the nature of homogeneous bodies and was already recognized by Walter Burley, who declared, without further discussion or amplification, that two unequal bodies of homogeneous composition (*equalis commixtionis*) would fall with equal speed in a vacuum.[127] It is probable that Burley's rationale for this daring statement was much the same as that given by Thomas Bradwardine (ca. 1290–1349) at approximately the same time. In his *Tractatus de proportionibus*, perhaps the most significant and influential medieval treatise on mathematical theories of motion, Bradwardine declared that "all mixed bodies of similar composition will move at equal speed in a vacuum" because "in all such cases the moving powers bear the same proportion to their resistances."[128] Years later, Albert of Saxony adopted the same conclusion with much the same justification.[129]

When examined against the background of the development of the concept of a mixed body, derivation of this significant conclusion seems almost inevitable. If differently proportioned mixed bodies of two to four elements were assumed to fall with different speeds in a vacuum, it seemed plausible to suppose that identically proportioned, homogeneous mixed bodies ought to fall or rise with equal speeds, regardless of differences in size. Because lightness and heaviness were the causes of natural motion in mixed bodies – and not mere size – two homogeneous bodies of unequal size were identical in all the properties that determined speed. It was logical to infer their equality of velocities in a void, although as we saw, not in a plenum, where the resistance of the external medium was conceived (at least by Albert of Saxony) as altering the total equality of motive force and resistance (see note 129).

Among those who adopted the conclusion of equality of fall in a void for unequal, homogeneous mixed bodies, only Thomas Bradwardine felt the need to explain how this radical departure from Aristotle's physics might be conceived. In his monumental *Tractatus de proportionibus*, in which he developed the powerful concept of "ratio of ratios,"[130] Bradwardine made use of a widely accepted distinction between intensive and extensive measure, as for example, between intensity of heat (temperature) and quantity of heat, and also between specific weight (an intensive factor) and gross weight (an extensive factor).[131] It was this kind of distinction that Bradwardine applied to ratios of force and internal resistance in mixed bodies. Concerning resistance, he explained:

It is not inconsistent for the same [agent] to have an identical proportion *qualitatively*, i.e. in its power of acting, with both the whole and the part [of its resistance], but not so *quantitatively*. For although the whole and the part are unequal in quantity, they can however be equal in the quality of resisting. And therefore just as they do not differ in their quality of resisting, so their movements through media do not differ in quality of motion (which is swiftness and slowness), but rather in quantity of motion, which is [in] length or brevity of time.[132]

Similarly, motive forces

can be proportional to their resistances *qualitatively*, that is, in their power of acting. From such proportionality arises equality of motions *qualitatively*, that is, in swiftness and slowness.[133]

In arriving at equality of speed for two unequal but homogeneous mixed bodies in the void, Bradwardine has treated force (heaviness) and internal resistance (lightness) intensively, or qualitatively, rather than quantitatively. Qualitatively, a ratio of force to internal resistance, or heavy to light, is the same for any unit or quantity of homogeneous matter. Whatever the size of the two unequal, homogeneous mixed bodies, in each of them the ratio of force to resistance per unit of matter is equal. Now because "from such proportionality arises equality of motions *qualitatively*, that is, in swiftness and slowness," it follows – in accordance with the universally accepted axiom of medieval physics that equal ratios of force to resistance produce equal speeds – that these two unequal, homogeneous bodies would fall with equal speeds in an extended vacuum.[134]

The significance of this concept must not be underestimated. The behavior of mixed bodies in a vacuum as it came to be understood and accepted by some fourteenth-century authors was totally subversive of at least two of Aristotle's most basic arguments against the existence of void space. Whatever the difficulties for the natural motion of elemental bodies in a vacuum, the case for mixed bodies seemed relatively straightforward. With light and heavy elements functioning as motive forces and internal resistances, it was obvious that they would move with finite, successive motions in a vacuum, thus producing a plausible counterinstance to Aristotle's claim that motions in a void would be instantaneous. But not only would mixed bodies move with finite speeds, they would also move with different speeds, depending on the ratio of the degrees of heaviness to lightness. Where the ratios differed, so did the speeds. Here was a consequence that shattered Aristotle's other argument that in the absence of a material medium, as in a void, all bodies of whatever weight would fall with equal speeds. With no material medium to cleave, Aristotle could find no justification for a heavier body falling faster than a lighter body and concluded that they would fall with equal speed, which he thought absurd. For Burley, Bradwardine, and others, however, equality of fall was a logical and intelligible consequence of the doctrine of mixed bodies[135] because the latter contained their own motive forces and internal resistances. When applied to motion in a vacuum, the doctrine of mixed bodies, with its associated concept of internal motive and resistive forces, contradicted Aristotle's most effective arguments against the vacuum. Whereas none accepted the real existence of vacuum, it had become apparent that if such an entity did exist, the motion of mixed bodies (and for some, elemental bodies as well) would be finite and successive and, depending on circumstances, different or equal in velocity. Motion in a vacuum had become a reasonable and plausible concept.[136]

8. MOTION IN A VACUUM IN THE SIXTEENTH AND SEVENTEENTH CENTURIES: GALILEO AND THE MEDIEVAL TRADITION

Centuries of medieval discussion of the conditions of motion in a hypothetical vacuum were continued by Aristotelian commentators on the *Physics* well into the seventeenth century. The Coimbra Jesuits (*Conimbricenses*) and the Barefoot Brothers of the College of Alcalá de Henarez (*Complutenses*), with their collective commentaries in the sixteenth century, as well as various seventeenth-century commentators such as Bartholomeus Amicus, the Scotist commentators Bartholomeus Mastrius and Bonaventura Bellutus, and the Thomist John of Saint Thomas in his *Cursus Philosophicus Thomisticus*, all had occasion to consider motion in a vacuum. But despite greater organization of the topic and slight alterations and additions, the basic fund of arguments remained essentially what it had become by the end of the fourteenth century.[137] Not even the new availability in Latin translation of the *Physics* commentaries of Philoponus and Simplicius did much to alter the situation. Indeed, although Philoponus had made the celebrated observation that two unequal weights dropped from a given height would strike the ground at almost the same time, he denied this for the same bodies falling in a void, in which, he insisted, "the same space will . . . be traversed by the heavier body in shorter time and by the lighter body in longer time. . . ."[138] In comparing the fall of two unequal weights in a void, Philoponus adopted the very Aristotelian position he had refuted for fall in a material medium. Bodies of different weight of whatever composition would not fall with equal velocities in a vacuum. But if Philoponus did not distinguish between elemental and mixed bodies, and therefore did not arrive at the medieval conclusion that homogeneous mixed bodies would fall with equal speeds in a vacuum, he did attribute the differentiation of velocities in a vacuum to the gross weights of bodies,[139] a move that focused attention on the mobile itself, rather than on any resistive capacity of the vacuum or on tensions between contrary qualities such as heaviness and lightness. Nor did he distinguish between parts of a vacuum corresponding to the natural places of the elements. A vacuum and its parts were identical in behavior and properties. Although the appearance of the works of Philoponus did not introduce much that was radically new on the subject of motion in a void, his ideas both supplemented and conflicted with the anti-Aristotelian medieval tradition. Because of their antique Greek provenance, however, Philoponus' contributions gained added credence and authority.

Galileo's arguments and conclusions on the possibility of motion in a void must be viewed against the background just described. In his early career, he responded to traditional problems on motion raised in the works of Aristotle. Although he made virtually no mention of the large traditional commentary literature on the works of Aristotle, especially the *Physics*, Galileo clearly benefited from the centuries of medieval discussion that sought to justify finite motion in a vacuum. It is no exaggeration to interpret Galileo's contributions on the problem of finite motion in the void as a continuation of the late medieval tradition. But Galileo was no scholastic. Although he may have profited from the medieval legacy, which was still alive and well known in the sixteenth century, he departed from that tradition in significant ways. Even more than his predecessors, Galileo recognized the centrality of the problem of motion in a resistanceless void space. In what follows, we shall describe and identify the old and the new in his treatment of this fundamental theme.

It is typical of Galileo that his interest in the vacuum was not as a place or space, but only as a possible resistanceless backdrop for motion. He was primarily concerned with problems of motion and not with philosophical and metaphysical considerations about the existence and nature of vacuum. Notably absent in his reflections are the numerous troublesome problems that preoccupied scholastic commentators, as for example, whether a dimensional void space and a material body can occupy the same place simultaneously, and whether a vacuum is a separate space without substance or accident. He simply ignored the philosophical and theological aspects and ramifications of traditional scholasticism. Galileo was content merely to assume the hypothetical existence of vacuum and to explore the consequences of an assumed motion in the absence of all resistance. A vacuum was simply an entity devoid of matter and therefore unable to oppose resistance to any body that might be placed in it.

Galileo's most extensive discussion of motion in a void appears in his *De motu*, written about 1590 and unpublished in his lifetime.[140] Like so many of his medieval predecessors and sixteenth-century contemporaries, Galileo discussed all of Aristotle's arguments except the inertial consequence (see Chapter 1 and note 13). In the tenth chapter of *De motu*, Galileo argues "in opposition to Aristotle . . . that if there were a void, motion in it would not take place instantaneously, but in time."[141] In refuting Aristotle, Galileo did not invoke the *distantia terminorum* argument as justification for successive and finite motion. Instead, he developed the position associated first with Avempace in the Middle Ages and then also with Philoponus in the sixteenth century, namely that "natural speed" is a function of "natural weight" and that both are completely effective only in a void because a material medium serves only to retard the natural speed of the natural weight of the body. As Galileo explained:

to put it briefly, my whole point is this. Suppose there is a heavy body *a*, whose proper and natural weight is 1000. Its weight in any plenum whatever will be less than 1000, and therefore the speed of its motion in any plenum will be less than 1000. Thus if we assume a medium such that the weight of a volume of it equal to the volume of *a* is only 1, then the weight of *a* in this medium will be 999. Therefore its speed too will be 999. And the speed of *a* will be 1000 only in a medium in which its weight is 1000, and that will be nowhere except in a void.[142]

Galileo saw this as a refutation of Aristotle's argument because it could now be readily seen "that motion in a void does not have to be instantaneous."[143]

In this example, we see that the determinant of velocity for Galileo is the effective, rather than the gross, weight of a body. It is the comparison of an equal volume of body with an equal volume of medium that determines the velocity of a falling body. Already in the eighth chapter of *De motu*, Galileo had argued that homogeneous bodies of unequal size and therefore unequal weight would fall in a uniform medium with equal speeds. The essence of his argument emerged in a refutation of the Aristotelian contention that the doubling of a body's weight results in a doubling of its speed. Galileo argued that

if we suppose that bodies *a* and *b* are equal and very close to each other, all will agree that they will move with equal speed. And if we imagine that they are joined together while moving, why, I ask, will they double the speed of their motion, as Aristotle held, or increase their speed at all? Let us then consider it sufficiently corroborated that there is no reason *per se* why bodies of the same material should move [in natural motion] with unequal velocities, but every reason why they should move with equal velocity.[144]

This conclusion, first formulated for fall in a medium,[145] was later extended to cover fall in the void when Galileo declared that "we can show that bodies of the same material but of different size move with the same speed in a void."[146]

For Galileo, the effective weight, or heaviness, of a body in a medium was dependent on the difference in the specific weights of the body and the medium through which it fell. Galileo explained effective specific weight when he declared that "in a plenum, such as that which surrounds us, things do not weigh their proper and natural weight, but they will always be lighter to the extent that they are in a heavier medium. Indeed, a body will be lighter by an amount equal to the weight, in a void, of a volume of the medium equal to the volume of the body."[147] Thus a difference in specific weights determined the velocity of a body, so that the free fall of a body is representable as

$V \propto$ *specific weight of body* – *specific weight of medium*, where V is speed; and the velocity of a rising body as

$V \propto$ *specific weight of medium* – *specific weight of body*.

Because the specific weight of a vacuum is zero, bodies could only fall – not rise – in empty space with speeds that are directly proportional to their specific weights.[148] Obviously, then, bodies of different specific weights will fall with different speeds in a vacuum,[149] and bodies of the same specific weight will fall with equal speeds. Consequently, if two bodies of unequal size have the same specific weights, they must necessarily fall with equal speeds in the same medium and with equal but greater speeds in a void.

Galileo thus arrived at the same conclusion as had Burley, Bradwardine, and Albert of Saxony more than two centuries earlier: Unequal, homogeneous bodies of like composition will fall in a void with the same speed. A comparison with Bradwardine, who alone in the Middle Ages provided reasons for his conclusion, is instructive. It is clear that both Galileo and Bradwardine employed intensive rather than gross or quantitative factors. Whereas Bradwardine employed a ratio of force to internal resistance per unit of matter, Galileo used weight per unit volume of homogeneous matter. Equality of the respective intensive factors in bodies of similar composition but unequal size would produce equality of speed.

Not only did Galileo adopt the medieval conclusion of equality of fall of unequal, homogeneous bodies in a void, but he probably would have agreed with Nicholas Bonetus that projectile motion in a vacuum could be achieved by a self-expending impetus.[150] Despite the obvious and undeniable connections that link Galileo to the long medieval and Renaissance scholastic tradition in favor of finite motion in a void, there are fundamental differences that set Galileo apart from his predecessors, who were essentially Aristotelians on the basic principles of physics. With Galileo we have moved outside of the Aristotelian framework and into early modern physics.

Whereas Galileo's medieval predecessors assumed that heaviness and lightness were absolute properties of bodies and conceived those qualities in mixed bodies as contraries that functioned as motive force and internal resistance, Galileo completely abandoned the concept of absolute heaviness and lightness. He insisted that everything has weight and that in any comparison of bodies one will be heavier, equal to, or less heavy than another. Void alone is weightless – and it is not a substance.[151] Abandoning absolute heaviness and lightness, Galileo used the notion of relative density in the precise form of specific weight, and thus had a means of determining true weights in a void and effective weights in media.

In abandoning absolute heaviness and lightness, Galileo avoided some of the major, though often ignored, difficulties that derived from the application of this pair of contraries to motion in void and plenum. Whereas medieval Aristotelians explained direction of motion in terms of absolute lightness and heaviness functioning as motive qualities, Galileo relied on the relation between the specific weights of body and medium, the latter equal to zero in a void. Gone is the crucial medieval distinction between pure, or simple, elemental bodies and mixed bodies. With his concept of specific weight, Galileo treated *all bodies alike*. Whatever a body's material composition, its specific weight could be determined and all the appropriate comparisons made in plenum and void. In effect, Galileo replaced the simple and mixed bodies of the Middle Ages with bodies conceived as homogeneous, Archimedian magnitudes. The medieval concept of internal resistance was thus rendered meaningless. In Galileo's scheme, bodies that fell in a void did so with speeds that were directly proportional to their specific weights. In the medieval conception, mixed bodies could move naturally up or down in a vacuum regardless of their absolute weights. Direction of motion was determined by the dominance of lightness or heaviness. A material mixed body that had components of earth and water but was yet dominated by the combined lightness of its light elements would, despite the possession of a certain determinate weight, move upward in a void toward the region of fire. For Galileo such a body could only move downward. But the reason such a mixed body, despite its weight, would rise toward the region of fire in the medieval interpretation follows from the Aristotelian doctrine of natural place. The medieval vacuum was usually assumed to lie between the moon and earth. Despite the postulated annihilation of all matter in this region – earth, water, air, and fire – the vacuum of each elemental region continued to exert an influence. Thus the vacuum of fire or vacuum of air, as Albert of Saxony referred to the void regions formerly occupied by each of these elements (see above, Section 7c) continued to affect the behavior of material elements. A body dominated by the combined lightness of fire and air, for example, would rise upward in the void if located in the former regions of water or earth. With Galileo's reliance on specific weight and density, the doctrine of natural place was virtually abandoned as a factor in the determination of the direction of motion. For Galileo, void space was, by implication – for he did not discuss the problem – truly homogeneous and without any capacity to influence the behavior of material bodies.

In the long, controversial history of the problem of the possibility of finite motion in a void, Galileo played a decisive and crucial role. He restored the tradition usually associated with the names of Philoponus and Avempace, namely that velocity is determined by the *difference* – not the ratio, as Aristotle and most medieval scholastics believed – between the weight of a body and the resistant medium through which it falls. When that resistant medium is removed, as in a vacuum, a body will fall with a natural, finite velocity, a velocity that Philoponus presumed to be proportional to the gross weight of the body and that Avempace may have taken to vary as the "limitation of the agent" (*limitatio agentis*), an opinion reported in the Middle Ages but always rejected.[152] During the Middle Ages, motion was never conceived as arising from a simple subtractive relation between motive power and resistant medium. The resistance of a material medium was something to be divided into – not subtracted from – the motive power. Motion of any elemental or mixed body, whether in plenum or void, was the consequence of a ratio of motive force to resistance. The total resistance against a mixed body falling in a plenum was the sum of its internal resistance plus the resistive capacity of the external plenum. This total resistance was then divided into the motive force. (For an example from Albert of Saxony, see note 129 of this chapter.) In a vacuum, the internal resistance alone was divided into the motive force. The universally accepted physical and mathematical representation of motion as arising from a ratio of motive force to resistance[153] virtually compelled medieval natural philosophers to accept Aristotle's position that finite motion without resistance was impossible. For Aristotle that resistance could only be associated with a material medium. By accepting the possibility of finite motion in a vacuum, medieval scholastics sought to identify resistances in the vacuum itself (the *distantia terminorum*), or in the body, or in both.

With Galileo resistance became utterly superfluous, a mere retardant to motion. Galileo upheld the more fruitful subtractive law of Philoponus and Avempace. But whereas neither Philoponus nor Avempace had provided a physical mechanism that would explain how the subtraction of all or part of a material, resistant medium could produce a natural, finite velocity in void or plenum, Galileo made good this serious deficiency in *De motu*. By introducing specific weight as the criterion for measuring the difference between body and medium,[154] Galileo, and indeed Benedetti before him,[155] made the subtractive law intelligible.

Specific weight provided Galileo with a consistent standard of comparison for motions in void and plenum, a consistency that was nowhere in evidence in the medieval theory of mixed bodies involving ratios of motive force to internal resistance. By adjusting the numerical values assigned the individual elements in a mixed body, one could generate strange results, as Albert of Saxony demonstrated when he argued that two unequal, homogeneous bodies that fell with equal speeds in a void would fall with unequal speeds in a homogeneous, material medium; that the same mixed body could fall more quickly in a plenum than in a vacuum; and that one and the same mixed body could fall with a greater velocity at the beginning of its motion than at the end.[156] These conclusions were in conflict with Aristotelian physics and impossible to reconcile with it. While that physics survived, discussions of motion in a void would be of an ad hoc and anomalous nature. Galileo changed this situation. From his earliest discussion in De motu to the final pronouncement in the Discorsi that all bodies of whatever weight and composition would fall in a vacuum with equal speed, Galileo made motion in a void an integral part of a new non-Aristotelian physics. And yet Galileo's path was prepared by his medieval predecessors. Their numerous arguments, fashioned over the centuries, made the possibility of finite motion in a void plausible and intelligible. Not only were these arguments available to Galileo, they made his own proposals far less radical than they might otherwise have appeared.

But if quite a few of Galileo's medieval forerunners embraced the idea that motion in a void space could be finite, they were unable to accept the possibility that a real, extended vacuum could exist anywhere in the finite cosmos. On this vital point, they were in complete agreement with Aristotle. The world was a material plenum, and nature abhorred a vacuum. The hypothetical void of the discussions on motion was a natural impossibility. Let us now examine the basis for this unyielding medieval belief.

4

Nature's abhorrence of a vacuum

1. FORMULATION OF THE CONCEPT

Few dicta are more inextricably linked with the Middle Ages than the declaration that "nature abhors a vacuum." Although the full significance of this famous principle would be described and explicated only in the fourteenth century, it had already emerged in the thirteenth, when expressions such as *natura abhorret vacuum, horror vacui*, and *fuga vacui* began to appear.[1] The origin of the principle is, however, unknown. But already in the first half of the twelfth century, Adelard of Bath expressed the fundamental idea of nature's resistance to a vacuum. In denying that magic plays any role in the failure of water to pour out of the holes in the bottom of a clepsydra when the holes in the top are stopped up, Adelard explained this strange phenomenon in cosmic terms by observing that

the body of this sensible universe is composed of four elements; they are so closely bound together by natural affection, that just as none of them would exist without the other, so no place is empty of them. Hence it happens, that as soon as one of them leaves its position, another immediately takes its place; nor is this again able to leave its position, until another which it regards with special affection is able to succeed it. When, therefore, the entrance is closed to that which is to come in, it will be all in vain that you open an exit for the water, unless you give an entrance to the air. . . . Hence it happens that if there be no opening in the upper part of the vessel, and an opening be made at the lower end, it is only after an interval, and with a sort of murmuring, that the liquid comes forth.[2]

Matter that departs one place in a cosmic plenum must always be replaced by contiguous and adjacent matter. Adelard's account emphasized in positive terms the natural affection of matter rather than nature's desire to avoid the formation of a vacuum at all cost. With Adelard we do not yet have an explicit declaration that nature abhors a vacuum.

A quite similar idea was presented by the Greek pneumaticist Philo of

Byzantium (fl. ca. 250 B.C.), whose *Liber de ingeniis spiritualibus* (*Book on Pneumatic Devices*) had been translated from Arabic to Latin no later than the thirteenth century.[3] In general agreement with Aristotle (see below), Philo believed that air and water (and also air and fire) were commingled and continuous and that water would follow air as if glued to it with birdlime.[4] Thus if air rose, it would draw up the water below it. Philo thus denied the existence of natural or artificial separate vacua, for "if air is evacuated, any of the bodies that are mixed with it succeed it immediately because by their very natures they are impelled [to do so]."[5] The idea that water would follow air upward because they were continuous was also adopted by Peter of Abano. In his *Conciliator differentiarum*, Peter argued that air and water shared a common surface so that one was inclined toward the other. Indeed, if air were removed from above water, the latter, because it normally shares a common surface with air, would by a violent motion rise upward to fill the space left vacant by the air. Although Peter may have believed that nature abhors a vacuum, he made no appeal to such a principle. On the contrary, he explained further that if water or earth were withdrawn from below air, the latter "tends naturally [to move downward] into its place by the inclination of its heaviness *and not by the necessity [of avoiding] a vacuum.*"[6]

With Philo, Adelard, and Peter of Abano, water and air can move in directions contrary to their natural inclinations because of some natural and common affinity with the element that lies naturally above them. If nature's abhorrence of a vacuum was an operative principle in their thought, it was only tacit. A more fundamental and direct source was Averroes' commentary on Aristotle's *De caelo*. There can be little doubt that Aristotle's vigorous attacks against the existence of any kind of vacuum constituted the broad background from which the principle that nature abhors a vacuum was ultimately derived.[7] But Aristotle himself formulated no equivalent of the medieval principle. And yet one passage of his may be singled out as of particular significance because it provided the point of departure for the commentary by Averroes. In *De caelo* (4.5.312b.4–12), Aristotle declared that "air will not move upwards into the place of fire, if the fire is removed, except by force in the same way as water is drawn up (σπᾶται) when its surface is amalgamated with that of air and the upward suction (σπάσῃ) acts more swiftly than its own downward tendency. Nor will water move into the place of air, except as just described."[8] In his explication, Averroes argued[9] that in order to prevent formation of a vacuum, air would descend (when the water below was removed) and water would rise (when the air above was removed). To forestall formation of vacua, both elements would thus act contrary to their natural tendencies. As reinforcement for his position, Averroes also mentioned the opinion of Alexander of Aphrodisias that when air contracts and occupies a smaller volume, it attracts water upward to prevent formation of a vacuum.[10]

Taken together, the ideas of Adelard, Philo, and Averroes clearly express the concept of nature's abhorrence of a vacuum. Only a name or phrase was needed to identify it for convenient reference, and at least three such were supplied in the thirteenth century. Underlying these shorthand expressions was the assumption that matter necessarily and inevitably rushed in to fill places or spaces that were in danger of becoming void and would do so even if required to move in directions contrary to its natural inclinations. Air would descend and water would rise.

But what was there in the nature of matter, or in the structure of the cosmos, that enabled an element to move in a direction that was contrary to its natural tendency? What caused air to descend in order to prevent formation of a vacuum, and why did water rise? In brief, how could nature's abhorrence of a vacuum be explained?

2. UNIVERSAL AND PARTICULAR NATURES

The major response to this fundamental question had already been proposed in the thirteenth century by Roger Bacon,[11] Johannes Quidort of Paris,[12] and the anonymous author of the *Summa philosophiae*, falsely ascribed to Robert Grosseteste.[13] They assumed that all bodies are influenced by at least two natures, particular and universal. By virtue of its particular nature (*natura particularis* or *specialis*) a body moves to its natural place, a heavy body down and a light body up.[14] In the ordinary course of events, bodies behave in accordance with their particular natures. But matter is also influenced by a universal nature (*natura universalis*) whose origin is the celestial region whence this universal power or influence – sometimes described as a "celestial" or "supercelestial agent" (*agens celeste* or *superceleste*)[15] or "celestial force" (*virtus caelestis*)[16] – is diffused throughout the sublunar realm. In the material plenum of the Aristotelian cosmos, the universal nature was conceived as the vigilant guardian of material continuity, a veritable universal regulative power.[17] Indeed an inexorable tendency to preserve and maintain the continuity of that plenum was its most essential characteristic.[18] Whenever formation of a vacuum was imminent and danger of separation threatened cosmic continuity, the omnipresent universal nature would, if necessary, cause bodies to act contrary to their particular natural tendencies: Heavy bodies would rise and light bodies fall. To preserve the continuity of the universe, the heavens themselves would even transform their natural circular motion to rectilinear motion. Thus if fire were suddenly to descend from its natural place below the concave surface of the lunar sphere, John Dumbleton believed that the heavens would immediately move down in a straight line to prevent formation of the dreaded vacuum.[19] Such behavior, which was contrary to Aristotelian physics, was not, however, caused by fear of a vacuum. As Bacon put it, "that a vacuum not occur" (*ne fiat vacuum*) is not a cause. A mere negation cannot be the cause of a positive action.[20] The cause of the contrary behavior of bodies is the universal nature, or agent, diffused everywhere in the sublunar world. It was a power capable of acting externally on bodies, or was perhaps something absorbed by all bodies and therefore an inherent property of matter.[21]

Perhaps ultimately derived from the pseudo-Aristotelian *Liber de causis*, which had been translated by Gerard of Cremona from Arabic to Latin in the twelfth century,[22] the distinction between particular and universal natures was widely accepted by the end of the fourteenth century. By this time it was sometimes encapsulated, without elaboration, in different versions of the expression "nature abhors a vacuum."

3. THE EVIDENCE FOR AND AGAINST NATURE'S ABHORRENCE OF AN INTERSTITIAL VACUUM (*VACUUM IMBIBITUM*)

As a fundamental medieval principle, the concept of a universal nature was often used to explain certain observations and experiences that were thought to illustrate nature's abhorrence of a vacuum. Largely inherited from Greek and Arabic sources, these "experiences" (*experientia*) or "experiments" (*experimenta*), as they were frequently designated, represented the empirical basis of the belief that nature would not tolerate the existence of vacua. They were usually grouped in clusters of three to five and were included in commentaries or questions on the fourth book of the *Physics*, either in some variant of the question "whether the existence of a vacuum is possible"[23] or occasionally in that part of the fourth book where Aristotle discussed rarefaction and condensation.[24] Ironically, these experiments were usually presented as initial evidence for the existence of vacua, only to be revealed subsequently (in the same question) as instances of nature's abhorrence of a vacuum.[25] As the oft-cited evidential basis for the rejection of the existence of vacua, a number of these experiences must be described to convey an appreciation for the cumulative impact they had on medieval physical thought.

Scholastics were concerned about two kinds of intramundane vacua, small and large. The first of these was conceived as a multitude of minute, interstitial, void pores lying between particles of matter and designated by expressions such as *vacuum imbibitum*,[26] *vacuum interceptum*,[27] *vacuum diffusum*,[28] *vacuum infusum*,[29] and *vacuum immixtum*;[30] the second, commonly referred to as *vacuum separatum*,[31] was described as an extended, separately existing space.[32]

a. THE PROBLEM OF CONDENSATION AND RAREFACTION

Because interstitial vacua were held to lie between the particles of material bodies and were conceived by the ancient Greek atomists as fundamental for the explanation of condensation and rarefaction, the *vacuum imbibitum* was usually discussed in the context of condensation and rarefaction in the commentaries and questions on the fourth book of Aristotle's *Physics*.[33] Two experiences that were commonly presented as *prima facie* evidence in favor of the existence of microvacua were drawn from Aristotle. In the more significant and widely used of the two, Aristotle explained that a proof offered by those who believe in the existence of void "is what happens to ashes, which absorb as much water as the empty vessel."[34] Although Aristotle's response[35] was not wholly ignored, Latin scholastics tended to follow the reply made by Averroes, who, while admitting that he had never witnessed the phenomenon in question, denied the existence of interstitial vacua and assumed instead that the ashes corrupted some portion of the water.[36] In effect, there was less water when the vessel contained ashes than when the water was poured into a previously empty vessel filled only with air.

One scholastic who considered this argument was John Buridan. On the assumption that a vessel filled initially with ashes can receive a volume of water equal to the volume of the vessel, Buridan first presented the interpretation of the vacuists in the form of a natural *reductio ad absurdum*: Either the water is received into the vacuities between the parts of the ashes or the water penetrates the ashes so that two bodies occupy the same place simultaneously.[37] Because the latter alternative is impossible, the existence of vacua must be accepted. In rejecting this interpretation, Buridan based his reply on the argument of Averroes, whom he mentioned. If the ashes are hot and dry, as Buridan seemed to assume, they will cause a great part, or at least some part, of the water to evaporate and presumably be transformed to air, in which state the water departs the vessel.[38] Moreover, some of the water will merely replace the air that occupied the pores between the discontinuous matter of the ashes. For these reasons, Buridan, who was obviously unable to arrive at any precise quantitative determinations, suggested the possibility, which he apparently adopted, that "the vessel could receive more water than if the ashes were not there."[39] Buridan thus expressed the common medieval position that in a vessel partially or wholly filled with ashes there would be less water than in a vessel devoid of ashes.[40] However small the matter of the ashes might be, it does occupy some portion of the vessel's space. For if this were untrue and a vessel with ashes could receive as much water as when it had no ashes, one might well conclude, as did John of Jandun, that the ashes are completely void and occupy no part of space. It would then follow that "whatever part of the space it [the ashes] occupies, either the water is not received in it, or two bodies will be in the same place simultaneously."[41]

The second experience reported by Aristotle, to which atomists appealed in defense of the existence of interstitial vacua, involved wine, the wine skins used to decant the wine, and a wine cask. As Aristotle described it, "people say that a cask will hold the wine which formerly filled it, along with the skins into which the wine has been decanted, which implies that the compressed body contracts into the voids present in it."[42] Unlike the illustration with the ashes, Aristotle did not trouble to refute this claim, perhaps because it appeared manifestly false, as John of Jandun suggested.[43] For whatever doubts one might have about the tenuous ashes, there could be no question about the substantiality of a wine skin, which must of necessity occupy a part of the space in the wine cask, unless one conceded the impossible, namely that two bodies could occupy the same place simultaneously. Jandun allowed that there might appear to be as much wine in a cask with a skin as in one without, "but if the skin were removed and [the wine] measured carefully, there would be no equality."[44]

For the reasons cited above, interstitial vacua were almost unanimously rejected during the Middle Ages. The readily observable phenomena of condensation and rarefaction thus had to be explained within the cosmic frame of a material plenum. Not surprisingly, the basis of the arguments and justifications were supplied by Aristotle.

With the occurrence of condensation and rarefaction taken as indubitable,[45] Aristotle sought to explain these contrary phenomena in a manner consistent with his physical and metaphysical principles. A given quantity of matter, presumably prime matter, could at different times assume contrary qualities, such as hot and cold, black and white, and so forth. Although a body might at one time be hot, it was also simultaneously potentially cold, and would indeed become cold when that potentiality was actualized. The same reasoning was applied to condensation and rarefaction. Because dense and rare were contrary qualities, one and the same quantity of matter could occupy different volumes at different times without the addition or loss of matter. Two transformations of this type were distinguished in elemental bodies.[46] One type was the change from air to water, or vice versa, a process that involved the corruption of one substance and the generation of another, where the same fixed quantity of matter served as substratum for both elements. Here a given volume of water could be substantially changed to a larger volume of air, or the process could be reversed. By contrast to the substantial transformation involved in the first change, the second change involved only a simple alteration of a single element, as when a given quantity of matter, say air, becomes one size and then another. That it could now be of one volume and later another is explained by the fact that it is always potentially another size. In both types of change, however, the most significant datum is that *no matter is gained or lost*. Condensation and rarefaction are thus distinguished from diminution and increase, where the quantity of matter always varies.[47]

It would appear that in Aristotle's interpretation an underlying quantitatively fixed material substratum is capable of "stretching" (i.e., rarefying) and contracting (i.e., condensing) over a range of volumes without breaks in its continuity.[48] The assumption of vacuous pores was thus unnecessary. But if Aristotle spoke of stretching and contracting, as if the material substratum were some elastic substance, he also defined rare and dense by the proximity of material parts, declaring in the *Categories* that "a thing is dense owing to the fact that its parts are closely combined with one another; rare, because there are interstices between the parts."[49] The parts of a given quantity of matter may occupy smaller or greater volumes according to whether they lie more or less close together. But it is the same matter that contracts and expands and the same matter that takes on the opposite qualities rare and dense. Moreover, when a body contracts and becomes dense, it also becomes heavy; and when it expands and rarefies, it becomes light.[50]

If Aristotle succeeded in eliminating the void as an explanation for condensation and rarefaction, his own explanation brought to the fore new and far more troublesome problems. For one thing, a given quantity of air would weigh more when condensed and encompassed in a smaller volume than when rarefied and spread over a larger volume. Indeed an inverse relationship seems to obtain between heaviness and volume. But how can one and the same quantity of matter vary its heaviness with its volume? Aristotle offered no explanation. And when does an alteration of air from a larger to a smaller volume become a corruption of air and a generation of water? How are we to conceive a fixed and determinate quantity of matter occupying different volumes and places without any break in continuity? Is matter really elastic, and therefore capable of stretching over a range of volumes? On all these questions Aristotle was silent. Even the category to which rare and dense belonged was left unclear. Did this vital pair of opposites belong to the category of quality, quantity, or place? Or did they perhaps belong to the category of local motion?[51]

Thus did Aristotle bequeath to the Latin West the plenist response to the vacuist position concerning the fundamental phenomena of condensation and rarefaction. His interpretations, along with Averroes' comments and elaborations, served as the basis of the plenist position until challenged in the sixteenth century and subsequently abandoned in the seventeenth with the advent of the mechanical philosophy. The nature of the medieval treatment of the problem of condensation and rarefaction in a material plenum cannot be examined here.[52] For our purpose, it suffices to know that although interesting and stimulating discussions were formulated, the response lay within the overall Aristotelian framework. And yet there was at least one remarkable scholastic, Nicholas of Autrecourt, who was prepared to defend the vacuist position against Aristotle and his followers.

b. NICHOLAS OF AUTRECOURT'S DEFENSE
OF THE INTERSTITIAL VACUUM

Of Nicholas of Autrecourt's major treatises, only the *Exigit ordo executionis* survived the public burning of his works in 1347. Fortunately it preserved a substantial discussion on vacuum.[53] As an anti-Aristotelian, Autrecourt was convinced that Aristotle's numerous demonstrations of alleged truths were no more probable, and indeed were frequently less plausible, than other possible demonstrations.[54] Although Autrecourt agreed with Aristotle that the *vacuum separatum* was indeed impossible,[55] he sought to show that the *vacuum interceptum*, or interstitial vacuum, was a more plausible explanation of rare and dense than that provided by Aristotle.[56] In Autrecourt's opinion, the arguments that Aristotle had adduced against internal vacua in the fourth book of the *Physics*, presumably including the experiences of the ashes and the wine skin, were perhaps helpful in explaining why people assumed the existence of interstitial vacua but were inconclusive as arguments against their existence.[57]

In attacking Aristotle, Autrecourt first denied Aristotle's conception of dense and rare as derived from the generation of some new quality. Dense and rare are the result of the local motion of parts of bodies. The condensation or contraction of a body "occurs only by the retreat [or compacting] of bodies, as in wool; or because the parts come together, that is, because more [parts] are related more closely than before." Conversely, rarefaction will not occur "unless the parts of this body are more separated than before. Thus 'dense' and 'rare' do not come to be except by the local motion of the parts."[58] It is thus on the motion of particles, a mechanical phenomenon, that Autrecourt rested his case against the qualitative explanation of the Aristotelians.

Like the Aristotelians, Autrecourt explained the apparent increase of new wine in a jar by rarefaction, but a rarefaction that results "not because of the generation of a new quality . . . , nor because of the advent of new bodies, but only, it seems, because the parts, having [previously] come together as if there were a certain trampling down, now separate and are more distant from each other."[59] Autrecourt also attacked the widely discussed ashes example in the version given by Averroes.[60] Not only was Averroes criticized for failing to observe and experience the phenomenon personally, but his explanation to account for the diminished quantity of ashes that is found upon drying was also incorrect. Instead of assuming that the smaller quantity of ashes results from a material corruption, Averroes should have followed the ancients and explained the diminution by a greater proximity of the particles of ash.

Whatever difficulties might be associated with an atomic theory and interparticulate vacua, Autrecourt was convinced that far greater absurdities would result from the assumptions of the Aristotelians.[61] For if the world is a material plenum, as Aristotle would have it, then (1) two bodies would be able to exist simultaneously in the same place; or (2) the movement of any one thing would necessitate the instantaneous movement of every other thing in the universe; and (3) rectilinear motion would be impossible. If vacua are denied and body a moves and changes its place, it will either come to occupy the place of another body, b, in which event two bodies would occupy the same place simultaneously, or body b would vacate its own place to make room for body a, thus raising the same problem for body b: Either it occupies a place simultaneously with body c, or c vacates its place and the same problem arises ad infinitum. From the motion of a single body, then, conditions (1) or (2) above must result; that is, either two bodies occupy the same place simultaneously, or when one body is moved all other things in the universe must also be moved, because every body must vacate and yield its place to another body. Although Autrecourt did not draw the further consequence, it is obvious that this potentially infinite process must either occur in an instant or every motion would involve an infinite process.

Rectilinear motion would also be impossible in a material plenum where all motion must be conceived as if it occurred in a circle in terms of the mutual replacement of contiguous bodies. To illustrate the dilemma, Autrecourt drew upon "a certain worthy master" and argued that if rectilinear motion is conceived as a process in which body a in place A moves to place B directly in front of it, and body b in place B moves to place A, and so on,[62] straight-line motion would be impossible because the body or bodies directly behind a would occupy place A before b could reach it, because the body behind a has immediate access to place A, whereas body b must come around a as they exchange places. Because a chain reaction would arise that involved all the bodies behind a, body b would be prevented from exchanging places with a. Under these conditions, rectilinear motion could not occur. Moreover, if a body is to be moved, it must yield to a push. But if bodies *a* and *b* are held to exchange places simultaneously, how can it be determined which body is being pushed and moved? And if simultaneity is denied and it is assumed that *a* enters place *B* in an instant other than that in which *b* enters place *A*, or vice versa, these two bodies would occupy the same place simultaneously. For should *a* enter *B* in the first instant, than just prior to *b*'s departure, *a* and *b* together would occupy *B* simultaneously, even if only momentarily.

On the assumption of interparticulate vacua, however, Autrecourt believed that the difficulties and absurdities described above would be avoided. Motion of a body through air, for example, involves compression of the air in the sense that some fixed number of particles would occupy a smaller volume. They will condense to make room for a moving body, or perhaps the atoms of the moving body itself might draw together. Condensation or contraction would not involve an infinite process but would only proceed sufficiently far to make room for the moving body. Not only would simultaneous occupation of the same place be avoided, but so would the possibility that every motion would generate an instantaneous and infinite sequence of other motions.

Readily conceding that Aristotle's most subtle argument against the void was the assumption that local motion in a vacuum would occur in an instant,[63] Autrecourt denied the claim by appeal to the *distantia terminorum* argument: Even in a vacuum the prior parts of its dimension must be traversed before the posterior parts (see Chapter 3, Section 3). Motion of particles in interstitial vacua would also be temporal. But would such motions be comparable in the same manner as are motions through resistant media? Arguing in the affirmative, Autrecourt, who assumed that both space and time are composed of indivisible units, insisted that when a body or atom moves through three units of space in three units of time, and another body or atom traverses three units of space in six temporal instants, the first is moved with twice the speed of the second because it rests in each unit of space for only one instant, whereas the second rests in each for two instants.[64] But how do atomic bodies come to rest in a vacuum? Autrecourt explained that rest is either an intrinsic natural property of bodies or is caused by the resistance of an external medium the parts of which do not readily compress. Because a resistant external medium is absent from a vacuum, he concluded that rest is an intrinsic property of atoms but offered no explanation as to why one atom should rest in a unit of space for only one instant and another for two or more temporal units.

Because of its relative brevity, Nicholas of Autrecourt's defense of physical atomism and interparticulate vacua left many unanswered problems and raised questions that cried out for discussion and analysis. But the fact that he proposed such a system as a serious alternative to the Aristotelian physics of the plenum is a tribute to his courage and intellectual acumen. It is especially commendable because his primary knowledge of atomism was likely to have been derived from hostile accounts, notably those of Aristotle and Moses Maimonides.[65] And yet in one vital respect, Autrecourt differed not at all from his immediate predecessors and contemporaries. For like them, he believed that nature abhorred the *vacuum separatum*, or extended vacuum. Indeed this conviction affected his explanation of atomism, as when he sought to analyze the behavior of inflated bladders that were only slightly compressible. He argued that the particles of air in the inflated bladder compress not at all, or only to a certain hypothetical critical point, so that formation of a vacuum may be prevented. Thus despite the existence of interstitial vacua and the capacity for further compression, the atoms of air naturally cease to compress for fear that a separate vacuum would be formed in the bladder. Even in his explanation of the increase in the size of a body, there was the suggestion that the separation of atoms is controlled by nature's universal fear of extended vacua. If so bold an intellect as Nicholas of Autrecourt could not be tempted into a defense of intramundane *vacua separata*, it comes as no surprise to learn that his fellow scholastics found that potential temptation easily resistible. In fact they were pleased to cite, and even occasionally to devise, numerous experiences and experiments to demonstrate nature's uncompromising abhorrence of the *vacuum separatum*. We must now describe and evaluate the most significant of these experiments.

4. THE EVIDENCE FOR NATURE'S ABHORRENCE OF A SEPARATE VACUUM (*VACUUM SEPARATUM*)

a. THE BURNING CANDLE IN AN ENCLOSED VESSEL

Among the most striking illustrations that nature abhorred a vacuum were those employing fire and heat. In this genre, an experiment of special interest was that of the burning candle the bottom part of which is placed in a bowl or dish filled with water. Shortly after a glass vessel with a narrow orifice is inverted over the upper portion of the burning candle and placed in contact with the surface of the water (Figure 1), the water in the bowl rises into the orifice of the inverted glass vessel. Thus contrary to its natural tendency to remain at rest in the bowl or dish, the water rises. But why should the water rise under these circumstances? Philo of Byzantium, whose *Liber de ingeniis spiritualibus* included this example, argued that the flame of the candle destroyed the ambient air, whereupon the water below ascended and replaced the quantity of air destroyed. The cause of the water's elevation was the need to prevent formation of

Figure 1. Experiment to show nature's abhorrence of a separate vacuum: the burning candle in an enclosed vessel. (Reproduced from Liber Philonis de ingeniis spiritualibus, *in Schmidt (ed.),* Heronis Alexandrini Opera, I, p. 476.)

a vacuum following destruction of the air.[66] Although Philo's important description was probably known to some in the Latin Middle Ages, the candle experiment probably received its widest dissemination through Averroes in the latter's commentary on Aristotle's *De caelo*. In the course of a complicated discussion about the relative degrees of heaviness and lightness possessed by all elements except fire,[67] Averroes observed that Themistius disagreed with Aristotle and Alexander of Aphrodisias that air moves more easily to the place of water than the motion of water to the place of air. When a reed (*canna*) is placed in water and the air is sucked out, the water rises, according to Themistius, because the surfaces of the water and air are moist and united so that as the air is sucked upward it drags the water violently with it. But neither air nor water can rise and also draw earth upward. According to Themistius, the earth's dryness makes it impossible for its surface to unite with the moist surfaces of air or water. Thus the water at the bottom of a heated vessel will presumably rise because the ascending air will draw it up violently.

At this point, Averroes introduced the candle experiment by noting that the commentators do not explain why, under certain circumstances, the water in a heated vessel rises but the air does not. For if we place a candle in a vessel and, immediately after covering the orifice of the vessel, we invert it over the surface of water and remove the cover, the water will rise into the vessel, although no additional air has entered the vessel and the enclosed air did not and could not rise anywhere.[68]

If Averroes knew the Arabic translation of Philo's treatise,[69] he certainly did not follow the latter's description, which, as we saw, makes no mention of inverting the vessel with the candle over the surface of water.[70] Averroes had thus supplied a second version of the candle experiment to the Latin Middle Ages. But in rejecting Themistius' explanation that the air above the water draws the latter up by attraction, Averroes offered three possible alternatives. Fundamental among these is that the fire of the candle will transmute some of the surrounding air to fire, which will then rise to the top of the vessel. If the vessel is upright and the flame of the candle is near the orifice, the air-turned-fire will rise to the orifice and come to rest against the lid. The remainder of the air, after following the fire as far as possible, will come to rest violently below the fire. The vessel with the candle, however, is immediately inverted over water and the lid is removed. As this happens, the air-turned-fire will rise to the top – that is, the true bottom – of the vessel, as will the newly created fire formed from the air around the flame as the stop is removed and before the flame is extinguished. Because the orifice is now plunged below the surface of the water and no new air can enter, the water will rise to fill the places vacated by the fiery matter that has risen to the top (i.e., bottom) of the inverted vessel. Formation of a vacuum is thus prevented.[71]

But Averroes looked with favor upon a second cause proposed by Alexander, one that presumably operated simultaneously with the first. When the fiery part of the air cools as the vessel is inverted over the water, the air within the vessel contracts and occupies a smaller place, thus causing the water to rise and immediately fill the potentially void space. The rising water thus prevents formation of a vacuum.[72] A third operative cause was viewed as a consequence of the first two. The motions generated by the actions of the first two causes were such that they communicated a strong motion to the water itself.[73]

In Averroes' quite detailed discussion, the underlying concern is that no vacuum be permitted to form. His elaborate explanations were devised toward this end. Despite their familiarity with Averroes' intricate account, medieval Latin scholastics largely ignored the details and usually presented their discussions in terms of nature's abhorrence of a vacuum, preferring to invoke the popular distinction between particular and universal natures. Thus, in his account, Johannes Canonicus assumed that the rising water would extinguish the flame,[74] an action that would produce a vacuum were it not for the simultaneous activity of the air and water within the enclosed space of a vessel. On the assumption that a given quantity of fire occupies a much greater volume than the same quantity of the matter of air or water, the place of the extinguished flame and the fiery matter that surrounded it could be filled only by rarefaction of the remaining air and the simultaneous ascent of the water. Thus where Alexander of Aphrodisias and Averroes had conceived a contraction of air in terms of a cooling action, Canonicus was convinced that the air had to expand to fill an even greater space than before. And apparently it required assistance, so that Canonicus further explained the ascent of the water by the action of the universal nature, which sought to prevent formation of a vacuum by counteracting and temporarily nullifying the particular nature of the water.[75]

Using the action of heat but not the candle, Friedrich Sunczel in the fifteenth century offered a variation of the candle experiment. He assumed that a glass vessel with a long neck was heated over a fire and then inverted over a surface of cold water. As the air cooled and contracted, the water ascended into the vessel to prevent formation of a vacuum.[76]

b. LIQUIDS IN REEDS, SIPHONS, AND SEALED VESSELS

The role of fire and heat in the experiments just described produced a variety of explanations about the behavior of fire, air, and water and their nearly simultaneous interactions in the closed system of vessels.[77] A much simpler phenomenon, in which heat played no part, was that involving siphons or reeds. According to Philo, if one orally sucks out the air from the long, narrow, siphonlike mouth of a vessel used for wine tasting, the liquid in the vessel will follow the air upward as if glued to it,[78] a conception that may have been based on Aristotle's description in *De caelo* quoted earlier (see Section 1). In commenting on this very passage, Simplicius stressed the same attractive, or drawing, power of air on water when the two were in contact. William of Moerbeke's 1271 translation of Simplicius' commentary on *De caelo* furnished a detailed description of the way in which water and blood were drawn upward in siphons and cupping-glasses, respectively.[79] And as we have already seen, Averroes, in his description of the opinions of Themistius, mentioned that when air is drawn through a reed, water follows it upward with a violent motion. Like Philo and Simplicius, Averroes emphasized the union of the surfaces of air and water, a circumstance that enabled the former to drag the latter upward as it ascended.

Although these treatises were available, Latin scholastics again chose to interpret the action of the rising water solely in terms of prevention of a vacuum and largely ignored the relationships between the contiguous surfaces. This tendency is already evident around 1230, in a commentary on Sacrobosco's *Sphere* attributed to Michael Scot.[80] Here we are told that when the air has been extracted from a reed (*canna*), one end of which is in the mouth and the other end in water, the water will act contrary to its nature and ascend in the reed to prevent a vacuum. As our author explained, "it is more possible that a heavy body rise upward than that a void place should be left behind" as the air is withdrawn.[81] Drawing upon a treatise *On Emptiness and Void* (*Tractatus de inani et vacuo*), Marsilius of Inghen described the action of a curved siphon in which "a vessel is made with two arms, one longer than the other, and the shorter is assumed to be in water." If the air is then extracted from the longer arm outside the vessel, "the water rises through the shorter arm which it would not do unless it were preventing a vacuum."[82] Repeating the universal conviction that "common nature" (*natura communis*) abhors a vacuum because of its potentially disruptive action, Friedrich Sunczel observed, among the numerous experiences that revealed the impossibility of the existence of a vacuum, that a heavy body would rise to prevent its formation, as when "you find that water is drawn and sucked upward in reeds." Conversely, light bodies descend to achieve the same end, as evidenced by the presence of air and exhalations in the concavities of the earth, so that "light rests under heavy."[83]

Nature might, of course, have achieved the same objective quite differently. In the sixteenth century, Luis Coronel (d. 1531), a master at the University of Paris, observed that rather than allow water to rise as the air is exhausted in a siphon or reed, nature might have prevented formation of a vacuum by causing the siphon to break apart, thus allowing air to enter from outside. This course of action might appear to some as less absurd than the ascent of the water, which violated a fundamental natural principle that heavy bodies cannot rise toward the natural place of a lighter body.[84] Indeed, in certain circumstances nature seemed to choose this very course of action, as when the water in a siphon, or sealed vessel, freezes and the vessel breaks apart.

The typical medieval explanation of this oft-mentioned example relied on two assumptions: (1) that no additional matter could enter the closed vessel and (2) that when water freezes in a sealed vessel, its volume diminishes as it condenses and contracts. Because the water will no longer occupy all the space in the sealed vessel and no additional matter can enter, a vacuum will inevitably occur. Fortunately, nature intervenes and ruptures the vessel at the appropriate moment. [85] But the observed fact that water in a siphon does indeed rise to prevent formation of a vacuum was sufficient indication for Coronel that there is less resistance to the ascent of water than to the destruction of the siphon. "And since nature always favors the less difficult course, it proceeds through the ascent of the water rather than by the breaking of the body [of the siphon]."[86] When the water freezes, however, and is not only rendered incapable of rising to prevent a vacuum but actually shrinks in volume to produce one, nature then achieves its end by shattering the vessel. Thus the manner in which a vacuum can be prevented may depend on the physical state of the water enclosed in a vessel or siphon. Whatever that state may be, nature will act as it must to prevent disruption of the continuity of matter.

Water in an enclosed vessel was assumed occasionally to diminish in volume without freezing. Roger Bacon, for example, assumed that a copper vessel was completely filled with water and tightly closed. At the end of a year, inspection would reveal that the water had diminished in volume, although nothing had entered or departed. Under these circumstances, it seemed that a vacuum had been produced in the sealed vessel. Bacon, of course, denied this consequence and explained the phenomenon by the generation of air from water, because an element can be generated from any other element by virtue of their common matter *and* by means of a universal agent (*agens universale*), which is of a celestial nature. It is the universal agent that actually causes the air to be generated from the water in order to prevent formation of a vacuum.[87]

Altering the conditions, Bacon confronted another contrived situation. What would happen if a vessel were only partially filled with water, and somehow the water departed and nothing could enter to replace it? The sides of the vessel would immediately come together to prevent the generation of a vacuum. But the contact of the sides and the consequent destruction of the vessel would be caused not by the universal nature or the particular nature but by the fact that "there would be nothing between them which could constitute a distance, since nothing would be there and 'nothing' does not make a distance."[88] The void produced under these conditions is here interpreted as a nothing that is a negation and therefore incapable of functioning as a distance or extension that could separate anything. (But see Chapter 2, Section 1, where Bacon seemed to treat nothing as something.) Hence the sides of the vessel will necessarily come together.

c. THE BELLOWS

One of a number of devices that captured the attention of medieval discussants of void was the common bellows. As Buridan explained, its operation could show "that we cannot separate one body from another unless another body intervenes."[89] By completely stopping up all the openings of a bellows after all the air was expelled and its sides had collapsed and come into contact, Buridan argued that "we could never separate their surfaces" because a vacuum would then exist.[90] As a measure of nature's resistance to this possibility, Buridan declared that "not even twenty horses could do it if ten were to pull on one side and ten on the other."[91] Those sides were separable only if "something were forced or pierced through and another body could come between the surfaces."

In yet another version, the mouth, or nozzle, of a bellows was inserted into the single opening of a compact, sturdy vessel. With the bellows-vessel system completely closed off so that nothing could enter from out-side, John of Jandun then assumed that "the sides of the bellows are elevated and separated." Two possibilities now present themselves, each seemingly productive of a vacuum. Either the air from the vessel will enter the space between the sides of the bellows, in which event the space of the vessel will be emptied of air and a vacuum will exist; or the air will remain in the vessel and the inner concavity of the bellows will remain void.[92] To counter these potentially dreadful consequences, Jandun as-sumed that "universal nature" (*natura universalis*) would prevent the separation of the sides of the bellows."[93]

In denying that the sides of the bellows would be separated for fear of a vacuum, Jandun and others who agreed with him signified their doubt that the air within the vessel could disseminate and rarefy sufficiently to fill the bellows as well. The safer course was to deny the possibility of separating the sides of the bellows.

d. THE CLEPSYDRA

In the medieval literature on nature's abhorrence of a vacuum, the two most popular demonstrations were the clepsydra and the separation of two plane surfaces. They represent a fitting conclusion to our discussion.[94]

Basically a decanting vessel, the clepsydra is characterized by a narrow, open neck and wide body with small holes at the bottom, which is the part first submerged into the liquid to be decanted, usually water or wine. When all the air in the vessel is expelled and replaced by the incoming, rising water, the narrow orifice at the top is stopped up, usually by cov-ering it with the thumb. Upon lifting the vessel from the water, one ob-serves that the water remains in the now elevated clepsydra despite the expectation that it would fall through the tiny holes at the bottom.[95] Al-ready in the first half of the twelfth century, Adelard of Bath had not only described the clepsydra without naming it[96] but also attributed its seemingly magical behavior to a natural affection existing among the four elements that constituted the universe, an affection so powerful "that as soon as one of them leaves its position, another immediately takes its place; nor is this again able to leave its position, until another which it regards with special affection is able to succeed it."[97] By the thirteenth century, however, the wave of translations had made available explana-tions of the water's behavior in which it was assumed that while one's finger stopped up the narrow upper orifice, the water was unable to de-part lest a vacuum be left in the clepsydra. Despite its natural tendency to fall, the water remained in the clepsydra to prevent formation of a vacuum.[98]

Medieval discussions of the clepsydra usually involved a description of the device, though often without employing the term, and an explanation of the liquid's behavior, as when Marsilius of Inghen, after subsuming the clepsydra experiment under the class of actions in which "a heavy body existing upwards without any obstacle [to its descent] does not descend because a vacuum might occur," declared that "if a vessel having many small openings below and a large opening above was obstructed, the water would not descend through the lower openings because there would not be any other means of preventing a vacuum."[99] With wine as his liquid, John Buridan first described the standard action of the clepsydra[100] and followed it with a variation in which the vessel is half-filled with wine and half-filled with air before its upper orifice is stopped up. The wine will now descend through the opening below[101] because the trapped air can rarefy and expand to fill the space vacated by the departing wine.[102]

What happens in the vessel is but the counterpart of rectilinear motion in the world at large. When a body moves rectilinearly, it causes a condensation of matter in the direction of its motion and a simultaneous rarefaction in its wake, a process that could continue all the way to the heaven itself, that is, all the way to the concave surface of the lunar sphere. Were something like this to occur, the heaven itself would move inward to fill a potentially void space. Generally Buridan argued that if something dense is generated from something rare, the bodies surrounding it, including the heaven itself, must move toward that place to prevent a vacuum unless a compensatory alteration from dense to rare occurs elsewhere simultaneously.[103] So potentially devastating were the consequences that even Nicholas of Autrecourt, defender of the interstitial vacuum, believed firmly that nature truly abhorred the separate vacuum "which would be left behind in the vessel after the water had been removed."[104]

But what enabled nature to prevent formation of a vacuum in the clepsydra? We have already described responses that emphasized either a special affinity between the elements (Adelard of Bath) or a "mode of fullness" in the universe. Explanations based on matter's universal and particular natures ultimately proved most popular in explaining the liquid's behavior in the clepsydra. Roger Bacon and Walter Burley have left two of the most interesting interpretations.

Both of these distinguished authors were convinced that prevention of a vacuum was the primary reason why the water in the clepsydra remained suspended, contrary to its natural inclination to fall. They also denied that this contradiction of Aristotelian physics could be causally explained by appeal to the mere negation "that a vacuum not occur" (see Section 2 and note 20). A positive cause was required, and it was at hand in the universal nature or agent. (For a discussion of the universal nature, see Section 2.) Bacon argued[105] that even if water were to fall to its natural place, the universal nature would act to prevent formation of a vacuum by causing the surfaces of the clepsydra, which are themselves continuous, to collapse and meet. In preventing a vacuum, nature must therefore choose between two unpleasant and unnatural options. Either it compels the water to remain at rest suspended in the clepsydra or, should it permit the water to descend, it must collapse the clepsydra and join its interior surfaces. The latter alternative, however, is more disruptive of nature, because the collapse of the vessel would follow an apparent momentary formation of a vacuum in the vessel and thus produce two unnatural events, namely a vacuum that exists for only a moment *and* the collapse of a vessel.[106] Thus did Bacon make explicit what was frequently only implied in the medieval theory of a universal nature: Material continuity is always preserved in the least disruptive manner possible.

An apparent inconsistency in the behavior of the universal nature or agent was reported and subsequently dismissed by Walter Burley. In preventing formation of a vacuum in a clepsydra, the universal agent causes two distinct actions: (1) it forces water to rise in the clepsydra as air departs through the uppermost orifice, and when the upper orifice is stopped up (2) it prevents the descent of the water through the several holes at the bottom. As a celestial force, the universal agent must be assumed to act in a uniform and regular manner. And yet an examination of the phenomena that it is alleged to cause reveals seeming discrepancies. In order to prevent a vacuum, it sometimes causes water to rise and at other times to rest. Moreover, when the single orifice at the top of the clepsydra is stopped up, the water remains suspended, whereas when the orifice is left open, the water descends. Thus despite the assumption of uniform behavior for the universal nature, water sometimes remains suspended in the clepsydra and at other times descends. Is it therefore necessary to postulate another particular cause to explain the seemingly conflicting actions of the universal agent? If so, our problem returns, for we must then account for the manner in which the new cause generates the same actions.[107]

Responding to this dilemma, Burley reaffirmed the primacy of the universal celestial agent, which not only governs and controls all terrestrial elements[108] but, appearances notwithstanding, also acts uniformly on the orifice of the clepsydra, whether it is stopped up or not. Its steady and uniform behavior is not, however, to be interpreted by the contrary effects produced but must be understood with respect to a single primary effect, the purpose of which is to preserve the material fullness and perfection of the universe, namely its continuity, "because if a vacuum exists in any part of the universe, then a certain part of the universe, required for its perfection, would be lacking and the universe would not be perfect." Because nature (i.e., the universal nature, or agent) abhors a vacuum, we see that the contrary behavior of the water in the clepsydra contributes to one and the same end: preservation of material continuity. When the orifice is stopped up, the water is suspended in the clepsydra because its particular nature has been rendered inoperative by the direct intervention of the universal agent, which achieves its primary effect – to prevent formation of a vacuum – in this way. But when the upper orifice is left open, the primary effect is attained by the ordinary operation of the water's particular nature without the intervention of the universal agent. The water descends naturally because in departing the clepsydra it is replaced instantaneously by air entering the vessel through the open orifice.[109] That a celestial agent should be capable of producing two such contrary effects on sublunar bodies was, in Burley's judgment, no absurdity. After all, the sun causes both generation and corruption.[110]

Material continuity, and therefore universal perfection, was to be preserved at all cost. Nature's efforts to achieve this continuity were encapsulated in the principle that nature abhors a vacuum, which for Burley was an end or goal that functioned as a final cause, determining the behavior of the universal agent in causing the water to rest in the clepsydra.[111] To medieval scholastics it was virtually self-evident, as one author put it, that "unless the whole inferior [i.e., sublunar] world is conjoined [and united without intervening vacua], it cannot be governed by the superior [i.e., supralunar] world, as Aristotle teaches in the *Meteorologica*, Bk. 1, ch.2. Therefore *nature abhors a vacuum*, for which reason it confers on particular things a universal inclination to be joined together and united."[112] Far better that a body be kept from its natural place than that nature should suffer disruption of its material continuity. For "it is repugnant that a void should exist," but "it is not repugnant that a body should be found outside its proper place. It is more natural, then, that a body be moved in order to remain in immediate contact with another body rather than gain its proper place."[113] In medieval physics, an agent was held to act on a patient by immediate contact or through intervening matter. The existence of vacua would have rendered the transmission of causal effects impossible. Considering the threat that the existence of vacua posed to the Aristotelian physics of the plenum, it is hardly surprising that universal natures and agents were readily invoked. For they did not merely "save the phenomena" but the very world itself.

e. THE SEPARATION OF TWO SURFACES
IN DIRECT CONTACT

Originally proposed as evidence in support of the existence of a vacuum, the final experiment to be considered here was even more widely mentioned than the clepsydra. In *De rerum natura*, I 385–397, Lucretius argued that "if two bodies suddenly spring apart from contact on a broad surface, all the intervening space must be void until it is occupied by air. However quickly the air rushes in all round, the entire space cannot be filled instantaneously. The air must occupy one spot after another until it has taken possession of the whole space."[114] Although *De rerum natura* was virtually unknown until the fifteenth century[115] and is therefore an unlikely candidate for the direct source of this oft-cited medieval example, scholastics were well aware of it and, contrary to Lucretius' claim, sought to demonstrate that the separation of the surfaces did not imply formation of a vacuum, however small or momentary.

Whatever the source of the medieval tradition of the Lucretian argument,[116] it had already been received by the mid-thirteenth century, when Roger Bacon and Pseudo-Siger of Brabant asked whether or not a vacuum would actually form on the separation of two plane surfaces. These authors stand at or near the beginning of a traditional concern for a proper response to this question, which would eventually engage Galileo's attention in the *Discourses Concerning Two New Sciences*. With at least one notable exception (Blasius of Parma, whose views will be discussed below), there was general agreement with Lucretius: A momentary vacuum would indeed form if two plane surfaces, initially in uniform mutual contact with no air or material medium intervening, were separated in such a manner that they remained continuously parallel.[117] In rushing to fill the totally empty space between the now separated surfaces, the air would require some time, however small, to move from the outer perimeters to the innermost parts of the surfaces. Until filled by the inrushing air, a momentary vacuum would inevitably form there.[118]

Faced with this grim prospect but firmly convinced that formation of a natural vacuum was impossible, many came to argue that for fear of allowing a vacuum, two plane surfaces could not undergo parallel and uniform separation of all parts simultaneously. Separation could occur only gradually and successively, minute part after minute part. As each part separated, air would immediately and instantaneously enter to prevent formation of even a momentary vacuum. Such a gradual and successive part-after-part separation could occur only if one surface were at least slightly inclined to the other. Thus did Bacon insist that

two round tables could not be separated if one were above the other [i.e., in contact] unless there was an inclination of some part. Hence it would be necessary that some part be inclined before it could be raised, for otherwise a void would be produced. This is the result of [the action of] a universal nature. Those who respond in this way answer well, since it is impossible that it be raised in this way [without inclination]. This is obvious in water, for if a [plane] glass is placed in water and is inclined no more one way than another, no man in the world could lift it.[119]

Whether the plane surfaces were both assumed hard, or one hard and one soft, as with glass and water, Bacon's claim was obviously untestable. Every instance in which a plane glass is actually lifted from the surface of water, and generally every case involving the separation of two plane surfaces, would be interpreted ipso facto as a direct consequence of the inclination of surfaces, however slight or undectable. Should such a separation fail to occur, Bacon would merely declare that the surfaces were not inclined and nature operated to prevent a vacuum. In the absence of any proper means of determining whether any two surfaces were truly inclined or not, Bacon used his theory to make the determination after the fact.[120]

Perhaps it was with Bacon in mind that Walter Burley formulated a response to Bacon's widely received argument. In Burley's judgment, it was utterly irrelevant whether two plane surfaces, or tables, in direct contact were conceived to separate simultaneously and parallel or successively and mutually inclined. Either way, air would enter successively and require some time, however small, to fill the whole space between the surfaces. According to Burley, the assumption of an inclination of the planes and successive, part-by-part separation reduces the scope of the problem but offers no solution. Instead of explaining how the total space intervening between the surfaces might be filled, one must now explain how each of a series of minute parts is filled, where each minute part poses the same problem as does the whole intervening space. For as each part is in turn separated, the inrushing air first occupies the periphery of that part and only subsequently arrives at its interior in order to fill a vacuum that would already have formed.[121] But Burley went further and challenged the very foundation of Bacon's argument, which assumed that a part-by-part separation could occur. For if the surfaces in question are continua, and if the motion of a continuum is one, then "if any part is moved, the whole [continuous surface] is moved. Hence if such a body is elevated, a vacuum will form before the air can fill the inner parts of the space [between the separated surfaces]."[122] To those who believed they could escape the dilemma of the possible formation of a vacuum by denying that two perfectly plane bodies could actually be found in nature, Burley replied that it mattered not at all whether the surfaces were plane or not. For the same difficulty arises when two nonplane bodies are brought into contact, presumably because even if the two surfaces are rough and uneven, there will be at least one, and very likely more, points of actual contact. The formation of a momentary void would occur at each of those points of contact in the same manner as already described for perfectly plane surfaces.[123]

Although Burley's solution to the problem will be discussed below, the criticisms he formulated were often ignored as many during the Middle

Ages adopted a position similar to that of Roger Bacon and John of Jandun that separation without creation of a vacuum could occur only by the disengagement of successive parts (rather than simultaneous and parallel separation of the total surfaces), which presupposed an initial inclination of the surfaces. A striking departure from this tradition was represented by Blasius of Parma, who insisted that parallel, simultaneous separation of surfaces was indeed possible without formation of a vacuum or the need to invoke instantaneous motion. Nor would the converse of separation involve these absurdities and impossibilities, namely the situation in which two plane surfaces approach although remaining parallel until contact. The latter problem is considered first in the first conclusion of the third of five articles that constitute Blasius' *Questio de tactu corporum durorum*.[124]

If two plane, circular surfaces[125] approach each other continually while remaining parallel, Blasius argued that the intervening air would depart *successively and simultaneously* – though not instantaneously – without leaving behind a vacuum in the center just prior to contact.

The way this happens is as follows: let *b* be air near the center, and *a* be the air near the circumference. Now I say that when air *a* (that near the circumference) moves away, it actually moves outside [the circumference], and a measurable amount of time elapses. But whatever the given velocity with which air *a* might be moved toward the outside, the air in the center can be moved 100, or even 1000, times quicker by a supposition [given above]. [126] Therefore these airs, *a* and *b*, will be moved to the outside in the same time.[127]

At this point, Blasius raised an obvious objection. If one and the same motive force propels all the intervening air, the air near the center ought not to move beyond the circumference as quickly as the air near the circumference.[128] To this, Blasius replied that

although this central air and this circumferential air are moved by the same mover, as [for example] by a plane stone which is moved downward toward another [plane stone], yet they are not moved with the same ratio because the air situated at the circumference has air external to it that continually resists its motion more and more because of its greater and greater condensation. But the central air has a continually smaller and smaller resistance because the air adjacent to it becomes continually less and less and rarer and rarer. And this is clear to the understanding so that although this and that [air] are moved by the same mover, they are not moved in the same ratio, as is obvious.[129]

By this explanation, Blasius sought to provide a physical mechanism that would somehow render intelligible the ultimate departure of the intervening air as the surfaces approached and eventually came into contact. The differential speeds of departure, quicker near the center and slower near the circumference, arise from condensation just beyond the surfaces and a consequent continuous rarefaction near the center. At the last moment, just prior to the contact of the approaching surfaces, all of the rarefying air would somehow clear the intervening space simultaneously but successively, thus avoiding the double absurdities of instantaneous motion and formation of a vacuum. As it stands, however, Blasius could avoid one or the other, but not both. Either all of the continually rarefying air departs instantaneously and no void occurs; or the air near the center departs successively, but as it reaches and joins the denser air near the circumference, its continuity is broken and a vacuum is left behind just before all the air simultaneously clears the area between the circular surfaces, which then immediately come into contact.

Blasius appears to have been aware of the inadequacy of this description, which made no physical sense of the last moment before contact of the surfaces. At best, it was a vain effort to enable the reader to picture how the air's departure might be both successive and simultaneous while avoiding instantaneous motion and the formation of a vacuum. But his real explanation – or, more properly, the basis for such an explanation – must be sought elsewhere, in the second conclusion of the third article, where Blasius considered the more traditional problem of the separation of surfaces. But if the problem was traditional, Blasius' solution was not.

In an extraordinary treatment of the problem of the contact and separation of plates, Blasius drew upon earlier medieval discussions that were concerned with first and last instants, as well as extrinsic and intrinsic boundaries of continuous physical processes and magnitudes.[130] As the circular surfaces approach, the air becomes rarer and rarer but never separates at any point to allow formation of a vacuum.[131] As long as the surfaces approach but do not meet, the air will continue to rarefy and yet fully occupy the diminishing intervening space. Hence there is no last assignable instant in which rarefaction ceases prior to contact of the surfaces. Therefore no vacuum can occur. And because the surfaces are assumed to fit together perfectly on contact with nothing else intervening, no vacuum can occur after the departure of the air.

Of crucial significance is the interpretation to be placed upon the notion of contact between the surfaces. Actual contact must be construed as the last move in the completion of the process of motion of the surfaces but as lying outside that process and actually serving as an extrinsic boundary to it. In effect, there is no last instant in which the surfaces are separated, although there is a first instant in which they are in contact. And if there is no last instant in which the surfaces are separated, there can be no last instant in which the rarefied air departs from the intervening space. Therefore no vacuum occurs. Contact of the surfaces not only terminates their movement but also signifies the total absence of intervening air – without, however, the existence of a vacuum because the surfaces are assumed to fit perfectly, thus excluding the possibility of unoccupied spaces.

With the plane surfaces now in uniform contact, Blasius, in the second conclusion of the third article, explained how separation could occur although the surfaces remain parallel and both vacua and instantaneous motion are avoided.

It is possible that two hard and plane bodies could be raised [or separated] while remaining parallel to each other. The conclusion is proved because if this did not occur, the reason for its non-occurrence would be *either* because a vacuum would be left *or* motion would take place in an instant, as is commonly said [or claimed]. But I declare that this would not happen because when you ask about the air existing in [the region around] the circumference *either* it is moved as quickly to the center [of the space between the surfaces] as to the mid-point of the radius [of the same place] *or* it is moved first to the mid-point of the radius and then to the center. I say that it is moved first to the mid-point of the radius and then to the center. But when you conclude, therefore, that a vacuum must then exist at the center, I deny the consequence because whenever there is air at the mid-point of the radius, there was previously other air in the center. Nevertheless, I wish [to argue] that all air that is in the center was first in the mid-point of the radius. Hence there was no first air that was first in the center just as there is no first [or initial] distance by which these plane bodies are now separated, as one supposition says.[132] Thus there is no air which entered [between the plane surfaces] first, so that any air that is now in the center was previously in the mid-point of the radius. And so no absurdity follows.[133]

In order to demonstrate the claim that no vacuum will form between the surfaces of two hard, plane bodies that are separated although remaining parallel to each other, Blasius had to show that air always exists between the surfaces after separation. To do this, he relied on the fourth supposition, which denies that a first moment of separation can be determined. In effect, there is a last moment, or instant, of contact but no first instant of separation. For if a first instant of separation exists, there must also be a minimum distance of separation. But given any initial distance of separation, one can always argue that the surfaces must have been previously separated by half that distance, and so on. Hence there can be no initial distance of separation and consequently no first instant of separation. But if there is no first distance and no first instant of separation, it follows that for any moment chosen after separation, air will fully occupy the intervening space associated with that particular moment, and no vacuum can occur. For if an alleged first instant of separation is arbitrarily chosen and it is argued that the air must first have reached the space corresponding to the midpoint of the radius of the circular surfaces and only thereafter the point corresponding to their centers, one could then, of course, plausibly claim that a momentary vacuum must have existed at the center prior to the arrival of the "first" air as it moves through the successive positions from circumference to center. But Blasius would counter any such move by denying that one could select a first instant of separation, without which there could be no initial entry of first air. Already prior to any first instant that might be chosen after separation, another earlier instant could be selected, which guarantees that air would have entered and fully occupied the intervening space prior to the occurrence of the initially chosen instant. Because this argument can be employed for any instant that might be chosen after separation, there is no moment when a vacuum could have existed after separation.

Blasius further insisted that whenever air enters between the separated surfaces, it does so successively, moving first to the midpoint of the radius and only then to the center. The entering air will not, therefore, fill the intervening space instantaneously. Because the moment of entry cannot be determined and because for any selected moment after separation it can be argued that the air already occupies the intervening space, the problem of instantaneous motion does not arise. In this way, Blasius not only demonstrated the nonoccurrence of vacuum but also avoided the absurdity of instantaneous motion. No choice need be made between them, as if somehow one were a lesser absurdity than the other. By resort to an argument based upon a denial of a first instant of separation, Blasius could demonstrate the logical impossibility of a vacuum, which, in turn, eliminated the problem of instantaneous motion.

As significant and interesting as was Blasius of Parma's logical solution to the problem of the separation of two surfaces, the major physical resolution of the dilemma it posed came in a strikingly different manner: by denying the claim that two physical surfaces could come into direct contact. This solution was derived from Aristotle, who in *De anima*[134] had argued that surfaces could not come into direct contact in air because the air itself would intervene, just as in water direct contact between two surfaces is prevented by the mediation of the water itself, as evidenced by the wet surfaces of bodies brought together under water. In the Middle Ages, the rather obvious inference was drawn that because a material medium actually intervened between two surfaces brought into seemingly direct contact, their separation could not produce a vacuum. Some, for example, Roger Bacon and John of Jandun, considered this argument inconclusive on the grounds that it did not include those cases in which a single surface came into contact with the surface of air or water. Under these circumstances, neither air nor water intervened, and a momentary

vacuum would presumably occur after separation.[135] To avoid this unacceptable consequence, we saw that Bacon and Jandun assumed that
separation without formation of a vacuum was possible only when the
surfaces were mutually inclined. But Aristotle's solution proved attractive
and formed the basis of the other major medieval response that denied
formation of void space following the separation of two surfaces. One of
the most noteworthy defenses of Aristotle's position was formulated by
Walter Burley.

In support of the claim that two plane surfaces, or tables, could not
come into direct contact, Burley assumed that the approach of any two
plane surfaces was quicker than the escape of the air, water, or fire in the
intervening space, from which he concluded that as the two surfaces came
to rest, and seemingly into contact, a thin film of air, water, or fire would
always be trapped between them. Upon separation, formation of a vacuum
would be avoided because the intervening film of matter would instantly
expand and fill the widening space until the surrounding medium – air,
water, or fire – arrived.[136] But what would happen if the air or fire trapped
between the surfaces prior to separation was already rarefied to its ultimate
limit – that is, just short of a break in continuity? Under these conditions,
a vacuum would inevitably form immediately after the separation of the
surfaces because the trapped air, already rarefied to its limit, could not
expand to fill the space between the now separated surfaces.[137] Only by
the unacceptable assumption of an instantaneous motion of the surrounding medium could the vacuum be avoided.

If such conditions actually obtained in nature and rarest fire or air intervened as described, Burley believed that the two surfaces would be inseparable unless one of them subsequently bent or folded.[138] Indeed whatever the original nature of the two surfaces or tables, whether plane or
nonplane, separation could not occur "because no intervening body could
be rarefied to fill the place [that would be left void. Hence] one table can
in no way be raised [or separated] from another, whether uniformly or
non-uniformly." For the given conditions, separation could occur only by
sliding or pulling one surface over the other.[139]

Burley took cognizance of yet another seemingly troublesome problem.
If one great stone were superposed over another with air assumed to lie
between, would not the downward thrust of the upper stone overcome the
weak resistance of the intervening air and bring the two surfaces into
direct contact?[140] On the assumption that it does, how would the air between the approaching surfaces escape just prior to contact? In coping
with this problem, Burley faced the same dilemma as those who believed
in the possibility of direct contact and had to explain how the intervening
matter escaped just prior to contact. The dilemma itself is expressed in
terms of a stone coming into contact with a watery surface. At the com-

pletion of its motion toward the water's surface, all parts of the stone's undersurface should strike the water simultaneously "because the whole stone is brought into contact with the water simultaneously." But the air between the surfaces of stone and water "does not withdraw as quickly from [under the surface of] the middle [of the stone] as [it does] from the outer parts of that space. Therefore a vacuum is left there [under the middle of the stone]."[141] According to Burley, some avoided this unacceptable consequence and denied formation of a vacuum by insisting that all of the intervening air withdraws simultaneously, though successively.[142] Therefore all parts of the two surfaces would meet simultaneously.[143]

Against possibly damaging arguments of this kind, Burley stood firm and doggedly reiterated his conviction that the plane surfaces of any two bodies could not come into contact but would always remain separated by a thin layer of a material medium such as air, water, or fire. With respect to the example of the superposition of one stone over another,[144] Burley insisted that whatever the heaviness of the uppermost stone, every part of it would tend to move downward uniformly and equally toward the stone directly underneath. If, then, it were possible for two surfaces to come into direct contact without an intervening medium, every part, and therefore the whole, of the undersurface of the descending uppermost stone should come into contact with the upper surface of the lower stone simultaneously. "But this could not happen unless the intervening air withdrew beyond the circumference [of the stones] according to all its parts. And so it [i.e., all of the intervening air] would withdraw without a motion [i.e., instantaneously], which is impossible. Thus however much a great stone is superposed, it could not descend through the intervening air because it is then possible that all the intervening air would withdraw instantaneously."

Perhaps because he was less than convinced of the persuasiveness of his response, Burley sought to reinforce his position by invoking the universal celestial agent. For if we assume that

a stone can be of such heaviness that air will not suffice to resist it, it must be said that although air, in its proper [and ordinary] power, is insufficient for resisting the stone, yet, nonetheless, by the power of the superior [celestial] agent, which seeks to preserve the perfection and plenitude of the universe, the moved air suffices to resist the stone, however heavy it may be. Hence the intervening air does not yield to the stone because it [i.e., the stone] is held back by the superior [celestial] agent, which powerfully seeks to prevent a vacuum.[145]

Of course, the heavier the solid body that presses down, the closer it will approximate actual contact with another body. But although air is sometimes denser than at other times, it will always intervene.[146]

According to Burley himself, opponents of his position seem to have derived at least one patently absurd and laughable consequence from the denial of direct contact between surfaces. If a material medium always intervened between any two surfaces, it would follow that animals could not be said to walk on earth but only on air or water, either of which must lie between an animal's foot and the earth.[147] Although initially Burley seemed to deny the absurdity of this consequence, he conceded that in common language animals are indeed said to walk on earth, not air, for "an animal is said to walk around on that which supports it; the earth is such a support, but not air. Thus it is said that an animal walks on earth and not on air because earth sustains and supports, but air does not."[148] At this point, Burley seemed at best to have confused his readers and at worst to have blatantly contradicted himself and allowed direct contact between surfaces. If the latter, then formation of a momentary vacuum, which Burley had previously rejected as impossible, would now be unavoidable as the intervening air withdraws just prior to direct contact between the surfaces of foot and earth. Whether Burley intended to concede so much is difficult to determine on the basis of his all too brief discussion. Perhaps he only wished to make the point that although there is a cushion of air between the surfaces of an animal's foot and the earth, it is not the air that actually supports the animal – for things fall through air and are not supported by it – but ultimately the earth itself, on which the intervening air rests. At this point in his tangled argument, Burley would have been well advised to invoke his universal celestial agent.

5. THE SIGNIFICANCE AND BRIEF SUBSEQUENT HISTORY OF THE EXPERIMENTS AND EXPERIENCES DEMONSTRATING NATURE'S ABHORRENCE OF A VACUUM

The experiments and experiences described in the preceding section, as well as others, were interpreted throughout the Middle Ages from the standpoint of Aristotelian physics – that is, from the assumption that the world is a material plenum in which void spaces, naturally or artificially created, are impossible. Although a variety of interpretations were proposed for the different examples, only a few basic explanations were involved. The most obvious explanation was rarefaction and condensation, without which changes of dimension and size in a plenum would be impossible. When the matter that originally filled a sealed vessel, or closed system, was reduced or contracted in volume, it was often assumed that the remaining matter, or the contracted matter, instantly rarefied to prevent formation of a vacuum. Thus in the candle experiment, the part of the air that is left after the consumptive action of the fire or the cooling action of the external air was assumed to expand to fill the potential vacuum that would otherwise form. Similarly, the potential vacuum in a sealed vessel filled with water that is subsequently frozen and erroneously thought to contract in volume would be filled by a subtle vapor derived from the rarefaction of the ice. Where additional matter could enter a vessel in which a potential vacuum might occur, it always did so, as in the candle and siphon experiments, where water ascended to perform this vital function. Here, however, it was not rarefaction and condensation that were thought to be operative but the principle of material continuity, which caused the air and water to adhere to each other in such a manner that when the former rose, it pulled the latter up with it to prevent formation of a vacuum. Whatever the faults and limitations of the explanatory appeals made to the mechanism of condensation and rarefaction or to the principle of material continuity, they were consonant with Aristotelian physics or at least had a foundation in specific discussions by Aristotle.

But other illustrations and experiences drove plenists to more extreme ad hoc explanations. Clever examples were devised that seemed to exclude the possibility of appeal either to condensation and rarefaction or the usual sense of the principle of material continuity. In this regard, severe challenges were posed by the bellows, the separation of surfaces, and the clepsydra. The sides of a deflated bellows completely emptied of air had to remain in contact and inseparable regardless of the forces that might attempt to separate them. If the sides parted, this was a sure sign either that the bellows had not been totally evacuated and sufficient air remained within to expand and fill the inner concavity, or that the surface of the bellows had been ruptured. Material continuity was thus preserved, as it had to be in a plenist natural philosophy, but at a considerable price: the otherwise inexplicable behavior of the bellows itself. Even greater ingenuity and ad hocness were required for examples involving the separation of surfaces. On the assumption that such surfaces could make direct contact without an intervening material medium, one could avoid the vacuum and the associated absurdity of instantaneous velocity only by insisting that the plates could be separated only gradually, part by part. Even this solution was untenable, as Walter Burley demonstrated. There was, of course, the logical construction of Blasius of Parma. But because it followed in the tradition of paradox and philosophical puzzle associated with Zeno of Elea, it could hardly be expected to play a significant role in the history of our problem. The alternative explanation, in which either air, water, or fire was always assumed to intervene between any two plane surfaces, however close they might appear, avoided the twin pitfalls of vacuum and instantaneous motion but, as we have seen, it placed an insupportable physical and intellectual burden on the thin film of intervening matter.

It was in this desperate situation that Walter Burley, like so many others, resorted to the universal nature, or celestial agent. The thin film of air lying beneath the surface of the heavy stone would not be compelled to depart and leave behind a vacuum because the universal nature would intervene and support the weight of the stone to prevent so dreadful a consequence. The universal nature was the ultimate explanation, a *deus ex machina* to resolve all difficulties. Nowhere were its services required more than in the clepsydra. With the upper orifice stopped up, the water could not depart without leaving a vacuum behind. No physical explanation could be adapted to explain this unnatural behavior. Such a cosmic condition could be explained only by invocation of a universal nature, which had only to determine the least disruptive means of achieving this essential goal. In the clepsydra, it was better for nature to cause the water to remain elevated than to allow it to descend naturally and prevent the vacuum by collapsing the sides of the vessel. But in the case of water frozen in a sealed vessel, the universal nature occasionally chose to smash the vessel, as all could have observed. Whatever the mechanism selected by the universal nature, its omnipresence within the world was sufficient guarantee that if the ordinary processes of nature were unable to prevent formation of a vacuum, its enormous, if not absolute, power was more than equal to the task.

In the course of the sixteenth century, the traditional medieval plenist interpretations of the experiences and examples described above underwent a striking transformation. With the introduction and availability of previously unknown or little known ancient texts that assumed the existence of vacua, such as Lucretius' *De rerum natura* and Hero's *Pneumatica*,[149] and the gradual development of a strong anti-Aristotelian movement, fear of the dreaded vacuum diminished and in some instances vanished entirely. A number of examples previously interpreted from a plenist conception of the universe, with its fundamental assumption of nature's abhorrence of a vacuum, were now thought to represent solid evidence for the existence of artificially created vacua.[150] Although sixteenth-century thinkers no more performed the experiments they discussed than did their medieval predecessors, they viewed them in a quite different light. Thus Bernardino Telesio believed that if a bellows were constructed with "thick and heavy" rather than "loose and thin" materials, separation of its sides after the air had been evacuated and a vacuum left behind would indeed be possible.[151] Francesco Patrizi argued similarly that if the water that filled a pouch were squeezed out and the mouth of the pouch bound tightly as its sides lay flattened against each other, the sides of the pouch could subsequently be separated and an empty space would truly lie between them.[152]

Adopting the traditionally false belief that water contracts upon freezing, both Telesio and Patrizi were convinced that in a vessel filled with water that is subsequently frozen, a void space will appear where the water has contracted. As with Telesio's bellows illustration, we must suppose that the vessel is sufficiently strong to resist fracturing.[153]

Although Telesio and Patrizi accepted the traditional interpretations of the clepsydra experiment, they suggested modifications that would have permitted formation of a vacuum. By using honey or some other liquid denser than water and by enlarging one of the holes in the bottom of the clepsydra, Telesio argued that the heavier liquid would indeed flow out of the enlarged hole and leave a vacuum behind.[154]

Perhaps the most striking evidence for the possibility of creating an artificial vacuum was proposed by Adrian Turnèbe. Turnèbe described an experiment, probably drawn ultimately from Hero of Alexandria's *Pneumatica*, in which a siphon is inserted through a specially made hole in a glass ball. After the hole is caulked, the air can be exhausted from the ball by a sucking action of the mouth. Evidence that a vacuum remains can be obtained by covering the siphon with a finger and inverting the whole apparatus into water, following which "the liquid would enter when you would remove your finger; and it would be drawn into the interior of the ball."[155]

Attempts to explain the consequences of the separation of two surfaces also continued beyond the Middle Ages and, as with the other experiments and examples, formation of a vacuum came to be readily conceded. Galileo himself may be seen to mark a culmination in this long tradition, as evidenced by his significant discussion in the First Day of the *Discourses Concerning Two New Sciences*. Although Galileo adopted the traditional judgment that nature abhors a vacuum, he concluded from the actions of hard, polished surfaces in a variety of circumstances that a momentary vacuum would indeed form upon the separation of such surfaces. In the *Two New Sciences*, the problem of separation arises in the context of resistance to fracture and breakage offered by hard bodies. One cause of this resistance is said to be the void itself. Salviati explained that

We may see whenever we wish that two slabs of marble, metal, or glass, exquisitely smoothed, cleaned, and polished and placed one on the other, move effortlessly by sliding, a sure argument that nothing gluey joins them. But if we want to separate them while keeping them parallel, we meet with resistance; for the upper slab in being raised draws the other with it, and holds it permanently even if it is large and heavy. This clearly shows nature's horror at being forced to allow, even for a brief time, the void space that must exist between the slabs before the running together of parts of the surrounding air shall occupy and fill that space.[156]

Salviati further explained that if the slabs or surfaces were not perfectly clean and their contact was imperfect, air would lie between them. Under these circumstances, when separation is slow, the resistance we would feel derives from the weight of the upper slab. But when separation is effected by a sudden pull, the lower slab will be lifted momentarily, only to fall back as the intervening air expands and prevents formation of a vacuum. Because these actions are intended to thwart formation of a momentary vacuum, Salviati concluded that nature abhors a vacuum and will prevent it whether the slabs or plates are in perfect or imperfect contact.

Sagredo, however, suggested that a momentary vacuum would indeed form, from which he inferred that, contrary to the common contention of Aristotelian philosophers, motion in a void would not be instantaneous.[157]

For if it were, the two surfaces would be separated without any resistance whatever, the same instant of time sufficing for their separation and for the running together of the surrounding air to fill the void that might [otherwise] remain between them. Thus, from the following of the upper slab by the lower, it is deduced that motion in a void would not be instantaneous. It is then further deduced that some void indeed does remain between the surfaces, at least for a very brief time; that is, for as long as the time consumed by the ambient air in running to fill this void. For if no void existed there, neither would there be any need on the part of the ambient air of running together, or of any other motion.[158]

Because Sargredo's argument was left standing, it seemed that Galileo agreed with the common medieval opinion (Blasius of Parma, at least, excepted) that separation of plates or surfaces in perfect contact would produce a momentary vacuum.[159] But where *all* his medieval predecessors devised explanations that would avoid or explain away this disastrous consequence, Galileo accepted it and saw no dire effects resulting therefrom. Although he believed that nature abhorred and resisted formation of extended vacua of whatever size, it was not always able to prevent their occurrence. On all such occasions, however, nature sought to fill them as quickly as possible.

By interpretations such as these, in which it was assumed that artificial *vacua separata* could be made without disastrous consequences to the universe, the medieval principle that nature abhors a vacuum was undermined and eventually abandoned. Medieval examples that were ostensibly favorable to the existence of separate, extended, artificial vacua had always been interpreted from the standpoint of a plenist conception of the world and then rejected. This attitude changed in the sixteenth century, when many of these same examples were found supportive of the vacuist position. Acceptance of interstitial vacua and growing acknowledgment that *vacua separata* could be created artificially were the antecedent conditions for the denouement in the seventeenth century when, with scorn and ridicule, Blaise Pascal rendered untenable the principle that nature abhors a vacuum.[160] Although many of the most significant physical scientists of the seventeenth century were convinced that *vacua separata* existed naturally in the universe, even those who denied the claim were not likely to take refuge in nature's abhorrence of a vacuum. Aside from a brilliant but uninfluential defense of interstitial vacua by Nicholas of Autrecourt, medieval natural philosophers rejected the natural or artificial existence of all vacua within the cosmos. Although they were prepared to speculate at considerable length about the conditions and possibilities of finite motion in hypothetical *vacua separata*, the consequences of real empty space were too destructive of all that had come to represent the medieval world view to win even the most hesitant and tenuous acceptance. But if intramundane vacua could win no place in the real scheme of things during the Middle Ages, the history of the concept of extramundane space, which will constitute the second part of this study on space and vacuum, would be quite different.

PART II

Infinite void space beyond the world

Largely a theological problem

Medieval concern for the various problems associated with the possibility of void space within the world was, as we have now seen, both extensive and intensive. This is hardly surprising when it is realized that every commentator on Aristotle's *Physics* had to confront a variety of arguments against the possible existence of empty space. With regard to the possibility that a vacuum might exist beyond the boundaries of the cosmos, the situation was quite different. For although, as will be seen, Aristotle had occasion in *De caelo* to reject extracosmic space, he considered the matter only briefly, and scholastic commentators on *De caelo* rarely used Aristotle's discussion as a point of departure for further speculation. In fact, brief discussions on extracosmic void were just as likely to turn up in commentaries on the *Physics*. Although Aristotle's arguments for the rejection of extracosmic void were to play a central role, the probable reason why the subject of extracosmic void did not become a regularly discussed theme in the Aristotelian commentary literature is the theological nature of the problem as it emerged in the Middle Ages during the thirteenth century. Although the historical roots of the concept of extracosmic void were both secular and theological, speculation about it became significant in the late Middle Ages only after it became enmeshed in theological debates about God's location, His absolute power, and the conditions that obtained before the creation. Medieval theologians, and very few of them, would thus develop the greatest interest in the possible existence of such a space. For the professional Aristotelian natural philosophers, who were the teaching masters in the medieval universities and without formal training in theology, the concept was not only cosmologically unacceptable but potentially dangerous because of its obvious theological implications and pitfalls. It was thus left to a small number of medieval theologians to lay the foundations for the more substantial scholastic discussions of the sixteenth and seventeenth centuries. To comprehend and appreciate the late medieval developments, which would, as will be seen, have a considerable impact on spatial conceptions of the seventeenth century, it is first essential to describe the secular and theological origins of those developments. At the very least, three significant intellectual currents must be distinguished. The first debated the probability and

103

reality of the existence of an infinite vacuum beyond the world; the second involved a theological condemnation of 219 propositions in 1277 by the bishop of Paris; and the third was concerned with conditions prior to creation, namely the possible need for a pre-creation void space in which the world had to be located and with the location of God Himself. We shall now describe these in turn.

5

The historical roots of the medieval concept of an infinite, extracosmic void space

1. ARISTOTLE'S REJECTION OF EXTRACOSMIC VOID AND THE REACTION IN GREEK ANTIQUITY

The idea of extracosmic void space reached the Latin West from a number of sources during the Middle Ages. As with so much else, it was Aristotle who conveyed the concept in the form that would be most widely known and that would be central to all subsequent discussion.[1] In the course of rejecting the existence of a plurality of worlds in *De caelo*, Aristotle declared categorically that "neither place, nor void, nor time" can exist "outside the heaven."[2] He had earlier argued that no bodily mass could come into being beyond the heavens, or outermost circumference of the universe, and inferred from this that neither place nor vacuum could exist there because "in every place a body can be present"[3] (no body, therefore no place) and because "void is said to be that in which the presence of body, though not actual, is possible"[4] (no possibility of body, therefore no vacuum). Aristotle concluded that absolutely nothing existed beyond the universe, a nothing that was best characterized as a privation.[5] His denial of extracosmic existence to place, void, time, and body was frequently repeated. Special reliance was placed upon the necessary connection between body and void. By definition, vacuum was conceived as a place devoid of body but capable of receiving it. It was usually concluded that the existence of a vacuum would be impossible if no body could possibly occupy it. The existence of extracosmic void was thus inextricably linked with the occupation of it by a body. Rejection of extracosmic void was often made to depend on the impossibility of any body existing beyond the world. Without the possibility of occupation by a body, the existence of vacuum beyond the world was deemed impossible.

At least one scholastic was prepared to alter Aristotle's definition to allow for the existence of extracosmic void. In the third of five ways of conceiving a vacuum, Roger Bacon defined vacuum as "a space in which there is absolutely no body, nor is there any natural aptitude for receiving any body; but to assume [vacuum] in this way, [is to assume it] beyond the heaven."[6] By altering Aristotle's definition, Bacon was able to formulate another that did allow for the possibility of a vacuum beyond the world but nowhere else. By accepting Aristotle's argument that no body could possibly exist beyond the world, and by defining one type of vacuum as that which could not possibly receive a body, Bacon devised a conceptual void space that could exist only beyond the world.[7]

Bacon's extraordinary move, proposed in a bare three lines without elaboration, is one that found no imitators or followers. It is obvious, however, that Bacon went beyond Aristotle in allowing something besides total privation to exist outside the world. Perhaps he felt the kind of intuitive reaction against Aristotle's denial of extracosmic existence that appeared very early on. Indeed, even before Aristotle's time, as he himself informed us, the Pythagoreans had proclaimed that what lies outside the heavens is infinite,[8] by which they appear to have understood an infinite extent of air, with the air apparently conceived as equivalent to void.[9]

Armed with a proper concept of void, the basis for a more fundamental reaction to the denial of extracosmic existence would yet be provided by a Pythagorean. Assuming void as empty space devoid of air and matter, Archytas of Tarentum (first half of the fourth century A.D.), Plato's contemporary, declared:

If I am at the extremity of the heaven of the fixed stars, can I stretch outwards my hand or staff? It is absurd to suppose that I could not; and if I can, what is outside must be either body or space. We may then in the same way get to the outside of that again, and so on; and if there is always a new place to which the staff may be held out, this clearly involves extension without limit.[10]

Although this fragment, as preserved in Simplicius' *Commentary on Aristotle's Physics*, was unknown in the Middle Ages,[11] it formed the basis for one of the most telling arguments against Aristotle's finite cosmos. It is obvious that Archytas' objective was to argue for "extension without limit" without determining whether it was body or void that lay beyond the world. In a crucial passage from Simplicius' *Commentary on De caelo*, which was known in the Middle Ages in the translation by William of Moerbeke from Greek to Latin in 1271, we learn that the Stoics decided in favor of infinite void. The argument is basically the same as Archytas', except for a seemingly arbitrary assumption that eventually one would exhaust whatever matter lies beyond the world and reach the beginning of an infinite void.[12] The assumption of an infinite extracosmic void by the Stoics was not, however, as arbitrary as would appear from the passage reported by Simplicius. Ironically, the Stoics followed Aristotelian physics and cosmology in their major outlines, agreeing with Aristotle that the cosmos is a finite sphere without any vacua whatever (it was filled with *pneuma*) and insisting that all existent matter and reality were contained within it. It was thus plausible for them to infer that if anything existed beyond our world, it would be void space. But there is a problem here. Because the Stoics accepted Aristotle's definition of void as that "in which the presence of body though not actual is possible"[13] and because they also agreed with Aristotle that nothing existed outside the cosmos, should they not have followed Aristotle and concluded that because no matter could exist beyond the world, it was impossible for void to receive body and therefore impossible that an extracosmic void exist? It was a different conception of "possible" that enabled the Stoics to diverge from Aristotle. Where for Aristotle "a proposition is possible only if it becomes actual at some time . . . to the Stoics this condition is not necessary; a proposition is possible if nothing external prevents it from being true. Consequently, the condition beyond the periphery of the cosmos satisfies the definition of void, even if body *never* comes to occupy it."[14]

Void space was thus conceived as a three-dimensional receptacle for the finite cosmos. Indeed that appeared to be its only function, because interaction between cosmos and void was denied on the grounds that void has no properties of its own and can in no way affect the material world, which constitutes a closed system that cannot be dissipated into void.[15]

But granted that the world is surrounded by a vacuum, why, it was asked, must it be infinite? Cleomedes provided a typical Stoic response when he argued that because no body could exist beyond the physical world, no material substance could limit void; and because it was absurd to suppose that void could limit void, or that void should terminate at one point rather than another – a clear violation of the principle of sufficient reason – infinite void space seemed an irresistible conclusion.[16] One could also have applied Archytas' argument to the extracosmic void itself to demonstrate its infinity.[17] Thus was born the concept of a spherical, finite cosmos sealed off from but surrounded by an infinite void space.[18] It would find occasional favor in a quite different context in the Middle Ages and gain in popularity in the sixteenth and seventeenth centuries until abandoned in the eighteenth.

The infinite space that surrounded the world was the product of cosmological and physical controversy and had nothing to do with any alleged application of Euclidean geometric space to the physical world (see Chapter 2, Section 2a). From the earliest beginnings, associated with the name of Archytas of Tarentum, all the way to the Scientific Revolution of the seventeenth century, those who fashioned the concept of a dimensional, infinite space paid no homage to Euclid. When Pierre Gassendi argued in behalf of a three-dimensional void space, his supportive appeal to the ancients did not include Euclid but rather Epicurus and Nemesius.[19]

2. THE CONDEMNATION OF 1277

The Stoic argument in favor of an infinite, extracosmic void as reported to the Middle Ages in Simplicius' commentary on De caelo was perhaps the most explicit defense of this opinion that the Middle Ages inherited from antiquity.[20] But because Simplicius himself rejected the argument, and because the entire weight of Aristotelian physics and cosmology stood against it, extracosmic void might have received much less attention than it did, were it not for the Condemnation of 1277 some six years after Moerbeke made Simplicius' commentary on De caelo available in Latin translation.

The condemnation of 219 diverse articles in theology and natural philosophy by the bishop of Paris in 1277 was a major event in the history of medieval natural philosophy.[21] Whatever doctrinal and philosophical disputes or personal and group animosities may have induced bishop Étienne Tempier to promulgate the sweeping condemnation,[22] its most significant general result was an emphasis on God's absolute power (potentia Dei absoluta) to do whatever He pleased short of a logical contradiction. Although the doctrine of God's absolute power was hardly new in the thirteenth century,[23] the introduction into the Latin West of Greco-Arabic physics and natural philosophy, with their independent and often deterministic philosophical and scientific explanatory principles, conferred on that doctrine a new and more significant status.[24] After 1277, appeals to God's absolute power were frequently introduced into discussions of Aristotelian physics and cosmology, which had previously been largely restricted to a consideration of natural causes and principles. Whether by implication or explicit statement, many of the condemned articles asserted God's infinite and absolute creative and causative power against those who would circumscribe it by the principles of natural philosophy. Nowhere is the spirit of the condemnation better revealed than in article 147, which condemned the opinion "that the absolutely impossible cannot be done by God or another agent," where "impossible is understood according to nature."[25]

The condemnation was in effect throughout the fourteenth century.[26] Many of the most significant scholastic authors had occasion to allude to or explicitly cite one or more of its articles.[27] The supernatural alternatives considered in the aftermath of the condemnation conditioned scholastics to contemplate physical possibilities outside the ken of Aristotelian natural philosophy and frequently in direct conflict with it. Its most characteristic feature, God's absolute power, was invoked in a variety of hypothetical situations. Indeed hypothetical possibilities derived from supernatural actions became a characteristic feature of late medieval scholastic thought.

In its impact on physical thought, no area was more affected by the condemnation and its central idea of God's absolute power than the concept of extracosmic vacuum. Here two articles played a paramount role: the thirty-fourth, which denied that the First Cause, or God, could produce more than one world,[28] and the forty-ninth, which denied that God could move the world with a rectilinear motion simply because a vacuum would be left behind.[29] After 1277, scholastics at Paris were compelled to concede that God could create as many worlds as He pleased and to allow that He could move our entire spherical universe rectilinearly even though a vacuum remained in the place it vacated.

In exploring the consequences of these possibilities, as we shall see below, concepts and ideas contrary to Aristotelian physics and cosmology were often found plausible rather than impossible. Articles 34 and 49 made it appear reasonable to assume that an extracosmic void space existed beyond our world. For if God did make other worlds, it was assumed that void space would intervene between them; and if God moved the world rectilinearly, it was further assumed not only that a void space would be left behind but also that the world itself moved into and out of other empty spaces that lay beyond.

Although no articles of the condemnation concerned vacua within the cosmos itself, it seemed obvious that if God could create or allow a vacuum beyond the world, He surely could do the same within the world. And so it was, as we have already seen, that God was frequently imagined to annihilate all or part of the matter within the material plenum of our world.[30] Hypothetical problems that might arise from such supernatural actions were often discussed. Analyses of a variety of "thought experiments" conceived within intracosmic and extracosmic vacua in the late Middle Ages were often made in terms of Aristotelian principles even though the conditions imagined were contrary to fact and impossible within Aristotelian natural philosophy. God's absolute power to create and annihilate matter and vacua within and beyond our world, as well as to move the world itself, was, as we shall see, discussed into the sixteenth and seventeenth centuries by scholastic and nonscholastic authors, among whom we may mention Francisco Suarez, Pierre Gassendi, John Locke, Thomas Hobbes, and Isaac Newton.

3. DID A VOID SPACE EXIST BEFORE THE CREATION OF THE WORLD?

Few topics in the thirteenth century were more controversial than that of the eternity of the world. As the most prestigious supporters of the world's eternity – and by eternity is meant without beginning or end – Aristotle and Averroes had placed Christian scholastics in a harsh dilemma: either believe that the world was uncreated and without beginning, which was contrary to traditional interpretations of Scripture,[31] or accept the existence of a pre-creation void space in which the world was created. Aristotle, who first formulated these alternatives in his *De caelo*, declared that "nothing is generated in an absolute sense" and then went on to explain that

generation of any body is impossible unless we can posit a void free of body; for the place which has held, since its inception, the thing now in course of generation, must previously have contained only emptiness without any body at all. One body can be generated from another, e.g., fire from air, but from no preexisting magnitude nothing can be generated.[32]

Thus if a body had actually been created from nothing – a possibility that Aristotle expressly denied – the space that now contains it must previously have been void, and therefore temporally prior to the body it now contains.

In his lengthy commentary on this passage, Averroes extended Aristotle's reasoning to the whole world, arguing that if the world were generated in time, it must have been generated in a place that is either a vacuum or the terminus of a containing body. The latter is impossible because, prior to the creation of the world, no containing body could have existed. Therefore, it followed that if the world was created, a vacuum must have previously existed in order to receive it.[33] Indeed all who believed the world was created, including the Muslim *Loquentes*, or Mutakallimun, who assumed its creation from nothing, accepted the existence of a vacuum prior to the creation of the world.[34] For Aristotle and Averroes, who denied on numerous grounds the possibility of the existence of void space and the generation of something from nothing, the choice was obvious: The world is ungenerated and eternal.

Neither of the alternatives proposed by Aristotle and Averroes was found acceptable in the Latin West,[35] at least not after the Condemnation of 1277, which, in damning both options, listed numerous objectionable articles about the eternity of the world[36] and actually condemned the necessary existence of an independent pre-creation void space.[37] That an uncreated world should cause grave concern to Christians is hardly surprising, but it is not readily apparent why a pre-creation void space should have been viewed with sufficient alarm to warrant condemnation.

The explanation lies in the possibility that a pre-creation void space could be conceived as an entity of eternal duration independent of God. Christian anxieties on this point were admirably expressed by Thomas Bradwardine, who declared that the properties or characteristics of such a pre-creation place or space could have "no positive nature, for otherwise there would be a certain positive nature which is not God, nor from God . . . ; such a nature would be coeternal with God, something no Christian can accept."[38] It was fear of this consequence, in Bradwardine's view, that prompted the bishop of Paris to condemn the opinion that many things are eternal.[39] Thus was a powerful argument formulated against the existence of an independent, eternal, void space, the kind advocated by atomists and Stoics. On theological grounds, it was essential to deny the possibility of an uncreated, eternal, and infinite void space that was completely independent of God.[40]

But if the dilemma posed by Aristotle and Averroes was rejected by denying both the eternality of the world and the existence of an independent void space that preceded its creation, the question remained as to whether the world required a container or place, and if so, what might be identified as functioning in this capacity.

Respondents to this question during the Middle Ages would have had to consider whether the world was to be conceived as temporally infinite and uncreated or temporally finite and created. If the first, or Aristotelian, position were adopted, the place under discussion would inevitably involve Aristotle's definition of place as the innermost, immobile surface of a containing body. Here the problem was one of conveying a sense of place to a world that had neither beginning nor end. Because all admitted that no material body existed beyond our finite world, determining a sense of place for the last material sphere, or for the world itself, posed a serious problem and was regularly discussed in commentaries on Aristotle's doctrine of place in the fourth book of the *Physics*. In the history of medieval spatial concepts, however, the problem of determining a physical sense of place for the world played no role.[41]

On the assumption, however, that the world is temporally finite and created, an assumption that was taken as scriptural truth in the Middle Ages, the problem of the place of the world took on a quite different and totally un-Aristotelian complexion. Because our finite world was the first physical creation of an all-powerful deity, it was natural to inquire whether the world was created in something nonphysical – that is, was it placed or located in something that was already in existence or not? One response, especially popular in the thirteenth century, was to deny that God had need of a pre-creation place in which to locate the world. "When God made the world," one author explained, "He made the place or dimension to receive the world. And so there was neither place, nor dimensions, nor are there now – beyond the heaven."[42] Whatever is meant by "place" in this context, it is clear that it was created simultaneously with the world and did not precede it. By the fourteenth century, however, the problem of a pre-creation void emerged in a context quite different from that described earlier. By then, some were convinced that an eternal, uncreated, pre-creation void space could exist only if it were in some sense associated with God Himself, for this was the only plausible way that a Christian could attribute eternity to space. The discussion of extracosmic space thus came to focus on its relationship to God. As space became linked with God, the problem of God's location became profoundly important to the concept of space itself. While it was usually assumed that God is omnipresent in the world He created, it seemed natural to inquire about His whereabouts before creation. Because God is an immutable being, His place or location must be permanent, unchanging, and eternal. Thus if He is in the world now, He must also have been here before the world. But is the place of the world itself the extent of God's omnipresence? Or does God extend beyond the world of His creation? How should one respond to the question "where is God?" Consciously or not, many responses had spatial connotations.

4. WHERE IS GOD?

Already in pagan Greek thought, the barest beginnings of a connection between God and space are detectable in certain Stoic authors, who identified their ubiquitous, intramundane *pneuma* with the divine spirit.[43] Stoic contributions, however, were limited by the fact that their infinite, three-dimensional, extracosmic void space was assumed to be totally devoid of body and spirit.

It was thus the Judeo-Christian tradition, not pagan Greek thought, that would provide the basic framework of ideas necessary for the developments of interest in this study. Probably during the lifetime of Christ, and certainly before Christianity was more than a local phenomenon, Philo Judaeus (ca. 20 B.C.–ca. A.D. 45), or Philo of Alexandria, as he was also known, identified God and space unequivocally when he explained that God Himself is called a place by reason of His containing things, and being contained by nothing whatever, and being a place for all to flee into, *and because He is Himself the space which holds Him; for He is that which He Himself has occupied*, and naught encloses Him but Himself. I, mark you, am not a place, but in a place; and each thing likewise that exists; for that which is contained is different from that which contains it, and the Deity, being contained by nothing, is of necessity Itself Its own place.[44]

In a later rabbinic tradition, God was described as the place of the world, by which was usually meant "the omnipresence of God within a universe from which He is separated and which He transcends."[45]

Similar remarks appear in early Latin Christian authors. Saint Cyprian (ca. 200–258) would declare that God is "one and diffused everywhere,"[46] and Arnobius of Sicca (ca. 260–ca. 327) would say of Him: "Thou art the first cause, the place and space of things created, the basis of all things whatsoever they be."[47] In *De trinitate*, Boethius (ca. 480–524/525) asserted that God is "everywhere but in no place."[48] From such brief remarks and others that might be included, it is reasonable to infer that no well-developed ideas on the nature of space were held by these early representatives of Christian thought. And yet their opinions were frequently cited as the problem of space grew in importance between the Middle Ages and the sixteenth and seventeenth centuries. As authors in these later periods sought respectability and support for their interpretations, they made much of what early Christians said about the relationship of God and space. The brief statements of Cyprian and Arnobius, for example, were invoked by Joseph Raphson in his *De spatio reali* (1702) as if they had foreshadowed his own opinions elaborated over some ninety pages.[49] Cyprian's one-line remark was interpreted as a declaration that "[God is] everywhere through the whole world and beyond the world through an immense vacuum. . . ." Throughout the period covered by this study, the discovery of profound truths and insights buried in brief, enigmatic, and often cryptic statements by ancient authors was commonplace, a state of affairs that was especially true for the subject of space.[50]

In this connection, Saint Augustine (354–430) was a special favorite. As the early Christian who played the most significant seminal role in the development of ideas involving God and space, it is ironic that Saint Augustine explicitly rejected the existence of an infinite, extracosmic space. To the taunting query of the Manichaeans, "Where was God before the heaven and earth existed?", Augustine's reply that God was only "in Himself" (*in seipso*) and not in another[51] would have a profound and enduring, though ambiguous, effect on the subsequent history of the problem of God's relationship to extracosmic space. Because Augustine said nothing about God being in a pre-creation void, but only that He was in Himself, many would take this as sufficient reason to reject extracosmic void. In truth, as we shall see, Augustine's God is in Himself dictum was readily accommodated to belief in and rejection of extracosmic void (see Chapter 7, notes 110, 125, and 143) and was frequently cited between the fourteenth and seventeenth centuries.

Despite the denial of extracosmic void space, Augustine was one of at least two major sources for an idea that would prove crucial to the relationship between God and void space. In *The City of God*, book 11, chapter 5,[52] Augustine took issue with certain non-Christians who shared with Christians a belief in God as a spiritual being and creator of all things. Noting with approval that these men rightly held that God, the divine substance, cannot be limited but is "spiritually present everywhere," Augustine went on to say that if infinite spaces existed beyond our unique world, these non-Christians would be committed to a belief in the omnipresence of the divine substance in those infinite spaces because no reason could be adduced for confining God to our finite world. Fortunately, they were not compelled to extend the presence of the Creator beyond the cosmos because "they maintain that there is but one world, of vast material bulk, indeed, yet finite and in its own determinate position." Indeed they would probably argue "that the thoughts of men are idle when they conceive infinite places, since there is no place beside the world, . . ."[53]

The idea that if spaces existed beyond the world, God, or spirit, would be present may have been derived by Augustine from the Hermetic treatise *Asclepius* (or *De aeterno verbo*, as it was called in the Middle Ages), a work familiar to Augustine[54] and well known in the Middle Ages. Speaking to Asclepius, Hermes Trismegistus declared:

But as to void, which most people think to be a thing of great importance, I hold that no such thing as void exists, or can have existed in the past, or ever will exist. For all the several parts of the Kosmos are wholly filled with bodies of various qualities and forms, each having its own shape and magnitude; and thus the Kosmos as a whole is full and complete. . . . And the like holds good of what is called "the extramundane," if indeed any such thing exists; for I hold that not even the region outside the Kosmos is void, seeing that it is filled with things apprehensible by thought alone, that is, with things of like nature with its own divine being. . . .[55]

And so Asclepius, you must not call anything void, without saying what the thing in question is void of, as when you say that a thing is void of fire or water or the like. For it is possible for a thing to be void of such things as these, and it may consequently come to *seem* void; but the thing that seems void, however small it be, cannot possibly be empty of spirit and of air.[56]

Despite a basic agreement with Aristotle that an actual void could exist neither within nor beyond the world, Hermes has here expressed a thought that was potentially subversive of Aristotelian cosmology. If void did exist beyond the cosmos, it would surely not be void of spiritual substances "apprehensible by thought alone," although it might well be devoid of visible physical bodies.[57] Here may have been born the idea of an extra-mundane space filled with spirit but empty of matter,[58] a concept whose reality is actually denied by Hermes. Saint Augustine and the author of *Asclepius* were thus agreed that the world is finite with nothing existing beyond, neither space nor void. But if something did exist beyond, they also concurred that the divine spiritual substance would of necessity be omnipresent in it.

The actual existence of such a spirit-filled extra-mundane void space would be affirmed in the fourteenth century. By that time, not only were the ideas and concepts described in this chapter in full effect and readily available, but the problem of God's "spatiality" and its relationship to the world He created had become a regular feature of medieval theological discussion. The locus of these discussions was Peter Lombard's *Sentences* (*Sententiae*), the most famous theological treatise of the Middle Ages and one on which all theological students had to comment before receipt of the theological degree. In book 1, distinction 37 of this twelfth-century work, Peter described the ways in which God is said to be in things ("quibus modis dicatur Deus esse in rebus").[59] Under this heading, theologians regularly considered God's habitat prior to the creation, the ways in which God could be said to be in a place, and whether He could be moved in any way. Incorporating numerous citations from scriptural and patristic sources, especially Saint Augustine, book 1, distinction 37, became one of the significant places, particularly in the sixteenth and seventeenth centuries, where the possible existence of extracosmic space and God's relationship to it could be appropriately considered.[60] We must now turn to the medieval problem of extracosmic void space.

6

Late medieval conceptions of extracosmic ("imaginary") void space

With perhaps a few minor exceptions,[1] there was little serious discussion of the possibility of extracosmic void prior to the Condemnation of 1277. When the problem did arise, Aristotle's rejection was usually adopted with little elaboration, as in the anonymous *Liber sex principiorum*,[2] falsely ascribed to Gilbertus Porretanus; in Johannes de Sacrobosco's *Sphere*;[3] and in Robert Grosseteste's *Commentary on the Physics*.[4] Even Thomas Aquinas found little occasion to discuss the possibility seriously, perhaps because he thought extracosmic void an untenable suggestion.

After 1277, however, the situation altered dramatically and the possibility of the existence of extracosmic space came to be discussed in two interrelated, though distinguishable, contexts. In the first context, the primary concern was with the possible existence of void space independent of God but assumed to have been created by Him before, during, or after the creation of the world. The possibility that God could create finite vacua at will was regularly conceded after 1277,[5] though His ability to make an infinite vacuum was, as will be seen, seriously questioned. In the second context, extracosmic void space was not assumed independent of God but was conceived to be in some sense associated, as a property or attribute, with God's omnipresent immensity. Although not all discussions of extracosmic void can be fitted neatly into one or the other of these two contexts – Jean de Ripa's treatment of the subject, for example, seemed to fall into both – it is convenient to consider them separately.

The first approach described above was concerned with God's absolute power and the void spaces and bodies He might create beyond the world. In contemplating the cosmological consequences of these hypothetical and imaginary divine acts – which all had to consider as possible – any scholastic master, whether theologically trained or not, could legitimately and appropriately discuss the issues. It was different with the second context, in which void space became enmeshed with God's omnipresence and immensity and was thereby converted to a theological problem suitable for discussion only by theologians, and not by the masters in the arts faculty,[6] who outnumbered the theologians many times over. In what follows, I shall arbitrarily designate the first category of discussion "the God-created, independent, separate, extracosmic void" and the second "the God-filled, dependent extracosmic void." Although these classifications are arbitrary, the distinction is real, as will be made clear below. Before I attempt to describe and analyze these two aspects of the history of the concept of space, the reader should be made aware that the extracosmic void we shall be concerned with was frequently characterized as "imaginary space" (*spatium imaginarium*). Because the meaning of the term "imaginary" in this expression is hardly obvious, and came to signify many things, it is advisable at this point to inquire into its range of meanings from the Middle Ages to the seventeenth century.

1. THE MEANINGS OF THE TERM "IMAGINARY" IN THE EXPRESSION "IMAGINARY SPACE"

As with so much else concerning void space, it was Aristotle who probably furnished one of the important foundational conceptions in the complex of meanings that would eventually cluster around the expression "imaginary space."

In the fifth and most important of five reasons explaining why people come to believe in the existence of the infinite, Aristotle pointed to the fact that *in our thoughts*, certain things appear to be inexhaustible, or without end, as with number, mathematical measures or magnitudes, "and what is outside the heaven."[7] In the discussion that follows, Aristotle made it clear that for some people what lies "outside the heaven" is an infinite void and place in which an infinite body or infinite worlds might exist.[8] To indicate our imaginary conception of the endless and seemingly inexhaustible nature of the extracosmic, at least one of the medieval Latin translations of the *Physics* used the term *existimatio*, the same term found in the Latin translation of Averroes' *Physics* commentary, where Averroes declared that what lies outside the heaven "is thought" or perhaps "imagined" (*existimatio*) to be an infinite vacuum.[9]

Much more interesting, however, is Averroes' discussion in *De caelo*, book 1, comment 92, where he expressly linked extracosmic void and imagination in the context of Aristotle's rejection of the existence of a plurality of worlds. Averroes said that some may argue that if the world were created from an exemplar, it is necessary to assume the existence of a plurality of copies. For the exemplar of the world would be like a form, and the world itself would be a particular instance of that exemplar – that is, it would be form in matter. Now, as Aristotle explained, "it is universal in our experience that, among things whose substance is bound up with matter, there are many – indeed an infinite number – of particulars similar in form, so that there either are or can be many worlds."[10] Although Averroes, like Aristotle, would reject a plurality of worlds,[11] he accepted Aristotle's generalization that experience indicates that a plurality of individuals will exist when form is associated with matter. Therefore, it is not surprising that some people believe in the existence of a plurality of worlds. In this, we are assisted by our "imaginative power" (*virtus imaginativa*).[12] For we can imagine that beyond the world a vacuum exists that not only could contain another world but in which another world might indeed exist. The kind of imagination involved here is no different from that in which we see innumerable separate individuals in the air around us. Because air is imperceptible to the sight, it is as if each individual thing existed in a vacuum; and yet outside each individual thing, there is another individual. The same reasoning would apply to a plurality of worlds in a vacuum beyond our world. Indeed, "since the imagination (*imaginatio*) is assimilated to the understanding (*intellectus*)," the Muslim *Loquentes*, or Mutakallimun, believed that we must first understand that the vacuum exists before we can say that the world now exists after nonexistence. The need for more than one instantiation of a world composed of form and matter demands the existence of a vacuum to contain those other worlds. Because those other worlds are created, our understanding (*intellectus*) informs us that the existence of a vacuum beyond our world is necessary in order to receive them. But it is our imaginative power (*virtus imaginativa*), based on experiences of the kind described above, namely the independent distribution of things in air, that enables us to conceive, or imagine, that extracosmic vacuum[13] in which our world, and perhaps others, must lie.

The association between imagination and extracosmic space continued in the Latin West. Already in Robert Grosseteste's summary of Aristotle's fifth reason for believing in an infinite, we are told that "imagination (*ymaginacio*) assumes an infinite space" and that "imagination (*ymaginacio*) says" that infinite space lies beyond the heaven. (For the Latin text, see note 4.) Aristotle, Averroes, and Grosseteste all rejected the existence of extracosmic infinite void, but they recognized that some people could not imagine an end to extension and mistakenly inferred the existence of an infinite space beyond the world. Thomas Aquinas and Pseudo-Siger of Brabant, who also denied extracosmic space, emphasized that only in the imagination was it possible to conceive such an entity. Denying that anything exists beyond the world, Aquinas explained that "when we speak of nothing being beyond the heavens, the term 'beyond' betokens merely an imaginary place (*locum imaginatum*) in a picture we can form of other dimensions stretching beyond those of the heavens."[14] To Pseudo-Siger of Brabant, the term "beyond" (*extra*) could signify either a true place or one that is imaginary (*secundum imaginationem*). For we can perceive something by our imagination only if it is in a place. Presumably, then, if we wish to imagine a body beyond the world, we must first imagine a place for it, even though no such place exists there.[15] Thus the term "beyond" (*extra*) leads us to assume or expect something "out there" beyond the world. And even though on other grounds we are convinced that no place, void space, or body lies beyond, yet by extrapolation from mundane experience we can imagine spatial dimensions extending ad indefinitum beyond the world, or imagine bodies, and therefore places, beyond the last sphere.[16] The strong intuitive sense that spatiality of some kind must exist beyond the finite world occasionally proved overriding, as when Nicole Oresme declared that the "human understanding consents naturally that beyond the heavens and world, which is not infinite, there is some space, whatever it may be; and one could not readily conceive the contrary."[17] But as with some Arabs earlier, reason and truth also convinced Oresme of the real existence of extracosmic space.[18]

What Oresme sensed so irresistibly in the fourteenth century, what Archytas and the ancient Stoics gave utterance to in Greek antiquity and al-Rāzī in the Arab Middle Ages, was underscored in the twentieth century by no less an authority than Sir James Jeans, who observed that "hard though it may be to imagine space extending for ever, it is harder to imagine a barrier of something different from space which could prevent our imaginations from passing into further space beyond."[19] Until the formulation of the theory of relativity early in this century, few scientists would have denied the endlessness of spatial extent. But in the Middle Ages it was quite otherwise. With the rare and extraordinary exception of Hasdai Crescas[20] and certain theologians, including Oresme, who developed a special tradition to be discussed shortly, Aristotle's powerful influence made the Stoic argument and the acceptance of extracosmic space eminently resistible.

Indeed, certain scholastic authors warned against relying on or trusting the imagination. Henry of Ghent was especially upset by those who seemed incapable of accepting the truth of a demonstration unless it was in conformity with their imaginations. Despite Aristotle's demonstration that neither body nor void nor time could exist beyond the world, there were those who denied this because they failed to realize that their imaginations were limited to finite quantity and assumed, instead, that everything beyond the world is infinite.[21] Similarly distrustful of the imagination was John Buridan, who believed it responsible for the false assumption that space exists everywhere, an assumption that was as untenable as the belief that the sun is no larger than a horse simply because it appears so to the senses. In such cases, the understanding (*intellectus*) must correct the errant sense appearance and imagination.[22] Others, by contrast, used sense experience as the very basis for rejecting extracosmic void, as did Franciscus de Mayronis (d. ca. 1328), who denied extracosmic void because it is not manifested to us in our world. Thus it is not like air, which though invisible is yet made apparent to us in an inflated skin, from which its departure can be sensed.[23]

Although a number of scholastic authors readily conceded the human capacity to imagine an indefinitely extended space beyond the world – indeed some, like Oresme, would consider its denial contrary to human understanding – few were prepared to accept its reality. They would probably have been receptive to the cautionary warnings of Henry of Ghent and John Buridan and subjected their imaginations to the guidance of reason and understanding, a procedure that would have vindicated Aristotle's emphatic denial of void, place, and body beyond the world. For those who followed this path, the spatial extension that is imaginable beyond the world would have been nothing more than a mental fiction, an interpretation of "imaginary" that was congenial to all who upheld the strict Aristotelian position.

But if *reason* could induce Henry of Ghent and John Buridan to reject the existence of extracosmic space as a mental fiction conjured up by the undisciplined imagination, *that very same reason* would prompt Nicole Oresme to argue for the actual existence of an incorporeal space beyond the world. In Oresme's contrary view, reliance on sense perception would lead inevitably to rejection of extracosmic space because our senses cannot comprehend or perceive such an entity.[24] Thus, if Oresme had used the term "imaginary," which he did not, to describe a really existent extracosmic space, its imaginary character would have derived solely from the inability of our senses to perceive it. Its reality is affirmed by reason and understanding alone.

As the problem of extracosmic space became more widely and intensively discussed in the sixteenth and seventeenth centuries, earlier suggestions were made more explicit and new ones emerged. For Thomas Bradwardine, Franciscus Toletus (1532–1596) and Francisco Suarez (1548–1617), imaginary space was equivalent to vacuum, or void space. (All three will be discussed below.) To the Coimbra Jesuit commentators on Aristotle's *Physics*, space was imaginary because its dimensions were imagined "in a certain relationship corresponding to the real and positive dimensions of bodies."[25] By the second half of the seventeenth century, Otto von Guericke (1602–1686) presented a variety of meanings gathered from his considerable familiarity with the literature on extracosmic space. He recorded that some thought that the term "imaginary" in the expression "imaginary space" signified nothing; or something devoid of all reality; or the negation of all being; or something merely fictitious. Yet others "understand imaginary space as some merely possible, immense, corporeal mass that is diffused everywhere into infinity; or [imaginary space is] a possible location of such a corporeal mass [or quantity]." And finally, there are those who conceived imaginary space as "God Himself, who, in accordance with His immensity is necessarily everywhere, or infinitely diffused."[26] It is obvious that numerous meanings of the term "imaginary" were developed over the centuries, ranging from a mental fiction to an actually existent reality. The various significations attributed to imaginary space conceived as an existent reality were historically of considerable importance. Indeed much of this study is devoted to an elucidation of those meanings. But now let us return to the first two contexts of discussion distinguished earlier.

2. THE GOD-CREATED, INDEPENDENT, SEPARATE, EXTRACOSMIC VOID SPACE

The impact of particular articles of the Condemnation of 1277, as well as the general emphasis on God's absolute power, which was one of the major consequences of the condemnation, fostered an attitude in which it became commonplace to concede that God could create a finite, extracosmic void. As part of the general scenario, it was further imagined that God created one or more bodies in that void, so that discussion could then center on the mutual relationship of those bodies to each other, or to our world, or to the void space in which they had been placed. Despite the frequency with which this theme was considered, no one during the Middle Ages came to believe that God had actually created a three-dimensional finite or infinite vacuum outside the world. Buridan, for example, argued that a supernaturally existing extracosmic, infinite space was an untenable assumption because "we ought not to posit things that are not apparent to us by sense, experience, natural reason, or by the authority of Sacred Scripture. But in none of these ways does it appear to us that there is an infinite space beyond the world."[27] Perhaps Buridan's sentiments were representative of those who rejected not only infinite extracosmic space but also any finite void space beyond the world. In denying the reality of spaces, however, Buridan and other scholastics were compelled to concede that if God wished, He could at the very least create finite, extracosmic spaces.[28] On the assumption that He did create such a space or spaces, these same scholastics found themselves reflecting on the nature of that space and the status of bodies created within it. Although these problems were often discussed in the context of Aristotelian science, Aristotle's opinions were of little help because his definition of place assumed that the body in place was always surrounded by another body, the innermost, immobile surface of which served as the place of the contained body. To those who might be asked to describe the place of a single, unsurrounded body created beyond the world, Aristotle's definition was useless. New departures were called for and occasionally produced, as when John Buridan inquired about the status of a body located beyond the world. On the assumption that God may not have created any extracosmic space, Buridan concluded that although a body located beyond the world could not be in a place, because it was contained by no other body, it was, paradoxically, in a space. The space Buridan had in mind was not a dimensional vacuum but the dimension of the body itself. Employing a version of the old Stoic argument derived from Simplicius, Buridan explained that if a man could thrust his arm beyond the last sphere, "it would not be valid to say that he could not place or raise his arm there [simply] because no space exists into which he could extend his hand. *For I say that space is nothing but a dimension of body and your space the dimension of your body.* And before you raise your arm outside this [last] sphere nothing would be there; but after your arm has been raised, a space would be there, namely the dimension of your arm."[29] Similarly, if God created a bean beyond the world, the magnitude of the bean would be its own space[30] because no separate space lies beyond the world for it to occupy. Indeed, as with the bean so with the whole world. For just "as all the parts of the world are spaces" and dimensions, the space of the whole world would be the aggregate of the spaces of its material parts.[31]

The space that Buridan described here was destined for a significant history. Described as "internal space" by Franciscus Toletus and "real space" by Francisco Suarez, it would contribute to the spatial conception adopted by Descartes, its most famous proponent. (On internal space, see Chapter 2, Section 2a.) Compelled to consider the status of bodies in extracosmic void, Buridan applied to them the usual conception of measurement in the plenum of the Aristotelian world, in which he defined "the space [or distance] between me and you" as "nothing but the magnitude of the intervening air or of another natural body, if one should intervene."[32] Thus even if God created a bean beyond the world or allowed an arm to extend beyond the last sphere, the magnitude intervening between the termini of arm or bean constituted its own space or distance.

But where no matter or corporeal extension intervened between two bodies, or termini – that is, where a vacuum separated them – most scholastics would have denied the existence of an intervening distance or space. If, for example, God created a body in a vacuum beyond the world, Marsilius of Inghen was prepared to argue that no measurable distance would separate the body from the world. Indeed world and body would be in contact. Thus did Marsilius give expression to the medieval conviction that nature abhors a vacuum. Just as the surfaces of a clepsydra would instantly collapse and come into direct contact when the water within flowed out, so the extracosmic body comes into immediate contact with the convex surface of the last celestial sphere.[33] Marsilius allowed that God could cause a stone to separate from the convex surface of the last sphere only if He also created intervening bodies. With only a void space between them, however, there is nothing to function as a corporeal dimension or distance of separation, and the bodies would come into contact.[34] Indeed if God created three spherical worlds in mutual contact, no measurable distances could be said to intervene in the vacua that lie between their convex surfaces. In such a configuration of worlds, distances could be measured only curvilinearly between the points of any particular convex surface, because these distances would be separated by the continuous matter of the surface.[35]

a. IS EXTRACOSMIC, CREATED VOID DIMENSIONAL AND ARE MEASUREMENTS POSSIBLE WITHIN IT?

It is appropriate at this point, and perhaps even essential, to inquire about the kind of extracosmic void Marsilius and others may have had in mind. Was it three-dimensional or simply nothing, where "nothing" is conceived as without accident or incorporeal substance? It is evident that many who confronted this issue denied dimensionality and all other attributes to extracosmic vacuum. Denial of dimensionality is not derived from any direct statements to that effect, but rather from a denial of the possibility of direct distance measurements in such emptiness.[36] In the material plenum of the Aristotelian physical world, intervening dimensions appeared inseparable from the material quantity of bodies. Because material body was everywhere, the measurement of any distance or separation between any two bodies or termini implied the necessary existence of matter between those bodies or termini. Without intervening bodies, no measurement was possible. Buridan gave expression to this fundamental Aristotelian concept when he declared that "the space [or distance] between me and you is nothing but the magnitude of the intervening air or of another natural body, if one should intervene." (For the text and references, see note 32.) It was on this basis, as we saw, that Buridan argued that a body created in the region beyond the world constituted its own space (see Chapter 1, Section 2a, and above). It is also the reason why Marsilius of Inghen, Albert of Saxony (see note 34), and very likely Buridan were agreed that no distance measurements could occur in extracosmic void. Bodies placed beyond the world would come into direct contact and lack any measurable separation. Already in the thirteenth century, Roger Bacon and Richard of Middleton had expressed similar opinions about vacua within the world. According to Bacon, the sides of a totally evacuated vessel would collapse and come into contact because nothing cannot constitute a distance (see Chapter 4, Section 4b). Imagining that God had annihilated every created thing between the immobile earth and the heaven, Richard of Middleton concluded that under these circumstances no distance would exist between heaven and earth unless God subsequently created a separate dimension, causing a distance to exist between them.[37] For the scholastics mentioned here, all distances in a vacuum would be equal because nonexistent. They confronted a situation in which bodies were imagined created in a hypothetical void beyond the world in which no distances could be measured and therefore no dimensions assigned. Such a void would be like a pure nothing except that it could somehow receive bodies.[38]

Not all scholastics shared this interpretation of extracosmic void. At least two, Henry of Ghent (d. 1293) and Jean de Ripa (d. ca. 1370), made startling departures from the more usual attitude just described. Henry of Ghent distinguished two ways of measuring distances. The first is the traditional conception of measurement *per se* and is applicable only to the plenum of our world because it requires the intervention of corporeal, material dimensions.[39] But where a positive, material distance measurable by the first way is absent, a measurement *per accidens* may be possible. Here the measurement is made by virtue of positive dimensions that exist in something that is not itself a plenum, namely a three-dimensional vacuum. Because this three-dimensional vacuum lies between two distinct and separate bodies, the distance between them can be measured and the void itself would be something that can receive bodies. That a vacuum intervening between two bodies can actually separate the bodies and measure the distance of separation is shown by an example. Henry imagined that a three-foot wall lies alongside a vacuum that separates two bodies, *A* and *B* (Figure 2). Because, by assumption, the wall between the bodies

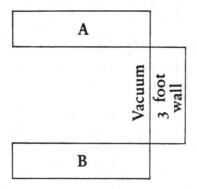

Figure 2. *Example proposed by Henry of Ghent to show measurement of a vacuum* per accidens. *See text for explanation.*

is three feet in length, it follows that body A, above the vacuum, must be separated from body B, below the vacuum, by a distance of three feet. And because the wall measures the vacuum, the latter must also be three feet in length and, by virtue of its measurability, capable of receiving bodies. Although Henry insisted that the intervening vacuum is absolutely nothing in itself (*secundum se omnino nihil est*), it can obviously function as if it were something *per accidens* – that is, it can serve to separate bodies by a measurable distance, which, in turn, indicates that it can receive bodies. Despite the lack of positive existence, a vacuum can be considered a positive dimension. For Henry, then, measurements over an intervening vacuum are no different than measurements over an intervening plenum. But because the latter has positive existence, measurements through it are characterized as *per se*, whereas measurements over or through a vacuum, which is absolutely nothing, are described as *per accidens*.[40]

On the basis of the distinction between *per se* and *per accidens* measurements, Henry argued that if God created a body beyond the world, or even another world, that is not in direct contact with the outermost celestial sphere of our world, we may infer the existence of an intervening vacuum, the distance of which can be measured. However, if God chose to create a body beyond our world that is actually in direct contact with our world, such a body would be neither in a plenum nor in a vacuum because no distance, *per se* or *per accidens*, separates it from the world. A body created beyond the world but in direct contact with it would be in a "pure nothing" (*purum nihil*), the same pure nothing in which the world itself was created.[41]

In the fourteenth century, Jean de Ripa sought to demonstrate that the concepts of place and measurement were also applicable to vacuum, or imaginary space. To achieve his objective, he devised four paradoxes[42] that revealed serious difficulties with the Aristotelian conception of place and measurement as confined to a plenum. The first two arguments concern the surrounding place itself, which was defined as the inner surface of the containing body. Here, De Ripa assumed that he is in a place surrounded by air. If the air moves and varies around his body, as it will, the immediate container, or place, surrounding his body will be continually altered. Because local motion is defined as successive changes of place, and his place is continually changing, it follows that he is in motion even while he is at rest. If we now assume that the place surrounding him remains constant as he approaches some fixed terminus, he would undergo no local motion because his place will not have changed.[43]

The final two arguments concern the intervening distance between two bodies and constitute an important statement on the medieval concept of measurement. On the assumption that distances are measurable only by means of intervening bodies, that is, only in a plenum, then if two bodies,

A and *B*, are separated by any distance whatever, the corruption or destruction of the intervening bodies would cause *A* and *B* to come into direct contact, even though they remain otherwise motionless in their original places. Finally, if the bodies intervening between *A* and *B* were destroyed successively, part by part, then even though neither *A* nor *B* moved, *A* would draw closer to *B* with the destruction of each part.

To avoid these absurdities, De Ripa appeared to suggest that the Aristotelian concepts of *place and measurement* are inadequate because they are exclusively designed to account for place and distance measurements in a plenum, and are therefore unable to assign meaning to places and intervening distances with respect to bodies or spiritual substances in a vacuum beyond the world. And yet, every body and spiritual substance located beyond the world or imagined there must be conceived to have some kind of place, even though it is surrounded by no other distinct and separate surface. And if each such entity can possess its own place, or *ubi definitivum* (see Section 2b), in empty space, then the distances between them ought to be measurable without the existence of intervening matter. For if the intervening matter between two bodies or places were destroyed, thus leaving behind an imaginary space, or vacuum, there is every reason to believe that the bodies and their intervening distance would remain unaltered, rather than come into contact, as would happen if measurements are assumed possible only in a material medium. The distance that now intervenes in the empty, or imaginary, space that separates bodies would surely be identical to the distance that previously separated them in the material plenum. It follows, then, that the distance between angels or bodies in imaginary void space is as measurable as a distance between two bodies in the positive places found only in a material plenum.[44] From this we may reasonably infer, though De Ripa did not, that imaginary void space is itself three-dimensional. Despite a radically different context of discussion (see Section 2b), Jean de Ripa arrived at the same conclusion as Henry of Ghent more than a half century earlier. Indeed, for De Ripa, measurements in void space were somehow even more fundamental than in a plenum.

The opinions of Henry of Ghent and Jean de Ripa were not destined for general acceptance. Neither of these scholastics considered the implications of a dimensional extracosmic vacuum. To assume a created, three-dimensional vacuum in which distance measurements were possible would have raised the old Aristotelian problem in which a dimensional body occupies a three-dimensional void space. Two equal three-dimensional entities could not simultaneously occupy the same location by natural means. Only supernatural action could accomplish this natural impossibility. In conformity with articles 139, 140, and 141 of the Condemnation of 1277, all scholastics had to concede that God could cause a dimension to exist by itself and several of them to exist simultaneously in the same place.[45] Although John Buridan readily conceded that God could make a spatial dimension separate from matter and devoid of accidents – that is, a dimensional vacuum – and that God could also cause several dimensional entities to exist simultaneously in the same place,[46] it certainly did not follow that He had ever done so or would ever choose to do so. (See Section 2a for the relevant passage and note 27.) Indeed we have already seen that Buridan considered separate spaces beyond the world superfluous (see Chapter 2, Section 2b). Because any bodies that might possibly exist beyond the world would possess their own internal space (see Chapter 2, Section 2a), they surely did not require a separate, external space. The creation of such spaces would, in any event, have been superfluous. For if extracosmic bodies required separate dimensional spaces as places, those spaces, in turn, would require separate dimensions, and so on, ad infinitum (see Chapter 2, Section 2b). To Buridan and many others, it appeared highly unlikely that God would have created a superfluous, separate, three-dimensional void space in which to locate the world or to place possible bodies beyond the world.

The widespread discussion of the possible existence of a separate void space during the Middle Ages was largely perpetuated by the impact of the Condemnation of 1277 with its general emphasis on God's absolute power and its specific mention of the possible divine creation of accidents without subjects, dimensions without matter, and several dimensions in the same place simultaneously. If the reality of a dimensional vacuum won few if any adherents, its *possible existence* as a creation of the divine power was doubted by none. But if God's power to create finite extracosmic vacua was never questioned, His capacity to create one of infinite dimension was seriously doubted. If the Condemnation of 1277 compelled assent to the supernatural creation of vacua both within and without the world, it was never intended to enlist support for the supernatural creation of an infinite dimensional void, which, as we shall now see, posed difficult problems.

b. CAN GOD CREATE AN ACTUAL INFINITE VACUUM?

Once again, it will prove convenient and instructive to turn to John Buridan, who in different works both denied and affirmed the possibility. In his *Questions on the Physics*, Buridan inquired "whether there is any infinite magnitude" and offered infinite space as the primary illustration of such a magnitude. Although he nowhere mentioned vacuum in this discussion and described infinite space as a body, there is little doubt that the space in question is a three-dimensional void. Buridan described the position of those who seek to establish that if any space existed beyond the world, it would, by the principle of sufficient reason, be infinite. For why should it be of one size rather than a greater size? Whatever the size of the finite space assumed, one could always properly ask whether another space lies beyond.[47] Thus, "if it could be shown that there is a space there, it ought to be conceded that it is infinite.[48] And that space which would have length, width, and depth, would be a body. Therefore, it ought to be conceded that there is an infinite body."[49]

In refuting this argument, Buridan, in conformity with the Condemnation of 1277, especially article 34 (see above, Ch. 6, Section 2a), dutifully allowed that "it must be believed on faith that God could form and create beyond this world other spheres and other worlds and any other finite bodies of whatever size He wishes." Indeed, for any finite body God creates beyond the world, He could create another twice as large, or 10 times as large, or 100 times as large, or as large as He pleases provided that it is a finite number of times greater. This much Buridan felt compelled to concede. But he apparently felt under no compulsion to believe that God could create an actually infinite magnitude or actually infinite void space. For although God could, by His absolute power, create a more perfect version of any creature, He could not create a creature of infinite perfection because such a creature would be as perfect as Himself. Indeed if God made any created thing so perfect that it could not be made more perfect, this would actually constitute a limit to His power. It is absurd, then, to suppose that God could create an actually infinite magnitude, or infinite void space. Such a creation would imply that He had reached the limit of His creative power and was consequently unable to create a yet larger space because, as Buridan explained, "it is absurd to believe that there is something greater than an actual infinite."[50]

But if, in his *Questions on the Physics*, Buridan saw the creation of an actual infinite as a threat to and limitation on God's absolute power, this concern is nowhere in evidence in his *Questions on De caelo*. There, in response to those who would argue that not even the omnipotent divine power could create an infinite space, Buridan declared unequivocally that on grounds of faith it must be conceded that God could indeed create an infinite mobile body and an infinite, immobile, three-dimensional space within which that body could be moved with a circular motion. The interpenetration of the two corporeal dimensions, body and space, can be easily achieved by the divine power.[51]

Whether Buridan's definitive opinion was to be found in his *Questions on De caelo* or the *Physics* may be indeterminable. But he surely exhibited the dilemma that the creation of actual infinites posed to fourteenth-century scholastics. As a nontheologian, however, Buridan was unable to pursue this important theological question freely.[52] Many of the theologians who were able to do so denied that God could create an actual infinite entity.[53] But a significant number, including Gregory of Rimini, John de Bassolis, Franciscus de Mayronnes, Robert Holkot, Nicholas Bonetus, Gerard of Bologna, Johannes Baconthorpe, and Paul of Perugia, thought otherwise and saw no logical contradiction in conceding that God could create one or more infinite entities.[54]

These "infinitists," as Anneliese Maier has appropriately labeled them,[55] generalized the domain of possible actual infinites to include not only powers, geometric magnitudes, numbers, and intensities but also quantities, including time and space.[56] But if space was a possible infinite, none of those mentioned above appear to have isolated it for special attention. One who did was the Franciscan theologian Jean de Ripa, who, about 1354 or 1355, took up the problem of extracosmic space in book I, distinction 37 of his *Sentence Commentary*, the standard locus for consideration of the manner in which God existed in all creatures and created things.[57]

When he proclaimed that "God is really present in an infinite imaginary vacuum beyond the world,"[58] De Ripa appeared to have assumed the actual, not merely possible, existence of an infinite void space (but see below). And in a manner reminiscent of Thomas Aquinas' attempted reconciliation of the concept of a creation with that of a world that may have had no beginning, De Ripa described the infinite void not only as eternal – that is, without beginning or end – but also as something created by God, for it flows from His presence as from a cause and is thus totally dependent on Him.[59] Because God is assumed to be omnipresent in an infinite void of His own creation, is it appropriate to identify and equate God's infinite immensity with that infinite empty space? Not in De Ripa's opinion. "The infinity of a whole possible vacuum or imaginary place," he declared, "is immensely exceeded by the real and present divine immensity."[60] In truth, God is "an intelligible infinite sphere."[61] that surrounds and circumscribes every imaginary place. Thus did De Ripa distinguish two infinites, the ordinary kind exemplified by imaginary void space and a single superinfinite equated with God's immensity.[62]

The distinction between infinities was a consequence of De Ripa's conviction that the hierarchy of creatable, spiritual substances had no theoretical finite limit, as was widely believed, but that God could intensify them all the way to actual infinity.[63] To grasp the significance of this concept for the problem of infinite space, it is essential to realize that in book I, distinction 37, De Ripa, like Peter Lombard, was concerned with the manner in which spiritual rather than corporeal substances are said to be in things, including places and spaces. Basing himself on Scripture and influenced by Saint Augustine, Peter Lombard had explained that something is locatable or circumscribable in two ways:[84] either because it is a dimension with length, depth, and width, and therefore constitutes an extension or distance that is presumably coextensive with its place, as would be the case with a body;[65] or, if it is not coextensive with its place, it can at least be delimited by a terminus or boundary. It is evident that the second mode applies to both corporeal and spiritual substances, for as Lombard explained, no created thing is everywhere but everything is somewhere.[66] Formalizing Lombard's distinction and terminology, it became customary to designate the place of a body as its *ubi circumscriptivum* and the place of a created spirit, such as an angel or soul, as its *ubi definitivum* (or *diffinitivum*).[67] It was thus the *ubi definitivum* of spiritual substances, rather than the *ubi circumscriptivum* of bodies, that was of fundamental concern to Jean de Ripa.[68]

Perhaps because the *ubi definitivum* applied to both corporeal and spiritual substances, De Ripa made much of the analogous manner in which bodies and spiritual substances occupied places. Just as bodies differ in dimension or extension and occupy greater or smaller places as their dimensions are greater or smaller, so do spiritual or intellectual substances differ hierarchically by intensive degrees of perfection; the greater the perfection, the greater the possible place that can be occupied or in which it can coexist.[69] For if the place of a spiritual substance were not proportioned to its intensive degree of perfection, it would follow that "any angel or soul could naturally coexist in an imaginary infinite vacuum beyond the world; or, if there were infinite worlds beyond this one, they [angels or souls] could be simultaneously equal to any of those worlds and could naturally coexist in any part [of those worlds], which no sane person would concede."[70]

Granted that created or creatable spiritual substances could be hierarchically arranged according to intensive degrees of perfection, was it possible for God to create a spiritual substance with an actual, infinitely intensive degree of perfection along with an infinite place or space that it would necessarily occupy?[71] Because the creation of such infinites did not involve a logical contradiction, De Ripa regarded the denial of God's ability to create them as tantamount to a rejection of His absolute power.[72] It seemed illogical to De Ripa that the great chain of creatable substances should culminate in a finite entity, as if God were incapable of proceeding to infinity. But if God could create an actual, infinitely intensive substance, so also would it be appropriate that this substance occupy a divinely created infinite place.[73] Thus did De Ripa assign to each spiritual substance a place appropriate to its degree of intensive perfection, a process that could culminate only with an infinite place or space being assigned to an infinitely intensive substance. But as we have already seen, these created infinites are in no way equal to God's infinity. We do justice to God's infinite and incomprehensible power only when we understand that not only can He create all manner of possible infinites but that He immeasurably exceeds and circumscribes them all. For only God can be called uncircumscribable.[74]

For Jean de Ripa, then, God is not only "really present in an infinite imaginary vacuum beyond the world" but He also immensely exceeds that infinite void space (see notes 58 and 60). But what is the nature of that imaginary infinite void in which the deity is omnipresent? Those who believe that only something positive and real – that is, a material body – can function as a position (situs), or location, for something else will judge it to be nothing, no more than a figment of the mind[75] in which no material or spiritual thing could be present. But the consequences of this assumption are untenable for both bodies and spiritual substances. Its untenability for bodies was demonstrated by De Ripa in the paradoxes described in Section 2a, in which the assumption of positive places yielded only absurdities with respect to motion and measurement. To these we may add one concerned with the world as a whole. If God created only our world, with no positive places beyond, it would follow that not even He could subsequently move the world with a rectilinear motion, because no positively created places would exist beyond the world into and out of which it could be moved. To argue this way, however, would be to deny God's power to move the world rectilinearly and thus to assume an article condemned at Paris in 1277.[76]

Similar difficulties would arise for spiritual creatures if all places were necessarily positive. Indeed the failure to assume the possibility of imaginary places would jeopardize God's absolute power. For if God destroyed everything except a single angel, that angel must be judged to be nowhere because it could be in no positive or corporeal place; nor could it be in itself because an angel does not by itself constitute a place or location.[77] Moreover, without positive places, that same angel would be unable to move from one place to another.[78] Serious problems also arise with regard to events prior to the creation of the world. Without the existence of imaginary space, any angel that God might have created before the existence of the world would have been created nowhere and could not have been separated by any distance from any other spiritual substance created before the world,[79] a position that was already shown to be untenable. Moreover, if only positive places existed, God Himself could be present only in real, positive places, from which it would follow that He is now present in the world but was nowhere present prior to its creation, when no positive places existed, a seemingly absurd consequence.[80] For all these reasons, De Ripa was convinced that God could be conceived as omnipresent in an infinite, imaginary void space that He had created from all eternity and in which He could act anywhere without suffering any change.[81] With the assumption of God's omnipresence in an infinite space, even before the creation of the world, De Ripa aligned himself with those who stood in opposition to a venerable tradition stemming from Saint Augustine and Peter Lombard, both of whom had argued that prior to the creation of the world, God was only in Himself and not in any space.[82]

Although imaginary space or place is not a positive entity it is obviously something, and something rather important. It functions as the *ubi definitivum* of all spiritual entities, such as angels or simple intellectual substances, and functioned in that capacity prior to the creation of the world. Such substances occupy limited places within it, places whose extent is always proportional to the intensive perfection of the particular spiritual being contained therein. It also would appear that any imaginary place is potentially capable of becoming a positive place or space when a body is created or placed in it. Although De Ripa omitted discussion of the precise differences between positive and imaginary places, the distinction between them seems to depend on the presence or absence of body, respectively.[83] Indeed the creation of the world itself might serve to exemplify the transformation of an imaginary to a positive place.

Thus did Jean de Ripa establish that imaginary space must of necessity be something. But what kind of something? How is it to be conceived? What are its properties? Because it is devoid of body and is explicitly described as a vacuum, our inquiry must focus on the nature of that vacuum. Although we know it is not a positive thing, it remains to be determined whether it is nonetheless real. We recall that De Ripa declared that "God is really present in an infinite imaginary vacuum beyond the world" (see above and note 58), from which it would appear to follow that imaginary space is something real, for how could God be in something unreal? And yet De Ripa nowhere described imaginary space as real or actual. To the contrary, he frequently referred to it as only "possible."[84] We are left with the impression that this infinite, imaginary space would actually exist if any one of the various possible intensive infinites or created spirits also existed. Because God could create any one of those intensive infinites, He could also create the infinite, imaginary space in which they would be located. Indeed He probably could also create it independently of all other intensive infinites.

But did He actually create such a space? By describing the mode of creation in Neoplatonic terms (see above and notes 59 and 82), it would seem that De Ripa thought He had. Even more significant was his unqualified declaration that "God is really present in an infinite imaginary vacuum beyond the world." Such a space would also seem essential for any of the spiritual creatures that might have been created before the creation of the world. And yet despite such impressive evidence, De Ripa often referred to extracosmic imaginary void as only possible rather than actual. Its merely possible existence is further emphasized by the manner in which De Ripa supported his claim that "the infinity of a whole possible vacuum or imaginary place is immensely exceeded by the real and present divine immensity." (For the Latin text, see note 60.) He explained that such an infinite, imaginary place or space (*situs ymaginarius*) would arise from the divine immensity as from a prior cause, which implies that the divine immensity or real, divine presence must exceed it.[85] Moreover, De Ripa allowed that such a space might even be three-dimensional because it could be constructed from an imaginary place of one cubic foot that is replicated an infinite number of times to produce "a whole possible, [infinite], imaginary place." But even this infinite volumetric space would be exceeded by the "divine immensity, which immensely super-exceeds every imaginary place."[86] For either God's divine and infinite immensity is exactly present in this infinite space and therefore "equal" to it with nothing extending beyond, or His immensity is greater than it.[87] If the first of these alternatives is true "and the whole [infinite] imaginary place is equal to the divine immensity," God would, in some way, be "circumscribable by some imaginary thing outside Himself" and "He would be able to be present in some imaginary place that is equal to Himself." That is, if God's divine immensity is equal only to an infinite imaginary space, one could just as well declare that God is contained by the equal space as vice versa. The repugnance of this consequence was intensified when De Ripa observed further that if God's immensity and an infinite imaginary space were truly equal, then it would also follow that some creature might also exist in an infinite imaginary void space, whether by natural or divine power. If this occurred, it would follow that a creature could be as spatially, or "locally" (*localiter*), immense as God Himself and would thus be uncircumscribed by God, consequences judged impossible by De Ripa. By assuming the other alternative, however, namely that God's immensity is greater than any infinite, imaginary space, the latter would lie within God's infinite immensity. But anything contained within an infinite is infinitely exceeded by that infinite. Therefore God's immensity infinitely exceeds any infinite, imaginary void space.

In juxtaposing God's immensity with a volumetric, dimensional, infinite space, the latter lying within the former, it would seem that at the very least Jean de Ripa allowed that if such a space existed, it might well be three-dimensional. Although this is undeniable, we have no particular warrant for supposing that De Ripa's imaginary space was conceived solely as a dimensional entity. Perhaps it was intended as just another illustration of the way in which God could employ His absolute power. Unfortunately, there is nothing in De Ripa's lengthy discussion that enables us to decide whether the possible imaginary spaces he discussed were to be taken dimensionally or nondimensionally. And this ambiguity, in turn, is perhaps connected with the overall uncertainty as to whether De Ripa conceived the infinite imaginary vacuum as an actual existent or merely as a conceptually real but only possible existent. For De Ripa, God could undoubtedly create an actually infinite, imaginary vacuum – to deny this would be to limit His absolute power – and He could undoubtedly occupy such a coeternal space. But whether intentionally or not, De Ripa left his final judgment as to the actual existence of an infinite imaginary vacuum shrouded in obscurity and uncertainty.

If De Ripa failed to furnish clear insight into the nature and existence of infinite imaginary space beyond the world, he was vigorous and unequivocal in his emphasis on God's uncircumscribability and the infinite difference between His immensity and the mere infinity of any possible or actual imaginary space in which God might be omnipresent. For it was obvious to him that "the divine immensity circumscribes every other imaginable finite or infinite [thing], not only causally but also by really coexisting with it and surrounding and circumscribing it, for which reason only God can be deservedly called uncircumscribable."[88] It was in defense of this profound position that De Ripa attacked an unnamed doctor (cited only as *iste doctor*), who was probably none other than Thomas Bradwardine, whose opinions on the relations between God and infinite space had probably been formulated but a short time before. According to De Ripa,[89] when the moderns hear or read what is said in the recent literature, namely "that God is in an infinite, imaginary vacuum beyond the world," from which it is then concluded that God can be truly and deservedly said to be immense and uncircumscribed, they take this statement at face value without realizing that "this doctor does not assume or hint that this is something peculiar to God Himself" but rather assumes that uncircumscribability is "common to God and to many possible creatures." Thus God is no more immense than other possible creatures and is no more uncircumscribable than they. De Ripa adds that "this doctor" has attempted in many ways to show that God could not produce something outside and independent of Himself that is intensively infinite and of infinite perfection. In De Ripa's view, this author errs by always assuming that every possible intensive infinite would be equal to the divine infinity, rather than infinitely less than it. On this assumption, any intellectual spirit that might be actually present in an infinite imaginary space beyond the heaven would be just as uncircumscribed spatially as God Himself.

3. THE GOD-FILLED, DEPENDENT, EXTRACOSMIC VOID SPACE

Thus did Jean de Ripa make mention of the major rival opinion that would prove far more significant and enduring than his own in the subsequent development of the concept of infinite space. As the author of that opinion, Thomas Bradwardine, some ten years earlier, had formulated ideas and arguments that joined God and space in a relationship strikingly different from De Ripa's. Possibly the first to attempt this difficult intellectual feat, Bradwardine dramatically encapsulated his major thoughts in five corollaries contained in his influential *De causa Dei contra Pelagium*, written perhaps in 1344.[90] Derived within the framework of a chapter titled "That God is not mutable in any way," the five corollaries,[91] consistent with God's immutability, declare:

1. First, that essentially and in presence, God is necessarily everywhere in the world and all its parts;
2. And also beyond the real world in a place, or in an imaginary infinite void.
3. And so truly can He be called immense and unlimited.
4. And so a reply seems to emerge to the questions of the gentiles and heretics – "Where is your God?" And, "where was God before the [creation of the] world?"
5. And it also seems obvious that a void can exist without body, but in no manner can it exist without God.[92]

In these five momentous corollaries, Bradwardine, perhaps for the first time unequivocally, linked the existence of an infinite, imaginary void space with God's omnipresence. To grasp their significance and meaning, we shall present Bradwardine's justifications and demonstrations for each and, where possible, attempt to determine, or at least suggest, the historical antecedents that produced them.

a. BRADWARDINE'S FIRST COROLLARY

The first corollary, *that God is everywhere inside the world*, was demonstrated by the assumption that God is not in any particular place in the world. However, because God is omnipotent, we must assume that He could be in any particular place of His choice. But He could not achieve this by means of any kind of motion because such a motion would have to be either the motion of a creature or of God Himself. It cannot be the motion of a creature, however, because God is necessarily present in every creature. On the implicit assumption that the world is a plenum, Bradwardine inferred that a creature has always existed in every particular place since the creation of the world. Therefore it follows that God has always been in every particular place. Nor indeed could God Himself move into any particular place from which He is alleged to be absent within the world. Such a motion would imply a change in God, who is immutable. Bradwardine thus concluded that God must always have been in that place and, by the same reasoning, in every other place in the created world.[93]

b. BRADWARDINE'S SECOND COROLLARY

Invoking the first corollary, Bradwardine demonstrated the second, *that God is everywhere outside the finite world in an infinite void*. He assumed that *A* is the fixed imaginary place of the world and *B* an imaginary, and quite distinct, place outside the world. He further assumed that God moves the world from *A* to *B* and that the world is now in place *B*.

In assuming these conditions, Bradwardine knowingly and deliberately violated two fundamental Aristotelian assumptions about the universe, namely that the world is immobile and that no void space, finite or infinite, exists beyond it. Both departures are justified by appeal to God's absolute power. Aristotelians argue for the world's immobility because

[they] assume that every local motion is necessarily upward, downward, or circular – i.e., away from the center [of the world], toward the center, or around the center. But they say that if this motion [from *A* to *B*] were assumed, it could not be any one of the ways just mentioned. For this reason, they say that it is impossible for the world to be moved.

But in Bradwardine's judgment, these Aristotelians

seriously diminish and mutilate the divine – indeed, omnipotent – power. For, in the beginning, God could have created this world in *B*. Why, then, is He unable to put it in *B* now? Furthermore, He can now create another world in *B*. Why, then, is He unable to put this world in *B*? This reply [that the world cannot be moved] is condemned by Stephen, Bishop of Paris, in these words: "That God could not move the heavens [that is, the world] with rectilinear motion; and the reason is that a vacuum would remain."[94] But this response does not avoid the difficulty. For it could be assumed that without [resort to] local motion, God could create another world in *B* and annihilate the world in *A*. [On this assumption,] the difficulty returns.[95]

Thus whether the world is moved from *A* to *B* by local motion or first annihilated in *A* and subsequently re-created in *B*, either of which alternatives is possible for God, the Aristotelian principle that the place of the world is immobile would be violated.

It was not only Aristotle himself who was subjected to severe criticism but also Aristotle's Christian followers. For although as Christians they accepted the creation of the world and thus departed from Aristotle's belief in its eternity, they appear to have committed themselves to a pre-creation void space of fixed size to which the created world had to conform, a position that restricts God's power. (On the pre-creation void, see Chapter 5, Section 3.) Their commitment to this position derived from their acceptance of Aristotle's rejection of the existence of void or place beyond the cosmos (see Chapter 5, Section 1, and notes 2–5). In adopting this opinion and denying the existence of extracosmic void, they also had to deny God's existence beyond the world and a fortiori to deny the possibility that the world could be moved from its present place. Without extracosmic void, the world must have been created in a void space of fixed size that, once filled by the created world, left no more void space in existence outside the world. Under these circumstances, not even God could move the world from its present sole possible place.

In Bradwardine's view, this constituted a grave restriction of God's omnipotence because it implied that God could not have altered the size of the world.[96] "For they [i.e., Christian Aristotelians] must say that God necessarily makes the world in place A, and that place A, and no other, existed before there was a world. But why this place and no other? Why was the world just this size and no greater or smaller?" But if God made the world, it must be conceded, argued Bradwardine, that He could have made the world any size whatever and as many places for it as He pleased. But if God did not fix the place, and therefore the size, of the world by His own free choice, then the world must have determined its own size. "If, however, the size is determined by the world itself – and not by God – what power determines this? What is the reason for it? What nature has fixed the world at this [particular size] and inviolately determined the limit beyond which it cannot pass?"[97] There is, of course, no such power, the very conception of which is absurd and contrary to the Christian faith. The existence of a void space that is neither God nor created by God could have no positive nature, for otherwise it would be coeternal with God (see Chapter 5, Section 3, and note 98), which is impossible, because the bishop of Paris condemned the existence of eternal things other than God.

Bradwardine thus rejected both alternatives of the dilemma posed by Aristotle and Averroes: either that the world is eternal or that an eternal pre-creation void place existed in which the world was created.[98] How Bradwardine resolved this problem will be seen below, but from the arguments just described, he concluded that not only could God move the world from its present place to any other place, but that He could also have created as many imaginary void places beyond the world as He pleased in order to make such motions possible. In fact, the creation of such vacua beyond the world was unnecessary because God is already omnipresent in an infinite, imaginary void space beyond the world, as the second corollary declared.

Bradwardine's demonstration of God's ubiquity in an infinite void rests ultimately on God's infinite perfection and power,[99] which enabled Him to be simultaneously in every part of the world when it was created[100] and to have preceded the world in the very place where He created it. Prior to the creation, God's presence depended only on Himself and not on the creation or any creature. That God must have been eternally in the place of the world and did not arrive there from elsewhere is obvious because any movement to the place of creation would constitute a mutation, and therefore an imperfection, in God's status.[101] Because God could have created the world anywhere He pleased, an infinite number of void places must be assumed in which He has existed eternally, from which it follows that He exists everywhere in an infinite imaginary void or place.[102] As further support for God's infinite ubiquity, Bradwardine declared that God is not confined to the place where He created the world because "it is more perfect to be everywhere in some place, and in many places simultaneously, than in a unique place only."[103] Without need of any creature and eternally at rest, "God is, therefore, necessarily, eternally, infinitely everywhere in an imaginary infinite place, and so truly omnipresent, just as He can be called omnipotent."[104] And because God created the world somewhere within this infinite imaginary space, it follows, though Bradwardine failed to make it explicit, that the place of the world and the infinite void beyond are homogeneous, a point that Giordano Bruno would make much of in deriving the infinity of the universe.[105]

c. BRADWARDINE'S THIRD COROLLARY AND THE MEDIEVAL DICTUM THAT "GOD IS AN INFINITE SPHERE WHOSE CENTER IS EVERYWHERE AND CIRCUMFERENCE NOWHERE"

From God's omnipresence is an infinite imaginary place or space, the third corollary – that God is truly "immense and unlimited" – follows immediately.[106] Of the numerous biblical, patristic, and pseudo-Hermetic quotations adduced by Bradwardine in support of God's uncircumscribable, infinite immensity,[107] two were frequently cited and are especially noteworthy. The first, drawn from the twelfth-century pseudo-Hermetic treatise the *Book of the XXIV Philosophers*, declared that "God is an infinite sphere whose center is everywhere and circumference nowhere."[108] Enormous significance has been attributed to this extraordinary metaphor, of which it has been said that it "presupposes an understanding of God and man which had to lead men beyond the medieval cosmos."[109] Originally applied to God, it inherently suggested an application to the cosmos itself and was so applied by Nicholas of Cusa (ca. 1401–1464), who transferred the metaphor from God to the universe and thereby "helped prepare the way for the new astronomy."[110] The "new astronomy" envisioned here is not the finite, heliocentric world of Copernicus but the infinite world of Bruno and seventeenth-century Newtonian physics. In extending the metaphor to the cosmos, Cusa conceived the universe as an "infinite unity" realized in a multiplicity that is "spread out in space and time."[111] Because an infinite universe spread out in space implies an infinite space, it follows that Cusa had derived that infinite space from the implication of the metaphor.

If we grant this much to Cusa, we must all the more concede priority to Thomas Bradwardine, who invoked the famous metaphor in a specific discussion of imaginary infinite vacuum some 100 years before Cusa. Whether Bradwardine was inspired to consider and proclaim an infinite extracosmic void after reading that God is an infinite sphere whose center is everywhere and circumference nowhere is probably indeterminable. But the nature of the metaphor and the context in which it was employed afford no good reasons for believing that it produced a new concept of infinite space. In fact, Bradwardine used the metaphor solely in support of the claim that God is truly "immense and unlimited."[112] It does not appear in any of the four other corollaries, which, for the history of spatial conceptions, were singly and collectively more important than the third corollary. To uphold the claim of God's immensity and unlimited nature, the historic description of the deity as an infinite sphere was but one of three definitions of God that Bradwardine derived from the *Book of the XXIV Philosophers*, to which he added others from biblical, patristic, and pseudo-Hermetic sources. Because Bradwardine used it nowhere else and employed numerous other metaphors in behalf of God's immense and unlimited nature, we have no grounds for attributing any unusual significance to the metaphor of the infinite sphere. As far as the history of medieval concepts of space and vacuum is concerned, it deserves no central place and played no significant role.

But what of the claims for Nicholas of Cusa? Perhaps he applied the metaphor to a real infinite space, where medieval scholastics did not. Curiously, although Cusa allowed that the world is not enclosed within any limits and is therefore not finite, he explicitly denied the infinity of the world.[113] Thus even if the world were in a space, the latter would not be infinite, as required by the metaphor. But if, for argument's sake, we concede that an unenclosed, unlimited universe is virtually equivalent to an actually infinite universe, it would still not follow that Cusa had enunciated a doctrine of real infinite space. For he may have had no concept of space whatever. Cusa seems not to have been concerned with the sorts of problems that are essential to the formulation of a spatial conception, as for example, whether or not the universe is an infinite plenum or perhaps a finite plenum surrounded by an infinite void; or whether it is a separate dimensional entity or something nondimensional; or whether it is equated with God's immensity; or something different from all of the above. Cusa simply did not discuss the nature of space.[114] Despite his argument for an unlimited universe, which may have influenced an occasional non-scholastic author, Cusa played no direct role in the development of a concept of infinite space and may be omitted from further consideration.

By applying the metaphor of the infinite sphere to the infinite cosmos, rather than restricting it to God or void space itself, Cusa, and after him Giordano Bruno, Pascal, and others, took a different approach than Brad-wardine, De Ripa, Nicole Oresme, and a number of early modern scho-lastics, who sharply distinguished between the finite cosmos and the infinite void space that lay beyond, which was not conceived as part of the cosmos but was associated with God Himself.[115] In scholastic thought, the pop-ular metaphor of God as infinite sphere was never explicitly transferred to the universe itself, and played no more than a supportive role for the claim of God's infinite immensity. And yet it was in the elaboration and explication of God's presence in an infinite void space beyond the world, rather than God's relationship to an infinite, material cosmos, that scho-lastics confronted many of the difficult issues about space per se. It was in this context that they forged new descriptive modes for spatial exis-tence, an existence that had to be compatible with the properties and at-tributes of an omnipresent deity. The fund of ideas and conceptions that were developed in this effort would prove useful to nonscholastics in the seventeenth century. These developments were thus not the consequence of a single brief, albeit striking, metaphor but rather the end product of a complex historical process involving debates about pre-creation voids, the possibility of extracosmic void, and the impact of the Condemnation of 1277, with its emphasis on God's absolute power and certain of its articles that were of particular relevance to arguments about space and void.

The second popular quotation in support of God's unlimited immensity was drawn from Saint Augustine's *Confessions*, book 7, chapter 5, where Augustine sought to compare the creature to the creator by imagining the Lord embracing the world "in every part and penetrating it, but re-maining everywhere infinite. It was like a sea, everywhere and in all di-rections spreading through immense space, simply an infinite sea. And it had in it a great sponge, which was finite, however, and this sponge was filled, of course, in every part with the immense sea."[116] In depicting the relationship between the finite world and God's infinite immensity by the figure of a sponge immersed in a boundless sea, Augustine had furnished a graphic model for those who would believe our world was immersed in an infinite void that was equated with God's immensity.

With God declared immense and unlimited by the third corollary and infinitely omnipresent in an imaginary void by the first two corollaries, Bradwardine sought to explain the nature of the divine extension. Although we may rightly call God "infinite, infinitely great, or of great magnitude," it would be improper to assign any attribution signifying actual extension.

For He is infinitely extended without extension and dimension. For truly, the whole of an infinite magnitude and imaginary extension, and any part of it, co-exist fully and simultaneously, for which reason He can be called immense, since He is unmeasured; nor is He measurable by any measure; and He is unlimited because nothing surrounds Him fully as a limit; nor, indeed, can He be limited by anything, but [rather] He limits, contains, and surrounds all things.[117]

d. BRADWARDINE'S FOURTH COROLLARY

On the basis of God's demonstrated omnipresence in an infinite, imaginary space beyond and within the world, the fourth corollary triumphantly proclaimed a reply to those old questions of heretics and Gentiles: "where is your God? And where was God before the [creation of the] world?" An obvious reply has been readied for both questions: God was, is, and always will be omnipresent in an infinite, imaginary void space. Thus did Bradwardine opt for a pre-creation void, the nature of which he described in the fifth corollary. (This was Bradwardine's reply to the problems raised in Chapter 5, Sections 3 and 4.)

e. BRADWARDINE'S FIFTH COROLLARY

In the final corollary, Bradwardine formulated the most fundamental characteristic of an imaginary infinite void. It was "a void" that "can exist without body, but in no manner can it exist without God." Bradwardine's inseparable association of God and void made his infinite vacuum radically different from that of the atomists and Stoics. Bradwardine was, of course, quite aware that vacua inside and outside the world had been rejected for a variety of reasons. For some, the claims of the atomists, Democritus and Leucippus, and the Pythagoreans that the existence of vacuum was essential to explain a variety of effects (for example, local motion, augmentation and diminution, condensation and rarefaction, alteration, the separation of things so that they might be distinguished from one another, and even the respiration of the heavens from the infinite void beyond) seemed an unnecessary and unwarranted assumption.[118] Others agreed with Aristotle when he rejected the existence of vacuum beyond the world, a vacuum that was conceived in terms of Aristotle's definition of place as the innermost, or ultimate, surface of the containing body and his definition of vacuum as a place that is naturally and potentially capable of being filled but that is not actually filled. Because no containing surface can exist beyond the world, neither can a vacuum exist there.[119]

Such arguments against the vacuum left Bradwardine unconvinced. They were neither necessary nor irrefutable. On the contrary, he was prepared to conclude that "by means of His absolute power, God could make a void anywhere He wishes, inside or outside the world. In fact, even now there is an imaginary void place beyond the world, [a place] which I say is void of body and of everything other than God, as the preceding arguments have urged."[120] In this final corollary, Bradwardine enunciated a new kind of void, one empty of everything except God and, because the latter is extensionless, perhaps also extensionless. Thus did Bradwardine affirm the actual existence of a "spirit-filled," imaginary infinite void space and make a reality of the suggestions of Saint Augustine and the anonymous author of the Hermetic *Asclepius*.[121]

f. BRADWARDINE'S LEGACY

With his five corollaries and accompanying arguments, Thomas Bradwardine established the basis for subsequent scholastic discussion of extramundane space. Rather than conceive imaginary infinite space as the creation of an omnipotent God whose own immensity "immensely exceeded" the infinite space of His own creation, as Jean de Ripa would have it, Bradwardine rejected the very idea that God could create an actual infinite and chose to conceive God's infinite immensity as somehow equivalent to, and therefore omnipresent in, imaginary infinite void space.[122] Although Bradwardine's opinions were all but forgotten until the publication of the *De causa Dei* in 1618, it was some version of Bradwardine's conception of the relationship between God and infinite space that was adopted and explicated by numerous scholastics during the next few centuries. After the fourteenth century, little if anything was heard about the creation of an actual infinite space.

In view of the theological nature of Bradwardine's version of a God-filled infinite space surrounding our finite cosmos, it comes as no surprise to discover that those who adopted some form of his conception were ipso facto concerned with the *divinization of space* rather than with its geometrization or physicalization. As God's infinite immensity, imaginary infinite space could be assigned only those properties that were compatible with His attributes. This is readily apparent from what has already been said. Despite our inclination to speak metaphorically and metaphysically of God as "infinite" or "infinitely great," Bradwardine insisted that God "is infinitely extended without extension and dimension" (see above).

Because infinite void space is identified with God's immensity,[123] that space would appear to lack extension. Although Bradwardine failed to make this explicit, an extensionless space gains further credibility from the fact that theological conditions required that infinite void space be denied any "positive nature" (see Chapter 5, Section 3) by means of which it might be construed as an entity independent of and separate from God.[124] To acquire a better appreciation for the problems Bradwardine faced and the solutions he proposed, it is useful at this point to describe the ideas enunciated some 100 years earlier by Richard Fishacre, the first Dominican graduated from Oxford University, who incepted as a master of theology there in 1236 and taught at Oxford until 1248, the year of his death.

During the period 1150–1250, especially the latter part, Christian theologians became concerned with the problem of divine infinity, that is, with the kinds of attributes that could be properly assigned to the deity.[125] It was this problem that Fishacre confronted in his commentary on the first book of the *Sentences* of Peter Lombard, composed shortly before 1245. Here, in book 1, distinction 2, question 2, Fishacre enumerated five ways in which we might conceive God to be infinite: intension, extension, duration, numerosity, and division. Of these, only extension will be discussed here.[126] On the implicit assumption that whatever is infinite in extension will occupy an infinite space (*infinitum spatium*), Fishacre distinguished two ways in which this can occur. In the first way, a greater – presumably finite – part of the infinitely extended thing will occupy a greater part of that infinite space. But God, who is of infinite simplicity, does not occupy a space in this way and is thus not extended or diffused. In the second way, an infinite extension occupies the infinite space such that the totality of the occupying thing is wholly in the least part of that space as it is in the whole of it. Thus if a created thing occupied this infinite space, or if this space were occupied by an infinity of worlds, God could exist wholly in each part, just as the whole soul is in every part of the body. It is in this second sense that God could exist wholly and totally in each part of that infinitely extended magnitude, just as the whole soul exists totally in every part of a finite body.[127] Thus a spirit is not extended in the same way as a body. A finite spirit is wholly in every part of a finite body and an infinite spirit, namely God, would be wholly in every part of an infinite magnitude, if such existed.[128]

In the final section of his discussion on infinite extension, Fishacre even argued that, in at least one important sense, we can know an infinite extension in the same way we can know a finite extension.[129] The latter is known to us in two ways, by its nature and measure. Fishacre believed we can know an infinite extent in the first, though not the second, way. If the infinite magnitude is assumed to be homogeneous and possessed of the same nature in every part, then we can know it. For if we know the smallest part of an infinite extension, namely the indivisible point, then we know the nature of the whole infinite magnitude because that nature flows from the point. We are thus able to understand the signification of the term "infinite" from the fact that our minds can understand the signification of the term "finite." But we cannot know the measure of an infinite magnitude because the infinite is unmeasured and immense.

What, then, is the possible relevance of Richard Fishacre's ideas on infinite space for Bradwardine? It surely does not lie in any assumption of the existence of a real infinite space, because Fishacre's discussion of infinite space and infinite magnitude is purely hypothetical. The former is presupposed for the existence of the latter, and both are introduced in order to conceive the manner in which God might be omnipresent in an infinite extension. (Compare to De Ripa's discussion in note 73 of this chapter.) Moreover, Fishacre does not suggest or imply that his infinite space is to be taken as a vacuum of which no mention is made. If Bradwardine knew Fishacre's discussion, he would thus have been aware of an argument that indicated how God might be conceived to occupy an infinite extension or space, namely that God would be totally and indivisibly in every part of it, just as the whole soul is in every part of a finite body. Although this may have served Fishacre's objective in providing a description of the mode of God's infinite ubiquity, it was of little use to Bradwardine because the manner in which the hypothetical, separate, infinite space came into existence was irrelevant to Fishacre. But this was precisely Bradwardine's major problem. Any existent infinite extension or space possessed of a positive nature distinct from God, as was Fishacre's hypothetical space, would of necessity be either coeternal with God, which no Christian could accept, or created by God, which Bradwardine believed impossible because God could not create an actual infinite. (See notes 98 and 122 of this chapter.) Although convinced that God was in an actual – not hypothetical – uncreated infinite imaginary, void space, Bradwardine was nevertheless unable to employ Fishacre's solution. He thus settled for a vague relationship between God and the infinite void that strongly suggests that the latter was to be taken as extensionless. Vague though it was, Bradwardine's association of God with an infinite void probably began the long process that would eventually enable Christians to accept the existence of an infinite void.

4. MEDIEVAL OPPOSITION TO A GOD-FILLED, INFINITE VOID

Although it is not unreasonable to expect that a few additional proponents of infinite extracosmic void space associated with God's immensity will be discovered, it is likely that Jean de Ripa, Thomas Bradwardine, and Nicole Oresme were its principal supporters.[130] It seems appropriate to inquire why the doctrine was accepted by so few theologians of the late thirteenth and fourteenth centuries. Most either ignored it or paid it scant attention, as for example, Thomas Aquinas, who, while acknowledging that one could imagine such a space, rejected it with virtually no discussion. (See above, Section 1 and note 14.) But around 1300, Richard of Middleton and Duns Scotus offered substantive reasons for their rejections, reasons that were often cited by scholastic opponents of extracosmic void in the sixteenth and seventeenth centuries. Because their arguments were formulated in the context of commentaries on book 1, distinction 37 of Peter Lombard's *Sentences* (see Chapter 5, Section 4), it is possible that these same arguments also served to deter acceptance of extracosmic void in the fourteenth century.

In considering whether God is outside of every place, Richard of Middleton distinguished two ways in which God could be conceived to be outside the world.[131] In the first, which Richard adopted, God's immensity (*immensitas*) is assumed to exceed our finite world infinitely and to be in no way dependent on or limited by that world. The possibility of other worlds and the assumption of God's absolute immutability form the foundation for Richard's derivation of God's infinite immensity. For if God created another world, He would be in that world as He is now in ours, because it would be an imperfection and a sign of mutability if God should have to move from our world to another that is newly created.[132] Because the same argument would apply to any other possible world, Richard concluded that God must be infinitely immense.[133] Indeed the famous metaphor of the pseudo-Hermetic *Book of the XXIV Philosophers* serves as further evidence of God's infinite immensity. If we understand that "center" signifies the creature and "circumference" the divine immensity, then the description of God as an intelligible sphere whose center is everywhere and circumference nowhere implies that the divine immensity is beyond the world and apparently of infinite extent.[134]

The second way in which God might exist beyond the world is in an infinite space, the existence of which Richard denied. The second way is therefore false.[135] Unfortunately, Richard offered no further justification for his rejection of God's omnipresence in an infinite extracosmic space. Indeed, we cannot be certain whether the infinite space he envisioned was separate and independent of God, who merely occupied it by virtue of His infinite immensity, or whether that space was inseparable from God and His actual immensity, as Bradwardine and Oresme would later propose. Such vagueness may only signify Richard's desire to deny either of these possible relationships, and generally to reject any linkage between God and infinite space of any kind. Whatever Richard may have meant, he was subsequently interpreted as an opponent of infinite extracosmic space, whether or not the latter was independent of God. Richard's legacy, then, was twofold. He insisted that (1) God's divine immensity extended infinitely beyond the world, even though nothing else did;[136] consequently, (2) he divorced God's infinite immensity from any association with infinite void space.[137] It remained for Bradwardine to unite them.

Duns Scotus' rejection of an infinite extracosmic void emerged in a dispute about a position taken earlier by Thomas Aquinas. The latter had argued that all actions, even God's, occur in the place where the effect is produced or through the mediation of some agent or medium. Thus if God were not in a place, He could not act there because action at a distance would be impossible.[138] God's omnipotence was in no way diminished by this because Aquinas, and those who followed him, argued that because God was everywhere, nothing was actually distant from Him. Thus if Aquinas had believed in the existence of extracosmic space, which he did not, he would undoubtedly have assumed God's omnipresence there, as well as in the world. Otherwise God would be unable to produce any effects there, an untenable position.[139]

Duns Scotus and his followers found nothing compelling in Aquinas' argument and assumed contrarily that God's presence in a place was not a necessary precondition for His acting in that place. In book 1, distinction 37 of his *Sentence Commentary*, the *locus classicus* of discussions on God's ubiquity, Scotus considered whether God's ubiquity with respect to His power (*potentia*) implies that He is also everywhere with respect to His essence (*essentia*). Or to put it another way, does God's omnipotence imply His immensity?[140] Scotus concluded that God's omnipotence depends on His will, not on His omnipresence or immensity. What God wills comes to be whether He is near or far from what He wills.[141] Because "action at a distance" follows from God's will, not His omnipresence, we must not imagine, Scotus continued, that before the creation of the world, God was necessarily omnipresent by essence in an infinite vacuum in order that He might actually be present in every possible place where He could have created the world.[142] Because God can create anything He wishes anywhere at all by merely willing it, His essential presence in the place where He created the world is unnecessary; and *a fortiori*, so is His essential omnipresence in an infinite void space. In sum, although natural agents must be present where they act, God need not.[143] An infinite extracosmic void was therefore superfluous and unnecessary for God to act wherever He pleased, and Scotus rejected it.

In combination with the Augustinian dictum that before the creation of the world God was in Himself (see Chapter 5, Section 4), from which many inferred that God was not in an infinite void space, and the fact that Thomas Aquinas and his many followers did not accept the omnipresence of God in such a space, the opinions of Richard of Middleton and Duns Scotus served to reinforce in specific terms the general medieval view that no God-filled space existed beyond the world. And yet the idea of such a space, especially in the form espoused by Thomas Bradwardine and Nicole Oresme, was somehow preserved into the sixteenth and seventeenth centuries, when it emerged as a widely and intensely discussed problem in scholastic commentaries at the very time when the Aristotelian physical system came under the final assault that would bring on its demise.

7

Extracosmic, infinite void space in sixteenth- and seventeenth-century scholastic thought

1. JOHN MAJOR AS A POSSIBLE LINK BETWEEN MEDIEVAL AND EARLY MODERN SCHOLASTICS

Although we have now seen that numerous late medieval authors had occasion to express an opinion on extracosmic void, few considered it within the theological context developed in the last chapter. This situation changed dramatically during the sixteenth century, when the relationship of God and a possible infinite void space came to be discussed at great length by numerous scholastic authors of major and minor significance. That the problem was a legacy of the late Middle Ages can scarcely be doubted. And yet the names of Bradwardine, De Ripa, and Oresme, who accepted the possibility or actuality of an infinite void space associated with God's immensity, go unmentioned in the great debates that developed in the sixteenth and seventeenth centuries. Despite the publication of Bradwardine's *De causa Dei* by Henry Savile in 1618, it played no apparent role in the sixteenth century, and no citation of it in the seventeenth has yet come to light.[1] But the ideas that Bradwardine, De Ripa, and Oresme expressed are much in evidence and may plausibly be assumed to form the ultimate basis for the elaborate and detailed discussions of the later period. But by whom were their ideas transmitted? At least two candidates are suggested by Gabriel Vasquez, who, as an opponent of the idea that God is omnipresent in an infinite void beyond the world, saw fit to describe and evaluate at some length the major arguments and authorities arrayed for and against that concept.

In disputation XIX of his *Disputationes Metaphysicae*, composed in the late sixteenth century, Vasquez declared that "Major and Caietanus taught that God is outside the world and is able to be in a vacuum."[2] The figures intended here are John Major (1469-1550) and Thomas de Vio, or Cardinal Cajetan (1469-1534). Of the two, only John Major, a Scottish theologian, logician, historian, and natural philosopher, will be considered here.[3] As a teacher for many years at the Collège de Montaigu of the University of Paris and one who was familiar with the significant problems of fourteenth-century scholastic thought, Major was surely aware of the problem of extracosmic space from the late medieval tradition. He expressed favorable opinions on the existence of such a space in at least two places, a treatise *On the Infinite* (*Propositum de infinito*)[4] and his *Sentence Commentary*, book 1, distinction 37,[5] the *locus classicus* of discussions on God and extracosmic space.

In the *Propositum de infinito*, published earlier than the *Sentence Commentary*, Major included a brief passage in which he declared that God is in every true or imaginary place and indeed exists beyond the heaven, or world, in an infinite, imaginary place or space. As proof of this, he argued that if God created a body beyond the world, He would surely be in that same place, and also remain there if He subsequently decided to destroy that body. Moreover, God was in an infinite imaginary place before the creation of the world. To deny this would imply that God was only in the place where He created the world, which is irrational, because it is necessary to allow that God can be in any possible place, as may be inferred from Solomon's statement (1 Kings 8:27) that if "the heaven and heaven of heavens cannot contain thee; how much less this house that I have builded?" To leave no doubt of his position, Major concluded that although "others may say that God is not outside the heaven, such talk must not be considered. I am not of their opinion."[6]

Pursuing the theme of extracosmic space in his *Sentence Commentary*, Major subsumed all under the heading of "Whether God is everywhere and in an infinite imaginary place beyond the heaven."[7] He enunciated his ideas in three conclusions: (1) "God is everywhere by essence, presence and power";[8] (2) "God can be outside the heaven by existing in the same place with a creature";[9] and (3) "it is probable that God is beyond the heaven in an imaginary infinite space."[10] In demonstrating these conclusions, Major not only incorporated most of what was said in the *Propositum de infinito* but added considerably to it, including the pseudo-Hermetic metaphor of God as sphere, the Augustinian dictum that God is in Himself (*in seipso*), and additional biblical authorities. Except for the Augustinian dictum, all but one of Major's nine biblical, Hermetic, and prayer citations are found in Bradwardine's *De causa Dei*, but not in book 1, distinction 37 of Peter Lombard's *Sentences*. This strongly suggests a continuous tradition from Bradwardine, or from one or more earlier authors, all the way to John Major and beyond.[11]

That God is everywhere by "essence, presence and power" (*essentia*,

presentia et potentia) is demonstrated not only by appeal to biblical authorities and the invocation of saints mentioned by Peter Lombard but also
by reason "because God is somewhere and there is no greater reason that
He should be in one place rather than another."[12] Nor would God be con-
taminated by the occupation of vile and evil places.[13] In defense of the
second conclusion, Major argued that an angel or body could coexist with
God outside the heaven. But God can subsequently destroy the angel or
body and remain by Himself in that same place or in an imaginary place.
For it would be amazing if God could not remain there after having destroyed a coexistent creature.[14]

Judging from the attention paid to it, the defense of the third conclusion was clearly Major's primary concern. That God is in an imaginary
infinite space beyond the heaven is declared on the basis of God's infinite
perfection, which signifies infinite strength that has existed through an
infinite time and therefore ought to imply that God should also exist
throughout an imaginary, infinite place.[15] Moreover, wherever God was
before, He must surely be there now. Hence if He was exactly where the
world is now, He is also presently there. But God could certainly create
another world concentric or eccentric to this world; indeed He could
create an infinite number of such worlds and be in each of them then
and now. Consequently, He could be in an imaginary, infinite place.[16]
Although the form of Major's arguments differs from that of his medieval
predecessors, there is yet a strong similarity rooted in God's infinite perfection and His ability to create other worlds, the places of which He
could occupy before and after their creation (see Chapter 6, Section 3b).

But with John Major, we find a significant new element that was lacking in fourteenth-century discussions about God and extracosmic space.
I refer here to the Augustinian dictum that before the creation of the
world, God was in Himself (see Chapter 5, Section 4). There were those
who argued that if God was in Himself before the creation of the world,
He could not be in a space beyond the world or indeed in any place whatever. Such a concept was unintelligible to Major, who could not imagine
that God was in no possible place. In support of his conviction that God
must have been, and must now be, in a place of some kind, Major invoked
Augustine's *City of God*, book 11, chapter 5, where, as we saw earlier (see
Chapter 5, Section 4 and note 53), Augustine had argued that if infinite
spaces existed beyond our world, one ought not to doubt that God would
be omnipresent therein. Although Major clearly misinterpreted Augustine
– who, in concluding the passage in question, denied unequivocally that
there are places beyond the world – his interpretation would frequently
be introduced in support of God's omnipresence in an infinite space beyond the world.[17]

The supernatural destruction of matter is also employed to advantage in the defense of God's ubiquity in an infinite void space beyond the world. If God were in a room filled with air, He could surely destroy the air, for it would be absurd to question this claim. Even with the air destroyed, however, God would still remain in the room. But if God can exist all alone in that void room, He could also exist beyond the heaven without any created thing coexisting with Him.[18]

At this point, Major mentioned the argument of "a certain doctor" (*doctor quidam*) that would seem to cast doubt on the analogical argument of God's presence in an evacuated room and His existence beyond the heaven. This "certain doctor," who, as a marginal annotation indicates, is none other than Saint Bonaventure, argued that if the impossible could be achieved and a place filled with matter could be completely evacuated, a "capacity" (*capacitas*) of the deprived place would remain. Now a "capacity" is something and God can be in it, but a "privation" (*privatio*) is nothing and God cannot be in nothing.[19]

To this Major responded that God could, if He wished, make a vacuum directly without first destroying the matter in an enclosed space. If this happened, the capacity of which our "certain doctor" spoke would not exist because the vacuum would not have been created by the supernatural destruction of matter. Despite the lack of a capacity in this instance, the vacuum would presumably be something because we have assumed that God created it. Moreover, whatever this "passive capacity" (*capacitas passiva*) might be, Major insisted that it could also be found beyond the heaven because that region is just as appropriate for having a place as is the room in which the vacuum was created. Thus God could be in a place beyond the heaven just as well as in a place in an evacuated room. Indeed "since God is posited to be outside the heaven by His power, why should anyone deny that He is [also] there by His essence?"[20] That God must be beyond the heaven by His essence gains support from Hermes Trismegistus, who declared that "God is a sphere whose center is everywhere and circumference nowhere," from which Major concluded that God's circumference is not terminated by the last heaven or outermost sphere of the world.[21]

Despite the claim of Gentile sources such as Hermes Trismegistus, what should one say to the argument that scriptural texts fail to indicate God's solitary existence beyond the heaven without the presence of any creature? Indeed "naturally speaking" there is no reason to believe that God exists beyond the heaven. Major simply admitted that naturally speaking it seems proper to deny that God exists beyond the heaven. But on supernatural grounds – and this is what Hermes probably intended – we ought to believe that by His essence God does exist beyond the heaven.[22]

John Major may well stand at the beginning of the final phase of the discussion about God's omnipresence in an imaginary, infinite void space.

And yet he failed to explain the relationship between God and extra-cosmic space. He did, however, take cognizance of the charge that if the third conclusion is true and God exists in an imaginary, infinite space beyond the heaven, it would seem to follow that God could create an actual infinite, that is, an actual infinite space. Were this true, the counter-argument continues, such an actual infinite would be as perfect as God.[23] As a firm believer in the possibility of an actual infinite,[24] Major conceded immediately that God could indeed create an actual infinite, but he simply denied that such an actual infinite would be equal in perfection to God or that it could offer infinite resistance to God. Its inferiority to God is made evident by the fact that God could corrupt and destroy it either instantly or successively.[25] Within this context, it is tempting to suppose that for Major the imaginary, infinite space in which God is omnipresent by His essence was God's own creation. But this is left vague and ambiguous and appears only in the context of a reply to a charge about the possible danger to God from an actual infinite. We cannot be confident that Major really thought of imaginary infinite space as something that God actually created and in which He is omnipresent. If Major truly believed this, he chose to make it known only by vague implication, without any noticeable effort to justify or defend his position.

2. IN DEFENSE OF GOD'S OMNIPRESENCE IN EXTRACOSMIC, INFINITE SPACE

If John Major was truly one of the significant initiators of the final phase of the controversy on extracosmic, infinite space in which God is omnipresent, the energy he devoted to an elucidation of the problem would prove minuscule by comparison with what was to follow. During the second half of the sixteenth century, and especially in the last decade of that century, lengthy and substantial analyses were composed. The relations between an omnipresent God and a possible imaginary infinite space had by then become a major issue to which many authors, mostly Jesuits, would turn their attention. What was thus begun would continue through the seventeenth century and even into the eighteenth until, for reasons that will be suggested in the next chapter, the problem was eventually resolved, or to be more precise, made to disappear. Because the opinions expressed were diverse, complex, and frequently lengthy, no convenient or easy summary description can be provided. To characterize the essential features of the interpretations developed during the sixteenth and seventeenth centuries, it will be advantageous and convenient to describe separately the conceptions of a few of the major scholastic discussants of God and extracosmic space. By using these conceptions as focal points for further elaboration and comparison, a reasonable sense of the scope and substance of this significant and troublesome issue may be conveyed. Of those who accepted some version of the opinion that God was omnipresent in an extracosmic infinite space, we shall concentrate here on the opinions of Francisco Suarez, Pedro Fonseca, and the Coimbra Jesuits at the close of the sixteenth century and Bartholomeus Amicus, Emanuel Maignan, and Franciscus Bona Spes in the seventeenth.

a. FRANCISCO SUAREZ

Among scholastic authors who concerned themselves with the problem of God and space, few if any ever reached the high level of lucid and intelligent analysis attained by Francisco Suarez (1548–1617). Although he did not incorporate his views within a commentary on book 1, distinction 37 of Peter Lombard's *Sentences*, his discussion falls squarely within the theological tradition. In Disputation 30 of his *Disputationes Metaphysicae*, which was devoted to the theme "What is God?", Suarez, in the seventh section of that Disputation, considered "whether it could be demonstrated that God is immense?"[26] Strictly speaking, Suarez understood by "immensity . . . a disposition toward a place, or the presence of God in all things."[27] Moreover, this immensity must be conceived as infinite because "every agent ought to be conjoined to the patient [or recipient] in which it operates. But God is a universal efficient agent for all things, and in all things that exist; therefore God is intimately present in all things," as numerous biblical passages attest.[28]

But does this infinite immensity of God imply the existence of an infinite, imaginary space? Suarez, who argued in the affirmative, knew well the powerful Scotist tradition against any attribution of spatial immensity to God (on this controversy, see Chapter 6, Section 4). For who would deny, the Scotists argued, that if God were confined to a single place and yet possessed of all His intensively infinite active power, He could not act on things at any distance from that place? It is therefore no diminution of God's perfection and power to assume that He is not immense in a spatial sense, but that He can achieve what He wills by action at a distance.[29] God's infinite power to act is quite different from His alleged immensity. They are indeed two different kinds of quantities. His infinite power to act is like an intensive quantity, whereas His immensity is like an infinite extension. But there is no good reason to assume that infinite immensity follows from His infinite power to act intensively.[30] For if whatever God can do by an alleged infinite immensity, with its implication of spatial extension, He can also do by His intensively infinite active power, the attribution of infinite immensity would appear superfluous. Nor indeed does the argument that God is in all things entitle us to infer His immensity. If all the things God created are in this finite world, then we may only infer that His immensity is confined to the finite world. Thus from God's action in creating our finite world, we may not properly infer His infinite immensity beyond that world, where no things exist in which He need be present.[31]

Even more to the point, however, were the questions raised by the Scotists about God's actual relationship to an alleged imaginary space. Why, they asked, should God be present in an imaginary space that is nothing in itself, for which very reason it is called "imaginary space"? How could God be present in such a space before the creation of the world? To assert that God is in such a space prior to the creation is no more appropriate than to claim that God was also in the things He would create before He created them, because those things are as much nothing as is this space! God's alleged presence in such an imaginary space is as imaginary as the space itself. Furthermore, in natural things the proximity of an agent to a patient or recipient, or the thing on which it acts, is something real and causes the actualization of a form from the patient's potentiality. But the presence of God as agent in an imaginary space, conceived as patient, is not real but only imaginary, as is the space itself. For not even God can produce an effect from the potentiality of a space that is nothing.[32]

To Suarez, the controversy seemed more verbal than substantial. He assumed that God's mode of existence was such that by means of His immensity, He could really and truly exist in anything He might wish to create and achieve this without any mutation. To conceive how, by means of His immutable immensity, God could exist in all things He chose to create, it is necessary to think of Him in relation to space, for God can exist in any *corporeal* spaces even if these are conceived to extend infinitely. Suarez explained:

We cannot conceive the disposition and immensity of the divine substance except by means of a certain extension, which, of necessity, we explain by means of a relation to bodies. And when we separate real bodies either from the thing itself, or in the mind, we necessarily perceive a certain space capable of being filled by certain bodies, [a space] in which the whole divine substance is present – the whole [divine substance] in the whole [space] and the whole [divine substance] in each of the particular parts of it [i.e., of the space]. And by this presence, we signify nothing other than the aforesaid disposition of the divine substance.[33]

In this momentous passage, Suarez made explicit what Bradwardine had only implied, namely that despite our theological conception of God as an unextended entity, a judgment shared by virtually all scholastic authors, we conceive the divine substance *as if it were extended*. Here indeed was a serious problem, one that Suarez boldly confronted by distinguishing between God's substance and the manner or mode in which He is present in something.[34]

The difficulty arises from the fact that our assumption of God's presence in divisible things, such as body and space, somehow suggests that God must also be divisible and composed of parts. Because no spiritual substance is divisible, this must surely be a confusion. To explain God's relationship to space, Suarez resorted to analogies involving angels and the soul. An angel can move from one part of space to another – that is, it can lose one part of space and acquire another – so that extension is clearly involved in the location of angels. But the extension is in the parts of space, not in the angel itself, which is indivisible.[35] Similarly, the human soul does not have any real parts in itself but is united with a person's whole body, which does have parts.[36] Generally, it is the bodies or spaces occupied by immaterial substances that involve the latter in a relationship with parts. Thus, if we understand God's omnipresence *as if He was, is, and will be forever diffused through the whole of an extended space*, the extension in all this is surely not on the part of God but arises from God's relation to the different parts of that space. For example, God's presence in relation to the center of the earth is considered different from His relationship to the heavens because heaven and earth are located in different places. And yet, because we rightly suppose that God is omnipresent and because we must necessarily distinguish between different parts of space, we are inevitably led to speak of God as if He were somehow extended by virtue of His presence in every part.[37]

But if God is indivisible and omnipresent, whereas space is divisible, does this not somehow imply that space is independent of God? For if space were God's immensity, as Suarez believed, how could space be divisible and God not? Perhaps one ought to conclude that God and space are separate. Suarez left no doubt, however, that space has no independence from God. In elucidating this point, he distinguished between *real space*, which exists wherever there is body, and *imaginary space*, which is a vacuum. That God has no need for a real space is illustrated by an example involving the body of Christ, assumed to exist outside the heaven. Insofar as Christ's body exists beyond the heaven, it "is really present there without any surrounding body, and it corporeally fills that space which it occupies." The space Christ "occupies" is not, however, something real and distinct from Christ's body. Rather, it is the presence of Christ's body that makes a real space, which prior to the presence of that body was void and nothing. But prior to the presence of Christ's body, God was Himself present in that very same empty space. Indeed that void space, in Suarez's words, "is not something really distinct from the substance of God."[38] Thus God's substance, or His immensity, is equated with an infinite, imaginary void space, which is apparently unextended and non-dimensional. Any part of that void space occupied by body is converted to real and positive space. But that positive space is not needed by God, because He did not have to create the material bodies that constitute such real spaces. Without those bodies, all that would remain is imaginary space, or God's infinite immensity.

The significant and crucial distinction between real and imaginary space, only hinted at by earlier authors such as Bradwardine and De Ripa and rarely made explicit even later, is here made perfectly clear. A real space exists wherever there is body; where body is absent a vacuum will exist, which is described as an imaginary space.[39] Imaginary space is converted to real space when occupied by body and reconverted to imaginary space when vacated by body. Suarez's real space, associated only with bodies, is virtually identical to the internal space of Buridan, Toletus, and Descartes, described earlier (on internal space, see Chapter 2, Section 2a). By making the real space of bodies internal and assigning positive properties to it, Suarez could preserve the nonpositive status of empty space and avoid assigning to it external, independent reality or divisibility. In this way, void space could function as God's unextended but infinite and insep-arable immensity. It was thus obvious for Suarez that "where we con-ceive space in the manner of a vacuum able to be filled by a body, there divine substance is more present and substantially fills the whole."[40] In-deed we can conceive spiritual things only in relation to material things, and where material body is lacking – as with a vacuum – we simply as-sume a disposition for it to be there, from which the presence of God may be inferred.

Up to this point, Suarez seems merely to have assumed God's omni-presence in an imaginary infinite space. But why should space be infinite, and why should God be omnipresent within it? Suarez provided exten-sive responses to these difficult questions. As an infinitely perfect being, God must have an infinite presence that cannot be confined to a finite place or space. God's immensity, which is unlimited, follows from His attribute of infinity.[41] Suarez argued that a finite substance is more per-fect to the extent that its presence extends over a greater sphere. Obviously, if a substance is infinitely perfect, its presence cannot be limited to a finite sphere, for otherwise a finite substance and an infinite substance would be equal in any finite sphere, even if the latter were infinitely increased (in infinitum).[42]

Moreover, because extent of presence is a kind of perfection, the divine substance ought to possess it in the highest degree and should therefore "be able to fill all bodies or spaces even if they are increased to infinity."[43] In yet a third argument for God's infinite omnipresence, Suarez argued that no limit can be set for God's presence. No reason can be offered as to why God should be confined or restricted to one part of a full or empty space rather than another, or why He should be present in only one or many possible worlds He might create. By a principle of sufficient reason, then, God should be infinitely omnipresent.[44]

God's infinite omnipresence was demanded for yet another reason. Suarez believed that every cause is conjoined to its effect, so that God, as agent, must be present wherever He acts to create effects. Because God could create an effect anywhere at all, it followed that He must be infinitely omnipresent. Thus a dependence relationship between cause and effect is offered as an essential reason for proclaiming God's necessary infinite omnipresence.[45] In the traditional controversy as to whether God had to be in the place where He caused an effect, Suarez aligned himself with Thomas Aquinas and against Duns Scotus, who had denied God's necessary presence in the places where He acts or might act, for which reason Scotus had rejected God's omnipresence in an infinite extracosmic space (see above, Chapter 6, Section 4, and Section 2a of this chapter).

b. PEDRO FONSECA AND THE COIMBRA JESUITS

Toward the end of the sixteenth century, the Jesuit fathers of the University of Coimbra (Portugal) would argue for the existence of an imaginary, infinite space in a commentary on Aristotle's *Physics*. Their opinions formed part of a noteworthy series of commentaries on the works of Aristotle in which the new humanism was combined with the traditional medieval scholastic method.[46] The *Conimbricenses*, the collective title under which the Coimbra Jesuits published, were inspired to produce their great commentaries under the influence of their fellow Jesuit, Pedro Fonseca (1528–1599), who, some years prior to the emergence of the new commentaries, had already considered the problem of space in his commentary on Aristotle's *Metaphysics*, perhaps first published in 1589.

In considering the problem "Whether place is a true species of quantity," Fonseca declared in book 5, chapter 13, question 7 "That place as space is not a true species of quantity."[47] Among those who espoused space as a three-dimensional quantity, Fonseca included Philoponus, Simplicius, Galen, and surprisingly, even Aristotle.[48] For most of these eminent authors, space is a continuous, divisible, immobile, and permanent external quantity. The externality of space is usually demonstrated from the fact that "it does not inhere in the bodies which occupy it, for otherwise either it would withdraw when those bodies withdraw, and therefore would not be fixed or immobile; *or* it would wander from one thing into another, which cannot be if it is an internal or inhering quantity."[49] Against a separate three-dimensional space, Fonseca brought the traditional arguments cited so often in the Middle Ages. If space is three-dimensional, it would be a body, from which it would follow that two bodies could exist simultaneously in the same place, which is naturally impossible. Moreover, if every quantity, or three-dimensional entity, needs a space, then the space itself would need a place, and so on, "so that there would be [an] infinite [number of] equal spaces in any same space, which is not only impossible, but also superfluous, since one suffices if it is not a three dimensional quantity, but rather [has] a certain aptitude for holding [a three-dimensional entity]."[50]

At this point, Fonseca mentioned how Philoponus sought to avoid the difficulties associated with a separate external space. By dividing quantity into material and immaterial, the latter identified with space, Philoponus argued that a material quantity could be contained by an equal immaterial quantity.[51] However, this immaterial space does not exist independently but inheres in body. Thus Philoponus could reject separate void space because his immaterial quantitative spaces were never empty of body. Fonseca judged Philoponus' distinction unacceptable. For why should body be a material quantity and space an immaterial quantity? And how is it possible for these two three-dimensional quantities to exist simultaneously in the same place? Moreover, why should this immaterial spatial quantity not require a space in which to exist, and so on, without end?[52] With these rather traditional arguments, Fonseca rejected the existence of real, separate external and internal spaces.

Because space is neither a separate external quantity nor an internal immaterial quantity inhering in bodies, Fonseca concluded that it belongs to no true species of quantity and is not a true being. For

if it were, it would either be an uncreated being, which happens only in God alone; or it would be a created being, which it cannot be since space could not begin to exist, for whatever it is now, there it always was and always will be necessarily. Furthermore, if it were a being, it would be either a substance or an accident. But it cannot be a substance because it is not incorporeal, as is clear since it is divisible; nor is it corporeal because it would [then] consist of matter and form which even its advocates concede it cannot be. Nor indeed can it be an accident, since it would [then] be either without a subject, or it would go from subject to subject, that is, it would pass from the located body into the succeeding located body; and also because space precedes the located thing, [it cannot be an accident, since] an accident does not precede its subject.[53]

If space is not a true and real quantity, what then is it? Fonseca believed it to be an imaginary entity, though not one that depends on our imaginations, as if it were a mental fiction. It is a truly existent entity but one that is difficult to explain and describe. We can, however, be certain that it always has existed, exists now, and always will exist, and that it is a container of bodies or has an aptitude for containing bodies. Moreover, it offers no resistance to the bodies it contains. A body of a certain volumetric measure will occupy a part of space equal to itself without opposition from that equal portion of space. This imaginary space must also be actually infinite (*actu infinita*) and exist beyond the heaven in every direction, because it can contain all bodies of whatever size that God could produce, even to infinity.[54]

But one might inquire as to where this spatial container, or nonresistant entity, is. It must surely be somewhere, and if somewhere, then also in something. For reasons already described, it is certainly not in the bodies that are themselves located, or in a place. Bodies cannot receive bodies. Fonseca concluded that space simply exists where it is and believed it silly to pursue the matter further.[55] And yet in a closing statement, he did pursue the matter a step further by declaring that "It is not surprising that they [i.e., the containers of bodies, or spaces] are not in something, nor somewhere, since pure negations (*negationes purae*) are not privations that would need a subject. . . ."[56] By characterizing imaginary space as a pure negation rather than a privation, Fonseca may have been the initiator of a significant move that would prove influential well into the seventeenth century. At long last, advocates of an infinite, extracosmic space had a description, strange though it was, of this elusive imaginary entity.

To grasp the significance of Fonseca's distinction between "privation" (*privatio*) and "pure negation" (*negatio pura*), it is useful to recall our earlier discussion of Gabriel Vasquez (see Chapter 2, Section 1 and the relevant notes), who considered privation and negation synonymous because privation is defined as "a negation with a certain aptitude, so that [although] this emptiness is not filled, it is nevertheless capable of being filled" (for the Latin text, see Chapter 2, note 27). Vasquez concluded that a separate vacuum is not the negation of any subject or body and is therefore absolutely nothing or, as he characterized it, a "pure negation" (*pura negatio*). A space that is described as a vacuum can have no possible existence[57] and therefore cannot qualify as a privation or negation. Whereas Fonseca adopted the same definition of privation as did Vasquez, and assumed with the latter that void space is not a mere privation or negation, he employed the expression "pure negation" (*negatio pura*) in a manner radically different from Vasquez. Where the latter invoked the concept of a pure negation to characterize void space as absolutely nonexistent and therefore not even a privation, Fonseca employed it to signify that void space was somehow a true existent because it was eternal and capable of receiving bodies. Although Fonseca realized the need to characterize void space with a new expression in order to distinguish it from privations or ordinary negations, he provided no further elaboration of this extraordinary conception. In the seventeenth century, however, it would receive considerable attention, especially by Bartholomeus Amicus, whose opinions will be discussed below.

When Fonseca's Jesuit colleagues at the University of Coimbra considered the problem of God and infinite space, they did so not in a commentary on the *Metaphysics*[58] in response to the question of whether place is a true species of quantity but in a commentary on the *Physics* in reply to the question "whether or not God exists beyond the heaven [or sky]."[59] Here we find most of the basic elements that became part of a standard scholastic discussion on the relations of God and space. Because the Conimbricenses were supporters of the idea that God exists beyond the world in an imaginary space, they first presented arguments in opposition to their own position, an opposition that was founded on the conviction that God's essence need not be omnipresent beyond the world, an opinion that the Conimbricenses attributed to Duns Scotus, Saint Bonaventure, Durandus, and Capreolus.[60] In presenting the case for those who rejected God's presence beyond the world, the Conimbricenses singled out five "most powerful arguments,"[61] some of which we have already described. Because of their significance, they will be paraphrased here in the order given.

1. Nothing exists beyond the heaven, but God cannot exist in nothing. Therefore God is not beyond the heaven.[62]
2. If God were in a space that we can imagine beyond the heaven, this imagined space would be nothing real or positive. Hence if God could be in a space that is unreal and without positive qualities, we could also say that God is in darkness, or in other privations, which is absurd.[63]
3. Because extracosmic space is only a conception of the mind, to say that God is in such a space is to declare that God is not in a real thing but exists only in a space that is conceived by the mind. Therefore, God does not really exist beyond the heaven.[64]
4. Substances devoid of matter – that is, spiritual substances – are in places only by their operation and not by their actual presence. But God does not even operate beyond the world, just as He did not operate in any way before He created the world. For as Saint Augustine explained in the *Confessions*, "Before God made heaven and earth, He did not make anything at all."[65]
5. Finally, the old Augustinian dictum is invoked that God was in Himself before the creation of the world and therefore not in an imaginary space.[66]

Against these opponents of God's omnipresence in an infinite space, the Conimbricenses argued in typical scholastic fashion by appeals to authority and reason. The array of authorities invoked in defense of God's infinite omnipresence in an imaginary space is impressive and extensive, embracing pagan and medieval Christian philosophers and theologians as well as Church Fathers and scriptural passages. Among those whose works are actually cited, or who are at least mentioned, we find Hermes Trismegistus, Plato, Aristotle, Clement of Alexandria, Eugubinus, Gregory Nazianzenus, Saint Basil, Saint Denis (Pseudo-Dionysius), Athanasius, Saint Hilary, Saint Ambrose, Marius Victorinus, Saint Augustine, William of Auxerre (Guillelmus Altissiodorensis), Thomas Aquinas, Marsilio Ficino, and John Major.[67] Two brief scriptural passages and an extract from a prayer to the Virgin, all of which, by the end of the sixteenth century, were routinely invoked in defense of God's extracosmic presence,[68] close the list. Although most of these authorities were frequently cited in discussions of God and extracosmic space during the sixteenth and seventeenth centuries, the most significant were the famous pseudo-Hermetic metaphor in the form "God is an intelligible sphere whose center is everywhere and circumference nowhere" ("Deus est intelligibilis sphaera cuius centrum est ubique, circumferentia nusquam") (see Chapter 6, note 108); and the two Augustinian passages in which God's relationship to the world is likened to an immense sea in which a sponge is immersed (see Chapter 6, Section 3c), and the oft-misunderstood passage in which Augustine was thought to assert that God existed in infinite spaces beyond the finite world (see Chapter 5, Section 4, and Chapter 7, note 17). Not invoked as frequently as the passages just cited, but worthy of special mention, is a statement by Hermes Trismegistus in the Hermetic treatise *Asclepius*. As reported by the Coimbra Jesuits, Hermes declared that "God who dwells above the summit of the highest heaven is present everywhere and from all around He watches all things. For there is beyond the sky a space (*spatium*) without stars removed from all things corporeal."[69]

In their appeal to reason, the Conimbricenses presented three arguments. The first depends on an analogy between God's immensity and His eternity. Because all divine attributes are equally perfect, it is assumed that God's immensity is related to places just as His eternity is related to times. Thus divine immensity will have the same relationship to the difference of places as divine eternity has to the variety of times. Now God exists through an infinite, imaginary time that is embraced totally and indivisibly by His eternity. Therefore God should be assumed to exist in an infinite, imaginary space by virtue of His immensity.[70]

Without mention of Anselm of Canterbury (1033–1109), the second argument reveals the influence, *mutatis mutandis*, of Anselm's famous ontological argument for the existence of God, applied here to demonstrate the necessity of God's extracosmic existence in infinite spaces. Assuming Anselm's definition of God "as that which nothing greater can be conceived," the Conimbricenses argued that if God does not exist beyond the heaven, then something greater than God could indeed be conceived, namely a being that not only existed where God did but also existed in the infinite spaces beyond. "It is therefore impossible," they concluded, "that God not actually exist beyond the world."[71] Corroboration for this conclusion is offered by appeal to the principle that the greater the nobility of an immaterial substance, the greater or fuller its sphere of existence. If it were infinitely perfect, its sphere of existence would be infinite – that is, it could not be limited by any spheres or spaces.[72]

The final appeal to reason is reminiscent of Bradwardine's discussion, because it justifies the assumption of God's infinite omnipresence on the basis of His immutability. If God created two bodies beyond the world and separated them by a spatial interval, not only must we assume that God is where those bodies are, but we cannot avoid the conclusion that He is also in the intervening space. For if God were not in that intervening space but assumed to be only in the separated bodies, the impossible consequence would follow that God is separated from Himself. Indeed if God were not omnipresent beyond the world, the movement of bodies from one part of space to another would imply that He moved with them and thereby suffered change of place and consequent mutation. To avoid such impossibilities, the Coimbra Jesuits insisted that God must be assumed omnipresent in an infinite space beyond the world.[73]

With God's existence and presence beyond the heaven or world demonstrated to their satisfaction, the Coimbra Jesuits, in the fourth and final article of the question, turned their attention to the difficult problem of the nature of imaginary space and how God exists in it.[74] Like their Jesuit colleagues Fonseca and Suarez, and very likely their medieval predecessors Bradwardine and Oresme, the Conimbricenses declared that "space [within and beyond the world] is not a true quantity possessed of three dimensions." For if it were dimensional, it could not receive bodies, "since several such dimensions cannot be in the same position simultaneously by [any actions of] the forces of nature (*naturae viribus*)."[75] Adopting another familiar and traditional argument, the Conimbricenses proclaimed that this space "cannot be another real and positive being, since, beside God, no such being could exist from eternity. But this space always existed and always ought to be."[76] Imaginary space is thus characterized as nondimensional, lacking real and positive properties or attributes, and yet is assumed to be an eternal, existent entity. Existence must be assigned to it because "by means of this thing [or space] itself bodies are received within the world without the action of the intellect; and they can [also] be received outside [or beyond] the world if they are created there by God." Despite its unreal dimensions, which prompts us to describe it as imaginary, space is not an entity of the reason alone. The characterization of spatial dimensions as imaginary does not derive from "fictions (*fictitiae*) or depend solely on a mental conception"; nor are they imaginary because they are "beyond the understanding." Those dimen-

sions are imaginary "because we imagine them in space, in a certain relation corresponding to the real and positive dimensions of bodies."[77] Like
Suarez, the Conimbricenses conceived space as imaginary because it can
receive bodies and seemingly prefigures the dimensions of those bodies.
We therefore imagine that it also has those dimensions, although more
careful consideration and analysis demonstrate otherwise.[78] As for God's
relation to this imaginary space, the Conimbricenses, once again reminiscent of Suarez, held that "God is actually in this imaginary space, not
as in some real being but through His immensity, which, because the
whole universality of the world cannot [accommodate it], must of necessity also exist in infinite spaces beyond the sky."[79]

True to scholastic tradition and practice, the Conimbricenses concluded
the question with a response to each of the five initial arguments that
denied God's extramundane existence (see above). The charge that "God
cannot be in nothing – that is, in a space which is not a real and positive
being" – is simply denied by appeal to God's absolute power, "for otherwise no stone could exist beyond the world by [an act of] the divine
power."[80] In response to the second argument, space is distinguished from
other privations (such as darkness) because it can receive bodies. Hence
God could be in a privation or negation such as space, but not in any
other kind.[81] To the third argument that space is a mental conception,
the reply reiterated the truly existential nature of space despite the absence
of real positive being.[82] The fourth position was repudiated by denying
that God is in a place at all, because He is not surrounded by space, presumably because His immensity is infinite and uncircumscribable. The
claim that a spiritual substance is in a place only by its operation is also
rejected by the counterclaim that a spiritual substance exists in a place by
its substance, albeit in a special way, which the authors decline to pursue
further.[83] The Augustinian dictum that God was in Himself before the
creation of the world, which is the fifth and final argument, is countered
by the statement that "although St. Augustine makes no mention of
imaginary space in his book [*Contra Maximum*], he, nevertheless, made
mention of it in the *City of God*, Bk. 11 [presumably ch. 5]," which the
Conimbricenses had already cited in their favor.[84]

c. GOD'S IMMENSITY AND SPACE AS PRIVATION OR NEGATION

Of the authors who accepted the existence of an infinite, imaginary space
and who have been considered here thus far, some, namely Bradwardine,
Oresme, Suarez, Fonseca, and the Coimbra Jesuits, conceived it as identical to God's immensity and also capable of receiving bodies. No such
identification was made by Jean de Ripa and John Major. As confirmed
"actual infinitists" – that is, believers in God's power to create an actual
infinite space – De Ripa proclaimed that God's infinite immensity was

infinitely greater than any possible infinite space (see above, Chapter 6, Section 2b), and Major rested content with the declaration that if God did create an actual infinite space, its perfection would be unequal to God's (see above, Chapter 7, Section 1). By conceding the possibility that God could create an actual infinite space, neither De Ripa nor Major could identify God's immensity with that created space. No created entity could be equated with God's immensity because the latter is an uncreated "internal" attribute or property that is part of God's very essence and perfection. The possible infinite space of which De Ripa and Major spoke must therefore be characterized as a possible created entity distinct from God's immensity. As such, it could have been conceived as an infinite, eternal, separate dimensional or volumetric being. The model for such a space can be composed from scattered statements in De Ripa's discussion, where he described the creation of an eternal, infinite, imaginary void space and then suggested a dimensionality for it.[85]

Whatever De Ripa and Major may have thought about the dimensional nature of infinite space, it is obvious that assigning dimensionality to space would have posed no insuperable obstacles for an actual infinitist. It was otherwise for those scholastics who identified infinite space with God's immensity and therefore denied a creation for it. To identify imaginary, infinite space with God's immensity and also to assign dimensionality to that space would have implied that God Himself was an actually extended, corporeal being. Although Benedict Spinoza, Isaac Newton, and others would do precisely this, such a move would have been completely unacceptable in medieval and early modern scholasticism.

God's immensity, and therefore imaginary infinite space, could not be described as dimensional. To avoid the path Spinoza would take, scholastics who identified imaginary infinite space with God's immensity were compelled to grope for some means of describing a nondimensional space that, by its very association with God, had to be conceived as an existent something. Despite obvious difficulties and perplexities, they were eventually led to describe it as some kind of negation. That this may not have occurred until the late sixteenth century is hardly surprising. Opponents of God's extracosmic presence had often insisted that God could not be in a privation or negation, that is, in nothing.[86] This was one of the two or three most basic arguments against the existence of an extracosmic infinite space. It is then readily understandable that Bradwardine, Oresme, and Suarez, for example, avoided altogether the attribution of privation or negation to extracosmic space. Even Fonseca sought to avoid the characterization of space as a mere negation. And yet, terms such as "privation" or "negation" seemed the only way to describe a dimensionless space that somehow had the capacity to receive any bodies that God might create in it. Thus it was, perhaps in desperation, that Fonseca boldly designated this strange entity as a "pure negation," namely a privation that required no subject (see above, Chapter 7, Section 2b). It was the Coimbra Jesuits, however, who confronted the issue squarely and abandoned the outward form of Fonseca's distinction, though not the substance of it. For them, space was indeed a negation, but one that differed from all other negations by virtue of its capacity to receive bodies. Their position was thus almost identical to Fonseca's without invocation of a special term such as "pure negation." God can indeed exist in a negation that is His own immensity, because He can create bodies that can be received in that immensity. The emphasis on space as some kind of negation capable of receiving bodies would be widely discussed and may have received its most extensive support from the seventeenth-century Jesuit author Bartholomeus Amicus, whose opinions we must now consider.

d. BARTHOLOMEUS AMICUS

Like the Coimbra Jesuits, Bartholomeus Amicus (1562–1649) expressed his opinions on extracosmic imaginary space in a combination exposition and *questiones* on Aristotle's *Physics*, but rather than use as his point of departure Aristotle's discussion of the Prime Mover in the concluding sections of the eighth book, as the Coimbra Jesuits did (see note 59 of this chapter), Amicus chose to present his opinions in the context of Aristotle's analysis of vacuum in the fourth book of the *Physics*. Amicus' unusually lengthy discourse on vacuum embraced five questions, which included virtually all the traditional topics accompanied by a host of different opinions from numerous ancient, medieval, and early modern authorities.[87] It is the fifth and final question, "On imaginary space" (*De spatio imaginario*), that will concern us here.[88]

With Bartholomeus Amicus, the problem of imaginary space may well have received its most extensive and detailed treatment. Although he omitted mention of the most significant medieval discussants, namely Bradwardine, De Ripa, and Oresme, he appears to have been familiar with almost all other relevant medieval and early modern scholastic discussions for and against the existence of imaginary space. From the variety and number of sources cited, we may reasonably infer that the possible existence and nature of imaginary space was ranked among the major themes of early modern scholastic thought. The origin of the problem, however, was assigned to antiquity when Amicus declared that "the ancients said that imaginary space is a body or quantity in which other bodies are received." But Aristotle had rejected this opinion in the fourth book of the *Physics*, where he "showed that such a space would be impenetrable with another body. Thus such a corporeal space could not possess the nature of a place." By the seventeenth century, then, the three-dimensional vacuum that Aristotle had conceived as a body and rejected was designated an "imaginary space," an expression neither he nor his opponents used. But even upon elimination of the interpretation Aristotle rejected, Amicus observed that "the controversy about what imaginary space is remains among the moderns."[89] Following the elaboration of three major opinions held by various moderns, who largely represent scholastic authors of the sixteenth and seventeenth centuries,[90] Amicus presented his own opinion within the context of five conclusions, in each of which he responded to common and anticipated objections.

In the first of his conclusions, Amicus denied that space could be formally a positive thing (*quid positivum*) with respect to essence or existence.[91] Here he agreed with the traditional scholastic view that if, as some believe, space is eternal, it could not be a positive thing because prior to the creation of the world, no positive thing other than God existed. Space could not be independent of God, for that would imply coeternality with God and suggest that space is also a God; nor could it have been created by God, because God created nothing positive prior to the creation of the world; nor, finally, could it be produced by another God, because a plurality of Gods is unacceptable to natural reason.[92]

Not only is the idea of a positive imaginary space offensive to our concept of the deity, but it cannot be accommodated to basic notions of metaphysics and physics. For a positive space would have to be either a substance or an accident. If a substance, it would be either spiritual or corporeal. It cannot be spiritual, because it cannot be diffused or extended as if it were a body; nor can it be corporeal, for then it would necessarily be a quantity and incapable of receiving another quantity. As for being an accident, this is excluded "because an accident cannot be without a subject."[93]

To those who would characterize imaginary space as an immaterial substance that is diffused, indivisible, and able to receive bodies, Amicus replied that such a space, which is apparently conceived as a separate entity, must have parts rather than be indivisible, because things are said to be in different parts of space and a body is said to pass from one part of space to another. Indeed Amicus insisted that an imaginary space that is an infinite and indivisible substance could only be God, which is not what the proponents of this opinion intended. If space were God, it would be divisible only virtually in the sense that the things or bodies that coexist with it are formally divisible, whereas space itself remains indivisible.[94]

Among those who interpreted space as an immaterial substance, there were some who conceived it as composed of formally extended parts rather than as an indivisible entity. In this opinion, which Amicus may have derived from John Philoponus and his followers, the parts of space are not impenetrable and can therefore receive bodies.[95] Because the sole impediment to penetration is quantity, the immaterial space envisioned here would lack quantity. Not only would such a space be repugnant to faith, because it must necessarily exist before the creation of the world, but as Amicus explained, it would also be contrary to the principles of natural philosophy because it assumes that a corporeal substance can exist without quantity. The present interpretation is also defective because it assumes that this space, which consists of parts, exists in itself and not in something else. If this were so, there would be just as much justification to assert that every body and every thing exists only in itself and not in an imaginary space. Hence the entire dispute about imaginary space would be superfluous and useless, because we can investigate the nature of that space only by the mode in which finite things exist in it. But if those finite bodies exist in themselves, and not in space, our inquiry is useless.

Satisfied with his demonstration that imaginary space cannot be a formally positive thing (*quid positivum*), Amicus in his second conclusion declared that imaginary space must not be thought of as a formally positive quantity that is real only in the intellect or understanding.[96] Here Amicus argued that we cannot properly conceptualize space in quantitative terms, even if we concede that it is not formally real or objective. As the most obvious point in favor of his second conclusion, Amicus observed first that space has the capacity to receive body, whereas quantity lacks such a capacity because it is impenetrable with respect to other bodies or quantities. Hence if space is conceptualized formally as a quantity, it ought rather to be characterized as possessing an actual incapacity to receive bodies.[97]

A second major defense of the second conclusion derives from the assumed incorruptibility of space, because the latter is not corrupted or destroyed when a quantity is produced in it. But if space is conceived as a positive quantity and another quantity were created in it, the space receiving that quantity would have to be assumed as destroyed, presumably on the basis of the principle of impenetrability, because two quantities cannot exist simultaneously in the same place. Thus space would be corruptible, contrary to the assumption of incorruptibility.[98]

With the complete denial of formal positive being to space, Amicus assumed in the third conclusion that "imaginary space is formally a negative thing (*quid negativum*) [and yet] at the same time [is also conceived as something] positive."[99] That space must be a negative thing followed for Amicus from the conviction that before the creation it existed from eternity and since the creation it exists outside the orb of the universe. However, all positive things outside of God have been created by God as efficient cause; because space is not created, it cannot be a positive thing. And on the assumption that a thing must be either negative or positive, the negative character of space may be inferred from the first conclusion, where it was shown that it is not a positive thing.[100] Despite the formally negative nature of imaginary space, we do indeed perceive it as if it were a positive quantity extended to infinity. Thus it seems to have the nature of a positive thing (*quid positivum*) even though its positive features are assigned to it only by the mind.[101] But imaginary space is not a real being even though it may be conceived in the mode of a positive extension. It is a negation that cannot be a true and real quantity except by informal conceptualization.[102]

But if imaginary space is not a real being or real quantity, but only a negation that is conceived in the manner of a positive extension, on what basis can this imaginary space be said to have objective existence outside the mind? Is it not, after all, merely a mental fiction? Resorting to philosophical realism, Amicus insisted that by virtue of its real essence, imaginary space must also have objective, though not real, existence. An object of the understanding must have objective existence and cannot be merely a mental conception.[103] It obviously lacks the real existence of a quantity or substance, because it is neither. But as a negative thing (*quid negativum*) capable of receiving bodies because, as we shall see in the next conclusion, it is the negation of resistance, Amicus insisted that something objective must correspond to our mental and intellectual conception of imaginary space.

In scholastic philosophy, a negation must be characterized as the negation of something. It is in his fourth conclusion that Amicus explained what imaginary space is the negation of. "This negation, which constitutes imaginary space," he declared, "is not a negation of body, or quantity, or of real space or place.[104] But it is the negation of a resistance for containing an extended thing."[105] Because body, quantity, and real space or place are all associated with body, Amicus had only to show that imaginary space is not the negation of body to demonstrate the first part of this conclusion. He therefore explained that the negation of body, which he probably assumed to be a vacuum, does not exist simultaneously with the body itself. By contrast, imaginary space exists simultaneously with the body that fills it, because a capacity – and space has a capacity for receiving bodies – can exist simultaneously with the thing for which it has the capacity.[106] The second part of the conclusion – that imaginary space is the negation of resistance – is confirmed by the fact that the opposite, or contrary, of receptivity of body is resistance to such receptivity. For "space is conceived in the mode of a negative being diffused through the whole [and] ordained for receiving bodies. But this negative being cannot be receptive unless it denies that which impedes reception." That space is the negation of resistance is inferred by Amicus from the observation that in receiving the body that fills it, space must coexist with body and therefore lacks resistance to the reception of body, which is the only possible opposite terminus to active resistance to the reception of body.[107] Citing

Fonseca and others, Amicus reiterated that space is not a positive real thing but only a nonresistant container for the reception of bodies. Despite its categorization as a negation, however, space is conceived as something that is extended in proportion to the size of the body it receives. A given space would therefore be receptive to a body that is exactly its size but would be resistant to every greater body.[108]

At this point, a significant problem confronted Amicus: Is imaginary space a vacuum? Because both are negations, perhaps they are one and the same. Amicus insisted, however, that such an identification is impossible because space and vacuum are different kinds of negations.

A vacuum is called a negation of the body that fills it, since the vacuum is destroyed when the body that fills it arrives. . . . But space cannot be called a negation of the body that fills it because it is not destroyed by the position of the body that fills it, as is obvious, for space is conceived as a receptacle and container of body; but in receiving [body], it is not destroyed by what it has received, but is rather perfected by it.[109]

Although he may not have been the first, Amicus here raised an issue that did not emerge in the authors discussed previously. His predecessors either identified and equated imaginary space with vacuum (for example, Bradwardine, De Ripa, Oresme, Suarez, and perhaps John Major)[110] or spoke only of imaginary space or place without explicitly introducing vacuum into their deliberations (for example, Fonseca and the Coimbra Jesuits). The latter group thus neither affirmed nor denied a possible identity between imaginary space and vacuum. By reason of the properties he assigned to imaginary space, however, Amicus was virtually compelled to conclude that it differed from vacuum. The passage cited above makes this readily apparent. Because of its eternality, imaginary space is not destroyed when occupied by body but coexists with it, whereas vacuum, which is conceived as the negation of body, is destroyed when filled by body.

But the differences are even more profound and depend on the restrictive concept of vacuum that Amicus employed. A vacuum can occur only between the concave sides of a body, from which it follows that a vacuum can be generated only where it is possible for bodies to exist. This accords with Aristotle's definition of vacuum as "that which the presence of body, though not actual, is possible" (see above, Chapter 2, Section 1 and note 6; Chapter 5, Section 1). Because for Aristotle the existence of body is possible only within the world and not beyond, he concluded that vacuum was impossible outside the cosmos (see above, Chapter 5, Section 1), and because the world is a plenum, it was also impossible within the world. Amicus undoubtedly accepted this but was prepared to argue that if matter could be annihilated within the world, *a vacuum would remain and not a space*, because vacuum was defined as the emptiness lying between the concavity of a surrounding body.[111] If, by supernatural means, some interior part of the material world were annihilated or otherwise removed, a vacuum surrounded by the remaining matter of the world would remain. Indeed Amicus imagined that God destroys all sublunar things but preserves the sides of the lunar orb. In the center of this void concavity, God places a stone. The emptiness between the stone and the sides of the lunar orb is not a space but a vacuum, because that void is not only surrounded by matter but is capable of being filled again and therefore also capable of being destroyed.[112]

Beyond the world, conditions are radically different. Not only would Amicus probably have accepted Aristotle's argument that void is impossible there because the existence of body is impossible, but the latter condition implies that no void concavities are possible. Without surrounding body, vacuum is impossible. The infinite unenclosed emptiness beyond the world is thus not a vacuum but a space that "is said to have existed before the creation of the world and is now outside the universe" (for the Latin, see the concluding quotation in note 111 of this chapter).

The implications of Amicus' distinction between space and vacuum for the world and the space beyond are clear, though not expressed. Because of the impossibility of the existence of body beyond the world and the concomitant lack of void concavities, only imaginary space, not vacuum, can exist there. Even if a stone were created supernaturally beyond the world and then annihilated, a vacuum would not result because no surrounding matter exists to form a concavity where the stone was annihilated. Within the world, however, the converse obtains. The annihilation or removal of matter here would always produce a concavity and hence a vacuum, but no imaginary space. It thus becomes apparent that for Amicus, the created world lies immobile within an eternal, infinite, imaginary space that surrounds it on all sides. That the world is not in a vacuum is obvious to Amicus from the fact that it is not located within a concavity, which could exist only if the world were surrounded by matter. But the world is surrounded and contained only by imaginary space and occupies a portion of that space equal to the length of its diameter.[113] That surrounding imaginary space does not, however, penetrate the world. Empty concavities within the world are conceived as vacua and not as parts of the infinite, imaginary space that surrounds the world. For if a part were removed or annihilated, a vacuum, not a space, would remain, because every potential empty space within the world is surrounded by the remaining matter of the world. In short, although the whole world is in an infinite, imaginary space, its parts are not.

In the fifth and final conclusion, Amicus declared the absolute independent existence of space. As a negation, it ought not to be conceived as something inhering in a subject, because there is no subject in which it can inhere or exist. Thus we must assume that it exists per se.[114] Here Amicus resorted to the terminology of Fonseca and identified imaginary space as a pure negation (*pura negatio*) rather than a simple negation. To be identified with the latter would make space a privation, "since privation is the negation of a form in a subject that has an aptitude for [that] form." But when that form is received by that subject, the negation is destroyed. The negation that is space, however, does not behave in this manner. It is not destroyed or negated when it receives body but coexists with it. Space is a pure negation because it negates resistance to the reception of bodies, a property that it possesses eternally.[115]

Although space is an independent, albeit negative, existent, we are well aware that it cannot exist independently of God and are not surprised to learn that Amicus identified it with the divine immensity. Indeed it is because of that very identification that we must assume the infinity of imaginary space.[116] As God's immobile[117] immensity, however, infinite space is not an attribute of the deity because, as we have already seen, it inheres in no subject and exists per se in complete independence. And yet, what could it mean to claim independent existence for infinite space and to maintain, as Amicus did, that this space was also coeternal and coexistent with God?[118] How could anything be eternal and independent of God? Indeed Amicus himself had denied the eternality of space as a positive entity independent of God (see above, note 92 of this chapter). But the imaginary space he conceived as a negation seemed to possess some level of existence, even if only the capacity to receive bodies without resistance. Could even such an entity be coeternal with, but independent of, God? Amicus replied in the affirmative. His defense of this extraordinary claim hinged on an interpretation of "independence" peculiar to negations. Negations can exist independently of God because they are free from God's action. They coexist with God improperly and from eternity as long as God refrains from action. Indeed all creatures have coexisted improperly with God from eternity. With the creation, however, those creatures abandoned the negative mode of existence and assumed the positive existence and nature of beings. Once these creatures became beings, their preservation and existence as positive beings depended on God's direct influence. In their former state as negations, however, they also depended on God, but that dependence was based on God's inaction, which alone preserved their nonexistent negative status. Imaginary space is a negation – indeed a negation of a negation (see above, note 107 of this chapter) – that God chose not to act upon, thus enabling it to coexist improperly with Him from eternity.[119] But how can bodies and positive things exist and be present in a negative, albeit existent, space? On this momentous question, Amicus conceded that a body cannot be present in imaginary space in any of the regular predicamental modes that Aristotle had distinguished as the only possible relationships that could be assigned or attributed to a substance.[120] Such relationships required positive termini as found in the categories of quantity, quality, place, relation, and so on. Because imaginary space is only a negative terminus, Amicus declared that bodies can only exist in and relate to it transcendentally.[121] Although Amicus and others were convinced that space must exist as the receptacle for bodies and things, they could not describe any of the possible relationships between space and body in the ordinary terms and concepts available in scholastic philosophy. Only by ultimate appeal to a transcendental relationship could they resolve their dilemma.

In responding to numerous traditional objections against the existence of imaginary space, Amicus had occasion to reinforce his opinions, interpret various scriptural and patristic authorities favorably, and provide additional minor insights into the nature of that space. Here we learn that space is called "imaginary" to distinguish it from what is real and positive. (On the different conceptions of imaginary space held in the sixteenth and seventeenth centuries, see above, Chapter 6, Section 1; for Suarez, see Chapter 7, Section 2a.) For although we conceive imaginary space as a positive thing, it has no positive nature (see above, n. 101 of this chapter). Despite the absence of a positive nature, space as a negation appears to have some degree of objective existence[122] even though that existence may only be improper.[123] To illustrate this, Amicus resorted to analogy: Just as not everything received in another thing is real and positive, as when blindness is received in the eye, so everything that is capable of receiving need not be positive, as with imaginary space.[124]

As part of his overall defense, Amicus found it necessary to consider the attitude of the Holy Fathers that before the creation of the world, God was in Himself and not in another place or container. Was this tantamount to rejection of extracosmic space by the Fathers, as many claimed? Not for Amicus, who insisted that "they [the Fathers] do not deny that He was in an imaginary space in which the world was produced and in another infinite space in which other worlds could be produced." Of course, God's presence in that imaginary space is negative because He does not fill it as if He were in a place or container.[125] God's presence beyond the world is, in Amicus' view, well attested to by scriptural and patristic sources.[126] Of these, two passages drawn from Saint Augustine, as we have already seen, were standard justifications for extracosmic space. The first of these from *The Confessions*, book 7, chapter 5, depicts the world as a sponge immersed in the infinite sea of God's immensity (see above, Chapter 6, Section 3c and note 116), whereas the second, from the *City of God*, book 11, chapter 5, denounces those who deny God's incorporeal presence in spaces outside the world (see above, Chapter 5, notes 52 and 53).

Few treatments of imaginary space were as thorough and significant as that formulated by Bartholomeus Amicus. His reasoned conclusions that imaginary space is a negation of resistance capable of receiving bodies and that space must be distinguished from vacuum form the foundation of his spatial theory. The latter distinction seems especially important because it made a permanent, immutable entity of space, in contrast to the potentially mutable and generable nature of vacuum. Although no other explicit discussion of the difference between space and vacuum has come to my attention (a possible exception is Otto von Guericke, who differentiated between commonly perceived vulgar three-dimensional space and the dimensionless infinite space that served as the universal container of all things; see below, Chapter 8, Section 4c), it is likely that other seventeenth-century scholastic and nonscholastic proponents of absolute, eternal, and immutable, infinite space implicitly assumed a similar distinction between space and vacuum.

That space somehow had to be a nondimensional entity was for Amicus partly a direct consequence of the Aristotelian principle that two corporeal, dimensional entities could not occupy the same place simultaneously. If space were dimensional, it would be unable to receive any bodies whatever. Because we perceive that bodies are indeed received in space, it must be concluded that space has no dimensional properties. Moreover, it was only as a nondimensional entity that space could be identified with God's infinite immensity, an identification that in and of itself would have compelled the rejection of dimensionality to space. The identification of space with God's immensity was probably instrumental in generating the further conclusion that "imaginary space, which is a pure negation, is diffused through the whole orb [of the world] and beyond and precedes every creature through eternity."[127] Amicus thus left no doubt that imaginary space penetrated the entire orb of the world and was nevertheless, by virtue of its negative, nondimensional status, able to receive and coexist with all bodies.

Lurking in this formulation, however, were some serious internal inconsistencies of which Amicus was apparently unaware. These difficulties arose because of Amicus' insistence that if some part of the internal body of the world were removed by supernatural action, as for example, the sphere of fire or air, the emptiness that remains would be a vacuum and not a space. But if this is so, in what sense can it be said that space penetrates the entire orb of the world? If a vacuum is left behind under the circumstances described, it is obvious that space is absent, as Amicus himself emphasized. It would appear, then, that space merely surrounds body but does not actually penetrate it. From this we may infer that if a part of the world that included a portion of the surface of the outermost sphere were destroyed by supernatural action, an imaginary space would remain, because the emptiness left behind under these conditions would not be completely surrounded by a containing surface and therefore could not be characterized as a vacuum. Only if the matter destroyed were wholly within the world would a vacuum, and not a space, be left behind. But in the latter case, no imaginary space would lie within the world, which contradicts Amicus' insistence that imaginary space penetrates the entire orb of the world, and presumably any body wherever it might be.

The difficulty inherent in this position appears to derive from the Aristotelian definitions and interpretations of place and vacuum, to which Amicus subscribed. Vacuum is defined in terms of place as a "place deprived of body."[128] Thus a vacuum is the negation of a plenum[129] in a certain circumscribed place. But what is a place? A place is the surface of a material container that is in contact with the body contained. When a body is destroyed within a place – that is, within the concavity of the surface of the containing body – a vacuum is left that is interpreted as a three-dimensional quantity, or real space.[130] By rigorously adhering to these definitions, Amicus was led to conclude that when a body within the world was destroyed, a three-dimensional vacuum was left behind, surrounded by the surface of the containing body. This vacuum could not be imaginary space because the latter is neither a quantity nor a dimensional entity. But why did Amicus not allow the simultaneous coexistence of vacuum and space? After all, we have already seen (note 129) that Amicus assumed the possible coexistence of vacuum and spiritual substances. Because imaginary space is equated with God's immensity and is therefore a spiritual substance, why could it not coexist with supernaturally created hypothetical vacua within the world? To allow this would have resolved the difficulty and upheld Amicus' claim that imaginary space truly penetrates the entire orb of the world. For some inexplicable reason, Amicus made no such move. Perhaps this omission should simply be attributed to his ignorance of the dilemma. As his last word on the problem, we must simply accept that Amicus believed that imaginary space penetrates the entire orb of the world and every body anywhere in the universe *and also held* that only vacua, and not imaginary spaces, are left when bodies within the world are supernaturally annihilated.

e. EMANUEL MAIGNAN: IMAGINARY SPACE
AS VIRTUAL EXTENSION

Although Bartholomeus Amicus assigned an independent, separate existence to imaginary space and conceded that we perceive it as if it were a positive quantity extending to infinity (see above, Chapter 7, Section 2d), he emphatically denied that it was a real quantity. The quantitative sense that could be attributed to imaginary space was a mere conceptual mode with no objective reality. For those who grappled with the problem of imaginary space, the mode of quantification that might be assigned to this elusive entity became central. Its association with God made the quantification problem a delicate one. To attribute real quantity to a space identified with God's immensity would have signified nothing less than the corporealization of God. It would have transformed the deity into a three-dimensional extension capable of division like any other corporeal extension. Such a move would surely have evoked theological censure. Short of attributing actual extension to God, scholastics sought to provide some sense of extension to imaginary space. Many, like Amicus, did so by assigning that extension to human modes of perception. Within the constraints just described, a significant step was taken some years after Amicus by Emanuel Maignan (1601–1676), a scholastic theologian and member of the Order of Minims.

By contrast with all the other early modern scholastics discussed thus far, Maignan was an accomplished experimental scientist who not only wrote a treatise on sundials[131] but, in his lengthy *Cursus philosophicus*, devoted sections to ballistics and the vacuum, where he described Pascal's recent experiments and concluded that the existence of macrovacua and microvacua was possible.[132] Observations with a microscope reinforced his convictions about the nature of matter and space.[133] Aristotelian though he was on many matters, Maignan was thus capable of adopting un-Aristotelian and even anti-Aristotelian positions.[134] His scientific interests acquainted him with the new scientific ideas that were sweeping Europe. Not only was he knowledgeable about the writings of Galileo and Descartes, attacking the latter's physics on strictly scientific grounds,[135] but he was personally acquainted with Athanasius Kircher and his fellow Minims, Marin Mersenne and Jean-François Niceron, all involved in the new science.[136]

As a scholastic, however, Maignan did not abandon his interest in traditional themes. Among those included in the *Cursus philosophicus* we find the doctrine of imaginary space. Under the heading of "Philosophy of Nature" (*Philosophia naturae*), Maignan devoted chapter 8 to the concept of place, which led him directly to a consideration of imaginary space. Like most of those who discussed imaginary space, Maignan contrasted real space to imaginary space by associating the former with the plenum of the material world and conceiving the latter as an invariable and immutable entity diffused everywhere without end.[137] In the absence of any specific identification of imaginary space with vacuum, it is probable that Maignan would have agreed with Amicus that imaginary space is not to be construed as an infinite vacuum (see above, Chapter 7, Section 2d). We conceive this imaginary space as if it had fixed, immutable termini within which all things exist and between which distances can be measured.[138]

Maignan's description of God's relationship to space was couched in terms typical of those who believed in its existence: "By force of His immensity, the best and greatest God is substantially not only in this part of imaginary space in which He created the world but also in extramundane spaces infinite on all sides."[139] That God must exist beyond the world by His immensity can be shown by faith and reason. Faith indeed teaches that God is immense because He cannot be encompassed by any measure. If God did not exist beyond the heaven, He would be as limited as a rational soul within a body whose quantity it cannot exceed.[140] That God actually exists in an extramundane space also appears evident to reason. God must have existed in the space where the world was created because the creation of the world presupposes God's presence, for otherwise God could not have made the world there.[141] Nor can it be denied that just as God made our world in this space, so also could He create other worlds in other spaces. Indeed He could also place our world in another space either by moving it there directly or by destroying it here and re-creating it there, from which it follows that God must exist in extramundane spaces that are identical by virtue of God's uniform immensity.[142] God's infinite ubiquity in extramundane space is thus compatible with the Augustinian dictum that before the creation of the world, God existed only in Himself. The deity's existence in imaginary space before the world's creation is equivalent to being in Himself because that space is His immensity, as will be seen below.[143]

If God actually exists in an imaginary infinite space within and beyond the world, how does He achieve this? Is He perhaps actually extended, like a body? As expected, Maignan denies this. In the course of his discussion, however, he distinguishes two kinds of extension, formal and virtual; the former is reserved for corporeal quantities and the latter for spiritual substances. Formal extension applies only to quantities that have impenetrable parts or parts that lie outside one another, which comes to the same thing. Indeed we say that a quantity is extended – that is, formally extended – because its parts lie one outside the other. When the body is taken as a whole, the totality of its mutually related parts, or its formal extension, may be taken as equivalent to an internal extension. By contrast to things that have such parts are those "simple things" that lack parts. Without parts, a "simple thing" (res simplex), such as a spiritual substance, can have no proper extension because it has no parts that can lie one outside the other. Consequently, a simple thing cannot be properly in a place. And yet it must be somewhere, and if somewhere, it must also be accorded some sense of extension. To account for this, Maignan formulated the concept of "virtual extension" (extensio virtualis). A simple thing can be in a place "by its substance, so that the whole corresponds indivisibly to the whole extension of the place and to its particular parts."[144] It is in this way that God is in the whole universe.

Maignan's description of virtual extension is probably the equivalent of the medieval concept of *ubi definitivum*, because in both cases a spiritual substance is said to be wholly in the whole place and wholly in every part of it. (For this fundamental medieval concept describing the mode of God's ubiquitous omnipresence, see above, Chapter 6, Section 2b, and notes 67 and 127.) By force of His immensity, God is not limited to a determinate place, "or (1) to the narrow confines of the universe, but just as He is within the limits [of the universe], so also He exists beyond it [and] (2) He is not only in some determinate and definite breadth of extramundane space, but is also without limit in all things and exists in every part in the infinitely stretching spaces."[145] God achieves existence in these "infinitely stretching spaces" by means of His immensity, "or, which is the same thing, virtual extension," which, in turn, is equivalent to His substance.[146] Not only is this imaginary space potentially infinite in the sense that one can always imagine more of it beyond any arbitrarily chosen limit, but it is also actually infinite because, by a single act of the intellect, we can conceive an extension without end in which God exists everywhere.[147]

In grappling with the difficulties of imaginary space, Maignan found it convenient, as had many other scholastics, to relate imaginary space analogically with imaginary time. Just as God's infinite immensity forms the basis of our belief that He must exist beyond the real space of the world, so is God's eternity the foundation of our belief that He existed in imaginary time, before the beginning of the real time associated with the creation of the world.[148] Maignan conceded that he used the expression "imaginary space" as a convenient means of conveying the idea that God exists beyond the real space of the world. The statement "to exist truly outside real space" is equivalent to the statement "to exist truly in extramundane imaginary space."[149] Because God's immensity is infinite, it must exist beyond the real space of our cosmos. To describe that extracosmic region, we may use the expression "imaginary space" with the proviso that it not be conceived as a separate container in which God's immensity exists, because God cannot be contained by anything else.[150]

Whatever Maignan may have had in mind when he used the expression "imaginary space," he did not identify it with vacuum, perhaps revealing here the influence of Fonseca, the Coimbra Jesuits, and Bartholomeus Amicus (see above, Chapter 7, Section 2d). Nor did he characterize imaginary space as a negation or pure negation, as did the Jesuits just mentioned.[151] As God's immensity, Maignan conceived imaginary space as a virtual extension, seeming thereby to attribute to it a somewhat more positive existence than did his predecessors.[152] But no more than they did he wish, or dare, to assign a positive or actual extension to God's immensity or imaginary space. By employing a quantitative term like "extension" in his basic description of imaginary space, Maignan went as far as he could in expressing the strong dimensional sense inherent in terms such as "immensity," "space," and "extramundane." To have gone further would have risked the transformation of God into a three-dimensional being. That move was consciously avoided within the scholastic tradition. In avoiding it, scholastics had devised a variety of terms and concepts that seemed to imply extension while denying it. The danger of attributing dimensionality to God shaped the scholastic literature on infinite space and made it difficult, tortuous, and often incomprehensible.

f. FRANCISCUS BONA SPES: IMAGINARY SPACE AS TRUE SPACE

If Emanuel Maignan may be taken to mark a culmination of the "quantification" of God's immensity and imaginary space within the scholastic theological and physical tradition, another significant move was made by an otherwise unknown Reformed Carmelite, Franciscus Bona Spes, who presented his opinions "On Imaginary Spaces" in little more than one double-column page.[153] Two questions, or doubts (*dubia*), form the basis of his discussion: (1) "What are imaginary spaces?" ("*Quid sint spatia imaginaria*") and (2) "Whether God is actually in imaginary spaces" ("*An Deus sit actu in spatiis imaginariis*"). Because these spaces are imaginary, Bona Spes, in his response to the first question, insisted that imaginary spaces can be neither real and positive nor indestructible negations of real places. For if our imaginations were rendered inoperative, the allegedly indestructible negations of real places would also become impossible.[154] Furthermore, imaginary spaces conceived as negations imply a contradiction. If such spaces are anything at all, they must exist in some sense. But if they exist, they cannot be negations, because a negation signifies the denial of actual existence. Consequently, imaginary spaces conceived as negations would both exist and not exist simultaneously.[155] Bona Spes concluded that imaginary spaces are purely imaginary in the sense that they have no existence whatever. Despite their characterization as imaginary, theologians and philosophers commonly attributed to these spaces such properties as infinite diffusion beyond the heavens, eternity, and a capacity to receive bodies because of a lack of resistance. But this served only to confuse imaginary spaces with real possible spaces.[156] Although imaginary spaces are truly imaginary in the sense of being nonexistent, we accord them some kind of existence because we imagine them in a manner similar to that in which an obscure mist or vapor might be conceived to be diffused into infinity. As reinforcement for his claim that imaginary spaces are nonexistent, Bona Spes invoked Saint Augustine, whom he correctly interpreted as rejecting extracosmic spaces when Augustine proclaimed that "the thoughts of men are idle when they conceive infinite places."[157]

As a consequence of his denial of the existence of imaginary spaces, Bona Spes, in response to the second question, aligned himself with those who rejected the actual presence of God in imaginary spaces (he mentions Bonaventure, Scotus, Vasquez, Capreolus, Oviedus, Fromondus, and "many others") and against those who accepted it (Suarez, Major, Domingo Soto, Hurtado, Arriaga, Fonseca, Comptonus, and "many others").[158] He singled out two major arguments that were common to his opponents. The first attributes extracosmic presence to God by virtue of His immensity and supports this by appeal to Hermes Trismegistus, who described God as "an intelligible sphere whose center is everywhere and circumference nowhere."[159] From this they apparently inferred that God must exist in an imaginary space. The second argument holds that if no extracosmic imaginary space exists, God could not be present in any new world that He might produce beyond the heavens, which seems absurd. Or if He were present in that world, the partisans of imaginary space asked whether or not He would be in the intervening space between these worlds. An affirmative reply signified that God is indeed in imaginary spaces; a negative response implied that God would be as divided as the separate and distinct worlds to which He would be confined, which is absurd.[160]

By way of response to these arguments, Bona Spes introduced his own opinion, which is both radically different from all the interpretations described thus far and yet surprisingly similar. God would indeed be present in any new world He produced beyond ours and would also be present in the distances intervening between them. *But He would not accomplish this by existing in an imaginary space.* For God is Himself a real immense space that is indistinct from His own divine immensity.[161] In support of this opinion, Bona Spes could do no better than invoke Saint Augustine's oft-cited remark in the *Confessions* (book 7, chapter 5): The world in God is like a sponge in an immense ocean. Beyond that sponge, God is extended everywhere by His substance.[162]

Reacting against those who also justified the existence of imaginary spaces by analogy with imaginary times, assuming that just as the latter existed before the creation of the world, so must the former,[163] Bona Spes replied by rejecting the existence of imaginary times, thus destroying the analogy. Those alleged imaginary times are as chimerical as imaginary spaces. For just as God Himself is the real space of His immensity, so before the world's creation was He also the real time of His eternity, existing in that eternity by Himself.[164]

Thus did Franciscus Bona Spes dispense with the concept of imaginary space. But unlike so many of those who also rejected it, Bona Spes not only accepted God's infinite omnipresence beyond the world but also assumed the existence of an infinite space, one, however, that was not imaginary – for such a space could not exist – but real. That real space is nothing other than God's immensity. Radical as this change might appear at first glance, from another vantage point it may be construed as a minor terminological alteration: For the widely used expression "imaginary space," Bona Spes substituted the expression "real space." In most other respects, his real space is identical with the imaginary space of his opponents. Both were conceived as equivalent to God's immensity and therefore really indistinct from Him. Moreover, it is probable that Bona Spes would also have ascribed to real space most of the usual properties attributed to imaginary space, namely infinity, eternity, and the capacity to receive bodies without resistance – though, as we have seen, he definitely rejected the characterization of space as a negation.

To interpret the contribution of Bona Spes in this narrow manner, however, would constitute a gross misconception of the significance of his contribution to the literature on infinite space. Implicit in Bona Spes' attack on the concept of imaginary space is the sense that despite the identification of imaginary space with God's immensity, proponents of it had hypostasized imaginary space into a separate existent. Not only was the attribution of negation or pure negation to imaginary space contradictory, but proponents of that space had somehow ascribed to it a separate existence capable of receiving bodies. In almost nominalistic fashion, Bona Spes viewed this transformation with alarm. For him, the concept of imaginary space was not only truly imaginary in the sense of nonexistent, but also superfluous. If space is to be identified with God's immensity, it must be real, and not construed as some kind of hypostasized imaginary void or negation. As long as space was conceived as God's immensity, it could not, for Bona Spes, be imaginary or a negation; it could only be real. For what could be more real than God's immensity?

With the rejection of imaginary space and the identification of real space with God's omnipresent immensity, we may appropriately conclude our study of the scholastic inquiry into the relations between God and space. Whether that space was conceived as imaginary or real, however, all scholastics were as one in denying to it any real dimensional extension. To have conferred dimensionality on space conceived as God's immensity or substance would have been tantamount to a transformation of the deity into a three-dimensional being.[165] It would have made of God a divisible magnitude and destroyed His divine immutability. Within the scholastic tradition, the nexus of space and God's immensity was a sufficient deterrent to the postulation of a three-dimensional space. Outside that tradition, however, another fundamental interpretation would emerge. Although many would retain the long-standing association between God and space, that relationship would be transformed by the attribution of real extension to infinite space. That transformation and the problems associated with it must now be described.

8

Infinite space in nonscholastic thought during the sixteenth and seventeenth centuries

1. THE IMPACT OF THE
NEW GREEK TREATISES

As nonscholastic interpretations and approaches to nature gained adherents during the sixteenth and seventeenth centuries, many of whom would be among the leading scientific and philosophical thinkers of the period, a common attitude toward scholastic authors and their works developed. When not characterized by downright hostility and contempt, this attitude is best described as indifference. With regard to discussions of space and vacuum generally and infinite space in particular, indifference was usually manifested by silence. Except for occasional mention of an opinion or attitude of the "schools" or "schoolmen," and even one specific and respectful citation of scholastic sources,[1] nonscholastic authors chose to document and support their varied arguments about space, whether associated with God or not, with ancient Greek authors (Plato, Proclus, Epicurus, the Stoics, Hero, Simplicius, Plutarch, Philoponus, Hermes Trismegistus, etc.), the Church Fathers, Cabbalists,[2] and nonscholastic predecessors and contemporaries.[3] Silence about, or contempt for, the large and detailed scholastic literature on infinite space does not and cannot legitimize the inference that scholastic ideas about space and God played no role in shaping nonscholastic interpretations and opinions.[4] It only makes the determination of such influences difficult to demonstrate and document. As the previous chapters have shown, the problem of infinite space and its relationship to the deity had engaged the attention of scholastic authors since the fourteenth century. Whether scholastic ideas on this significant theme influenced nonscholastics in the sixteenth and seventeenth centuries will be considered in different places in the two concluding chapters. Thus far, it is fair to say that possible influences have either been ignored or decided arbitrarily in the negative.

The reaction against scholastic Aristotelianism was catalyzed by the availability in the sixteenth and seventeenth centuries of Greek and Latin treatises that had been unknown or ignored during the Middle Ages. Works by Plato, Lucretius, Plotinus, Proclus, Cleomedes, Hero, Plutarch, Simplicius, Philoponus, Hermes Trismegistus, and others quickly took their places alongside, and often replaced, long-standing and revered authorities.[5] Whether taken independently or as part of the Neoplatonic tradition, or even as the major link in the *pia philosophia*, that great chain of pagan and Christian philosophers and theologians stretching from Zoroaster to Ficino and beyond,[6] Plato was the central figure in the powerful eclectic philosophies that were developed in opposition to the dominant Aristotelian natural philosophy and cosmology of medieval and early modern scholasticism. Rival cosmologies also appeared to challenge the restricted conception of the Aristotelian cosmos. Two were of overriding significance. The first and more dramatic was the atomist cosmology described in the *De rerum natura* of Lucretius. Previously known only through the biased reports of Aristotle, Lucretius offered a rigorous defense of the concept of innumerable worlds randomly dispersed throughout the endless extent of an infinite void.[7] The second was of Stoic origin and, though less dramatic, was more popular. As reported by Simplicius, the Stoics conceived a unique, finite, spherical cosmos filled everywhere with matter but surrounded by an infinite three-dimensional vacuum devoid of worlds and any matter whatever.[8] Also available was the *Physics* commentary of John Philoponus, which repudiated Aristotle's attribution of corporeality to void space.[9] When the scholastic tradition, as described in the preceding chapters, is added to these sources, we can appreciate the emergence of spatial and cosmological conceptions that differed considerably, and even radically, from those expounded in the Middle Ages. Despite occasional mention, the Copernican heliocentric system would play but a small and negligible role. Whether the earth turned around the sun or vice versa made no difference to those who sought to characterize the elusive nature of space or who were concerned with what, if anything, might lie beyond our world.

2. THE INFINITE UNIVERSE OF LUCRETIUS AND THE GREEK ATOMISTS: GIORDANO BRUNO

Of the two major theories of infinite space described above, few opted for the Lucretian atomistic version with its countless worlds scattered through an infinite void space. In a society in which the doctrinal tenets of Christianity still exerted a powerful influence, the Lucretian cosmology, which denied creation and Providence and proclaimed a purposeless universe, could only repel or frighten potential adherents. But it did not frighten Giordano Bruno, who adapted the Lucretian essentials to his own extraordinary conception of God and creation. Falling just short of pantheism,[10] Bruno denied a creation from nothing and the finite duration of the universe. The latter is coeternal with God and is but an aspect of the deity. Because He is in everything – both matter and form – that emanates from Him, God is not logically antecedent to the universe, although He may be conceived as its source.[11]

Adopting the principle of plenitude, Bruno insisted that God had no choice but to create all possible things to the maximum capacity of His infinite power.[12] It was therefore essential that an infinite universe with an infinity of worlds exist. The cosmologies of Aristotle and the Stoics therefore had to be rejected: the former because it assumed the existence of only one world with nothing whatever beyond, neither time, nor void, nor place, nor matter; the latter because despite its infinite void space it contained only a single finite world, which was as nothing in that endless emptiness. Bruno could see no reason whatever for God to restrict His creative power to the finite.

Divine goodness can indeed be communicated to infinite things and can be infinitely diffused; why then should we wish to assert that it would choose to be scarce and to reduce itself to naught – for every finite thing is as naught in relation to the infinite?[13]

Why should God "be determined as the limit of the convexity of a sphere" when He could be "the undetermined limit of the boundless?"[14]

It was imperative, therefore, that Bruno expose the inadequacies of the Aristotelian and Stoic cosmologies. In his *De l'infinito universo et mondi (On the Infinite Universe and Worlds)*, which is a detailed refutation of Aristotle's *De caelo*,[15] Bruno sought to counter Aristotle's assertion of the finitude of the world and his denial of anything beyond, whether void or plenum. If nothing lies beyond the world, where is the world and where is the universe? asked Philotheo, who was Bruno himself. Aristotle's reply, Philotheo explained, is that the world "is in itself. The convex surface of the primal heaven is universal space, which being the primal container is by naught contained."[16] Such a conception is tantamount to the claim that the heaven and world are nowhere.[17] To those who would agree with Aristotle and conclude that "where naught is, and nothing existeth, there can be no question of position in space nor of beyond or outside,"[18] Philotheo countered that these are meaningless words because it is "impossible that I can with any true meaning assert that there existeth such a surface, boundary or limit, beyond which is neither body, nor empty space, even though God be there."[19] Bruno thus followed the tradition of those who, whether scholastic or nonscholastic, found it inconceivable on intuitive and rational grounds that something, either void or plenum, should not lie beyond our world. (See the discussion in Chapter 6, Section 1, especially Oresme's statement.)

One might well ask, and Bruno did, whether God himself, who exists beyond our world, does not perhaps serve as the container of all things in our world. If so, the need for extramundane void space or matter to perform that function would be obviated. Unable to conceive how a non-dimensional divinity could serve as a container for dimensional bodies, Bruno rejected the suggestion.[20] Besides, it would be beneath God's dignity to fill space or function as the boundary of a body.

The grave deficiencies of Aristotle's cosmology were not, however, remedied by the Stoics, who assumed the existence of an infinite three-dimensional void beyond Aristotle's finite spherical world. Although this cosmology led the Stoics to make a proper distinction between the finite *world* and an infinite *universe* – the former representing "all that which is filled and doth constitute a solid body," the latter embracing "not merely the world but also the void, the empty space beyond the world"[21] – it also produced a situation in which the void "hath no measure and no outer limit, though it hath an inner; and this is harder to imagine than is an infinite or immense universe."[22] Although Bruno considered the imposition of an inner limit to infinite space unjustifiable, the absence of body or worlds within the Stoic extracosmic infinite void was far more disturbing. Rigorously applying the principle of plenitude, Bruno insisted that infinite space is essential not for the mere "exaltation of size or of corporeal extent, but rather for the exaltation of corporeal natures and species, because infinite perfection is far better presented in innumerable individuals than in those which are numbered and finite."[23] An infinitely perfect God would surely not restrict His creative powers to the finite. And so Bruno not only rejected Aristotle and the Stoics but adopted the basic cosmological frame of Democritus and Epicurus, whose ideas he probably derived from Lucretius:[24] Worlds infinite in number are distributed through a universe of infinite extent.

But Bruno's worlds are not those of his bold atomist predecessors, nor do they resemble the Aristotelian cosmos. A world is not an ordered aggregate of planets and spheres enclosed by an outer surface. For Bruno every celestial body – earth, moon, sun, and so on – is a world moving by itself through its own course without attachment to an invisible sphere.[25] These infinitely many and diverse worlds[26] represent the perfect unity for Bruno because they are encompassed within a single infinite universe,[27] which is made possible by an infinite space, the nature of which we must now consider.

Because his striking cosmology was founded on an infinite universe,

Bruno's concept of an infinite space has been much discussed. Despite the attention it has received, the properties of that space have been surprisingly neglected.[28] These properties – fifteen in all – were formulated by Bruno in his most mature cosmological treatise, the *De immenso et innumerabilibus*, first published in 1591.[29] In this significant section,[30] Bruno first mentioned John Philoponus as a Peripatetic who did not acquiesce in Aristotle's antivacuist position but rather boldly defended its existence. For Bruno, as for Philoponus, "space is a certain continuous three dimensional physical quantity in which the magnitude of bodies is received."[31] To comprehend Bruno's deeper conception of the space described here in general terms, we must cite the fifteen properties that he attributed to it.[32]

1. Space is a quantity, which is obvious from the differences in equality between the bodies contained and comprehended by it.
2. Space is a *continuous quantity*. Indeed it is the most continuous of all physical magnitudes, because the actual division of it is impossible.
3. Among all things physical, space is necessarily first by nature. Not only does it precede the bodies that occupy it, but it remains immobile as bodies succeed each other within it. Moreover, when all bodies have withdrawn from it, space will remain alone by itself.
4. Space is also physical because it obviously cannot be separated from the existence of natural things. Despite its physical nature, however, space is neither matter nor form nor any composite of these.[33] Because it is neither of these fundamental entities, space must be considered to be beyond or apart from the natural (*praeternaturale*) and to precede it (*antenaturale*).
5. Space receives all things indifferently. The difficulties that may arise in the withdrawal and succession of bodies must be attributed to the bodies and not to the space.
6. Space is neither active nor passive and receives neither forms nor qualities.
7. Space does not intermingle or mix with anything else; nor does it yield to bodies. Only a body can yield to a body, with space functioning only as a receptacle or support for their mutual actions.[34]
8. In what may well be the most extraordinary attribute of space, Bruno, tracing the consequences of the seventh property, described space as impenetrable (*impenetrabile*). Only discontinuous magnitudes are penetrable because their parts are capable of variations in distance. Because space is absolutely indivisible (as we learn from the second property), it has no parts.[35] For Bruno, then, bodies seem to move into or through the separations that exist between parts of discontinuous matter.[36]
9. Space itself cannot be formed or shaped; only matter is alterable.
10. Because there should not be a space for a space or a place for a place, space itself is unlocatable.[37]

11. Space must be understood as outside all things because all things have a limit and shape.

12. Space is uncomprehended (*incomprehensum*) or unperceived because it is not confined within anything so that it can be grasped or comprehended.

13. Space is equal to the thing located in it, because that located thing cannot be of greater size (*aequalius*) than the space in which it is.

14. Space does not exist outside of bodies by the mere imagination or by thought (*cogitatio*) alone, because we are unable to conceive bodies unless they are somewhere. Although we cannot abstract body from space, we can infer space from body.

15. Space is neither a substance nor an accident because things are not made from it nor is it in things. Rather, space is that in which things are locally. It is a nature that exists "before the things located in it, with the things located in it, and after the things located in it."[38]

From the properties just described and from statements made elsewhere, it is evident that for Bruno space is essentially an infinite, homogeneous,[39] immobile, physical, three-dimensional, continuous, and independent quantity that precedes, contains, and receives all things indifferently, despite the further assumption of its impenetrability. In what would prove a significant move, though Bruno was not the first to make it, space was said to be neither a substance nor an accident.[40] It was thus removed from the categories distinguished by Aristotle as necessarily applicable to all things. Moreover, space could neither affect nor be affected by anything. Its role was apparently that of a pure container. Many of these properties and descriptions came to be widely accepted in the course of the sixteenth and seventeenth centuries. One that was not, however, is impenetrability. Here Bruno stood with a small minority. For the most part, his scholastic and nonscholastic contemporaries and successors assumed it as virtually self-evident that space has the property of yielding to bodies without resistance. Bruno, however, placed primacy on the continuity and indivisibility of space. The latter could have no parts because an actual infinite could not be divided into finite parts. The conception of bodies somehow penetrating and being in a three-dimensional space seemed to violate the latter's indivisibility and disrupt its continuity. To guarantee the preservation of indivisibility and absolute continuity, Bruno deemed it essential to assign the further attribute of impenetrability. But how can bodies move in and through an impenetrable space? Had Bruno not raised here the medieval dilemma posed by Aristotle's conception of a dimensional vacuum as an impenetrable body?[41]

Perhaps Bruno would have approved the approach adopted a century later by Joseph Raphson, who also assumed the impenetrability of space[42] but allowed that space could penetrate bodies.[43] The act of penetration was thus assigned to space rather than body. Although it was fairly common in the sixteenth and seventeenth centuries to allow that space penetrated body, it was unusual to assign to space the property of impenetrability, because impenetrability might somehow imply resistance to the occupation of space by body or an inability of space to yield to body. It was therefore more usual to assume that as space penetrated body, it also simultaneously yielded to it. With Raphson, and perhaps also Bruno, the property of impenetrability was intended to preserve indivisibility at the expense of the property of yielding, a move that would, in any event, produce the same results. Unfortunately, where Raphson was quite clear, Bruno was not. In *De l'infinito universo et mondi*, where Bruno assigned to space the capacity to penetrate everything, he did not attribute impenetrability to it; and, contrarily, in *De immenso*, where he assigned the property of impenetrability to space, he did not also declare or imply that space can penetrate bodies, although he did assume, as always, that bodies occupy space and that somehow the latter can contain bodies. Because all the essential ingredients of Raphson's interpretation appear disparately in Bruno, it is not implausible to assume that Raphson's conception was shared implicitly by Bruno. (For more on the relationship of indivisibility, penetration, and the notion of spatial parts, see below, Sections 4f, h.)

If we turn now to Bruno's *De l'infinito universo et mondi*, published in 1584, seven years before *De immenso*, the difficulties of the latter treatise vanish. Here there is no talk of impenetrability or a space that is not divisible into parts. Instead there is a confusion as to whether Bruno's infinite space is void or an ethereal plenum, or whether indeed these are one and the same.

By void Bruno did not mean a pure nothing, but rather "that which is not corporeal nor doth offer sensible resistance is wont, if it hath dimension, to be named Void, since we do not usually understand as corporeal that which hath not the property of offering resistance."[44] Thus void is a three-dimensional entity that offers no resistance to the reception of bodies. But is this the void of the Greek atomists as described by Lucretius? Is it a void in which completely empty three-dimensional space exists side by side with material bodies? Despite his admiration for the ancient atomists, such a conception would have been unacceptable to Bruno. The principle of plenitude alone would probably have led him to a full space. For as Philotheo, Bruno's spokesman, expressed it, "we shall find the Plenum not merely reasonable but inevitable."[45] The plenum Bruno envisioned was that of an infinite material ether that offered no resistance to the motion and position of bodies and that completely penetrated them while simultaneously receiving their qualities.[46]

Because the properties of the infinite void and the infinite ether were basically identical – that is, both were three-dimensional and could receive bodies without resistance – and because the latter was conceived as contained by and coextensive with the former, the problem arises as to which of the two Bruno intended as true space. As with so much else in Bruno, no obvious answer is possible. Of the two, however, ether is the more plausible choice. For although Bruno could declare that "there is a single general space, a single vast immensity which we may freely call void,"[47] it was more usual for him to speak of space as ether or matter, as when he proclaimed that "beyond the world is Space which is ultimately no other than Matter."[48] Occasionally, Bruno even spoke as if void and material ether are different ways of referring to one and the same three-dimensional space that underlies the universe.[49] But the infinite void of which Bruno spoke is clearly subordinate to the infinite ether that fills it. For Bruno, the universe is unequivocally a plenum filled by a ubiquitous ether. The void serves merely to contain that infinite ether in a manner similar to the way Philoponus conceived the finite material world to be contained in a never-empty three-dimensional vacuum.[50]

In truth, Bruno's infinite void was even more superfluous for his infinite universe than Philoponus' finite void was for his finite world. Because the ether is contained in the void and is coextensive with it, bodies and worlds actually move and subsist in the material ether rather than the empty void. Furthermore, the ether has the same basic properties as void, namely tridimensionality and absence of resistance. Thus the void seems to function only as a coextensive container of the ether,[51] whereas the latter serves as the direct medium in which bodies move and subsist. Because the ether seems able to perform all the functions of void, the latter appears superfluous. As an admirer of the ancient atomists, why did Bruno not rest content with an infinite void free of a material ether? His infinity of worlds could just as well have been scattered through an infinite void as through an infinite ether. But purely empty space would have violated Bruno's principle of plenitude. Mere emptiness would have detracted from the creative principle. Hence Bruno filled his infinite void with an infinite material ether and assigned to that ether the significant function of binding together the elements of the universe,[52] a function that no purely empty space could have performed. Indeed Bruno's ether is rather complex. Not only does it function as a void, but in its material aspect it is like a tenuous, infinite substance designated as ether in its purest state but conceived as "air" when it is close by and external to us and as "spirit" when part of the composition of bodies.[53] In light of the obvious superiority of ether over void and their obvious duplication of spatial functions, we may well inquire why Bruno bothered to introduce infinite void at all? Why should he have needlessly violated Ockham's razor by the addition of a superfluous and useless empty container? Perhaps it is best here to invoke Bruno's enormous admiration and respect for Lucretius and the ancient atomists. Superfluous or not, the infinite void enabled Bruno to pay respectful attention to atomist cosmology without truly adopting it. By declaring an infinite void and then filling it with an infinite ether that directly rendered it superfluous, Bruno could incorporate the empty space of ancient atomism into the ethereal space of his new cosmology.

By 1591, Bruno had come to emphasize the absolute indivisibility and continuity of space, a space that could not be divided into parts and therefore had to be assumed impenetrable. Thus was the resistanceless, yielding space of *De l'infinito* replaced by the impenetrable space of *De immenso*.

a. THE RELATIONS OF GOD AND SPACE IN BRUNO'S THOUGHT

Although God's powers and nature were of considerable concern to Bruno, he had virtually no interest in the relationship between the divine power and space. Bruno's thoughts on this, one of the central themes of this study, must be inferred from general cosmological concepts and statements about the deity. Let us recall (Section 2) that for Bruno God is not prior to the universe but coeternal with it. Thus form and matter, which constitute the substance of the world, are coeternal with God even though they emanate from Him directly and coequally.[54] As a basic entity in the universe, space would also seem to be eternal and to bear some relationship to God. On these issues, however, Bruno was silent. To acquire any sense of a Brunonian conception of the relationship of God and space, we must turn to some of the fifteen spatial properties described in *De immenso* (see above). Here Bruno's negative assertions about space are perhaps the most interesting. Space is neither matter nor form nor any composite of them (Property 4); nor is it a substance or an accident (Property 15). It appears, therefore, that space is none of the things that have emanated from God. In fact, space is a wholly independent physical quantity on which Bruno conferred seemingly contradictory properties, namely the capacity to receive all things indifferently (Property 5) and the property of impenetrability (Property 8), wherein it would seem incapable of receiving anything of a physical nature. Whatever the resolution of this difficulty, it is clear that Bruno's space is an independent entity said to be first by its nature (Property 3). (For an attempted explanation, see below, Sections 4f, h.) Indeed it is "beyond" or "distinct from the natural" (*praeternaturale*), or nature, and "prior to the natural" (*antenaturale*), or nature. Moreover, it is homogeneous and, because of its absolute indivisibility, is also without parts. From such properties, we may infer at the very least that space is an uncreated and eternal physical quantity. Except for God, it must be conceived as prior to everything, first among uncreated and eternal things such as matter and form. But the latter are described as emanations from God and are therefore in some sense causally dependent on Him. (See Chapter 6, note 59, for Thomas Aquinas' similar reconciliation of creation and eternality.) Because space is not so characterized, we may properly infer that it is eternal and uncreated, lacking even the special sense of creation assigned to matter and form.

But how, if at all, might space be related to God? Is space a noncreating, passive entity coeternal with but independent of God, an entity functioning solely as a container and receptacle for the material universe? Or is space to be identified with God Himself, perhaps as the latter's immensity or as an attribute of the divine substance? Of these two possibilities, the second is incompatible with Bruno's metaphysics, in which God is identified as a substance and His effects as accidents.[55] But as we have already seen, space is neither a substance nor an accident (see Chapter 1 and note 38 of this chapter). Thus space is neither God Himself (substance) nor any attribute of God as, for example, His immensity. But neither can space be an effect of God, because God's effects are accidents. And because Bruno also expressly denied that space is matter or form (see above, Property 4), both of which are coequal, direct emanations from God (though also coeternal with Him), it follows that space could not have been created by God.

We are thus left with the first alternative, which is blatantly heretical. For how could a positive physical quantity be uncreated, eternal, and totally independent of God? But the consequences of Bruno's description of space and the properties he assigned it lead inevitably to an infinite space that is coeternal with but wholly independent of God, who appears to have utilized it merely as the container of His infinite universe. The medieval fear that void space would be interpreted as an eternal, uncreated positive entity independent of God (see Chapter 6, Section 3b) was realized in the metaphysics and cosmology of Giordano Bruno.

But God did not merely locate His universe in Bruno's independent infinite space; He also located Himself within it by virtue of His omnipresence. It is natural to inquire whether God's omnipresence in an independent, three-dimensional, infinite space implies His actual physical extension. On this question, Bruno's response was unequivocal: "God is totally infinite, for He is everywhere in the whole universe and in each of its parts, infinitely and totally."[56] Being wholly in every part, God remains unextended and therefore indivisible. On this momentous point, Bruno followed the medieval explanation of God's omnipresence.[57] But it does seem that the space God occupied was not of His own making. And so we return to the astonishing implications of Bruno's spatial and metaphysical doctrines. Was Bruno aware of them? Did he realize that the infinite space he described and the properties he attributed to it were such as to exclude it from God's creation, or at least what passed for a creation in Bruno's system? Probably not. Bruno was apparently as unaware of the heretical implications of his thoughts on space as have been his readers to date. Heresy must first be detected before the charge of it can be made. Nevertheless, within the context of medieval and early modern thought, the implications, though not the intent, of Bruno's thought were heretical. Uncreated, eternal entities independent of God were theologically unacceptable throughout the period with which this study is concerned.

3. FINITE VOID SPACE AND THE INFLUENCE OF JOHN PHILOPONUS

a. BERNARDINO TELESIO

Although Bruno's bold and radical cosmology was sufficiently well known in the seventeenth century, his influence was small. Few would adopt his infinite universe with its infinity of worlds.[58] Even less significant was his doctrine of infinite space, the precise nature of which is obscure and ill defined. It was not Bruno's cosmology, or generally that of the ancient atomists, that would provide the framework within which nonscholastics would debate the nature and properties of infinite space. That role would fall to the Stoic universe with its finite and unique world surrounded by an infinite three-dimensional vacuum, which a few would fill with incorporeal ether or light. Most of the major nonscholastic discussants of infinite space during the sixteenth and seventeenth centuries – a list that would include Francesco Patrizi, Pierre Gassendi, Henry More, Isaac Barrow, Otto von Guericke, John Locke, and Joseph Raphson – worked within the framework of Stoic cosmology centered by then on a Copernican rather than an Aristotelian world.

Even before Francesco Patrizi published what is perhaps the first major nonscholastic version of the Stoic universe, the influence of newly available Greek treatises had begun to alter spatial concepts in the sixteenth century. Among such works, the commentary on Aristotle's *Physics* by John Philoponus was of special significance. Although he denied extracosmic void, Philoponus, as we saw (see Chapter 2, Section 3a), assumed the existence of a three-dimensional void that was coextensive with the world it contained and never empty of it in part or whole. Void space could not exist independently of body. As early as 1520, Giovanni Francesco Pico della Mirandola (1469–1533) upheld this opinion in his *Examen vanitatis*,[59] where Philoponus is mentioned by name; and by 1565, without mention of Philoponus, Bernardino Telesio (1509–1588), in his *De rerum natura juxta propria principia*,[60] nonetheless adopted and defended the former's interpretation of void space. Like Philoponus, Pico and Telesio confined their claims for the vacuum to our finite cosmos, with no concern for the possibility of an extracosmic void. Of these two similar positions, only Telesio's, which is the more significant, will be described here.

In the *De rerum natura*, Telesio conceived space as something absolutely incorporeal, as if it were nonbeing. Because of its incorporeality, it offers no resistance to bodies that enter it. Indeed it is by virtue of that same incorporeality and lack of resistance that void space can simultaneously coexist with the bodies that successively come to occupy it.[61] Although Telesio did not explicitly attribute tridimensionality to space, dimensionality seems implied by the claim for coexistence of body and incorporeal space, with the latter penetrating the former.[62]

Despite a separate and independent existence, Telesio insisted, as did Philoponus, that void space is always occupied by body.[63] Vacua do not exist naturally. Because the cosmos is actually a plenum, it follows that bodies moving through the void are actually penetrating a plenum. Bodies move by displacing other bodies in the cosmic void. In this circumstance, Telesio found the basis for the repudiation of Aristotle's claim that motion through a void would be instantaneous (see Chapter 1). Obviously, if that void is always filled with resistant matter, resistance would guarantee a finite, temporal interval for the traversal of any distance.[64]

As the receptacle of the material cosmos, incorporeal void space is absolutely immobile[65] and homogeneous.[66] Indeed it is wholly different from all things, being neither similar to nor dissimilar from anything else nor the contrary of anything.[67] In these suggestive thoughts, did Telesio wish to imply that void space is so unlike anything else that it is neither a substance nor an accident? We have already emphasized (see Chapter 1 and Chapter 3, Section 4) that in the Aristotelian tradition, three-dimensional void space was classified as a body and its existence rejected. As a body, it could not receive and coexist simultaneously with other bodies. Telesio had already denied that space is a body, but now he seemed to go further and suggest that because of its dissimilarity to everything else, it may not even be an accident. Indeed void space is not even the contrary of anything and could not therefore be conceived as the privation or contrary of body, which was one of the ways vacuum was described during the Middle Ages and even later (see Chapter 2, Section 1). Despite the vagueness of his account, Telesio indicated that space must be described in terms wholly outside the traditional substance-accident categories. Although, as will be seen below, medieval antecedents for such a move are clearly detectable within a theological context by a few authors who themselves denied the existence of a separate void space, Telesio appeared to mark the beginning of a significant development wherein those who assumed the real – not hypothetical – existence of an independent and separate void space would characterize it as something outside of, and prior to, things that are substances and accidents. The explicit and unequivocal move in this direction may have first been made by Francesco Patrizi, although scholastics had similar thoughts at approximately the same time (see below, Section 4a). Before we turn to Patrizi, however, it will prove instructive to describe the spatial conceptions of Tommaso Campanella (1586–1639), who not only followed in the broad path established by Telesio[68] but also developed opinions that went far beyond, although they seem thus far to have been largely ignored or misunderstood.[69]

b. TOMMASO CAMPANELLA

To grasp the essential features of Campanella's spatial doctrine, it is imperative to view it within the larger context of his hierarchically structured Neoplatonic metaphysical and cosmological schema. Although bits and pieces of Campanella's system can be found in numerous works, the major source is one of his last completed treatises, the *Universalis philosophiae seu metaphysicarum rerum iuxta propria dogmata* (Paris, 1638).[70] Here Campanella distinguished five kinds of worlds:[71]

1. The archetypal world (*mundus archetypus*) is actually God, who is within and without all things; before and after all things; and simultaneously with all things. Indeed He is infinite and everywhere. As an "immense and infinite" entity, the archetypal world exceeds "by far the limits and numbers of the corporeal world" and is "more real, better, and innumerable in its being, preceding and excelling the corporeal worlds without limit."[72]

2. The second world is that of mind (*mundus mentalis*), which includes both angelic and human minds.

3. The third is the mathematical world, or universal space (*mundus mathematicus, hoc est, spatium universale*), which constitutes the inner foundation of bodies because it is "prior to, after, and simultaneous with bodies and beyond and below [them]."[73] It is this spatial world, also assumed to be perpetual and unchangeable, that shall be considered in greater detail below.

4. The material, or fourth, world (*mundus materialis*) is the world of bodily masses (*moles corporea*), which encloses and sustains all forms.

5. The fifth and last world appears to be our particular world, consisting of finite things formed into a system (*mundus situalis*) by the constant struggle between hot and cold as they battle for possession of the bodily masses of the material world.[74]

All worlds are related hierarchically.[75] The four below the archetypal world participate in the latter, which is God. More specifically, our particular world (*mundus situalis*) is in the material world of bodily masses (*mundus materialis*) and vice versa. Similarly, the material world is in the mathematical world, or space (*mundus mathematicus seu spatium*), and vice versa. The same reciprocal relationships obtain between the mathematical and mental worlds (*mundus mentalis*) and, finally, between the mental and archetypal worlds. Universal space, or the mathematical world, is thus most intimately related to the qualitatively superior mental world directly above and the inferior material world immediately below. Space participates in the mental world and could not exist without it; similarly, matter participates in space, which is its foundation.[76] With this in mind, we can better appreciate Campanella's conception of universal space.

The mathematical world, or universal space, is God's direct creation. Indeed it is God's first creation, because it "precedes all beings, if not in time, at least in origin and nature."[77] Space is therefore "the place of all things that are sustained by the divinity," for "in him we live, move, and have our being."[78] For Campanella, space was the "first substance" (*substantia prima*) and matter the "second substance" (*substantia secunda*).[79] As a created entity, space cannot be God, as some Arabs have argued.[80] From this it is obvious, though left unmentioned, that space is not God's immensity. Indeed this is evident from the fact that Campanella, contrary to a common misconception, assumed that God's created space was finite, not infinite.[81] Well aware of the opinion that postulated an infinite space beyond the world, Campanella rejected it and assumed, with Telesio, the finitude of space.[82] Only God, not space, is infinite. And yet if space is finite and God infinite, does this not imply that the latter is somehow beyond or outside space? Campanella conceded the extraspatial existence of God, but he insisted that God is beyond space only "by nature" (*natura*) and not in the sense that He is in another place or space beyond the created finite space, "for indeed there is no place and space outside place and space, just as there is no humanity outside man, nor linearity outside lines."[83]

If space is finite, what then are its properties and nature? As with Telesio, it is three-dimensional,[84] incorporeal, immobile, homogeneous,[85] and suitable for receiving bodies, which would not be possible if, as Aristotle held, space itself were a body.[86] Because it is neither body nor dimensionless, incorporeal substance, Campanella classified space as intermediate between divine and corporeal beings.[87] It is a three-dimensional incorporeality. But how can we know that it is three-dimensional? The mind determines this from the observed dimensions of bodies.[88] Those bodies could not possess dimensions unless they existed in a three-dimensional space. "For a body could not have length unless it is in a space with length; nor depth unless in [a space] with depth; nor width unless in [a space with] width."[89] In the hierarchy of Campanella's worlds, the observed dimensions of bodies depend on the immediately preceding superior world of space, namely the mathematical world. But the dimensions of space, in turn, appear not to inhere objectively in or be integral to space itself. Those dimensions are conferred on space by the mind (*mens*), which "divides space [and makes] a line, a surface, and a depth, since it is in a metaphysical world of a higher order."[90] The superior mental world (*mundus mentalis*) thus imposes all mathematical figures – points, lines, and figures – on space, or the *mundus mathematicus*.[91] A line, for example, is not in space itself but in the mind that is thinking the line.[92] Indeed space itself "is unable to be extended, or widened, or come to a point, or be encompassed (*ingrossari*) by the composition of lines and surfaces. But we think about it in this way by [means of] the intellect [or understanding]."[93] If space possesses immaterial dimensions and seems divisible, it is not because of the bodies within it but by virtue of the superior world of mind, which imposes dimensions upon it. The observed dimensions of bodies are thus not actually and objectively in bodies, nor indeed are they objectively in space itself. They are imposed on space by mind, wherein they exist as ideas. The seeming dimensionality of space and bodies derives from the hierarchical relationships of the mental, spatial (or mathematical), and corporeal worlds. How incredibly different was Campanella's conception of space from that of his illustrious Italian predecessor, Francesco Patrizi, who insisted that all dimensions were objectively and actually infinitely extended in space and for whom the role of mind was confined to the perception of finite segments of these infinitely many objective infinite extensions.[94]

For Campanella the "true" nature of space seemed to presuppose a dimensionality imposed by the mind. And yet it is clear that he also thought of space as something outside the human mind. Unlike body, the parts of which can be separated and be mutually distant, space itself cannot be divided or separated. It is indivisible and unalterable.[95] Thus although bodies in space are divisible and alterable, space itself is not, even though the dimensions of body really belong to space, which, in turn, receives them from mind. For Campanella, space can be indivisible because dimensionality is not integral to it, but is imposed by mind. Whatever "objective" space might be without dimension as an inherent property is unclear, but we can be certain that Campanella was prepared to describe it as indivisible.[96]

As if the concept of space we have just described were not sufficiently unusual and even extraordinary, Campanella assigned one more basic property to it that would make his spatial theory virtually unique in the seventeenth century. He endowed space with sense and feeling! For as Campanella expressed it, "even in matter we find appetite and sense, so why not in space as well?"[97] The world for Campanella is a large spherical[98] animal wherein all the beings that constitute it, including matter and space, possess the capacity to sense and feel.[99] As an animal with feeling and sensation, the world seeks consciously to preserve its continuity. For

just as our own arm does not wish to be separated from the elbow, nor the elbow from the upper arm, nor the head from the neck, nor the leg from the thigh, but all oppose and hate division, so the whole world abhors division – which division occurs when any empty space or vacuum intervenes between any of its parts.[100]

Although convinced that artificial vacua could be produced,[101] Campanella appears initially to have adopted the traditional medieval position that nature abhors a vacuum. The material world works to preserve its continuity and thus immediately fills all potential vacua.

But in the *Del senso delle cose e della magia*, the very treatise in which he presented the traditional interpretation, Campanella suggested another way to explain the cosmic plenum and material continuity. Rather than resort to matter's abhorrence of a vacuum, Campanella assumed an affinity between space and matter. He declared that in an earlier draft of the *Del senso delle cose*, which had been stolen, he had explained the lack of natural vacua by the far greater fear that bodies had for a vacuum than for their own contraries. So pervasive was this fear that bodies would rather be destroyed by their contraries than permit existence of a vacuum within the world.[102] The same overwhelming desire for material continuity also seemed a suitable explanation as to "why the sky, although encompassed by space, does not expand into that space – because . . . it enjoys the mutual contact of its parts and it fears the nothingness that it might suffer by spreading."[103] But although this was a possible explanation for the lack of natural vacua within the world, Campanella explained in book 1, chapter 12 of the same *Del senso delle cose e della magia* that he was

now more inclined to say that there is pleasure in the contact between bodies and space because of that desire and love for expanded existence of which I spoke.[104]

Earlier, in the very same chapter, Campanella did indeed speak of the desire for expanded existence on the part of both space and matter. As a divine creation, space attracts to itself things that desire to occupy it, whereas material bodies take pleasure in expanding themselves to fill space. "We may conjecture further," Campanella continued,

that since the air hurries in order to prohibit a vacuum, there is joy felt in filling a void; and that the rush is not so much to prohibit the vacuum as it is to spread out in space; for the love of expanding oneself, multiplying oneself, and living full lives in spacious existence obtains in all things that multiply themselves, generate, and expand.[105]

The attractive power that space exerts on bodies led Campanella to conclude

that it is not bodies that give unity to the world, since the world is composed of contraries, but rather space, which interposes itself even between separate bodies and binds the world together.[106]

Thus did Campanella abandon the powerful and negative medieval tradition that nature abhors a vacuum and seeks at any cost to preserve its material continuity. (See Chapter 4, Sections 1, 2, and 4d, for medieval conceptions of nature's overriding concern to preserve material continuity and avoid a vacuum.) In its place, he substituted a generally positive theory in which potentially empty spaces are usually filled for reasons of joy and pleasure. Although he could never decide whether bodies filled space because of the powerful attraction of the space itself or whether they spontaneously and willingly rushed into spaces either because they wished to avoid division[107] or because of their irresistible, innate urge to spread and extend themselves,[108] Campanella was convinced that both bodies and space took pleasure in mutual contact. Space wished to be occupied, and matter was always pleased to oblige when no other obstacles interfered.

In light of Campanella's unusual spatial theory, we may properly infer that he did not believe in the existence of extracosmic space, contrary to what may be suggested by the passage cited above. Empty spaces within or without the world would be filled with bodies in ways, and for reasons, already described. The explanation that matter fears the disruption of its continuity and the consequent void that would result from its own expansion may be said to represent the medieval tradition that Campanella abandoned. If there were empty spaces beyond the world, both the space and the matter would eagerly and joyfully seek mutual contact. Campanella rejected the existence of natural vacua within and without the world precisely because he was convinced that matter sought to extend itself into a space that was eager to receive it. No spaces would be left empty. But what if there were not enough matter to fill the spatial expanse that God created? Campanella resolved this potential dilemma by the assumption that God had filled his finite created space with a formless matter, or body, that is capable of receiving many forms.[109] Campanella thus filled his empty space permanently with matter, as did his Italian predecessors Bruno, Telesio, and Patrizi (for Telesio, see above, Section 3a). The superior world of indivisible space, the *mundus mathematicus*, must precede the next inferior world of divisible body (*mundus ma-*

terialis).[110] The latter cannot endure without the former. Because the dimensions of bodies are derived from the space of the mathematical world (and ultimately from the mental world),[111] space penetrates all things and is penetrated by all things,[112] for "space is more intimate to body than body to itself."[113] Indeed it is space itself "which interposes itself even between separate bodies and binds the world together."[114] The unity of the world depends upon space, not bodies.

With such properties and functions assigned to space, the very existence of which Aristotle had denied, it comes as no surprise that Campanella excluded it from the traditional substance–accident categories and was thus in agreement with the likes of Telesio, Patrizi, and eventually Gassendi (see below, Section 4a, for Patrizi, and Section 4b, for Gassendi). Although space can be considered a substance in the literal sense of an essential substratum in which matter and bodies subsist and without which they cannot exist,[115] it is not a substance in the Aristotelian sense of a substratum in which accidents and qualities inhere. Nor indeed does space have any contraries.[116] Its perplexing and elusive nature may have prompted Campanella to exclude it from his own revision of the Aristotelian categories.[117] Now that space had achieved independent status, it seemed to defy easy classification. At the very least, however, it was accorded logical and natural, if not temporal, priority over created things, all of which it was assumed to contain.

Telesio and Campanella represent the two most significant proponents of a finite, spherical, physical world that was assumed to occupy a coextensive, independent void space that was, however, never empty of matter. Although preserving the finitude of the world, Francesco Patrizi would extend space itself to infinity and provide the fundamental model for the major spatial theories of the seventeenth century.

4. INFINITE SPACE IN THE STOIC TRADITION

a. FRANCESCO PATRIZI

With Francesco Patrizi (1529–1597) we have perhaps the earliest European proponent of the Stoic conception of the universe: a finite material world surrounded by an infinite extended void space. As a vigorous anti-Aristotelian – he was a dedicated Platonist – Patrizi sought to replace the four elements of Aristotle with four basic principles of which space was the first, followed by light (*lumen*), heat (*calor*), and fluidity (*fluor*). If we exclude the innumerable scholastic commentaries and *questiones* on Aristotle's discussion of place and vacuum in the fourth book of the latter's *Physics*, Patrizi's claim, made at the conclusion of his two chapters on space, that his was the first *systematic* description of space ever formulated seems tenable and reasonable.[118] His first publication on space was devoted solely to that topic. Giving it the Lucretian title *De rerum natura*, Patrizi divided his treatise into two books, the first on physical space (*De spacio physico*), the second on mathematical space (*De spacio mathematico*).[119] As the full title indicates, these were only the first two books on "the nature of things." When Patrizi published the whole of it in 1591, this time under the title *Nova de universis philosophia* (*New Philosophy on the Universes*), he reprinted the two books on space as the opening chapters of the fourth or final part, titled "Pancosmia."[120]

In the opening lines of his treatise, Patrizi proclaimed space as the first of God's creations. That space must be prior to all created things is virtually self-evident, for is it not "that which all other things required for their existence, and could not exist without, but which could itself exist without any other things, and needed none of them for its own existence?"[121] With this dramatic introduction, we stand on the threshold of a monumental break with the scholastic tradition and its concerns about a separate space. The space God has created first of all things will later be described as infinite in extent. God has thus created an actual infinite! [122] Where medieval scholastics might have conceded at most that God *could possibly create* an actual infinite space, none was so bold as to declare that He had done so.[123] If infinite space existed, it was as God's immensity and not as His separate creation. For Patrizi, however, not only is God said to have created an actual infinite space but the declaration is worthy of no further comment, as if the creation of an actual infinite were a commonplace without theological implication. And yet Patrizi did seem to locate God in the space of His own creation. For if God is indivisible, as He is, then He must exist in indivisible space. But even if God exists *nowhere* (*nullibi*), we cannot help but conceive this in some spatial sense; and if God exists somewhere (*alicubi*), either in the heaven or beyond, He is certainly in space; and if He exists everywhere (*ubique*), He cannot but be in space.[124] Patrizi would thus appear to have located God in the space that He Himself created, leaving us to ponder where God was before He created His own space. Was He in Himself in some Augustinian sense? Or is God's space different from the space He created for all subsequently created things? If not, how is God omnipresent in an infinite three-dimensional space? In truth, Patrizi appears not to have been much concerned with such questions, which seem to arise only incidentally. The theological implications of God's relationship to space, which so exercised scholastic authors, were of little interest to Patrizi.

The characteristics of space that Patrizi would formulate were derived from ideas and concepts that became available with the influx of Greek treatises in the fifteenth and sixteenth centuries, treatises that had not been available to medieval scholastics. For those who were eager and ready to abandon the traditional scholastic cosmology and physics, the new literature presented exciting alternatives or conferred respectability on well-known but previously rejected opinions. Spatial concepts and allusions were now available from works by Plato, Plotinus, Cleomedes, Plutarch, Hermes Trismegistus, Philoponus, Simplicius, Diogenes Laertius, Sextus Empiricus, and Lucretius. Preserved in some of these treatises were ideas on space cited from yet other authors whose works were lost. All this was now available to the growing body of anti-Aristotelian natural philosophers. Patrizi would use it to break completely with Aristotelian scholasticism.

Like some of the medieval scholastics whose ideas we have previously described, Patrizi adopted a Stoic cosmological frame in which a geocentric, finite, spherical cosmos is surrounded by an infinite void space.[125] On the strange assumption that an infinite magnitude could have a center, Patrizi located the center of infinite void space at the center of the world, justifying this move by the belief that all radii drawn from the center of the world would be equal, however far they might be extended.[126] The infinite space in the center of which the world is located is otherwise void, homogeneous, and immobile. It is, as we have seen, temporally prior to the world itself. Moreover, "if the world should be completely destroyed and become nothing, which some not at all obscure ancient philosophers asserted would happen, the Space in which the world is now contained, as *locus*, will remain entirely empty."[127] The existence of a finite world within infinite space prompted Patrizi to distinguish two kinds of space: (1) an infinite external space that surrounds the world and (2) the space of the world itself.[128] Because the space of the world forms part of the infinite space beyond, it is hardly surprising that the two spaces share common properties. For if the space of the world were empty, it would be void (*inane*) and thus be identical with the infinite space beyond. It is only the creation of the world in that space that sets it apart as a *locus*. A *locus* is a space filled with body. Thus space – that is, *empty* space – is prior to *locus*. And because unoccupied space is void (*vacuum*), it follows that "a vacuum is certainly prior to *locus*, and should be prior to it. But it is an [essential] attribute of Space to be a vacuum, hence Space is prior to *locus* both in nature and in time."[129] The major differences between the two spaces derive from the fact that one is filled with body (though not completely, because Patrizi assumed the existence of interstitial and even small separate vacua)[130] and the other is totally void. The space of the world, or *locus*, is finite because the world is finite. It thus differs from the infinite extracosmic void, which is both finite and infinite. It is finite because it terminates at the outermost boundary of the world and does not penetrate within (because the world prevents this); it is infinite because it recedes from the world to infinity in all directions.[131]

The true nature of infinite space is, however, conveyed in terms of the common properties that the two spaces share.

Neither of these two kinds of Space is a body. Each is capable of receiving a body. Each gives way to a body. Each is three-dimensional. Each can penetrate the dimensions of bodies. Neither offers any resistance to bodies and each cedes and leaves a *locus* for bodies in motion. And just as resistance (*resistentia, renitentia,* and *antitypia*) is the property of a body which makes it a natural body, so is a yielding offered to bodies and their motions the property of each kind of Space.[132]

By making the rigid distinction between body and three-dimensional space, Patrizi sided with John Philoponus against Aristotle and placed himself in a position to derive the remaining spatial properties. (See above, Chapter 2, Section 3d. Although Patrizi did not mention Philoponus in this discussion, he certainly knew his *Physics* commentary.) If void space is not a body, as it was usually conceived in the Middle Ages, then it need not have two of the three most fundamental properties of a body, namely resistance and impenetrability. But it could possess tridimensionality, the third essential property of bodies. With a tridimensionality that offers no resistance to the reception of bodies and can indeed penetrate bodies by yielding to them, space could coexist simultaneously with bodies and serve as their absolutely immobile container.[133] By making the assumption that space could simultaneously yield to and penetrate bodies, Patrizi clearly indicated that space is continuous, immobile, and homogeneous. For in yielding to a body without moving or offering resistance, space necessarily penetrates and coincides with that body rather than being displaced by it, as the opinion described earlier by Walter Burley would have it.[134] As the *locus* of that body, space continues to exist even as it coincides with the body that fills and occupies it.[135] Yielding for Patrizi did not signify the displacement of space by body as if body and space were mutually exclusive, a condition that would imply a discontinuous space. On the contrary, yielding (*cessio*) implied continuity of space, because in yielding to a body without resistance, the space simultaneously penetrated the body and coincided with it at every position in its path. For Patrizi, the interpenetration of body and space was an essential feature of the cosmos. It posed no problem because space, though dimensional, was not a body, offered no resistance to body, and could therefore coincide and coexist with it. (How Patrizi might have reconciled penetrability of space with its indivisibility is suggested below, note 218 and Section 4f).

Like Philoponus, Patrizi distinguished corporeal from incorporeal dimensions. Body and space are different three-dimensional entities. *Locus*, or space, is "evidently something incorporeal, having all the dimensions of a body, yet not a body."[136] Here, then, was Patrizi's reply to the old puzzle as to whether space is anything. The reply presupposes that place and space must exist, "For all things, whether corporeal or incorporeal, if they are not somewhere, are nowhere; and if they are nowhere they do not even exist. If they do not exist they are nothing."[137] Because Patrizi had already assumed that space is an entity created prior to everything else, its existence is undeniable by assumption. The question to be answered is not whether space exists, but what is it?

Now one might interpret space as an existent that is nothing, but Patrizi construed this as a contradiction. For "if anything is, it is a being (*ens*); if it is *not*, it is *nothing*."[138] Obviously, then, if space exists, it must be a being and therefore something. That "something," as we have seen, is an immobile, incorporeal, three-dimensional entity that is not a body but is capable of receiving and containing three-dimensional bodies. Patrizi resolved the old Aristotelian problem of the impossibility of interpenetration between a three-dimensional space conceived as a body and three-dimensional material body by simply denying that space is body.[139] He adopted the basic position formulated so effectively by John Philoponus against Aristotle. Patrizi's justification of the existence of a separate, three-dimensional, nonresistant, yielding space in which bodies could move and rest would win numerous adherents in the seventeenth century. The Aristotelian two-dimensional conception of place and the scholastic nondimensional space associated with God's immensity were abandoned. For those subsequently interested in God's omnipresence, however, a new problem had arisen: How could a presumably dimensionless God be omnipresent in an infinite three-dimensional space?

Following Patrizi, nonscholastics would usually conceive space as a distinct three-dimensional, infinite, immobile container surrounding a finite world. But rarely did they follow the Stoic model in its pristine form. Patrizi himself made radical alterations in the basically simple Stoic universe. Within the world, he assumed the existence of interstitial vacua[140] and thus abandoned the pure Stoic plenum. Of greater significance, Patrizi, like Bruno, would fill his infinite void with another substance, light. Upon creation of the infinite void space we have just described, God through His goodness chose not to leave it empty, and so filled it with pure light.[141] Why light? Because, as Patrizi explained, it is most like space. It is most simple (*simplicissima*); it can be diffused everywhere and fill the universe; it resists nothing and yields to everything and is, therefore, penetrated by everything. Although light is also incorporeal, as is space, it differs from space in one aspect: It is a body, a body in space.[142]

By arguing for an infinite, separate void space and then filling it with ether and light, Bruno and Patrizi reveal to us how difficult it was for anyone in the sixteenth and seventeenth centuries to accept the existence of pure void in the manner of the Greek atomists and Stoics. The existence of pure, unfilled spatial emptiness seemed an affront to divine creativity. Why should God leave vast extents of space empty of everything? The purpose of a spatial container is to contain. Hence Bruno and Patrizi, and others who followed, would fill that space. They would fill it, however, with things that were akin to space itself and that, above all, had the property of yielding to bodies without offering resistance.[143] Bruno's ether and Patrizi's light – and, for that matter, God's omnipresent immensity for scholastics and nonscholastics – met those conditions even as they did much else.

Despite his assumption that infinite space is filled with light and forms the *locus* of the world itself, Patrizi insisted that the essence of space is emptiness; to be filled or occupied is accidental to it.[144] What sort of thing is space if, being prior to all things, its essence is emptiness? Patrizi was quite emphatic as to what space is not. In words reminiscent of Telesio, he excluded any comparison of space with body by declaring that space is neither similar to nor the contrary of body.[145] Convinced that space is something, Patrizi inquired whether this implies that space is a substance or an accident as these were ordinarily understood within the context of the Aristotelian categories. Should this be so, then "if it is a substance, it is either some incorporeal thing or a body. If it is an accident, it is either a quantity, a quality, or some other such thing."[146] In fact, it is none of these. Because space is not a body and is not properly comparable to bodies, it cannot be described or characterized by any of the Aristotelian categories that are applicable only to bodies.[147] Moreover

granted that the categories serve well for worldly things (*in mundanis*); Space is not among worldly things (*de mundanis*), it is other than the world (*mundus*). It is the accident of no worldly thing (*mundanae*), whether body or not body, whether substance or accident – it is prior to them all. As all things come to be in it, so are they accidental to it; so that not only what are listed in the categories as accidents, but also what is there called substance, are for it accidents. Hence it must be philosophized (*philosophandum*) about in a different way from the categories.[148]

For all these reasons, then, Aristotle's traditional substance–accident categories, which were used to describe all things in the world, were inapplicable to space. In arriving at this conclusion, Patrizi may have initiated, or perhaps merely strengthened or made more explicit, a significant trend. Because Aristotle had rejected the existence of a distinct and separate void space, that space could not be assigned to any of the traditional categories.[149] With an exception that will be mentioned below, the usual medieval discussions on the hypothetical existence of void space omitted consideration of its ontological status. Acceptance of the existence of real void space in the sixteenth century changed this situation. It became essential to determine the real nature of space. In a world still dominated by Aristotelian thought, however, it was quite natural to inquire whether a truly existent space could be properly assigned to any of the Aristotelian categories. As early as 1565, Telesio declared, without elaboration, that space is dissimilar to everything else (see above, Section 3a), and in 1587, Patrizi specifically excluded space from the Aristotelian categories, a move that others would follow. At approximately the same time, scholastics also responded to the same problem in a similar manner. By 1589 (Chapter 7, note 47), Pedro Fonseca, using different arguments than Patrizi, denied that space is substance or accident (see Chapter 7, Section 2b). Other scholastics would adopt the same opinion.[150] Some, for example, Pitigianus, who denied the existence of void space, found the arguments that excluded space from the Aristotelian categories a convenient reinforcement for their belief in the nonexistence of space. After all, why should anyone suppose that something that was neither substance nor accident could exist? For others, both scholastic and nonscholastic, the conviction that space was neither substance nor accident did not signify nonexistence, but rather existence of a kind that differed from substance or accident. Indeed the conception of a three-dimensional space that was neither substance nor accident had already been imagined in the fourteenth century by Walter Burley and John Buridan. That space, however, was postulated as a supernatural creation because neither Burley nor Buridan believed in its natural existence.[151] No further description of such a space was forthcoming until the sixteenth century, when the real existence of a three-dimensional void space was postulated, albeit one that was filled with matter. Those who adopted a real three-dimensional space (and even some scholastics who thought of space as the nondimensional immensity of God) were convinced that it was not classifiable in terms of the traditional substance–accident categories. Obviously, such a space must differ radically from everything else. What could it be and how should it be described? A variety of nonscholastic and scholastic responses were made, some of which we have already seen. It is time now to consider one of the most significant reactions, that of Francesco Patrizi.

For Patrizi space is "substantial extension (*extensio hypostatica*), subsisting *per se*, inhering in nothing else."[152] As a substantial extension, space is, after all, a kind of substance, though not the substance of the categories "because it is not composed of matter and form. Neither is it a genus, for it is predicated neither of species nor of individual things.

It is a different sort of substance outside the table of categories."[153] Space is an absolutely homogeneous, immobile entity[154] that was conceived by Patrizi as a mean between body and incorporeal substance.

It is not a body, because it displays no resistance (*antitypas, aut resistens aut renitens*), nor is it ever an object of, or subject to, vision, touch, or any other sense. On the other hand, it is not incorporeal, being three-dimensional. It has length, breadth, and depth – not just one, two or several of these dimensions, but all of them. Therefore it is an incorporeal body and a corporeal non-body.[155]

Patrizi would reiterate the mediate nature of space in his section "On Mathematical Space," where he would describe mathematics as "the mean between the completely incorporeal and the completely corporeal . . . because Space is really a body that is incorporeal and an incorporeal that is a body."[156] In that same description of mathematical space, we are furnished further reasons why space must be a substance outside the table of categories. For Patrizi, space is an infinite continuum that cannot be divided by any force of nature or by any power of the human mind. Within infinite space itself are contained *in reality* (*re ipsa esse*) an infinity of lines, surfaces, bodies, and incorporeal qualities, which are infinite not only in number but also in extent. Despite the actual infinite extent of these mathematical magnitudes,[157] the human mind perceives them as finite extensions and assigns appropriate finite spaces to accommodate them.[158] Patrizi defined a *continuum* as that which has not been divided or cut and a *discrete* entity as that which has been divided or cut. As an absolute continuum, space is indivisible but contains within itself all lines, surfaces, bodies, and quantities that are infinite in number and extent. It is the human mind that then "cuts out" finite entities from within this infinite space. But these divisions are not actually made in the space itself, for it is impossible to divide absolute space either potentially or actually. Such divisions are only imagined in the mind.[159] On the basis of the nature of infinite space, Patrizi would argue that geometry, which is "the science of the continuous," is prior to arithmetic, "the science of the discrete."[160]

Because they were next in importance to the deity, Patrizi felt constrained to preserve those spatial properties that would confer upon space its near-divine status. Thus space had to be conceived as infinite, immobile, and immutable, from the last of which properties flowed the additional property of indivisibility. With such properties, we are no longer surprised that space was excluded from the Aristotelian table of categories.

b. PIERRE GASSENDI

In the course of the sixteenth century, Francesco Patrizi and Giordano Bruno had formulated two rather similar spatial paradigms set within two radically different cosmological systems of Greek origin, atomism and Stoicism. With their Greek models, they shared the concept of a three-dimensional infinite space as the ultimate container of all things, but unlike their respective Greek counterparts, both shunned absolute void and filled their spaces with matter, Bruno with ether and Patrizi with light. Of these two systems, Patrizi's would prove dominant in the seventeenth century, when his ideas and works were widely known by such luminaries as Gilbert, Bacon, Kepler, Fludd, Digby, Hobbes, and, of the greatest importance for the history of spatial theory, Pierre Gassendi and Henry More.[161] Because the latter two may have significantly influenced Newton,[162] their opinions on the nature of space must now be described.

Although he lacked the originality of Patrizi, Pierre Gassendi (1592–1655) played a more crucial role in the development of a spatial doctrine that would eventually form the basic frame for the cosmology and physics of the Newtonian Scientific Revolution.[163] To a surprising degree, he merged ideas from scholastic and nonscholastic sources and exercised an influence directly, or indirectly through Walter Charleton, on Pascal, Barrow, Newton, and Locke.[164] Not only was Gassendi's cosmology a hybrid drawn from Greek atomism and Stoicism, but his concept of space united scholastic concerns about God and infinite imaginary space with the by-then standard three-dimensional infinite space of nonscholastics, the specific properties of which he may have derived directly from Patrizi, and perhaps from Telesio and Campanella as well.[165]

Atomist though he was, Gassendi was only too well aware that full-blown Greek atomism, with its infinity of eternal atoms generating an infinity of worlds, was utterly unacceptable to the Church and its theologians. Not even the God-created version devised by Campanella would do. Gassendi's cosmic vision was much more modest and even medieval. Using as his alleged model the atomic theory of Guillaume de Conches, the twelfth-century Chartrain whom he called Aneponymus, Gassendi declared that

the idea that atoms are eternal and uncreated is to be rejected and also the idea that they are infinite in number and occur in any sort of shape; once this is done, it can be admitted that atoms are the primary form of matter, which God created finite from the beginning, which he formed into this visible world, which, finally, he ordained and permitted to undergo transformations out of which, in short, all the bodies which exist in the universe are composed.[166]

From a finite number of atoms, God thus created a single, finite world and placed it in an infinite three-dimensional void space. Here then is the basic Stoic universe: a finite world surrounded by an infinite three-dimensional void space. Gassendi's finite world was not, however, a Stoic

plenum but consisted of atoms and microvacua. He thus united fundamental elements of two Greek cosmologies, atomist and Stoic.

If the basic frame of Gassendi's universe was a combination of elements
drawn from these two potent Greek world views, most of his other spatial
ideas and arguments had been enunciated previously by scholastics and
nonscholastics discussed earlier in this volume. Gassendi's great contribution was to synthesize them into an intellectually appealing form that received wide dissemination, not only through his own works but also
through Walter Charleton's significant English-language summary of those
ideas in 1654.[167]

In order to establish a foundation for the attribution of certain basic
characteristics to space, Gassendi resorted to a favorite medieval and early
modern scholastic device: the imaginary annihilation of all or part of the
world. Because his initial objective was to demonstrate the very existence
of spatial dimensions, Gassendi imagined that God destroys all body and
matter below the lunar sphere but leaves the sphere itself intact. That
God can do this "no one would deny, except a man who denies God's
power."[168] With this seemingly contrary-to-fact – and even impossible –
condition accepted, Gassendi believed he could convince anyone of the
existence and nature of space.[169] Following the destruction of all three-
dimensional matter within the lunar sphere, we can easily imagine the
continued existence of dimensions where once the earth stood at the
center[170] as well as where the other surrounding elements – water, air,
and fire – formerly existed. The dimensions of the earth are now imagined
where it once was. Dimensionality exists even when body has been removed. It was evident to Gassendi that

wherever it is possible to conceive some interval, or distance, it is also possible
to conceive a dimension because that interval, or distance, is of a determinate
measure, or can be measured. Therefore, this is the nature of the dimensions that
we call incorporeal and spatial.[171]

By extending the scope of divine annihilations from the matter within
the lunar sphere to the entire world, Gassendi formulated his case for the
infinitude of incorporeal space. For if God should destroy the entire
world, the vacuum that remained would have "the same nature as the
vacuum that had existed beneath the moon"[172] and would be an incorporeal space. Gassendi then asked that we imagine the existence and subsequent destruction of worlds that are successively larger and larger. As
the worlds that God annihilates become progressively larger, so do their
corresponding incorporeal void spaces. Because this process could be carried on to infinity and produce a world of infinite extent, its subsequent

destruction would leave behind an infinite, incorporeal, three-dimensional void space.[173]

From these imaginary annihilations, Gassendi drew profound consequences. Because God could have created the world any size whatever all the way to an actually infinite body, and because the imagined destruction of a world, even an infinite world, would leave behind a three-dimensional incorporeal space, Gassendi concluded that "immense," presumably infinite, "spaces" (*spatia immensa*) must have existed before the creation of the world.[174] That endless space is also absolutely immobile, for if God were to move the world through it,[175] the space would remain motionless. As a motionless entity in which bodies can obviously coexist, space can offer no resistance to the bodies that occupy it and must therefore be incorporeal. The mode of spatial and bodily coexistence is thus one where space penetrates, or rests in, body.[176] As Gassendi put it, "space cannot act or suffer anything to happen to it, but merely lacks resistance (*repugnantia*), which allows other things to occupy it or pass through it."[177] Space is thus radically different from body because the former lacks the capacity to resist whereas our concept of the latter is "something that has dimensions and a capacity for resistance."[178] Wherever body exists, however, its corporeal dimensions always correspond to equal incorporeal dimensions that coexist with it.

The fundamental properties that Gassendi assigned to void space are much the same as those described earlier by Patrizi, to whom Gassendi openly acknowledged his debt.[179] But Gassendi derived one further fundamental idea from Patrizi: Space (like time) cannot be classified as a substance or accident, but lies outside the Aristotelian categories, which, in Gassendi's view, were based on the assumption that

all being is either substance or accident, and that all substance is either corporeal or incorporeal, and hence that all accident is either corporeal or incorporeal (since it pertains to substance, or a being having existence), and that of all the corporeal accidents the first is quantity of which place and time are species. From this you may understand that the common opinion holds place and time to be corporeal accidents, and consequently that if there were no bodies upon which they depended there would be neither place nor time.[180]

To this "common opinion," Gassendi, following Patrizi but with greater clarity and force, replied that "place and time do not depend upon bodies and are not corporeal accidents," for "even if there were no bodies, there would still remain both an unchanging place and an evolving time."[181] Space differs from all things that are ordinarily classified as substances or accidents. Indeed space and time must be added to, not included within, the categories of substance and accident, because every substance and accident must exist somewhere, in some place and some time at some moment. Space and time must therefore be added to substance and accident "as if to say that all being is either *substance* or *accident*, or *place*, in which all substances and all accidents exist, or *time*, in which all substances and all accidents endure." [182] Of these four primal entities, space and time are obviously more fundamental than substance and accident, for "even if the substance or [183] the accident should perish, the place would continue nonetheless to abide and the time would continue nonetheless to flow." [184] Moreover, because substance and accident are real, so also

space and time must be considered real things, or actual entities, for although they are not the same sort of things as substance and accident are commonly considered, they still actually exist and do not depend upon the mind like a chimera since space endures steadfastly and time flows on whether the mind thinks of them or not. [185]

In these momentous passages, Gassendi constructed the spatial framework of Newtonian physics and cosmology. Space is an absolutely immobile, homogeneous, inactive (resistanceless), and even indifferent, [186] three-dimensional infinite void that exists by itself whether or not bodies occupy all or part of it and whether or not minds perceive it. If Gassendi derived the most enduring elements of his spatial theory from Patrizi, he also diverged from the latter on two significant points that were central to the debates of the seventeenth century. For Gassendi, the space beyond the world is truly void of created things. Body, ether, and light are absent. The conception of Philoponus that void space must always be filled with matter, a view that dominated the Italian natural philosophers, was abandoned by Gassendi, who followed his Greek models – atomists and Stoics represented by Democritus, Epicurus (through Diogenes Laertius and Lucretius), and Cleomedes – as faithfully as was feasible and politic for a Christian, to say nothing of priest, in the seventeenth century. Gassendi was no Bruno or Campanella. [187] And so, having emptied the void of all created things, Gassendi filled it with God, the uncreated divine spirit. God is not only in Himself, where He was before He created the world, He is also everywhere. Although God is not an actually extended magnitude, like a body – indeed "the divine substance is supremely indivisible and whole at any time and at any place" [188] – yet we do imagine God *as if* He were infinitely extended, for

there is a kind of divine extension, which does not exist in one place only, but in many, indeed in all places. But let me add that since it follows from the perfection of the divine essence that it be eternal and immense, all time and all space are therefore connoted, without which neither eternity nor immensity could be understood. Therefore, God, indeed, both exists supremely in Himself and is infinitely perfect, but He also necessarily exists in all time and in every place. And when it is asked, "Where was God before He created the world?" one cannot deny that He was in Himself; but it must be conceded at the same time that He was everywhere, that is, in every place; that is, not only in that place in which the future world would be, but also in an infinity of other places. That God be in space is thought to be a characteristic external to His essence, but not with respect to His immensity, the conception of which necessarily involves the conception of space.[189]

Gassendi has here reconciled the two major interpretations involving the manner of God's omnipresence prior to the creation of the world. That God was truly in Himself prior to the creation is not to be construed as a finite limitation on His omnipresence, but it must be understood to signify that by His infinite immensity God was everywhere in an infinity of other places – that is, in an infinite space.[190]

But how, we must ask, is God related to the infinite space in which He is omnipresent? Did He create it? If not, is it His immensity or something independent of Him? For Gassendi, not only is space uncreated (*improductum*) but it is also independent (*independens*) of God, a blatantly heretical idea[191] proclaimed by Gassendi as if it were in no way offensive to the "sacred doctors" (*sacrorum doctorum*).[192] How could Gassendi have arrived at such a conclusion? To appreciate his rationale, it is necessary to recall that Gassendi had excluded space (and time) from the categories of substance and accident. What could be the true nature of a space that is neither a substance nor an accident? It is at this juncture that Gassendi invoked the concept of imaginary space and thereby injected the scholastic tradition into his overall spatial doctrine. By "space," Gassendi explained, "we do not mean anything but that space which is generally called imaginary and which the majority of sacred doctors admit exists beyond the world."[193] These same doctors do not, however, conceive extracosmic imaginary space as a chimera, or figment of the imagination, but call it imaginary "because we have an image of its dimensions by analogy to the dimensions that appear to our senses," a description of space virtually identical with that proposed by the Coimbra Jesuits toward the end of the sixteenth century and by Bartholomeus Amicus in the seventeenth.[194] We have recourse to analogy because the senses do not directly perceive space. Our analogy is based upon the hypothetically conceived divine annihilation of bodies described above, which Gassendi offered as convincing proof of the existence of an objective, albeit imaginary, three-dimensional space. There is, however, a radical difference between this imaginary space and all those positive created things classified as substances and accidents. Because space is neither substance nor accident, it is nothing positive (*nihil positivum*).[195] God, however, creates only positive things that are either substances or accidents; therefore God did not create space.[196] On the basis of this argument, Charleton,[197] and perhaps Gassendi himself, was convinced that charges of impiety had been avoided. Despite the fact that space is a real being, God did not create it because space is not a positive thing, being neither substance nor accident. Because no other creation could be invoked for it, infinite space must be uncreated and therefore coeternal with and independent of God. Gassendi's intent then was not to denigrate God but to defend, exalt, and safeguard His creative powers. He saw it as no compliment to the deity to assign to Him the creation of nonpositive entities such as space. God should be conceived as the creator of positive things only, and not of those extracategorial entities, space and time, that had emerged as independent beings only toward the end of the sixteenth century.

If Gassendi's conception of infinite space has been presented with reasonable accuracy, one cannot but judge it heretical within the context of medieval and early modern scholastic theology and Christian belief generally. Whatever his justification, Gassendi allowed space to be uncreated and independent of God, which unequivocally implied the existence of two eternal, uncreated beings, God and space, each independent of the other. Surely he would have been aware of the dangerous implications of his position. Perhaps Gassendi really meant to identify space with God's immensity and thus follow the path of the majority of scholastics who assumed the existence of an infinite extracosmic space. In this way, at least the objection to the uncreatedness, if not the independence, of space could be accounted for. Careful analysis of Gassendi's discussion offers no hope for such an interpretation. It reveals all too clearly that he did not believe space was God's immensity. For one thing, he distinguished between two kinds of incorporeal substance. God, intelligences, and human minds are incorporeal substances possessed of genuine, positive natures with faculties that can effect appropriate actions. They are not merely substances from which only incorporeal dimensions are absent, as in the case of space, which is incapable of acting or being acted upon and is the second kind of incorporeal substance.[198]

An even more important objection may be posed. If a three-dimensional space were God's immensity, we would have to assume that God Himself is a three-dimensional extended being! Now, we have already seen that Gassendi adopted the typical and traditional scholastic explanation for God's infinite, indivisible omnipresence, namely that God is totally in every part or point of space and therefore cannot be divided when space is divided (see above and note 188). God could thus be omnipresent in a three-dimensional space and yet remain indivisible, immutable, and dimensionally unextended. But if that three-dimensional space were actually God's immensity itself, God would be converted ipso facto into a three-dimensional extended being. Nowhere did Gassendi conceive of God as three-dimensional, and no evidence suggests that he equated God's immensity with a three-dimensional infinite space, a move that would subsequently be made in the seventeenth century.[199] The conclusion is unavoidable that for Gassendi space is a self-existent, incorporeal, nonpositive, uncreated entity that is independent of though coeternal with God, a conception that Isaac Barrow thought impious (see above, note 191 of this chapter), as indeed would most other Christians of the seventeenth century.

c. GASSENDI, OTTO VON GUERICKE, AND THE SCHOLASTIC TRADITION ON SPACE

Near the beginning of the discussion on Gassendi, his spatial conception was said to be a synthesis of the two major traditions on space in the seventeenth century, the Greek (derived directly from Lucretius, Diogenes Laertius, Cleomedes, Simplicius, and Francesco Patrizi, who added to the ancient heritage) and the scholastic. Because Gassendi formulated the spatial doctrine that would underlie the new physics and cosmology, it will be useful to identify briefly what he received from each tradition. From the first he derived the basic properties of space, namely its infinity, three-dimensionality, incorporeality, wholly passive nature including an incapacity to resist bodies and therefore an ability to coexist with material dimensions, and finally, from Patrizi, the exclusion of space from the traditional categories of substance and accident. As an anti-Aristotelian admirer of Greek atomist and Stoic thought, as well as of the ideas of the Italian natural philosophers, Gassendi eagerly relied on such sources, which should come as no surprise. But he also derived much from the scholastic tradition, which he had regularly criticized as part of his overall anti-Aristotelianism. Whatever the intensity and scope of Gassendi's hostility toward the physics and cosmology of scholastic Aristotelianism, he was well aware of the scholastic tradition on infinite, imaginary, extracosmic space[200] and recognized its potential utility as a support for the most radical feature of his doctrine, the uncreatedness of space and its independence from God.

We have already seen that the analogy between dimensions perceived by the senses and those conceived in imaginary space could have been derived from the Coimbra Jesuits or Bartholomeus Amicus (see Section 4b and note 194 of this chapter). The doctrine of divine annihilation was also a scholastic doctrine, as was the concept of God's omnipresence, by which the deity was said to be wholly in every part of space and therefore indivisible. Using the latter device, Gassendi could follow scholastic tradition and proclaim God as omnipresent in infinite imaginary void space.

Indeed that same scholastic tradition could even have furnished Gassendi with a number of crucial ideas that we have assumed were derived directly from Patrizi and the Greeks. In Chapter 7, we saw that prior to Gassendi, scholastic authors had already conceived an imaginary space that offered no resistance to bodies and indeed coexisted with them. (For Fonseca, see Chapter 7, Section 2b; for Amicus, Section 2d.) The exclusion of space from the categories of substance and accident was also unambiguously proclaimed by Pedro Fonseca at approximately the same time as Patrizi (see Chapter 7, note 53, and Section 4a of this chapter). Because Gassendi absorbed these ideas in conjunction with a concept of three-dimensional space, it seems far more likely that he derived them from Patrizi and the Greek tradition rather than from scholastics whose space was nondimensional. In certain significant respects, however, Gassendi differed radically from Patrizi and aligned himself with the scholastic tradition, namely in his assumptions that space is "nothing positive" (*nihil positivum*), uncreated, and independent of God. (For Patrizi, space was the first of God's creations; see Section 4a.)

Although all of these ideas appear in different parts of Chapter 7, Bartholomeus Amicus' discussion bears the closest resemblance to that of Gassendi. Not only did Amicus deny that space was a positive thing (*quid positivum*), but he described it as a pure negation (*pura negatio*) (see Chapter 7, Section 2d) that offered no resistance to the reception of bodies and was neither subject nor accident. Moreover, space was uncreated and independent of though coeternal with God, the very position Gassendi adopted. Their explanations of that independence differed, however. For Gassendi the uncreatedness of space followed from God's inability to create what lacked positive properties and was neither substance nor accident. Because it was nonetheless a real being and yet uncreated, it had to be independent of God. Amicus explained spatial independence from the very nature of a negation, which could remain independent only so long as God refrained from acting on it (see Chapter 7, Section 2d). God, however, possessed the power to act on negations, so that each negation, including space, was ultimately dependent on Him. The independence of space was thus strongly qualified and made dependent on the divine will. Gassendi attached no such qualifications to the independence of his space. Nevertheless, spatial independence for both Gassendi and Amicus derived ultimately from the negative nature of space, which paradoxically was also considered to have a real, objective existence. If Gassendi knew the work of Amicus, he would surely have included the latter in that group of unnamed doctors who did not object to the conception of space as uncreated and independent of God.

Gassendi's doctrine of space has much in common with what had been formulated within the scholastic tradition during the sixteenth and seventeenth centuries. Not only did scholastics have access to the same Greek texts that had become newly available in the sixteenth century, but they also had a long-standing tradition of anti-Aristotelian spatial ideas, especially on extracosmic space. Their ideas, however, were not harnessed to a three-dimensional space; Gassendi's was. With the notable exception of Otto von Guericke, Gassendi was perhaps the only other nonscholastic to incorporate scholastic ideas into his spatial theory. It is probably through Gassendi directly, or through his disciple, Walter Charleton, that scholastic influence was disseminated, an influence that is often apparent where mention is made of imaginary space or the relationship of God and space is at issue. Although Gassendi was a vital, if indirect, source of scholastic influence, Otto von Guericke (1602–1686) cited the spatial arguments and names of scholastic authors directly – for example, the Coimbra Jesuits, Athanasius Kircher, and Leonard Lessius[201] – and described in some detail the various meanings that had been enunciated for imaginary space (see Chapter 6, Section 1). He revealed direct scholastic influence perhaps more than any other nonscholastic of the seventeenth century. As the great experimentalist who demonstrated the elasticity of the air, its pressure, and the falsity of the traditional claim that nature abhors a vacuum, Guericke has received insufficient praise and attention.[202] Even less known are his opinions on the nature and extent of void space, to which he devoted the second of the seven books comprising his famous treatise *Nova Experimenta (ut vocantur) Magdeburgica de vacuo spatio* (*New Magdeburg Experiments* [*so-called*] *on Void Space*).[203] Despite his famous experiments, Guericke seems to have conceived absolute space as a nondimensional infinite being.

To comprehend Guericke's spatial theory, it is essential to realize that he placed all existent things in one of two exhaustive divisions. Every existent must be either an "Uncreated Something" (*Increatum aliquid*) or a "Created Something" (*Creatum aliquid*). No third division or category is possible because it could have no members.[204] What are these two divisions of being that together make up all existence? The Uncreated Something is "an infinite, immense, eternal, pre-existent and self-subsistent (*per subsistens*) essence which is other than [or different from] all created things and which contains all things but is [itself] contained by nothing."[205] By contrast, every Created Something lacks independence but gains its finite existence from the Uncreated in which it also subsists. Those who assume a third division called "Nothing" (*Nihil*) are mistaken because " 'nothing' is the negation of one thing and the affirmation of another."[206] Hence it is something and must belong to the Uncreated or Created, or perhaps to both. In different senses, Nothing can be assigned to both. Thus a chimera is said to be nothing, that is, nonexistent; however, to the extent that it is the mental conception of an animal, a chimera is not "absolutely Nothing" (*omnino Nihil*)[207] but has some degree of existence, that of a thought. Hence it qualifies as a Created Something.

But what of Aristotle's claim that nothing (*nihil*) exists beyond the world?[208] To which of the two divisions does this nothing belong? Von Guericke assigned it to the Uncreated. Thus if we ask what existed before the creation of the world, and one person replies an Uncreated Something and another Nothing, each would be correct because the Uncreated Something and Nothing are identical. When we say that heaven and earth, and all things, were made from Nothing, we mean they were made from the Uncreated Something. At this point, Guericke launched into what can only be described as a lyrical "Ode to Nothing" when he declared that

everything is in Nothing (*nihilo*) and if God should reduce the fabric of the world (*machinam mundi*), which he created, into Nothing (*nihilum*), nothing would remain of its place other than Nothing (*nihil*) (just as it was before the creation of the world), that is, the Uncreated (*Increatum*). For the Uncreated is that whose beginning does not pre-exist; and Nothing, we say, is that whose beginning does not pre-exist. Nothing contains all things. It is more precious than gold, without beginning and end, more joyous than the perception of bountiful light, more noble than the blood of kings, comparable to the heavens, higher than the stars, more powerful than a stroke of lightning, perfect and blessed in every way. Nothing always inspires. Where Nothing is, there ceases the jurisdiction of all kings. Nothing is without any mischief. According to Job (*Hiob*) the earth is suspended over Nothing. Nothing is outside the world. Nothing is everywhere· They say the vacuum (*vacuum*) is Nothing; and they say that imaginary space – and space itself – is Nothing.[209]

Thus did Otto von Guericke reduce Nothing, imaginary space, and space itself to one and the same thing. With his "Ode to Nothing," he served notice that his infinite space – for he would argue for its infinity – is not the inefficacious, impotent, and completely empty void of the atomists and Stoics. Space is now akin to the extracosmic, divine space of the scholastics. In view of the superlatives that Guericke heaped upon imaginary space, we are hardly surprised at his declaration that "the space which they call imaginary is true space."[210] Guericke thus abandoned the scholastic distinction between real and imaginary space[211] and conflated the two. But if imaginary space is a true and real space, it must be a positive and real thing,[212] which Guericke readily conceded. This conception bears a certain resemblance to the earlier views of Franciscus Bona Spes, who, it will be recalled (Chapter 7, Section 2f), rejected imaginary space because for him "imaginary" signified nonexistent. Only real space could exist, and its reality was assured because it was God's own infinite immensity. In almost all respects, the real space of Bona Spes is identical with the imaginary space that he rejected. Now Otto von Guericke also abandoned the distinction between imaginary and real space. Unlike Bona Spes, however, he did so not by abandoning imaginary space as a nonexistent negation but by the total identification of imaginary space with true and real space. Space is imaginary for Guericke because it is infinite and not directly perceptible. Nevertheless many things that are not directly perceptible are real and positive. For example, "it is necessary that one who never sees Rome, or a spirit, or an exotic animal, or some other thing, must imagine them."[213] The same can be said of the infinite, which, despite its incomprehensibility, "is grasped, at least in some way, by the imagination." Hence it "does not follow that Rome, or spirit, or the infinite, are not real (verum) or positive things."[214] Thus did Otto von Guericke equate, and indeed conflate, infinite imaginary space with real space.

But what is this space to which Guericke devoted so much attention? At the outset of a crucial chapter called "On Space" (book 2, chapter 4), Guericke attempted to answer this question. He explained[215] that he was not here concerned with the common, or "vulgar," conception of space as a three-dimensional magnitude. Rather, space was conceived as

the universal vessel or container of all things, which must not be conceived according to quantity, or length, width, and depth; nor is it to be considered with respect to any substance . . . nor through any diffusion, dilation, or expansion of itself. But [it is to be considered] only in so far as it is infinite and the container of all things, in which all things exist, live, and are moved and which supports no variation, alteration or mutation.[216]

Guericke thus distinguished two ways of conceiving space: the ordinary way as a three-dimensional quantity and another way as a universal container that is distinct from everything else and seems dimensionless. Although only the latter is relevant here, the distinction poses the question of whether Guericke believed in the dimensionality of space at all, a problem that will be considered below.

Except for the express denial of dimensionality, Guericke's conception of space as an infinite, universal vessel or container is similar to, though by no means identical with, previously described scholastic and nonscholastic descriptions of extracosmic space. As the universal container of all things, space, or imaginary space, is neither a substance nor an accident but is the container of all substances and bodies.[217] Like Patrizi, Gassendi, and numerous scholastics, Guericke conceived space as distinct from and prior to all substances and accidents. Indeed space can receive no forms, qualities, or accidents. Guericke, however, did not assign yielding to space as a property, although space appears to yield to body or substance because it is "permanent and immobile, indivisible, everywhere in all things,

[and] through all things whether corporeal or incorporeal,[218] with which nothing [else] intermingles here [in the world] or elsewhere, [whether space is] filled or empty." Space is not directly perceptible by the senses but is abstracted by the intellect from every material condition of nature. All things differ infinitely from space, and nothing can attain equality with it. The magnitude of all things seems derived from or measured with respect to it. Not only does space receive and penetrate all things, but it does so without action and passion – that is, without acting on anything or suffering any action. Space is also not locatable, because there can be no space of space. Contrary to Descartes's claim, then, body and space are not identical[219] but differ radically, as is also evident from experiments in which the totality of air is removed from a room, leaving behind an enclosed empty space. With the major exception that Guericke did not assign dimensionality to infinite imaginary space, and the minor difference that he denied to space the property of yielding, his description of spatial properties follows the tradition of Patrizi and Gassendi and of certain scholastics as well.

Because infinite extracosmic space, which is also the space of our world (see below and note 223), is identical with the Uncreated Something, imaginary space, and nothing, we are naturally led to inquire about Guericke's interpretation of the relationship between God and space. How, if at all, does a space that is equated with an Uncreated Something relate to God? Is it God's immensity or is it independent of God, as Gassendi conceived it? Although problems lurk in Guericke's resolution of this perennial and vexing question, there seems little doubt that he identified infinite void space with God's immensity and perhaps even with God Himself. Guericke's position on this matter emerges in the course of a summary account of the opinions of the ecclesiastics Jacques du Bois of Leyden and Athanasius Kircher, the eminent Jesuit scholar. In his *Dialogus Theologicus-Astronomicus*, published at Leyden in 1653 and directed against Galileo and all defenders of the heliocentric cosmology, Jacques du Bois had proclaimed the infinite omnipresence of God in an infinite void beyond the world. Du Bois should not have placed the divine essence in an infinite void, complained Guericke, but ought rather to have declared that " 'there is a place or space not *in which* the divine essence is, but which is itself the divine essence.' For God can be contained by no place (*ubi*) or vacuum or space, because He Himself is the place (*ubi*) for Himself, or the thing that is empty (*vacuum*) of every creature, or the space, or universal container of all things."[220] Here Guericke identified infinite space (or the Nothing, or Uncreated Something described earlier) with God Himself.

His criticism of Kircher differed somewhat. In a manner reminiscent of Bradwardine's fifth corollary (see Chapter 6, beginning of Section 3),

Kircher believed that even if a void could exist without body, which he denied, it could not exist without God. In the passage cited by Guericke, Kircher explained that "when you imagine this imaginary space beyond the world, do not imagine it as nothing, but conceive it as a fullness of the Divine Substance extended into infinity."[221] In Guericke's judgment, Kircher was wrong to say that "God fills all imaginary space, vacuum, or emptiness by His substance and presence" and simultaneously deny the existence of empty space. For "how can God fill what is not?" Rather, Kircher ought to have concluded with Lessius that " 'imaginary space, vacuum, or the Nothing beyond the world, is God Himself,' as he finally does when he announces that the space beyond the world is not Nothing, but is the fullness of the Divine Substance."[222] Not only are these authors right to suppose that the space beyond the world is the divine essence itself, but the same space, filled with the divine essence, is also found within the world, because the world is contained in this infinite space and even causes a slight diminution of it.[223]

But if space is God's immensity, or even God Himself, Guericke insisted that we must nevertheless understand that "the infinite essence of God is not contained in space, or vacuum, since God, who is present everywhere, is not contained in space, or vacuum, but is in Himself for Himself and is the space for Himself and is the thing that is void of every creature."[224] The Augustinian declaration that God is in Himself is, of course, compatible with all that has been said thus far. If infinite space is God's infinite immensity, one can rightly declare that space is in God and is for Him without implying that God's immensity is somehow more extensive than space.

If infinite space is God's immensity and the former is also equivalent to the Uncreated Something, or nothing, described earlier, we may rightly conclude that infinite imaginary space, or the Uncreated Something, is not independent of God in the manner assumed by Gassendi.[225] Indeed by all that has been said thus far, the Uncreated – or nothing, imaginary space, and vacuum, which are other names for it – is God Himself. The attributes ascribed to the Uncreated Something, or imaginary space, are those of the deity (see above). And as if to protect the nondimensionality of that deity, Guericke also insisted that infinite space must not be conceived as a three-dimensional magnitude. How then does this nondimensional imaginary space, which is God's infinite immensity, relate to that other three-dimensional space, which Guericke had described as the common or vulgar conception of space (see above)? Are there two different spaces? Or are they identical but conceptually different ways of perceiving the same thing? In either case, how can we reconcile nondimensionality and dimensionality? The difficulty is further compounded because in at least one place Guericke spoke of imaginary space as if it were an extended magnitude, though he left it uncertain as to whether this was his opinion or a sort of consensus derived from opinions he reported.[226] If this were really Guericke's opinion, God's immensity would be three-dimensional and He would be an extended magnitude, a conception Guericke clearly rejected in book 2 when he distinguished between imaginary space as nondimensional and space conceived in the common way. Despite difficulties of interpretation, there is little doubt that Guericke expressly conceived and proclaimed the nondimensionality of space insofar as the latter was a container or vessel. It is in its function as container that space is described as absolute, immobile, indivisible, and ultimately as nondimensional. Guericke clearly broke with the tradition derived from Philoponus, and accepted by Telesio, Patrizi, and Gassendi, that absolute space is an incorporeal, three-dimensional magnitude. It was the description of void space as an incorporeal magnitude that had won acceptance for the existence of a separate, three-dimensional space with which bodies could simultaneously coexist. Guericke's seeming rejection of this important solution to the old Aristotelian dilemma of the impossibility of the interpenetration of void and body is best understood as a consequence of his identification of space with God's immensity and a conscious, though unexpressed, desire to avoid the attribution of dimensionality to God. In this, Guericke must be seen as operating wholly within the scholastic tradition.

But what of the vulgar conception of space as three-dimensional? Although we cannot directly touch or see this space, Guericke held that we can perceive it with the eyes and mind operating simultaneously. For example,[227] if we imagine the air between two towers or mountains completely removed, our minds judge that the towers or mountains will not come into contact but will remain as they were. What is left after the evacuation of the air is pure empty space. Indeed Guericke believed that actual void – or pure empty space – could exist without the assumption of ethers or subtle fluids.[228] In book 3, chapter 9, he explained that vacuum must exist in the nature of things, for when the air in our world thins out to the point where it ceases to fill space, "there a pure space, void of all body, necessarily begins."[229] Is this pure space three-dimensional, as was the matter that filled it before evacuation or before it thinned out? Guericke provided no direct reply, but a careful analysis of his ideas suggests a negative response. If we recall that Guericke described the space within our world as identical with the space beyond our world (see above and note 223), and because the latter is dimensionless, so also must the "pure space, void of all body" be dimensionless.

In view of this conception, it is essential to contemplate the possibility that for Guericke the space we commonly perceive as three-dimensional is not a true representation of reality. Reflecting on the nature of space and its necessary relation to God, the mind arrives at a conclusion radically different from what is provided by ordinary perception. The available evidence indicates that Guericke believed space was nondimensional. In this he may have taken his cue from the Coimbra Jesuits, with whose discussion of imaginary space he was quite familiar. In their view, the dimensions of space are imaginary "because we imagine them in space, in a certain relation corresponding to the real and positive dimensions of bodies." (For this passage from the Coimbra Jesuits, see Chapter 7, Section 2b, and Chapter 7, notes 59 and 80, for the references to their *Physics* commentary, from whence it comes.) But they are not really there. This judgment of the Coimbra Jesuits, along with much else from them on imaginary space, was not only cited by Guericke in the *Experimenta nova* but also adopted by him.[230]

If this interpretation is plausible, there is no need to explain how absolute space could be both nondimensional, as God's immensity, and three-dimensional to our common perceptions. The latter would have no basis in reality. Thus when Guericke declared that "space must be understood in two ways: either with respect to the concept of the vulgar," which is three-dimensional, as he explained earlier, "or according as it is the universal container of all things,"[231] we should interpret this as a disjunction in which one or the other alternative, but not both, is true. On this basis, the available evidence points overwhelmingly toward the second alternative. In the absence of compelling evidence to the contrary, we must conclude that Otto von Guericke's infinite space was nondimensional, as was the God whose immensity it was. Without any doubt, his metaphysics of space was shaped by the scholastic tradition.[232]

d. EXTENDED SPACE AS GOD'S ATTRIBUTE: HENRY MORE

Despite Otto von Guericke's apparent rejection of infinite space as three-dimensional and his identification of it with God's immensity in the manner of numerous earlier scholastics, three-dimensional void space as described by Patrizi and Gassendi would eventually attract powerful supporters who would make it the absolute space of the new physics and cosmology. Although scholastics almost unanimously rejected an infinite three-dimensional space whereas numerous nonscholastics accepted it, there was otherwise a surprising degree of agreement on certain spatial properties such as infinity,[233] incorporeality, penetrability, indivisibility, lack of resistance, ability to coexist with bodies, homogeneity, immutability, and especially the exclusion of space from the traditional categories of substance and accident.[234] Although these properties were formulated by Patrizi and disseminated widely by Gassendi, there were significant disagreements among those who accepted some form of infinite space, disagreements that arose as a consequence of different perceptions of the relationship between God and space. Gassendi's own version, or at least the one that has been attributed to him here, was repugnant because it assumed an uncreated space that was coeternal with God and yet independent of Him, thus allowing for the simultaneous coexistence of two separate, actual infinites, which scholastics and nonscholastics would have found unacceptable. (For Guericke's statement, see note 225 of this chapter; for the medieval attitude, Section 3f and notes 98 and 122 of Chapter 6.)

With the exception of Gassendi, those who assumed extracosmic space related it to God in some way, usually as His immensity. As God's infinite immensity, space could be conceived as the instrumentality by which He achieved ubiquity and omnipresence. The mechanism for His omnipresence was not, however, infinite extension but rather the power to be wholly in every part of space simultaneously. Throughout the Middle Ages and most of the seventeenth century, scholastics and nonscholastics understood God's infinite omnipresence in this manner, a mode of explanation that was traceable to Peter Lombard, Saint John Damascene, and at least as far back as Saint Augustine and Plotinus (see Chapter 6, Section 3f, and Chapter 6, note 127). Whether space was conceived as nondimensional or dimensional, God's omnipresence could be explained by the "whole in every part" doctrine. For those who also believed that infinite space is God's immensity, the whole in every part doctrine was theologically acceptable only if space was assumed to be nondimensional, under which condition God would also be nondimensional, as theology required. For if space were assumed to be three-dimensional and also God's immensity, God would be transformed into a three-dimensional, extended being. Gassendi, who accepted a three-dimensional space and also invoked the whole in every part doctrine, avoided the imposition of dimensionality on the deity by a refusal to identify an extended space with God's immensity. He was undoubtedly aware of the consequences of such an action, as was Nicolas Malebranche (1638–1715), who in his *Entretiens sur la Métaphysique et sur la Religion (Dialogues on Metaphysics and on Religion)*, first published in 1688, declared, through his spokesman Theodore, that "The immensity of God is His substance itself spread out everywhere, and all of it is present everywhere, filling all places without local extension, and this I submit is quite incomprehensible." But "if you judge of the immensity of God by means of the idea of extension, you are giving God a corporeal extension," from which it would follow that "The substance of God will no longer be all of it wherever it is"[235] – by which Malebranche surely understood that the doctrine of the whole in every part could not apply to an actually extended God.

The steady and growing support that an infinite three-dimensional space had acquired during the seventeenth century is impressive. The arguments advanced in its behalf were satisfying and seemed to have overcome traditional Aristotelian and medieval objections. It was admirably suited for the new physics and cosmology that were emerging. But it raised serious theological problems that were ultimately incapable of satisfactory resolution. Gassendi's approach was unacceptable because it allowed for an uncreated three-dimensional infinite space that was wholly independent of God and coexistent with Him. Few, moreover, were willing – as was Patrizi – to assume the creation of an actually infinite three-dimensional space. As for those who identified infinite space with God's immensity, they had two options: either to assume the nondimensionality of space and thereby preserve the traditional nondimensionality of God Himself, as the scholastics did, or to assume the tridimensionality of infinite space and accept the inevitable consequence that God must be a three-dimensional being. Henry More would take this incredibly bold and unheard-of step in the 1660s, offering the definitive statement of his position in the *Enchiridion metaphysicum* of 1671.[236]

In an apparent defense of this radical move, More found it necessary to reject the traditional whole in every part doctrine, or "Holenmerism," as he called it, that was invariably invoked to explain how God could be omnipresent in the world or beyond in infinite space. (The term "Holenmerism" is from the Greek ολευμερῆ [see note 239]. For the background to the whole in every part doctrine, see Chapter 6, note 127.) As is well known, More insisted, in opposition to Descartes, that everything, whether corporeal or incorporeal, possesses extension.[237] Every spirit, therefore, including God, must be an extended being. To make this extraordinary conception acceptable, More found it necessary to destroy two significant opinions: (1) that of the Nullibists, which held that spirits are nowhere,[238] and (2) that of the Holenmerians, who, though conceding that spirits are somewhere, believed that

they are not only entirely or totally in their whole *Ubi* or *Place*, (in the most general sense of the Word) but are totally in every part or point thereof, and describe the peculiar Nature of a Spirit to be such, that it must be *Totus in toto & totus in qualibet sui parte.*[239]

In his assault on the venerable whole in every part doctrine, which was applied to the mode of God's omnipresence as well as the way a soul occupies its body, More devised a number of arguments centered on a diagrammatic representation of a solid body CDE (see Fig. 3).[240] The Holenmerians insisted that not only does a spirit or soul occupy the whole of body CDE, but it is also wholly in every part or point of that body, as, for example, wholly in point A, wholly in point B, and so on. They pre-

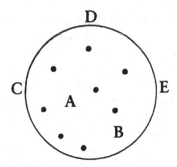

Figure 3. Diagrammatic representation of a solid body used by Henry More in his assault on the whole in every part doctrine. See text for explanation.

sented two major reasons in defense of their position. The first protects the absolute indivisibility of spirit. Without a doctrine of the whole in every part, a spirit would be divided when the body, or extended thing, it occupies is divided. And secondly, it is essential that the whole soul or spirit perceive what affects any part of it. Thus "whatever happens to it in C, or B, it presently perceives it in A, because the whole Soul being perfectly and entirely as well in C, or B, as in A, it is necessary that after what fashion soever C or B is affected, A should be affected after the same manner."[241]

Against these reasons in favor of the whole in every part doctrine, More raised a number of critical objections, the kind that must surely have occurred to many others who contemplated and even used this truly incomprehensible explanation. It remained for More, however, to make them explicit. Apart from his abhorrence of the notion that "a whole should not be divided into parts, but into wholes,"[242] More argued further that if the spirit, or soul, is totally in the whole body CDE and is totally in every point A or B, "it is manifest that they make one and the same thing many Thousand times greater or less than it self at the same time; which is impossible."[243] If the Holenmerians replied that the spirit or soul is really no greater in extent than any single physical point it occupies, but is yet capable of diffusing simultaneously throughout every other point of that body, or extension, More countered that if the spirit were all in one point, say A, nothing could be left over in point B or any other point. Under these circumstances, a spirit would be capable of such simultaneous diffusion and omnipresence only if it were also capable of "a stupendious velocity, such as it may be carried within one moment into all the parts of the Body, and so be present to them."[244] Such an instantaneous motion of spirit or soul was characterized by More as "outrageous" and "utterly impossible." Even if the whole of a spirit could be encompassed within a single point, it would surely be the smallest thing in the universe, which is ridiculous enough when applied to a created spirit but is "Reproachful and Blasphemous" when applied to "the *Majesty* and *Amplitude* of the *Divine Numen*."[245] With this powerful attack on an entrenched traditional conception of the way spirits occupy places, More may have had a decisive influence on Newton when the latter confronted the problem of God's relationship to space.

With the Nullibist and Holenmerian positions destroyed, only one plausible alternative remained: Spirit is extended like body. "For to take away all extension," as the Holenmerians did, "is to reduce a thing only to a Mathematical point, which is nothing else but pure Negation, or Nonentity."[246] The very essence of a thing, of being itself, whether spirit or body, is to have parts. But the properties of spirit were assumed the opposite of those possessed by body.[247] Where *body* is impenetrable (only to other bodies), physically divisible into parts, and devoid of perception, life and motion, *spirit* is penetrable (by other spirits), able to penetrate bodies (and other spirits), indivisible (or indiscerpible), and endowed with life and motion, with which last capability it could also contract and dilate itself. As one who believed that all being must be either a substance or an accident (that is, a mode of a substance), More classified spirit as immaterial substance because it subsisted by itself and needed no other thing for its existence.[248]

In view of the antithetical natures of spirit and body and the enormous significance More assigned to the realm of spirit, it comes as no surprise that he also filled the material world with a pervasive "Spirit of Nature" (or "Universal Soul of the World").[249] As an immaterial ether, was the Spirit of Nature, which More described as "the great Quartermaster-General of Divine Providence,"[250] the space of the world? No. Apparently coextensive with the finite, material world, the Spirit of Nature was conceived rather as an unconscious, nonmechanical, but real principle that held the world together and prevented formation of vacua within it. And like the bodies in that world, the Spirit of Nature also required a place. By 1671, however, when he published the *Enchiridion metaphysicum*, More joined numerous others who found unacceptable both Aristotle's claim that nothing whatever existed beyond our finite world and the Greek atomist belief in the existence of an infinity of worlds. Instead he adopted the popular Stoic interpretation of the universe and surrounded our unique, finite world with an infinite void space,[251] by which action he opposed not only Aristotle but also Descartes and the latter's assumption of a material world indefinite in extent.

More conceived this infinite void space as an absolute, immobile container into which and out of which bodies within the world moved. The existence of such a separate, three-dimensional spatial container was essential to account for the motion of a body from one place to another. Otherwise the place or space of a body, or its internal *locus*, would be the property of the body and move with it.[252] That a three-dimensional body could move through a three-dimensional space posed no more of a problem for More than it had earlier for Patrizi, Gassendi, and all who accepted the Philoponian distinction between material and immaterial extensions.[253] The mere possession of extension by space and matter did not make them equivalent, as Descartes would have it.[254] But if not equivalent, was space real and objective, or was it an imaginary mental concept, or "phantasm," as Thomas Hobbes had argued? Although Hobbes had differentiated between space and extended matter, he denied objective reality to space, which he described as a phantasm. In truth, Hobbes retained the scholastic distinction between real and imaginary space. Real space is the extension or magnitude of a body and has objective existence because it surely does not depend on our thought. Body, or magnitude, is, however, capable of producing effects on the mind. For if the world were annihilated, our memories of the magnitudes that formerly existed would remain and differ in no way from our perceptions prior to annihilation. Here was evidence that our mental concepts or phantasms of magnitudes are essentially devoid of external reality.[255] For Hobbes, the phantasm of former bodily magnitudes is imaginary space, which lacks real extension.[256] Although we can imagine the annihilation of the world with all its matter, we cannot imagine away space, because "space . . . is nothing but a phantasm, in the mind or the memory, of a body of such magnitude and such figure."[257]

More agreed with Hobbes that we cannot, as he put it, "disimagine" space.[258] But where Hobbes assigned real extension only to body and imaginary extension to space, More assigned extension and reality to both. The inner logic of his metaphysics, moreover, drove More to assign space to the realm of spirit. Because he regularly assumed that all existent being was divided into spirit or body, and because space was not body, as it was for Descartes, it had necessarily to be spirit. But what kind of spirit could an infinite, extended void space be? In his *Antidote Against Atheism* of 1662, More considered whether the idea of matter implies necessary existence. For is it not true that we cannot but imagine an extended space running infinitely in every direction? And if this space is extended and an entity – for it must be something if it is extended – will it not be corporeal matter? To reject this, which is nothing but Cartesianism, More suggested three possible replies as to what space might be other than corporeal matter.[259] Of these, the first and third conceive space as spirit, the first declaring that space is God's infinite immensity and the third that it is God Himself.[260] In a real sense, one may argue that More retained both of these opinions in the *Enchiridion metaphysicum*, some nine years later, because it is difficult to keep them apart. For even if one holds the weaker of these two relationships with the divine, namely that space is God's infinite immensity, that very claim implies that space is in some sense God, even if only as a manifestation or property. Space was indeed a property of the divine. As a three-dimensional void, space had to be the property of something. More was a firm believer in the subject-accident metaphysics of Aristotle and had assumed, as we saw, that every being had to be classified either as subject (i.e., substance) or accident (i.e., attribute). He thus broke with the tradition of Patrizi, Gassendi, and numerous scholastics who, as we saw, found it plausible, and even necessary, to exclude space (and time) from the categories of subject (substance) and accident.

For More "a real attribute of any subject can never be found anywhere but where some real subject supports it. But extension is a real attribute of a real subject . . . which is independent of our imagination."[261] Because extension can exist without matter, the subject that supports extension must be sought elsewhere than in matter. Besides matter there is only spirit. An infinitely extended attribute must therefore inhere in an infinite spirit. Because there is only one infinite spirit, infinitely extended space must inhere in God Himself. As evidence of this, More observed that the metaphysicians have attributed some twenty titles, or properties, to God that are identical to those "which fit the immobile extended [entity] or internal place (*locus*)."[262] Space and God are the same because each is "One, Simple, Immobile, Eternal, Complete, Independent, Existing in itself, Subsisting by itself, Incorruptible, Necessary, Immense, Uncreated, Uncircumscribed, Incomprehensible, Omnipresent, Incorporeal, All-penetrating, All-embracing, Being by its Essence, Actual Being, Pure Act."[263] Here we recognize many of the properties that had been attributed to God and space by both scholastics and nonscholastics for a number of centuries. In More's term-by-term analysis of these divine titles and attributes, we learn that as a spirit, space is homogeneous, without physical parts and therefore indivisible, "uncreated, because it is the first of all, for it is by itself (*a se*) and independent of anything else. And *Omnipresent* because it is immense or infinite. But *Incorporeal* because it penetrates matter, though it is a substance, that is, an in-itself subsisting being."[264] Although, in his enthusiasm, More made space both a substance and an attribute, he surely did not intend to identify or equate space with God, because the latter is much more than the twenty attributes assigned to Him. Those attributes do not "appear to concern the divine life and activity, but simply his bare essence and existence."[265] To arrive at a more complete picture of the divinity, those attributes must be linked with God's life and activity.

By his fusion of space and God in the *Enchiridion metaphysicum* of 1671, Henry More revolutionized the traditional concept of the deity. Although his divinization of space was hardly new, his attribution to God of incorporeal extension was. The consequence was an infinite, three-dimensional void space that was God's immensity, or His attribute, and the instrumentality of His infinite omnipresence. It was, as Bradwardine had insisted centuries before, a space that could be void of matter but in no way of God (see Chapter 6, beginning of Section 3). It was also a space in which God was no longer conceived as wholly in every part. Presumably God was now evenly distributed within His infinite space, however that might be interpreted. Did this signify that any division of space would somehow also divide God? Not at all, because More forestalled this possibility by the assumption of spatial indivisibility. For how could a simple space without parts be divisible? Moreover, those who adopted a three-dimensional space in the late sixteenth and seventeenth centuries assumed that immaterial extended space and material bodies could coexist because space always yielded to body, or, to put it another way, space always penetrated body. With such properties, it was easy for the likes of Patrizi, Gassendi, Campanella, and More to maintain the indivisibility of extended space, even though divisibility was always assumed without exception to be an essential property of corporeal extension, or bodies.

Despite the conscious and careful effort to make the dimensionality of God and spirit different from the dimensionality of body, there must have been something unsettling about More's conception of a three-dimensional God. No longer was God wholly in every part. Was He then less than wholly in every part? It certainly appeared so. But what interpretation could be attributed to such an omnipresence? What the traditional opinion lacked in intelligibility, it gained in the depiction of God in all His infinite majesty and power. By contrast, More's extended God appeared *prima facie* more intelligible but somehow seemed less impressive, having assumed the characteristics of an infinite body stripped of matter. Even if these issues could be ignored, acceptance of two types of extended beings, one divisible (corporeal, material extension), the other indivisible (incorporeal, spiritual extension) should have caused serious concern.

e. BEYOND HENRY MORE: BENEDICT SPINOZA AND THE CONFLATION OF GOD, SPACE, AND MATTER

If such problems troubled More's contemporaries and successors, their apprehensions and misgivings were not immediately apparent and would not become significant until the early eighteenth century. The idea that extended space is God's attribute or His substance won considerable acceptance, especially in England. That God must therefore be an extended dimensional being, which was a clear consequence of this doctrine, was proclaimed with less enthusiasm and often not at all. But the deification of space had captured the fancy of many during the second half of the seventeenth century. Before we turn our attention to More's impact in England and the opinions of Newton in particular, we must mention briefly the extraordinary, and basically anomalous, role of Benedict Spinoza (1632–1677). In his clash with Descartes, More attributed extension to matter and spirit and assigned space to the realm of extended spirit rather than extended matter, as Descartes had done. Here More drew upon the distinction between material and immaterial extension that had been formulated by Philoponus and had, as we have seen, played such a significant role in the development of spatial doctrine in the sixteenth and seventeenth centuries.[266] But if More and Descartes had radically different conceptions of spirit, they were yet agreed that the world was constituted of matter and spirit. By the attribution of extension to both spirit and matter, however, More greatly diminished the difference between them. Benedict Spinoza would eliminate it altogether when he made extension (space), matter, and God one.[267]

Despite the close relationship of his ideas and concepts with those of Descartes and More, Spinoza's extraordinary conclusions derive from his conviction that a material world could not originate from an immaterial God, nor from creation *ex nihilo*, nor indeed from matter that was coeternal with God. Spinoza resolved the dilemma by attributing materiality and extension to God and proclaiming Him as the sole existent substance.[268] As the single existent substance, God was infinite with an infinity of attributes, two of which, thought and extension, are known to us. Each of these infinite attributes was reducible to an infinity of "modes" all the way to particular finite modes, such as a single mind or a single extended body. They were all, however, merely aspects of God, the one and only existent substance.[269] From the introduction of the Greek concept of a separate, infinite, three-dimensional void space in the sixteenth century to Spinoza's *Ethics* in 1677, approximately 150 years, space had become indistinguishable from God Himself. Spinoza took the final step and conflated God, extension, matter, and space as one infinite, indivisible substance.[270] One could go no further and few, if any, would go as far. To identify and equate an infinite plenum with the deity was judged heretical and atheistic by most, for it destroyed the distinction between spirit and matter. Transmitted by Plato, Aristotle, Plotinus, and the Church Fathers, the dichotomy and hierarchy of spirit and matter constituted the foundation of Western philosophy and theology. This tradition Spinoza had repudiated. Those who thought space three-dimensional and believed it somehow had to be associated with God were well advised to avoid Spinozism and cling instead to some form of More's vision.[271] And this they did, especially in England.

Whether directly or indirectly, More seems to have influenced a number of his countrymen who gave thought to this issue in the seventeenth century. More's fellow Cambridge Platonist, Ralph Cudworth, reported an opinion, without disapproval, that appears to be More's. Gassendi's exclusion of space from the categories of substance and accident prompted Cudworth to declare that if space is "neither the *Extension* of *Body*, nor yet of *Substance Incorporeal*," it would be the extension or affection of nothing. Because this is impossible, and if "Space is a Nature distinct from *Body* and Positively Infinite," as the Democriteans and Epicurean atheists said, then there must be "some *Incorporeal Substance*, whose *Affection* its *Extension* is; and because there can be nothing *Infinite*, but only the *Deity*, that it is the *Infinite Extension* of an *Incorporeal Deity*; just as some *Learned Theists* and *Incorporealists* have asserted."[272] Whether More was one of the "learned Theists and Incorporealists" Cudworth had in mind is uncertain, but the description he gave fits More's interpretation. Because the argument is only one of many intended to demonstrate the existence of an incorporeal deity against the denials of the Democriteans and Epicureans, it is by no means clear whether Cudworth really believed that infinite space was "the *Infinite Extension* of an *Incorporeal Deity*."[273] But there can be no doubt that John Turner was convinced that God is an extended being when, in a sermon in 1683, he explained that God is an "Omnipresent *Extention*; an Extention not only of Attributes, but of Substance and of Nature."[274] Turner attacked the idea that God could operate where He is not present. As he conceived it, "to say that God can be present by any Virtue, or Power or Efficacy of the Divine Nature, where the real and local substance of the Divinity is not, is to affirm Transsubstantiation with the *Papists*, or to deny motion with *Zeno*."[275] Whether Turner reflected a certain space-mysticism that became popular in England as a result of More's ideas is unclear.[276] More's influence on some of the greater and lesser figures associated with the new science and philosophy, however, is highly probable and, in at least one instance, reasonably certain.

f. THE DIRECT INFLUENCE OF HENRY MORE: JOSEPH RAPHSON

That "reasonably certain" instance is Joseph Raphson, who in 1702 published an extraordinary Latin treatise on infinite space entitled *De spatio reali seu ente infinito* (*On Real Space or Infinite Being*).[277] It is abundantly evident that Raphson had read widely on the theme of infinite space. The first of his six chapters is an historical summary of ideas on infinite space from the ancients to Henry More, whom Raphson called "a most famous and praiseworthy man who reawakened this most ancient doctrine as if he had brought it to life from its ashes; and he has declared, augmented, and confirmed all these things in his *Metaphysics*."[278] Although More is presented as if he were the *terminus ad quem* of Raphson's history, Locke, Newton, and Leibniz are also mentioned, along with ideas from the various Greek philosophical schools (Platonic, Neoplatonic, Hermetic,[279] and Stoic), the Church Fathers, and the Cabbalists. Among individuals mention is made of Plato, Aristotle, Cleomedes, Simplicius, Iamblichus, Cyprianus, and Arnobius from the ancients and Ficino, Cornelius Agrippa, Patrizi, Descartes, Van Helmont, Gassendi, Hobbes, Spinoza, Otto von Guericke,[280] and Malebranche from the moderns. Even the scholastics are not wholly ignored, with the *De divinis perfectionibus* of the Jesuit Leonard Lessius cited, along with a number of general references to the "schools" (*scholae*) or to "scholastics" (*scholastici*).[281]

Like More, Raphson's cosmology was Stoic with a finite world, the limits of which could not, however, be determined by the human mind, surrounded by an infinite void.[282] In determining the status of that infinite void, Raphson allied himself with Henry More against Spinoza, whose assignation of matter as one of God's infinite attributes is rejected as false.[283] Extension and matter must be distinguished, with only the former attributable to God. In the fifth chapter, Raphson, the mathematician, derived the nature and existence of infinite space in the "geometric mode" (*more geometrico*), proceeding by definitions, axioms, postulates, demonstrations, propositions, and scholia.[284]

Space is defined as "the first extended thing in nature" (*extensum . . . intimum natura primum*), although it is the last thing to be arrived at by continuous division and separation. Its properties are those that had become widely accepted by the latter part of the seventeenth century and are, with a few noteworthy exceptions, traceable to Patrizi and Gassendi. Thus space is (1) absolutely indivisible, (2) absolutely immobile, (3) actually infinite, [285] (4) pure actuality, (5) all-containing and all-penetrating but, according to a corollary, impenetrable itself.[286] With this corollary, Raphson opposed those who would interpret the movement and location of bodies in indivisible space as a mutual and reciprocal interpenetration in which the body can be imagined to penetrate space because of the latter's lack of resistance and passivity while the space itself simultaneously penetrates the body.[287] Raphson would have none of this. For although the net result is the same, there is an enormous conceptual difference. To explain the coexistence of space and body by the penetration of indivisible space is to conceive that space as actually divisible. In a strict sense, the direction of penetration is not a matter of indifference but is determined of necessity by the need to preserve the property of indivisibility.

Most of the remaining properties agree with those adopted by Patrizi and Gassendi. Space is (6) incorporeal, (7) immutable, (8) one in itself, (9) eternal, for the infinite "cannot not be" (*non potest non esse*), (10) incomprehensible, (11) the most perfect of its kind, and (12) something without which extended things cannot live.

The thirteenth, and final, property is one that neither Patrizi nor Gassendi assumed but represents the foundation of Henry More's spatial theory: "space is an attribute (namely the immensity) of the First Cause."[288] Because space was defined as "the first extended thing in nature" and is here made the attribute of God, it follows that God must be an infinitely extended being. Raphson left no doubt about this. The existence of all things requires the direct presence of the First Cause.

But, how this essential and intimate presence can be explained in the hypothesis of the non-extension [of the First Cause] without a manifest contradiction has not yet been made clear; and it will never be possible to make it clear. Indeed, to be present by essence in places diverse and distant from each other, for instance in the globe of the Moon and in that of the earth, and also in the intermediate space, what else is it but, precisely, to extend oneself?[289]

Of course, God's extension is not the gross and crude extension of ordinary matter. And yet extension is itself a positive thing, even in matter. In God, however,

The infinite amplitude of extension expresses the immense diffusion of being in the First Cause, or its infinite and truly interminate essence. This [amplitude] is that originary *extensive* perfection, which we have found, so imperfectly counterfeited, in matter.[290]

The schoolmen were wrong to insist that extension is in God only transcendentally. For how can extension be derived in the world from an originative source that is itself unextended? This is no more plausible than the assumption that thinking beings could derive from an unthinking being.[291] For the remainder of the *De spatio reali*, Raphson presented an array of authorities in defense of the thesis that God is an infinitely extended being whose immensity is infinite void space.

g. WAS SPACE "GEOMETRIZED" IN THE SEVENTEENTH CENTURY?

Although we had earlier argued that no good reasons could be adduced for the assumption that Euclid himself presupposed the existence of a separate, three-dimensional, infinite void space in which to locate his geometrical figures (Chapter 2, Section 2a), Alexandre Koyré argued that acceptance of an infinite, three-dimensional void space in the seventeenth century was the consequence of a process he called "the geometrization of space,"[292] by which he presumably meant "the replacement of the Aristotelian conception of space – a differentiated set of innerworldly places – by that of Euclidean geometry – an essentially infinite and homogeneous extension – from now on considered as identical with the real space of the world."[293] Leaving aside the fact that Euclid's geometry probably presupposed no such space as described by Koyré, it is appropriate to inquire whether geometry played an instrumental role in the development and acceptance of the concept of a real, infinite, three-dimensional space. Of those who came to accept an extended, infinite void space, very few described it in terms that suggest geometry was a significant factor in their decision. Two who did were Francesco Patrizi, who published his views, and Isaac Newton, who did not.[294] We have already seen (Section 4a) that Patrizi assumed the real existence of lines, surfaces, and volumes in space and further assumed that each is infinite in extent. To do geometry, the mind must imagine, or "cut out," finite portions of these infinite figures. Newton's conception was strikingly similar, as will be seen below (Section 4l). All figures are actually in infinite space itself, an assumption that Newton apparently thought necessary to explain how figures of different shapes and magnitudes could actually move through or rest in space. Unlike Patrizi, however, who published his interpretation, Newton's conception appeared only in an early unpublished work (*De gravitatione*), probably completed between 1664 and 1668, while he was a student. (On the *De gravitatione*, see below, Section 4l and note 338.) Because Newton was a mathematician and Patrizi was not, the status of mathematician was not of itself crucial in the geometrization of space. Nor would all mathematicians geometrize space, as is borne out by Joseph Raphson and Isaac Barrow, the former of whom geometrized space whereas the latter did not. Raphson explained that he took abstract space as a genuine object, something that mathematicians had not done prior to his day.[295] All possible kinds of infinites, including extension, are subject to geometric calculations.[296] Some thirty-five years earlier, however, Raphson's fellow mathematician, Isaac Barrow, had argued for a space that was nondimensional, a space that had only a mere potentiality for receiving magnitudes. Although Barrow insisted that this space had to be compatible with the operational laws of geometry, the space he described is plainly nongeometrical.[297]

The mere endowment of infinite void with extension does not, and should not, imply any conscious geometrization of space. Mathematics was not the driving force behind the widespread acceptance of a dimensional space in the seventeenth century. The concept of an incorporeal extended space was Greek in origin. But Greeks such as John Philoponus, Cleomedes the

Stoic, and other atomists and Stoics, who described and formulated the ideas that would affect Western Europe in the sixteenth and seventeenth centuries, had developed their thoughts within a physical and cosmological, not mathematical, context. The subsequent descriptions of space and the controversies about its nature during the sixteenth and seventeenth centuries were largely of a physical, cosmological, and especially theological nature. From the fourteenth century on, space was gradually divinized and, despite the assumption of tridimensionality in the sixteenth and seventeenth centuries, the problem of God's relationship to that space remained central. In this movement, mathematics played a relatively small role.

h. "PARTS" IN AN INDIVISIBLE SPACE

But its role might have been larger. There were no inherent obstacles to the real mathematization, or geometrization, of space – not even an apparent paradox that underlies the whole history of spatial doctrine from the late sixteenth to the early eighteenth century. If space were geometrized, we might expect that in addition to tridimensionality, it would be necessary to assume *infinite divisibility*. And yet, as we have seen, Patrizi, Bruno, Campanella, Gassendi, More, Spinoza, Raphson, Newton, and others were as one in the assumption that *space is indivisible*.[298] But how could an indivisible space be reconciled with the needs of geometry for the infinite divisibility of *all* magnitudes? Moreover, as Raphson rightly saw, indivisibility implies impenetrability. Thus we have a space that is indivisible, impenetrable, and incorporeal. But how can bodies move through a space that is indivisible and impenetrable? Do we not also have here a resurrection of the old Aristotelian dilemma that plagued medieval physics: how to explain the hypothetical motion of bodies through an impenetrable void? (For the problem, see the beginning of Chapter 1 and all of Chapter 3 for attempts to cope with it.) The two situations differ radically.

Medieval scholastics interpreted intracosmic void space as body, from which its impenetrability followed. In the sixteenth and seventeenth centuries, a quite different conception of space was devised, one that was more akin to the extracosmic space of the medieval and early modern scholastics but having little in common with their intracosmic void. Obviously, if space is God's immensity, as it was for some, it could not be divisible. But even if not God's immensity, space was conceived as an immobile, homogeneous, absolutely continuous, infinite container that was completely unaffected by the bodies and spirits within it. When bodies in that space were imagined annihilated, as was often done, their spaces remained unaffected. If a body was imagined divided, the space it occupied was not similarly divided but remained as before. Space remained immobile as bodies moved from place to place. It could receive bodies and coexist with them by virtue of its incorporeality. In coexisting with bodies, however, space was not thought to be divided, because it remained completely unaffected by and independent of the bodies that occupied this or that part of it. The fact that bodies could occupy different parts of space also signifies that although indivisible, space was understood to have distinguishable parts.[299] The difference between divisible and indivisible magnitudes lies in the fact that the parts of the former are in principle separable and movable, whereas the parts of the latter are not. Space can be distinguished into parts that are contiguous, but those parts, like the whole, are immobile. If they were movable that magnitude would be a body, not a space.[300]

The coexistence of indivisible space with the bodies that occupy it involved some kind of instantaneous penetration. The manner in which this might be imagined to occur was susceptible to the following interpretations: (1) space penetrates body; (2) body penetrates space; and (3) space and body mutually, reciprocally, and simultaneously interpenetrate. Although all three interpretations are found, sometimes intermingled, the first is clearly the most consistent with the universally accepted property of indivisibility. If space is to retain that fundamental property, it would appear necessary to deny penetration, which implies division and disruption of continuity. Raphson's specific attribution of impenetrability to space was thus meant to guarantee space's inviolate indivisibility and, if properly understood, would probably have won the support of most seventeenth-century advocates of infinite space (see note 218 of this chapter). Ironically, in its relationship to space, it would be body, not space, that is penetrated. We can now also envision how that indivisible, infinite space could be reconciled with the infinite divisibility that was essential for geometry. Space, as we saw, has distinguishable but not separable or movable parts. A configuration of any size could thus be distinguished in space. Because that configuration – which is usually delineated by a body, however small or large – cannot be separated or moved from its location, the indivisibility and immutability of space are not affected. And yet any figure whatever can be distinguished in that indivisible space and even successively reduced or enlarged. By this means, paradoxically, the infinite divisibility required in geometry was reconciled, or made compatible, with the absolute indivisibility of infinite space.

Properties such as indivisibility, immobility, impenetrability with respect to body, homogeneity, and incorporeality served to differentiate and isolate space from body, which was composed of actually divisible, separable, and movable parts that could be transported from place to place or rest somewhere within the infinite expanse of space. Despite the often lavish descriptions of it, space had become a neutral entity that, to use Guericke's phrase, functioned primarily as the "universal container of all things."

i. SPACE AS PURE POTENTIALITY: THE INTERPRETATION OF ISAAC BARROW

The influence of Henry More on Joseph Raphson is unmistakable. Not only is More mentioned with praise and respect, but more significantly, his fundamental tenets form the foundation of Raphson's interpretation of the nature of space: (1) infinite, extended, void space is God's attribute and immensity, from which it follows that (2) God is also extended. When we turn to Isaac Barrow, John Locke, and Isaac Newton, on whose opinions we shall focus in the remainder of this chapter, the name of Henry More does not appear in their discussions of the nature of space. And yet his influence on all three has usually been assumed. Whereas Locke and especially Newton reveal the impact of More's ideas, the latter's alleged influence on Barrow is at best superficial but, in light of their nearly antithetical conceptions of space, probably nonexistent.[301]

What Barrow had to say about space is contained largely in the tenth lecture of his *Mathematical Lectures* (*Lectiones Mathematicae*).[302] Following a summary of arguments in defense of the Aristotelian and Cartesian thesis that space is equivalent to magnitude (body) and arguments in favor of a real space distinct from things, Barrow presented his own opinion. Space is not itself a magnitude but distinct from it, because space can exist where magnitude is not.[303] And yet Barrow allowed that space is not an actually existent thing, for it possesses no dimensions proper to itself.[304] But if space somehow exists as something distinct from magnitude but is yet not an actually existent something with dimensions of its own, what is it? How is this puzzle (*hic griphus*) to be resolved?

Barrow's resolution involved and required the characterization of space as "nothing other than pure potency, mere capacity, ponibility, or (please forgive the words), the interponibility of some magnitude."[305] To convey his meaning, Barrow employed a few illustrations,[306] the first of which involves the status of things prior to the creation of the world, when no body existed anywhere. And yet when God willed the creation of a body of some size, namely the world, it occupied a definite position, which could have occurred only if some capacity had already existed to receive it. With the world in existence, Barrow moved to his second example, which depends on the Stoic conception of the universe, namely a finite world surrounded by an infinite void. Because no body lies beyond our world, no actual dimension exists there. But if we acknowledge, as we must, the possibility that a body could be constituted beyond the world, a capacity to receive body must already exist there. That capacity is what Barrow meant by space. Hence an extracosmic space does indeed exist.

In the presumably later *Geometric Lectures* (*Lectiones Geometricae*), published in 1670, Barrow reiterated the existence of an infinite space prior to the world and affirmed the present existence of an infinite, extra-cosmic, void space, a space with which God Himself coexists, although the manner of that coexistence – whether space is God's attribute, or His creation, or whatever – is ignored.[307] Narrowing his illustrations still further, Barrow imagined that God has eliminated all the matter within an enclosure surrounded by four walls. Under these circumstances, a space will surely be left, because no one would deny that other bodies could be received there in the future.[308] As his final example, Barrow observed that if two magnitudes are contiguous, no space exists between them; but if they are separated by a distance, a space would exist between them simply because a body can now be received there.[309]

Despite its lack of dimension and positive attributes and its status as a mere potentiality capable only of the reception of body, space is no chimera or figment of the imagination.[310] It has the same kind of reality as "creability, sensibility, [and] mobility." By this Barrow meant, for example, that just as a body at rest has the capacity or potentiality for motion, so a space devoid of body has the capacity to receive body. With a nondimensional potentiality, or indefinite capacity, for the reception of magnitudes, space receives the dimensions of the magnitudes that occupy it; where it is unoccupied and in a state of potency, it possesses no dimensions of its own.

Anxious to protect the bastion of divine perfection, Barrow sought to allay fears that his conception of space would weaken or subvert that perfection. Space is no eternal, infinite, uncreated entity independent of God, and could not therefore imply the existence of anything besides substance or accident.[311] It is improper to hypostasize space as a separate existent outside the traditional Aristotelian categories of substance and accident, as Patrizi, Gassendi, and others, both scholastic and nonscholastic, had done. Even within those traditional categories, however, space is not a real substance or accident but simply *connotes the possibility and mode of a substance or accident*.[312] From this we may plausibly infer that space is not God's attribute or immensity, as with Henry More, and Barrow assigned to it no such glorious function. And yet God is not divorced from space. He is present wherever anything exists[313] and therefore coexists with space, a relationship not otherwise defined.

Barrow's conception of space can only be characterized as *sui generis*. As a nondimensional entity, it is akin to that of the scholastics and in no way properly relatable to the space conceived by Patrizi, Gassendi, or Henry More. It also varies greatly from the nondimensional space proposed by Otto von Guericke, who identified space with God's immensity and endowed it with lavish attributes, making it the most real of beings.

By contrast, Barrow's space is a rather shadowy thing, a mere potentiality barely possessed of existence. And though it was ontologically closer to the scholastic sense of space, Barrow ignored, perhaps deliberately, the intricate terminology that had been devised by scholastics over the centuries and that the reader has encountered in the preceding two chapters. In his concept of space, then, Barrow was not part of any larger movement or trend. He was uninfluenced by Henry More and would have no impact on Newton.

j. THE INDIRECT INFLUENCE OF HENRY MORE: JOHN LOCKE AND ISAAC NEWTON

If the influence of Henry More reached not to Isaac Barrow, it did affect two of the greatest English thinkers of the seventeenth century, the philosopher John Locke and the scientist Isaac Newton.

k. JOHN LOCKE

As one who consciously sought clear and distinct ideas, John Locke approached the subject of space with caution and restraint. In the evolution of his thought on space, he held at least three different interpretations: (1) that space is a mere capacity for the existence of body; (2) that space is a pure relation conceived as an abstraction from distances between real extremities; and (3) that space is an infinite, immovable absolute identified with the deity.

The first of these opinions is much the same as Barrow's and appeared in the entry to Locke's journal for Thursday, September 16, 1677, where he declared that "space in its self seems to be noe thing but a capacity or possibility for extended beings or bodys to be or exist. . . ."[314] Later in that same entry, Locke mentioned "imaginary spaces" that are "purely noething but meerly a possibility that body might there exist."[315] But then, in that very same sentence, he added that if a being does exist in this imaginary space, "it must be god whose being we thus make i.e. suppose extended." To Barrow's conception, Locke added elements of More's opinion, the third distinguished above, but the two are not really reconcilable. On the assumption that imaginary space is devoid of God, it is a pure "noething" that becomes an extended dimension when God, who is an extended being, is assumed therein. Two radically different opinions are represented here, and Locke would have had to choose between them before publication.

The second of his early opinions, in the journal entry for Monday, January 24, 1678, is compatible with the first and builds upon it. Space is now conceived as something abstracted from distances between real things. In its "confusd and generall sense," space still, as in the first opinion, "signifies noething but the existibility of body."[316] The notion of space becomes immediately more precise, however, when it is conceived as the distance between two real things. That the termini of the distance must be real is obvious to Locke, "for one cannot say nor conceive that there is any such thing as space or distance between something and noething or which is yet more absurd two noethings."[317]

Locke's final statement on the nature of space appeared in his *Essay Concerning Human Understanding* in 1690. Here he assumed the usual Stoic conception of the universe in which our finite world is surrounded by an infinite void.[318] By 1690 Locke's ideas about imaginary space had been transformed. No longer is imaginary space a pure nothing, a mere possibility for the existence of body. In the years between the journal entries described above and the *Essay*, Locke discovered the distinction between corporeal and incorporeal extension (see Chapter 2, Section 4) and rejected Descartes's conflation of space and body. Now he had only criticism for those who describe "imaginary space" beyond the universe "as if it were nothing because there is no Body existing in it,"[319] an attitude that Locke believed "ascribes a little too much to Matter,"[320] which is in no way necessary to the existence of space.[321] Space is now an immobile,[322] homogeneous entity that not only lacks resistance to motion,[323] but is also indivisible, for although it could be thought to have parts, those parts are not separable from one another.[324] And despite Locke's acknowledgment that we can have no positive idea of infinite duration or space, he assumed the infinity of space. To have a positive idea of quantity requires the possibility of addition between quantities, as in the addition of two days or two paces. But to have a positive idea of an infinite quantity, such as space, requires that we be able to add two infinites together and arrive at a greater infinite,[325] which is absurd. Our idea of infinite space is thus inevitably unclear,[326] something that few who proclaimed it would admit.

Despite the unclarity of our notion of infinite space, it is an idea no less clear than our idea of infinite duration. Those who think it obvious that God has existed from eternity but are doubtful that "infinite space is possessed by God's infinite Omnipresence"[327] are simply mistaken. The one is as "clear" as the other, and neither is truly intelligible. Thus did Locke advocate both God's eternity and His infinite omnipresence, using the former, which was accepted as virtually self-evident, to win support for the latter. Locke's firm belief in God's omnipresence in an infinite space was confirmed earlier in the *Essay* in the same manner, by linkage with God's eternity. For if everyone readily allows that God fills eternity, " 'tis hard to find a Reason, why any one should doubt, that he likewise fills Immensity: His infinite Being is certainly as boundless one way as another."[328] That infinite, extended void space might be God's immensity was, as we have seen, suggested by Locke as early as 1677 and left undecided. In the *Essay*, however, he seems to have opted in favor of space as God's immensity, as may be surmised from his declaration that "the boundless invariable Oceans of Duration and Expansion; which comprehend in them all finite Beings, and in their full Extent, belong only to the Deity."[329] If expansion belongs to God, it would follow that God Himself is an extended being. On this vital matter, Locke's position was the same as Henry More's.

Is it not plausible then to infer further that Locke also agreed with More that space is an attribute of God, and therefore an accident, or that it is the divine, extended substance, both of which claims were made by More? Fortunately, no inference is required because Locke provided us with a clear answer to a question that was bound to arise in the context of contemporary discussions about the nature of space. "If it be demanded (as usually it is) whether this *Space* void of *Body*, be *Substance* or *Accident*, I shall readily answer, I know not: nor shall be ashamed to own my Ignorance, till they that ask, shew me a clear distinct *Idea* of *Substance*."[330] That no clear idea of substance was available was made all too obvious by Locke, who observed that the term is applied in three radically different significations: "to the infinite incomprehensible GOD, to finite Spirit, and to Body."[331] We are no better off in our comprehension of accidents, which seem to be some type "of real Beings, that needed something to inhere in" and "were forced to find out the word *Substance*, to support them."[332] Of substance, then, Locke could only conclude that "we have no *Idea* of what it is, but only a confused obscure one of what it does."[333] The status of infinite void space is also left uncertain. Because Locke could not determine whether it was substance or accident, and yet acknowledged that infinite, extended space is something that belongs to God and is apparently God's immensity, space could be either substance or accident, as More believed; or it might be neither and perhaps lie wholly outside the Aristotelian categories in the manner advocated by Patrizi and Gassendi and those who followed their important tradition. Thus although Locke agreed with More that an infinite, extended, void space belonged in some sense to God and that God was therefore an extended being, he found no basis for identifying that space as either a substance or an accident, or as anything else for that matter.[334]

1. ISAAC NEWTON: FRUITION OF A LONG TRADITION

With Isaac Newton (1642–1727) the history of the spatial concepts that we have described thus far comes to full fruition. Prior to Newton, the doctrine of infinite void space played little role in science proper. As the chief architect of the Scientific Revolution, however, Newton would construct his new physics and cosmology within the frame of an infinite, absolute space. Such an achievement alone would entitle Newton to a major role in the history of spatial concepts. But even if his momentous contributions had been quite modest, or even minimal, his reflections on the nature of space would nonetheless demand our attention because they illustrate most of the major conflicts, confusions, and dilemmas that had developed since the sixteenth century.[335] Those reflections are embedded in both published and unpublished materials, the latter formidable in extent. Obvious though it may appear, it should be kept in mind that Newton's published ideas on space were in a form that he was presumably prepared to defend in public, whereas those in his unpublished manuscripts were, for whatever reason, not yet judged ready for public scrutiny. Certain significant aspects of Newton's thoughts on space appeared only in the unpublished manuscripts and never formed part of the public conception of his understanding. Although both sources are crucial for a proper appreciation of Newton's theory of space in all its aspects, much of what he thought about its true nature he apparently chose to leave unpublished. And what was published did not always bear his name, as in the Leibniz–Clarke correspondence, which nonetheless is legitimately assigned to Newton, who "participated closely in framing Clarke's replies to Leibniz."[336] In that famous exchange, as with his remarks on space in the General Scholium to the second edition of the *Principia* (1713), Newton's views on space were provoked by criticisms that were more than his prickly temperament could bear.[337] Unprovoked and left to himself, Newton might have said no more about space than was necessary for strictly scientific purposes – that is, little more than appears in the first edition of the *Principia*. His seeming reluctance to publish on space is perhaps best explained by the theological nature of the subject. During the seventeenth century, those who attempted to describe the properties of space usually found it essential to consider as well its ontological relationship to the deity. Newton may well, and wisely, have sought to avoid such a hazardous undertaking, but circumstances dictated otherwise when his sense of outrage caused him to enter the public arena and present his case.

Newton was concerned with the nature of space from his student days at Cambridge University. Sometime in that period, probably between 1664 and 1668, when he received the M.A. degree, Newton composed the *De gravitatione et aequipondio fluidorum* (*On the Gravity and Equilibrium of Fluids*), which remained unpublished.[338] In this treatise, he paid considerable attention to the subject of space. Although much of it would not appear in print, his ideas already revealed the powerful influence of Gassendi, whether through Charleton's *Physiologica-Gassendo-Charletoniana*, as is likely, or directly from the works of Gassendi.[339]

Like almost all who assumed the existence of infinite space in the seventeenth century, Newton accepted, and always retained, the Stoic conception of the universe with its finite world surrounded by an infinite space.[340] At the outset, Newton revealed Gassendi's influence when he denied that

space is either a substance or an accident, for "it has its own manner of
existence which fits neither substances nor accidents."[341] It is no substance
because it is not absolutely self-existent (*non absolute per se*) "but is as it
were an emanent effect of God, or a disposition of all being."[342] Moreover, space does not even act like a substance because it experiences no
motion and causes no sensations or perceptions. Nor is it an accident,
because space does not inhere in a subject. Indeed extension can exist
without a subject, for otherwise we could not "imagine spaces outside the
world or places empty of body." Newton also resorted to the popular
device of divine annihilation, when he concluded that space would not
disappear with any body that God might choose to destroy.[343] Indeed we
cannot imagine the nonexistence of space, although we might imagine
that nothing exists in that space.[344] One cannot but conclude that space
must exist and be independent of the things in it.

But if space is neither substance nor accident, it is surely not a nothing
(*nihil*), for the latter has no properties and cannot even be conceived.
Space, however, clearly has existence, and its ontological status seems to
place it somewhere between accident and substance, being more than an
accident but less than a full substance, though approaching "more nearly
to the nature of substance" (*et ad naturam substantiae magis accedit*).[345]
The most fundamental property of space is its tridimensionality – its
length, breadth, and depth – with which numerous of its other properties
are associated, namely immobility, indivisibility, lack of resistance, penetrability, homogeneity, eternality and immutability, and, of course, infinity.[346] Into this spatial frame, with its by-now familiar congeries of
properties, Newton, in a manner reminiscent of Francesco Patrizi but
even more explicit, set all possible geometrical figures. There are "everywhere all kinds of figures, everywhere spheres, cubes, triangles, straight
lines, everywhere circular, elliptical, parabolical and all other kinds of
figures, and those of all shapes and sizes, even though they are not dis-
closed to sight."[347] The invisible, immaterial existence in space of the
whole possible range of geometrical figures must be assumed because
material replications of any of them can pass through space and be contained therein. A material sphere, for example, can move through space
only because spherical spaces already exist there; a cubic body because
cubic spaces are already there to receive it; and so on. Newton's basis for
this unusual move seems to derive solely from our experience that all sorts
of different-shaped material objects are perceived to move through space
or to rest in it. Hence those spatial shapes must already be prefigured there.
The more usual interpretation, and presumably the one adopted by Gassendi, held that space was a uniform, homogeneous, incorporeal, extended
blank capable of receiving and coexisting with bodies of any shape. Of
course, Newton's assumption that all shapes are prefigured in space would

produce consequences that differed in no way from the assumption of space as a blank that contains no single shape but receives them all indifferently. If nothing else, Newton's choice of options revealed the mathematical orientation of his mind.

Newton was convinced of the infinity of space from the familiar Stoic argument that "we cannot imagine any limit anywhere without at the same time imagining that there is space beyond it."[348] Although we cannot imagine actual infinite extension, Newton believed that we can nevertheless understand it, because we can understand "that there exists a greater extension than any we can imagine."[349]

In this very early examination of the nature of space, Newton already revealed that intense interest in the relationship between space and God that would never wane. The spaces that are often thought to be nothing were for Newton "true spaces" (*imo vero sunt spatia*). For "although space may be empty of body, nevertheless it is not in itself a void; and *something* is there because spaces are there, although nothing more than that."[350] "Nothing more," that is, except God Himself, whose omnipresence Newton had already proclaimed a few paragraphs earlier, when he asserted that "the quantity of the existence of God was eternal, in relation to duration, and infinite in relation to the space in which he is present."[351] As the place in which God is omnipresent, space must be eternal. For if it were not coeternal with God but subsequent to Him, a time would have existed when God was nowhere, a thought repugnant to one who believed that "space is an effect arising from the first existence of being, because when any being is postulated, space is postulated."[352] The alternatives to an eternal, uncreated space are unacceptable, namely that God created our space but remained outside or that He created His own ubiquity.[353] But what about the problem raised by Descartes that an infinite space "would perhaps become God because of the perfection of infinity?"[354] In Descartes's concern there is an echo of the medieval uneasiness about actual infinites. (For a similar concern by Bradwardine, see Chapter 6, note 122.) Like Bradwardine, however, Newton denied that space is an actual infinite distinct from God. For "infinity is not perfection except when it is attributed to perfections."[355] But "infinity of extension is such a perfection with respect to being extended."[356] From these statements, it is perfectly proper to infer that Newton has here, inadvertently or not, made of infinite space an attribute of God. Because this infinity of extension does not exist by itself, it must be assigned to God, for to whom else can such an infinite perfection belong? But how could Newton conceive infinite extended space as an attribute of God in light of his earlier denial that space is either substance or accident? If space is God's attribute, does that not imply it is somehow an accident or a property of God? In responding to Descartes, Newton appears to have forgotten his earlier remarks and implicitly contradicted himself. However that may be, Newton never published these interesting, though somewhat confused, thoughts. When he did publish on the relationship of God and space, it was not to deny that space was either substance or accident but, on the contrary, to uphold space as a substance, if not an accident. Newton would never publicly identify himself with the position of Gassendi (and Patrizi), who distinguished space from the traditional categories of substance or accident. Newton's public utterances on the relationship of God and space would appear to link him clearly with Henry More. Does this signify, then, that for Newton, as for More, God is an extended and dimensional being? It would appear so. If Newton conceived infinite, extended, void space as God's attribute, it surely follows that God is an extended being. From his student days, as evidenced in *De gravitatione*, Newton adopted More's position. He would never abandon it. Indeed, as we shall argue below, he could not without also yielding up the only mechanism he had to account for God's infinite omnipresence. Thus what Newton ultimately derived from Gassendi in the *De gravitatione* was not the exclusion of space from the categories of substance and accident, to which he would never publicly commit himself, but the physical properties of infinite space, the infinite space he needed for the *Philosophiae Naturalis Principia Mathematica*.

The absolute infinite space that Newton proclaimed in the *Principia* is the space he required for the full realization of the first law of motion whereby "every body continues in its state of rest, or of uniform motion in a right line, unless it is compelled to change that state by forces impressed upon it."[357] In the scholium between the definitions and laws of motion, Newton explained that "I do not define time, space, place, and motion, as being well known to all."[358] Although space is left undefined, Newton did describe its properties, which are largely based on what he had already set down in *De gravitatione* and are ultimately traceable to Gassendi. "Absolute space, in its own nature, without relation to anything external," he declared, "remains always similar and immovable."[359] The order of the parts of space is immutable.[360] From such properties, the infinity of space may be inferred because "no other places are immovable but those that, from infinity to infinity, do all retain the same given position one to another."[361] Space also penetrates and permeates every body that occupies a part if it because "the place of the whole is the same as the sum of the places of the parts, and for that reason, it is internal, and in the whole body."[362] Going beyond the *De gravitatione*, however, Newton contrasted absolute, infinite space to relative space, which is "some movable dimension or measure of the absolute spaces; which our senses determine by its position to bodies; and which is commonly taken for immovable space . . . determined by its position in respect of the earth."[363]

And in a manner reminiscent of Patrizi, Newton further assumed the existence of an immobile point within absolute space that he identifies with the "common centre of gravity of the earth, the sun, and all the planets,"[364] that is, with the common center of the world. Where all other advocates of absolute infinite space were content to argue for it theologically, metaphysically, or by appeal to the impossibility of conceiving an end to space, Newton, the great experimentalist, sought to verify its existence experimentally by means of the rotation of a pail of water.[365] Unable to determine the issue kinematically, he believed it possible to resolve it dynamically. At best, the experiment could only confirm what Newton was already convinced of. Absolute infinite space was essential to his physics and cosmology and could not have been abandoned whatever the fate of the pail experiment.

Although there is a considerable similarity between Newton's spatial ideas in *De gravitatione* and the first edition of the *Principia*, there is yet much that separates them. In *De gravitatione*, Newton said as much, if not more, about the relationship of God to space than about the latter's purely physical properties, whereas in the *Principia* the name of God occurs but once,[366] and then without relevance to space. For his magnum opus, Newton had deliberately chosen to keep the *Book of Nature* distinct from the *Book of Scripture*.[367] Reluctance to reveal himself publicly on the interrelationship of God and space apparently persisted until 1706, although Newton's personal concern continued – as is evident, for example, when in the early 1690s he declared that space cannot be equated with God merely by virtue of its eternity and infinity. Instead God is omnipresent by reason "of the eternity and infinity of his space," which together render Him the most perfect being.[368]

In 1706, however, Newton went public when he added queries 17 to 23 to the Latin translation of his *Opticks* of 1704. By so doing, he set the stage for the famous Leibniz–Clarke correspondence, a controversy in which the relationship between God and space played a significant role and would involve Newton directly. In queries 17 to 23 and the Leibniz–Clarke correspondence, let me say it now, Newton appears to have held the position that God is literally an incorporeal, three-dimensional being actually possessed of length, breadth, and width and that His dimensionality is our absolute, three-dimensional, infinite space. Although he would never declare his position in this explicit manner, Newton was nonetheless a disciple of Henry More. A good sense of Newton's true sentiments is conveyed by David Gregory in a memorandum of December 21, 1705.[369] Plainly alluding to the new queries Newton was preparing to add to the 1706 Latin edition of the *Opticks*, Gregory declared that Newton had seriously thought of titling the last query "*What the space that is empty of bodies is filled with.*" Newton left no doubt as to what fills that space.

As Gregory related, "The plain truth is that he believes God to be omnipresent in the literal sense. And that as we are sensible of Objects where their images are brought within the brain, so God must be sensible of every thing being intimately present with every thing: for he supposes that as God is present in space where there is no body, he is present in space where a body is also present." Although Newton was apparently more forthright in private, we shall find more support for linking him to More than to any other contemporary opinion.

With the appearance of the Latin translation of the *Opticks* in 1706, Newton had finally committed himself publicly to God's omnipresence in infinite space. Contrasting the indirect mode of human perception by images to the direct manner in which God knows things, Newton, in query 20,[370] assumed that "there is a Being incorporeal, living, intelligent, omnipresent, who in infinite Space, *as it were* in his Sensory, sees the things themselves intimately, and thoroughly perceives them, and comprehends them wholly by their immediate presence to himself."[371] Not only does God perceive phenomena directly and immediately, whereas only images of things are carried into "our little sensoriums," but in query 23, Newton spoke of God's will acting in "his boundless uniform Sensorium," which, however, must not be conceived as God's organ because "He is an uniform Being, void of Organs, Members or Parts."[372] Indeed "God has no need of such Organs, he being everywhere present to the Things themselves."[373] From this we may conclude with David Gregory that "the plain truth is that he [Newton] believes God to be omnipresent in the literal sense." Infinite space may not be God's organ, but it is surely the place where He is dimensionally omnipresent, not figuratively but literally. And to make certain that he would not be misunderstood, Newton would declare it again even more emphatically in the General Scholium added to the second edition of the *Principia* in 1713. At first Newton appeared to deny that God's omnipresence is infinite space when he declared that God "is not duration or space, but he endures and is present." But in the next breath, he explained that God "endures forever and is everywhere present; and by existing always and everywhere, he constitutes duration and space."[374] In plain words, God is, among other things, also our space. A few lines below, all doubt was removed when Newton said of God:

He is omnipresent not *virtually* only, but also *substantially*; for virtue cannot subsist without substance. In him are all things contained and moved; yet neither affects the other: God suffers nothing from the motion of bodies; bodies find no resistance from the omnipresence of God. It is allowed by all that the Supreme God exists necessarily; and by the same necessity he exists *always* and *everywhere*.[375]

Although Newton went public with his views on God and space late in his career, with the queries to the 1706 Latin version of his *Opticks*, his private papers reveal a deeper and much earlier concern for the role of God in physics and cosmology.[376] By the time Newton wrote book II of the *Principia*, he had abandoned the notion that a material "aether" could provide causal explanations for physical phenomena and, as evidenced by unpublished remarks, had come to believe in the existence of "an infinite and omnipresent spirit in which matter is moved according to mathematical laws."[377] Here, then, was an immaterial aether, which Newton would substitute for the inadequate material aether. But "what could an immaterial aether be? To Newton, it was the infinite omnipotent God, who by His omnipotence is actively present throughout it."[378] God was conceived as an incorporeal aether "who could move bodies without offering resistance to them in turn," as the General Scholium would declare.[379] By the time of the Leibniz-Clarke correspondence in 1715–1716, Newton's physics and cosmology had come to depend on the efficacy of an infinite space that was literally God's three-dimensional omnipresence and that in scholastic terms would have been called God's immensity.

m. THE TUMULTUOUS CLIMAX:
NEWTON AND THE
LEIBNIZ–CLARKE CORRESPONDENCE

It is fitting and ironic that the climax of the tradition of a God-filled space should have seen Newton as its defender and champion locked in combat with that other giant of the age, Gottfried Leibniz (1646–1716). But even before Leibniz entered the fray against his old adversary, George Berkeley (1685–1753), in his *Treatise Concerning the Principles of Human Knowledge*, first issued in 1710, rejected Newton's belief in the objective, external existence of absolute and relative space.[380] Imagining the annihilation of everything in the world but his own body, a "pure space" would exist, declared Berkeley, only in the sense that "I conceive it possible for the limbs of my body to be moved on all sides without the least resistance; but if that too were annihilated then there could be no motion, and consequently no Space."[381] If without body no pure space could exist, it is "pernicious and absurd" to suppose

either that Real Space is God, or else that there is something beside God which is eternal, uncreated, infinite, indivisible, immutable. . . . It is certain that not a few divines, as well as philosophers of great note, have, from the difficulty they found in conceiving either limits or annihilation of space, concluded it must be *divine*. And some of late have set themselves particularly to shew that the incommunicable attributes of God agree to it.[382]

But it is clearly Leibniz who posed the greatest threat to Newton's conception of space and its relations to God. On the basis of queries 20 and 23 of the Latin translation of Newton's *Opticks*, in 1706, Leibniz, in a letter to Princess Caroline of Wales in 1715, effectively inaugurated the famous correspondence with Clarke by accusing Newton of the belief that "space is an organ, which God makes use of to perceive things by."[383] When Princess Caroline communicated parts of the letter to Samuel Clarke, a famous theologian and philosopher of the day and a trusted friend of Newton,[384] the first letter in the Leibniz-Clarke correspondence was history. By the time of Leibniz's death in November 1716, the total correspondence consisted of five letters from Leibniz to Clarke with five replies from Clarke. That Clarke's opinions accurately represented Newton's views and that portions of the correspondence may actually have been written by Newton has now been established.[385] When Clarke's replies to Leibniz are cited, it is a reasonable assumption that Newton's views are represented.

As the great champion of void space, Samuel Clarke, Newton's faithful spokesman, would include in his letters the salient properties of infinite space from both nonscholastic and scholastic sources. "Extra-mundane space (if the material world be finite in its dimension,) is not imaginary, but real,"[386] he declared. Those ancients who called it imaginary did not intend to signify by that term the unreality of space "but only that we are wholly ignorant what kinds of things are in that space."[387] Not only is it real, but infinite space "is one, absolutely and essentially indivisible."[388]

In his fourth letter, Leibniz had argued that "if space is a property or attribute, it must be the property of some substance. But what substance will that bounded empty space be an affection or property of, which the persons I am arguing with, suppose to be between two bodies?"[389] For if space is void, as Newton and Clarke believed, space would be an attribute without a subject, that is, "an extension without anything extended."[390] In his response, Clarke might have been paraphrasing Bradwardine from the 1618 edition of the *De causa Dei* (see Chapter 6, beginning of Section 3). "Void space," Clarke replied, "is not an attribute without a subject; because, by void space, we never mean space void of every thing, but void of body only. In all void space, God is certainly present. . . ."[391] But if God is present in void space, Clarke produced a syllogism to convince us that infinite space is not God ("Infinite space, is immensity: but immensity is not God: and therefore infinite space, is not God").[392] Here Clarke sought to repudiate the Spinozistic equation of space and God. Space is not equivalent to God. But Clarke would unhesitatingly declare space a property of God. For "space is not a substance but a property; and if it be a property of that which is necessary, it will consequently (as all other properties of that which is necessary must do,) exist more necessarily, (though it be not itself a substance,) than those substances themselves which are not necessary."[393] Nothing could be more obvious: Infinite space, which is immensity, is a property of God. Infinite space is therefore God's immensity, although, as we saw above, "immensity is not God," that is, not equivalent to God.

From these statements it is evident that Clarke, and therefore Newton, held that the substance of which space is an attribute is God Himself.[394] In effect, separate, absolute, infinite void space is God's immensity. But Leibniz would have none of this. He declared "space to be something merely relative, as time is; that I hold it to be an order of coexistences, as time is an order of successions. For space denotes, in terms of possibility, an order of things which exist at the same time, considered as existing together; without enquiring into their manner of existing. And when many things are seen together, one perceives that order of things among themselves."[395] If an absolute space exists independently of bodies, then every part of that homogeneous space will be identical with every other part. Under such circumstances, it would make no difference how God placed bodies in space. A particular fixed relationship between a group of bodies would obtain no matter where God placed them. Whether God placed them in an east-west orientation or vice versa would make no difference, which violates Leibniz's principle of sufficient reason. For "if space was an absolute being, there would something happen for which it would be impossible there should be a sufficient reason."[396]

To this, Clarke responded with an argument that might have come straight from Nicole Oresme's *Le Livre du ciel et du monde* in the fourteenth century.[397] For if space were as described by Leibniz, "nothing but the order of things coexisting; it would follow, that if God should remove in a straight line the whole material world entire, with any swiftness whatsoever; yet it would always continue in the same place."[398] The "absurdity" of Leibniz's position is twofold: (1) on the assumption that God moves the world with a rectilinear motion, he would have to deny its motion and (2) he would also have to hold that "nothing would receive any shock upon the most sudden stopping of that motion." Leibniz reiterated his position in the fourth letter.[399] The rectilinear movement of the world by God would affect no change whatever, for "two states indiscernible from each other, are the same state; and consequently, 'tis a change without any change." Moreover, such a motion would be pointless, and God does nothing in vain. But "two places, though exactly alike, are not the same place," Clarke patiently replied.[400] A man shut up in a cabin may not perceive the uniform motion of his ship, but the motion of that ship is nonetheless real and "upon a sudden stop, it would have other real effects; and so likewise would an indiscernible motion of the universe. To this argument, no answer has ever been given."[401] Leibniz replied by a reiteration of his previous position.[402] Because space is only the "order of the existence of things, observed as existing together," a single, finite, ma-

terial world moving forward rectilinearly in infinite empty space makes
no sense. A separate space does not exist and would have no purpose if it
did. "These are the imaginations of philosophers," Leibniz declared with
an obvious loss of patience and civility, "who have incomplete notions,
who make space an absolute reality. Mere mathematicians, who are only
taken up with the conceits of imagination, are apt to forge such notions;
but they are destroyed by superior reasons." By the time of his fifth reply,[403] Clarke must also have been weary of the issue. In an irrelevant
response, he pointed to Leibniz's mention of the motion of a finite, material universe, which was Clarke's own illustration, as evidence of Leibniz's concession that "there must necessarily be an empty extra-mundane
space." Where this particular argument might have gone in a sixth exchange does not bear thinking about.

What has just been described is all too typical of the Leibniz–Clarke
correspondence. It was less a genuine dialogue than two monologues in
tandem where, by some strange coincidence or prearranged harmony, the
letters of each correspondent contain identically numbered paragraphs
that frequently treat the same theme. One problem, however, in which
the same spirit prevails, is nonetheless of great importance for the insight
it provides not only into the particular views of Clarke, Newton, and Leibniz but also into one of the most basic notions about absolute space in the
seventeenth and early eighteenth centuries. The problem in question concerns the real nature of indivisible space and how it could be related to
God. We saw earlier (Section 4h) that in the seventeenth century an indivisible magnitude such as space was said to have parts that were distinguishable but not actually movable or separable from the whole. By contrast, a divisible magnitude, or body, had parts that were in principle
separable from the whole and movable. Let us now see how this distinction affected the Leibniz–Clarke correspondence.

In his third letter to Clarke,[404] Leibniz declared that those "gentlemen" who maintain "that space is a real absolute being" are involved in
"great difficulties" because such a space would have to be eternal and infinite. "Hence some have believed it to be God himself, or, one of his
attributes, his immensity. *But since space consists of parts, it is not a
thing which can belong to God.*"[405] If Leibniz was aware that Clarke,
Newton, and virtually all other proponents of absolute infinite space believed it indivisible, on what basis could he then accuse them of assuming
it divisible into parts and of assigning it to God as an attribute, knowing
full well that this would make God a divisible being? Is it plausible to
suggest, as did Koyré, that Leibniz had simply forgotten that those who
hold the absolutist conception of space "deny that space consists of parts
– *partes extra partes* – and assert, on the contrary, that it is indivisible?"[406]
I think not. Clarke, Newton, and Leibniz were all aware that, as assigned

to absolute space, indivisibility signified that space was distinguishable into parts that were, however, inseparable and immobile. If those parts were separable and mobile, we would have a body, not an indivisible space. When criticizing those gentlemen who believed that space is a real absolute being, Leibniz was not attacking the notion that if a real absolute space existed, its distinguishable parts would be immovable and inseparable, but was repudiating the broader notion that it could have distinguishable parts at all, whether separable or inseparable. For if parts are distinguishable in space, and space is assumed to be God's attribute, and therefore His immensity, God would have distinguishable parts even if those parts are immobile and inseparable. Therefore "space . . . is not a thing which can belong to God."

That Clarke and Newton distinguished parts in space is indubitable. When Leibniz insisted that "space being uniform, there can be neither any external nor internal reason, by which to distinguish its parts, and to make any choice between them,"[407] Clarke would declare that "two places, though exactly alike, are not the same place,"[408] thus simply reinforcing his earlier assertion that "different spaces are really different or distinct one from another, though they be perfectly alike."[409] To demonstrate that Clarke was committed to a divisible God when he attributed space to God, Leibniz had only to assert that Newton, Clarke, and all believers in an absolute, indivisible space accepted distinguishable parts in that space, which of course they did. In his reply, however, Clarke interpreted "parts" to mean movable and separable, rather than distinguishable. To suppose that indivisible, infinite space can be parted is, Clarke argued, "a contradiction in terms" because any partition of infinite space would yield another space. Consequently, space would be simultaneously parted and not parted.[410] And as if to remove any doubt that he intended movable parts, Clarke cited the fourth paragraph of his second reply, where, in defense of the indivisibility of space, he had declared that "to imagine its parts moved from each other, is to imagine them moved out of themselves."[411] Whatever the reason, Clarke misunderstood Leibniz's charge and responded to the wrong sense of "parts." Leibniz was not concerned in this critique with movable parts, but only with immobile, inseparable, distinguishable parts and what Clarke would call in his fourth reply, "essentially indiscernible" parts.[412] Leibniz did not accuse Clarke of the assumption that space was composed of movable and separable parts, although Clarke always responded as if he had. Leibniz had struck at the very foundation of the concept of spatial indivisibility as it was understood by advocates of absolute, infinite void space who also associated it with God. The traditional interpretation of God as an absolutely indivisible entity clashed with the conception of an absolute, infinite, extended, void space that was assumed indivisible and yet possessed of distinguishable, albeit inseparable and immobile, parts. Because this was indeed the opinion of More, Raphson, Clarke, and Newton, there was no satisfactory response to Leibniz's criticism and Clarke formulated none.

Although Newton never publicly admitted that God could possess distinguishable parts, and might have considered such a bald statement distasteful and even repugnant, his spatial doctrine and, as we shall see, his theology led him inexorably to it.[413] He was committed to a three-dimensional infinite void space that was the attribute or immensity of the deity. Because that space had distinguishable parts, God Himself must also be an extended, indivisible being with distinguishable parts. Of course, God was not to be conceived as an ordinary (corporeal) quantity because He has no separable, divisible parts. Newton may have sought to convey this in his brief *Avertissment au lecteur*, printed as part of Des Maizeaux's French preface to the 1720 edition of the Leibniz–Clarke correspondence. Although Newton's remarks are presented in Des Maizeaux's preface as if composed by Samuel Clarke, five drafts of it in Newton's own hand testify to his authorship.[414] Newton's intent was to clarify, in Clarke's name, the use of the terms "quality" and "property" in the Leibniz–Clarke correspondence. In one of the drafts, which differs little from the published preface, he declared that

wherever in the following papers, through unavoidable narrowness of language, infinite space or immensity & endless duration or eternity are spoken of as *Qualities* or *Properties* of the substance which is immense or eternal; the terms *Quality* & *Property* are not to be taken in that sense wherein they are vulgarly, by the writers of Logick & Metaphysicks applied to finite & created beings; for those writers consider space & duration as quantities & not as qualities.[415]

Here Newton sought to make clear that his conception of infinite space as a property or quality of God, that is, as God's immensity, was not to be construed as a corporeal quantity divisible into parts. Clarke, as we have seen, assumed in his correspondence with Leibniz that the latter had done just that. Because he was intimately involved in the correspondence and Clarke's views were his as well, Newton thought it wise to correct any misunderstandings about Clarke's use of the crucial terms "property" and "quality" as these had been applied in the relationship between infinite space and God's immensity. He was convinced, as was Clarke, that Leibniz had maliciously distorted Clarke's interpretation of part as applied to space, taking it to mean corporeal part, and therefore separable and movable, rather than merely distinguishable.

From both public and private statements cited above, Newton left no doubt that God is omnipresent in infinite space.[416] The same evidence indicates that, like Henry More, Newton believed God's ubiquity was achieved by His literal presence in every part of an infinite three-dimensional space,[417] from which it would seem to follow that God Himself is a three-dimensional being. Indeed that space must itself be an attribute or property of God, that is, God's infinite immensity. Submerged in Newton's often unclear, ambiguous, incomplete, and frequently revised statements and shifts of nuance is a fundamental assumption that God is a three-dimensional, extended, immaterial being. This basic belief, which Newton never made explicit, follows from his conception of infinite, extended void space as God's property, or immensity. The plausibility of this interpretation derives from the implausibilty and unfeasibility of three other possible positions that might be attributed to Newton: (1) that he was a Nullibist; (2) that he was a Holenmerist, as Henry More described those who assumed that God is wholly in every part; or (3) that he was an "operationalist," who believed that God can act a distance by His will rather than His actual presence.

Because Newton firmly believed that "the Maker and Lord of all things cannot be *never* and *nowhere*," he was plainly no Nullibist.[418] On the contrary, Newton believed that God is ubiquitous in an infinite extended void. But how can God be omnipresent and act everywhere in such a space? The second and third explanations attempt to cope with this formidable problem. In the second interpretation, God is said to be wholly in every part, which means He is omnipresent throughout the world and infinite space. This doctrine, which Henry More called "Holenmerism," is, of course, the traditional scholastic device for the explanation of God's infinite omnipresence.[419] Any attempt to foist this interpretation on Newton must be rejected. Nowhere, to my knowledge, did he mention or suggest such a mechanism for God's omnipresence. Moreover, because Newton was probably familiar with Henry More's *Enchiridion metaphysicum* and the powerful arguments against the whole in every part doctrine contained therein, it is not likely that he would have adopted it. To have accepted that traditional scholastic opinion would have required explanations to More's objections: How can a whole be divided only into wholes? How can one and the same thing be simultaneously many thousands of times greater or less than itself? And finally, if spirit, or God, is reduced to a single point, how could God be simultaneously in every other point of space? If the whole in every part doctrine of God's omnipresence ever occurred to Newton, it was surely found wanting and rejected.[420]

Nor did Newton accept the third option, that God could operate where He was not. God's actual presence is necessary for the efficacy of His actions. For as Newton explained in the General Scholium, God "is omnipresent not *virtually* only, but also *substantially*" (for the full citation, see above, Section 4l). He therefore does not act at a distance, a conviction

that Newton would reiterate in one of the five drafts to the *Avertissement
au lecteur* of 1720, where he would explain that God's ubiquity

renders him capable of acting & producing finite creatures wherever & whenever
he pleases & governing the whole Universe & that all this is consequent to his
being really necessarily & substantially Omnipresent & Eternal.[421]

In scholastic terms, Newton aligned himself with the numerous followers
of Thomas Aquinas, including Francisco Suarez, in opposition to Duns
Scotus and his disciples, who, as we saw, argued that God's presence was
in no way required to affect anything. His omnipotent will was sufficient.
For Duns Scotus – and for Leibniz – God could indeed act at a distance.[422]

If none of the three approaches just described was employed by Newton to explain the manner in which God could be omnipresent in infinite
space, only one plausible alternative remains: God is omnipresent because
He is actually a three-dimensional, extended being. Much of what Newton says about God and space is compatible with this interpretation but
with no other. Newton had even made God's literal omnipresence the
foundation of his physics, the basis for the maintenance of its mathematical
laws and therefore of lawful cosmic operation. To guarantee this state of
affairs, it was essential that God be literally everywhere, from which literal omnipresence it follows that God is an infinite extension with length,
breadth, and width. It was not, of course, something Newton could declare unequivocally. He was no Henry More and was all too aware of the
controversial nature of explanations that related God and space. Ultimately, however, he accepted More's interpretation because it was more
compatible with an infinite, extended void space than was its potential
rivals. Only by the assumption of God's literal omnipresence in an infinite space did Newton feel he could account for numerous phenomena that could not otherwise be explained by mechanical means. But of
all the plausible ways in which that omnipresence could be achieved,
I have sought to show that the assumption of an actually extended, three-
dimensional God was the only one acceptable to Newton, if for no other
reason than that it was the least objectionable. It was preferable to brazen,
or muddle, one's way through the formidable problems associated with
God's "parts," in a space with distinguishable but immovable parts, than
to cope with the even more calamitous problems associated with all the
other alternatives.

n. AFTERMATH

Newton's absolute, infinite, three-dimensional, homogeneous, indivisible,
immutable, void space, which offered no resistance to the bodies that
moved and rested in it, became the accepted space of Newtonian physics

and cosmology for some two centuries. It was, of course, the physical space inherited from the Greeks but given new significance in the early modern period by Patrizi and Gassendi. Despite criticisms of Newton's absolute space by those who emphasized the relativity of space or its psychological aspects, or even challenged its objective existence,[423] its triumph was complete by the end of the eighteenth century.[424] It was a triumph, however, only for the physical side of Newton's space. The God who filled it and whose property or attribute it was had vanished. The space that remained was more like Patrizi's than Newton's – a space that was neither substance nor accident nor God's immensity or property. True, Patrizi had made it God's creation, but that was of little consequence because God's presence and actions in that space were no longer required for the preservation of the universe and its mathematical laws of operation. Space was now a pure, infinite, three-dimensional container for all things and activities. Its divinity was gone. God's alleged relationship to space seemed beyond plausible and convincing explication and posed insuperable problems. In the end, it seemed best to let the spiritual and physical go their separate ways. Although Newton's God-filled space continued to arouse controversy among English divines and philosophers in the eighteenth century,[425] scientists gradually lost interest in the theological implications of a space that already possessed properties derived from the deity.[426] The properties remained with the space.[427] Only God departed.

PART III

Summary and reflections

9

Summary and reflections

Aristotle's denial of the existence of a separate void space must form the point of departure of any consideration of the history of spatial concepts from the Middle Ages to the Scientific Revolution. As an entity possessed of dimensions, such a space would have been equivalent to body and therefore unable to receive material bodies. But even if void space existed, finite and successive motion in it would be impossible, as Aristotle's numerous arguments had demonstrated. Although medieval and early modern scholastics would follow Aristotle and reject the actual existence of intracosmic void, they were nevertheless prepared to consider the behavior of bodies in hypothetical extended vacua. From such conjectures and speculations, many in the Middle Ages would conclude that if extended vacua could exist, motion in them would indeed be finite and successive, a conclusion derived from the widely accepted argument known as the *distantia terminorum*. On the basis of the Aristotelian assumption that extended void is like a body and therefore divisible into parts, the *distantia terminorum* argument provided a rationale and justification for finite and successive motion by demonstrating that successive parts of that vacuum could be traversed only sequentially and therefore of necessity in a finite period, however small. Widely accepted as kinematically sound, the *distantia terminorum* argument also demanded a causal explanation, which proved difficult to provide.

During the Middle Ages, it was assumed that successive and finite motions resulted only from resistance to bodies in motion, a resistance that could arise either from (or within) the mobile itself (*ex parte mobilis*) or from the resistance of the external medium through which the body moved (*ex parte medii*). In a vacuum, however, the problem of determining what might function as an external resistance (*ex parte medii*) posed insurmountable problems. Attention was consequently focused on the mobile itself, where two mechanisms were invoked: a self-expending impetus received within a body where the latter functioned as an overall resistance to the impetus; or an internal resistance that was derived from the medieval concept of a mixed body. Although internal resistances and mixed bodies would be swept away by Galileo, he made motion in a hypothetical void part of his new physics, employing certain arguments with a decidedly medieval flavor.

Despite widespread acceptance of the plausibility and intelligibility of finite, successive motion in hypothetical vacua, it was an almost universal conviction in the Middle Ages that nature abhorred a vacuum. Various arguments and experiments were proposed to confirm this truth and uphold the Aristotelian cosmic plenum. But if Aristotle's rejection of intracosmic vacua was almost universally sustained in medieval and early modern scholasticism, his denial of place and void beyond the world was widely repudiated. For a variety of significant reasons, the idea that an infinite void space existed beyond our spherical, finite cosmos took hold in the fourteenth century and, by the sixteenth and seventeenth centuries, won widespread acceptance. In scholastic thought, the nature of that extracosmic void space was much debated. By contrast to real space, the space associated with real bodies, infinite extracosmic void was called "imaginary" space, the space imagined in the absence of body. Of the greatest significance was the assumption that, although devoid of matter, imaginary space was filled with God and even identified as God's immensity, the means by which He achieved omnipresence. The ontological status of imaginary space was controversial and much discussed. Often characterized as a "pure negation" – to distinguish it from substance, accident, or privation – imaginary space was anything but a mental fiction or chimera. As God's immensity, it could hardly be a mere nothing or fiction. In fact, it was conceived as an existent entity, a something that could receive bodies without resistance. Because it was God's immensity, however, scholastics did not – and could not – attribute real dimensionality to imaginary space. Such an attribution would have converted God to an extended being and made of Him an incorporeal, albeit indivisible, body. And yet they came close to that fateful step, as can be seen by the descriptive terminology applied to imaginary space. The Coimbra Jesuits would liken it to the real and positive dimensions of body; Bartholomeus Amicus would conceive it in the mode of positive extension; and Emanuel Maignan would call it "virtual extension." As God's immensity, however, imaginary space could not be assigned real extension. The ambivalence was perhaps best expressed by Thomas Bradwardine in the fourteenth century when he declared that God "is infinitely extended without extension." The explanation of God's omnipresence in this infinite, though nondimensional, void space was by the whole in every part doctrine, whereby God was assumed to be wholly in every part or point of infinite space. By this device, God, as well as space, remained unextended and indivisible.

Under the influence of new Greek texts that became available in the late fifteenth and sixteenth centuries, anti-Aristotelians and nonscholastics would attribute real extension to infinite void space and thereby alter God's relationship to that space. With the adoption of the crucial distinction between material and immaterial dimensions, probably derived from John Philoponus, nonscholastics, especially Francesco Patrizi and Pierre Gassendi, believed they had overcome Aristotle's objection to the coexistence of body and extended void space. Like their scholastic counterparts, they adopted the Stoic universe, where a finite cosmos was immersed in and surrounded by an infinite extended void. Because it was thought capable of existence even if all the bodies in it were destroyed, some nonscholastics, notably Patrizi and Gassendi, excluded infinite void space from the traditional categories of substance and accident. In this they adopted a position similar to their scholastic contemporaries, who conceived space as a negation that was not classifiable within the traditional Aristotelian categories. For both Patrizi and Gassendi, space was conceived as an extended entity wholly independent of God, the former assuming it created, the latter uncreated. Neither of them, however, identified void space as God's immensity or attribute. To have done so would have transformed God into a three-dimensional being. That dramatic move was reserved for Henry More, who would eventually be followed by Newton himself.

Advocates of an infinite, extended void space were in general agreement on its physical properties. Space was homogeneous, immutable, continuous, indivisible, and capable of receiving bodies without offering any resistance or being affected in any way. Although most thought it also penetrable, some would characterize it as impenetrable. Both alternatives, however, appear reducible to much the same thing. Significant disagreement, however, centered on God's relationship to that space.

Patrizi and Gassendi represented one major interpretation of infinite void space, which they characterized as neither substance nor accident and therefore in no way God's attribute or immensity. Henry More charted a new course. Not only was space an attribute of God (though at times, More would call it a substance), but God Himself was an infinitely extended being. Indeed He was omnipresent by means of His three-dimensional space. Infinitely extended space is a property that inheres in God, the divine substance. More thus denied to space an independent status outside the categories of substance and accident. Without explicit mention of Henry More's spatial doctrine, Isaac Newton may be said to have chosen More's path rather than follow Patrizi and Gassendi. For Newton assumed the literal three-dimensional omnipresence of God in an infinite space that was God's property and, in effect, His immensity. Unable to explain all physical phenomena by mechanical means, Newton made God's actual spiritual omnipresence the foundation of physics and the guarantor of the efficacy of its mathematical laws. Leibniz objected to all these ideas in his correspondence with Samuel Clarke. Not only did he deny the existence of an external, absolute, infinite void space but he protested its identification with God's immensity. Because of the numerous and ultimately insoluble difficulties associated with an intimate relationship between God and space, Leibniz vigorously opposed it. In the end, he prevailed. Although among physicists his own relational theory of space was no serious competitor against Newton's infinite, extended void, the nexus between God and space, which so alarmed Leibniz in his correspondence with Clarke, would be broken by the end of the eighteenth century. Only the physical space would remain to furnish the container and frame for a universe operating under purely mechanical, Newtonian laws. The God on which Newton depended was elsewhere or nowhere.

In the historical developments described here, it was medieval and early modern scholastics who, for better or worse, introduced God into space in a manner more substantial than the vague, if dazzling, metaphors found in the earlier patristic, cabbalistic, and Hermetic traditions. Jean de Ripa, Thomas Bradwardine, and Nicole Oresme would proclaim an infinite, extracosmic void space with an omnipresent deity in the fourteenth century. Indeed Oresme called that infinite and indivisible space God's immensity. From then on, until the eighteenth century, scholastics and nonscholastics would attempt to explain and describe how God and space might be related. Scholastics linked a nondimensional space with an unextended God; others, such as Gassendi, would assume a three-dimensional space and an unextended God; and in the final stage, More and Newton would join an extended space to an extended God. Certain important nonscholastics, such as Gassendi and Guericke, revealed deep scholastic influences; others seem to have been indirectly affected, revealing an influence only in their use of scholastic terms and expressions that had become part of the language of philosophical and theological discourse. Scholastic ideas about space and God form an integral part of the history of spatial conceptions between the late sixteenth and eighteenth centuries, the period of the Scientific Revolution. From the assumption that infinite space is God's immensity, scholastics derived most of the same properties as did nonsholastics, and did so before the latter. As God's immensity, space had to be homogeneous, immutable, infinite, and capable of coexistence with bodies, which it received without offering resistance. Except for extension, the divinization of space in scholastic thought produced virtually all the properties that would be conferred on space during the course of the Scientific Revolution.

Scholastic contributions to the development of spatial doctrines in the period of the Scientific Revolution should occasion no great surprise. The unavoidable association of space with God, and therefore theology, made it natural for scholastics to explore the subject in depth. Aristotle's denial of void, place, and time beyond the world set the stage for an examination of God's relation to that region. In their investigation of the problem, scholastics formulated the initial association between God and extracosmic space. Without the fundamental assumption that God existed beyond the finite, physical world He created, the doctrine of infinite space as God's attribute or immensity would have been unfeasible in Western Europe. From the fourteenth century on, numerous medieval scholastics would associate extracosmic space with God's assumed extracosmic existence. Without the assumption of God beyond the world, the Stoic arguments about the necessity for extracosmic spatial existence would have proved of no avail against Aristotle. The existence of a separate extracosmic space without God or devoid of any relationship to God would have been unacceptable. Not even the introduction of Greek ideas about a separate three-dimensional void in the sixteenth century could have significantly altered the basic problem. A three-dimensional space might change the manner in which God's relationship to that space was perceived, but the need to explain that relationship was still the central problem for nonscholastics as well as scholastics. Not until the eighteenth century did space achieve independence from the divine omnipresence so that it might serve only the needs of physical science. Until then, nonscholastics grappled with a scholastic problem and often employed the same terminology and concepts. For all these reasons, the history of spatial doctrine must include the faceless scholastics as well as the major and minor figures who directly shaped the Scientific Revolution. The exclusion of scholastics from previous histories of space has limited our perspective and prevented genuine comprehension of the developments that eventually produced the fundamental frame of the Newtonian universe.

Space was perhaps the only scientific subject in which scholastics could hold their own. Mathematics played no role, and experiments on vacua were ambiguous and inconclusive. The tools of analysis in determining the existence and nature of space were metaphysics, theology, and thought-experiments, such as the imagined annihilation of all or part of the world. In these areas, scholastics had acquired great sophistication and a highly developed, subtle terminology. As nonscholastics groped along on the same impossible path, the experiences of their scholastic predecessors and contemporaries in elucidating the relations between God and space could hardly have been entirely unwelcome. In any event, it was scholastics who established the idea that God must bear an intimate relationship to space, an idea that remained viable well into the eighteenth century and played a central role in the scientific and theological thought of Isaac Newton himself.

Because God's eventual elimination from infinite space failed to affect the properties of space, does it not seem plausible to conclude that the intrusion of God into space was a mistake from the outset? Did it not delay for centuries the acceptance of a purely objective, scientific, physical space? Had the God-free, three-dimensional, infinite space of the Stoics and atomists been accepted immediately, a purely physical space would have been available at least two centuries earlier. In effect, we have already answered these hypothetical questions. No separate, external, infinite space could have been adopted in medieval and early modern Europe without an explanation of its relationship to God, on whom it had to depend in some way. For centuries, that relationship was explored, even as the physical properties of space received early and general acceptance. Before God could be removed from space, a general realization had to develop that the various mechanisms devised over the centuries to explain His omnipresence in infinite space were not only unsatisfactory but ultimately unintelligible. Only then could God be removed from space without a theological reaction and charges of atheism. After centuries of debate and controversy, this stage was finally reached in the eighteenth century, helped along, to be sure, by Berkeley and Leibniz. In a curious sense, one might argue that John Duns Scotus and Gottfried Leibniz triumphed over Thomas Aquinas and Isaac Newton. It was better to conceive God as a being capable of operating wherever He wished by His will alone rather than by His literal and actual presence. Better that God be in some sense transcendent rather than omnipresent, and therefore better that He be removed from space altogether. With God's departure, physical scientists finally had an infinite, three-dimensional, void frame within which they could study the motion of bodies without the need to do theology as well.

Notes

Scope of study

1 According to Aristotle, "Leucippus and his associate Democritus say that the full and the empty are the elements, calling the one being and the other non-being – the full and solid being being, the empty non-being (whence they say being no more is than non-being, because the solid no more is than the empty); and they make these the material causes of things." *Metaphysics* 1.4.985b.4–8. Trans. by W. D. Ross in *The Works of Aristotle, translated into English* under the editorship of J. A. Smith and W. D. Ross (12 vols., Oxford: Clarendon Press, 1908–1952), vol. 8. For a brief discussion, see Harry Austryn Wolfson, *The Philosophy of the Kalam* (Cambridge, Mass.: Harvard University Press, 1976), pp. 360–361, where Plutarch is cited as attributing to Democritus the statement that "something (τὸ δέν) is in no respect more existent than nothing (το μηδέν)."

2 Otto von Guericke, *Experimenta nova [ut vocantur] Magdeburgica de vacuo spatio* (Amsterdam, 1672, reprinted in facsimile, Aalen: Otto Zeller, 1962), p. 63, col. 2. For the complete passage, see below, Ch. 8, Sec. 4c, and n. 209.

3 See David Furley, "Lucretius," *Dictionary of Scientific Biography*, vol. 8 (New York: Scribner, 1973), p. 538, and below, Ch. 4, n. 115.

CHAPTER 1
Aristotle on void space

1 *Physics* 4.4.212a.5–6. Unless otherwise specified, quotations from Aristotle's *Physics* are from the translation by R. P. Hardie and R. K. Gaye in *The Works of Aristotle*, translated under the editorship of J. A. Smith and W. D. Ross, vol. 2 (1930). The bracketed addition is mine. Since Aristotle's own doctrine of place and its fate in the Middle Ages will not be considered here, see E. Grant, "The Medieval Doctrine of Place: Some Fundamental Problems and Solutions," in *Filosofia e scienze nella tarda scolastica: Studi in memoria di Anneliese Maier*, ed. Alfonso Maierù and Agostino Paravicini Bagliani (Rome: Edizioni di Storia e Letteratura; forthcoming, 1981).

2 *Physics* 4.4.211b.6–9 and 4.4.212a.3–5.

3 *Physics* 4.1.208b.25; see also 4.8.214b.18–19 and 4.7.213b.32.

4 *Physics*, 4.1.208b.34–209a.1.

5 *Physics* 4.1.209a.5–7; see also *Metaphysics* 13.2.1076b.1. Here and elsewhere in this volume where the problem of the impenetrability of bodies arises, I draw, often verbatim, on my article "The Principle of the Inpenetrability of Bodies in the History of Concepts of Separate Space from the Middle Ages to the Seventeenth Century," *Isis* 69 (1978), 551–571. For Aristotle's arguments, see 551–552.

6 For the entire argument, see *Physics* 4.8.216a.26–216b. 2.

7 See Grant, "Principle of the Impenetrability of Bodies," 552.

8 See *Physics* 4.8.216b.3–11.

9 *Physics* 4.8.214b.29–216a.20.

10 *Physics* 4.8.214b.30–33.

11 *Physics* 4.8.215a.5–12.

12 *Physics* 4.8.215a.13–18.

13 *Physics* 4.8.215a.19–21.

14 This interesting inertial consequence was rarely discussed in the Middle Ages. Reasons for its neglect are suggested in my article "Motion in the Void and the Principle of Inertia in the Middle Ages," *Isis* 55 (1964), 265–292. Although it is likely that Aristotle had an infinite void in mind (see my article, 265–266, n. 2), his arguments would also apply to an extended finite void within the cosmos. Moreover, it is likely that Aristotle conceived this motion as uniform and rectilinear. For why should a body moving through a void alter its speed or direction?

15 *Physics* 4.8.215a.22–24.

16 The first alternative appears to be the interpretation adopted by Averroes. Because differentiations of space do not exist in void, a body would move in whatever direction it is pushed. ("Et sic nullum motum movebitur propter aptationem ad aliquam partem aliam ab ea ad quam expellens expellit." Averroes, *Commentary on the Physics*, bk. 4, comment 70 in *Aristotelis opera cum Averroes commentariis* [9 vols. in 11 parts plus 3 supplementary vols.; Venice, Junctas ed., 1562–1574. Reprinted in facsimile Frankfurt: Minerva, 1962], vol. 4: *Aristotelis De physico auditu libri octo cum Averrois Cordubensis variis in eosdem commentariis*, fol. 157v, col. 1.) That Averroes considered this an illustration of violent motion in a void gains support from use of the term "expellatur" in text 70 of the *antiqua translatio* of Aristotle's *Physics*, which accompanies Averroes' commentary and was probably a translation of the Arabic text he used. In a second and later translation, which is printed along with the *antiqua translatio* in the Junctas edition that is cited in this volume, a body is said to be "carried" (*fertur*, corrected from *feretur*) in every direction. Thus if a body were merely placed, or released, in a void space, it would be capable of moving in any direction whatever, and perhaps, as the ultimate absurdity, Aristotle even intended that we understand the body as moving in all directions simultaneously, because all parts of void are identical and the body has no more reason to move in one direction than another. But if the body were pushed by an external force, Averroes' interpretation appears reasonable and might even be construed as an illustration of the inertial consequence whereby a body moves continually in one direction because void space lacks natural places toward which it could move naturally and come to rest.

17 *Physics* 4.8.215a.24–216a.11. Although I have evaluated this argument in my article "Motion in the Void," it will be discussed below in greater detail.

18 *Physics* 4.8.216a.12–20. Are the "equal velocities" to be interpreted as finite or infinite? Morris R. Cohen and Israel E. Drabkin (*A Source Book in Greek Science* 2d ed. [Cambridge, Mass.: Harvard University Press, 1958], p. 217, n. 1) argue that in this passage, equality of velocities is a consequence of the immediately preceding Aristotelian argument that all bodies of whatever size move with instantaneous velocity in the void. Therefore the equal velocities are infinite velocities. But the consequence of infinite speed is derived from a comparison of motions in plenum and void, whereas the present argument derives from a comparison of two bodies falling in the same medium. Here Aristotle sought to present another quite independent, absurd consequence of motion in a void. *Differences in finite speed* arise because a heavier body can cleave a medium more readily than a less heavy body. But if the same two unequal bodies were to fall in a void, where there is no medium to cleave, no cause or reason could be invoked to account for any differences in finite speed. Consequently, bodies of different weights will fall in a vacuum with equal *finite* speeds. From the standpoint of Aristotle's physics, equal finite speeds of unequal bodies are almost as absurd as instantaneous motion. I have discussed at length Aristotle's consequence of equality of fall in the void and its reception in the Middle Ages in "Bradwardine and Galileo: Equality of Velocities in the Void," *Archive for History of Exact Sciences*, 2, no. 4 (1965), 344–364.

We might also inquire whether the velocities in a vacuum are uniform or accelerated. One would suspect that Aristotle conceived them as uniform, because he would have had to invoke an additional cause to explain acceleration. In the Middle Ages, such motion was generally interpreted as uniform (see below, Ch. 3, n. 124).

19 *Physics* 4.1.209a.8–13.

20 Whether the point has dimension or not in the context of this discussion was not indicated by Aristotle.

21 *Physics* 4.1.209a.14–17. It is of interest that where Aristotle elsewhere sought to destroy the idea of void by conceiving it as if it were a body (see above), here his criticism depended on the denial of body to void.

22 *Physics* 4.1.209a.18–23.

23 *Physics* 4.1.209a.24–26; see also 4.4.211b.19–29.

24 Aristotle's true opinion of void was that it "seems to be a non-existent and a privation of being" (*Physics* 4.8.215a.10). Aristotle considered the meaning of privation in a number of places. Relevant to the void was his contrast of negation and privation when he said that "negation means just the absence of the thing in question, while in privation there is also employed an underlying nature of which privation is asserted" (*Metaphysics* 4.2.1004a.15–17; see also 5.22.1022b.22–1023a.6). Elsewhere (*Physics* 1.9.192a.5) he explained that "privation in its own nature is not-being."

25 See also Grant, "Motion in the Void," 286–287.

CHAPTER 2
Medieval conceptions of the nature and properties of void space

1 Among those who so characterized it, we mention Pseudo-Siger of Brabant (see Philippe Delhaye, *Siger de Brabant Questions sur la Physique d' Aristote* [Louvain: Editions de l'Institut Supérieur de Philosophie, 1941], bk. 4, question 7, pp. 153–154; for the passage, see below, this paragraph, and n. 4 of this chapter); John Buridan (see *Acutissimi philosophi reverendi Magistri Johannis Buridani subtilissime questiones super octo Phisicorum libros Aristotelis diligenter recognite et revise a Magistro Johanne Dullaert de Gandavo antea nusquam impresse* [Paris, 1509, reprinted in facsimile with the title *Johannes Buridanus, Kommentar zur Aristotelischen Physik* Frankfurt: Minerva, 1964], bk. 4, question 7, fol. 73r, col. 2); and Albert of Saxony (see *Questiones et decisiones physicales insignium virorum: Alberti de Saxonia in octo libros Physicorum; tres libros De celo et mundo . . . Recognitae rursus et emendatae summa accuratione et iudicio Magistri Georgii Lokert Scotia quo sunt Tractatus proportionum additi* [Paris, 1518], *Questions on the Physics*, bk. 4, question 1, fol. 43v, col. 1). I have also discussed this in my article "Place and Space in Medieval Physical Thought," in *Motion and Time, Space and Matter*, ed. by Peter K. Machamer and Robert G. Turnbull (Columbus, Ohio: Ohio State University Press, 1976), p. 138.

2 In mentioning the preexistent chaos proposed by Hesiod, Aristotle declared that Hesiod thus implied "that things need to have space first, because he thought, *with most people*, that everything is somewhere and in place." *Physics* 4.1.208b.33–35. The italics are mine.

3 *Categories* 5a.6–14.

4 Pseudo-Siger of Brabant, *Physics* (Delhaye ed.), bk. 4, question 7, p. 154. The attribution to Siger of Brabant of these *Questions on the Physics* (see above, n. 1) has been rejected, and Peter of Auvergne has been suggested as the possible author. See Albert Zimmermann (ed.), *Ein Kommentar zur Physik des Aristoteles aus der Pariser Artistenfakultät um 1273* (Berlin: Walter de Gruyter, 1968), p. xxxviii and n. 60, and Charles H. Lohr, S. J., "Medieval Latin Aristotle Commentaries, Authors: Narcissus – Richardus," *Traditio* 28 (1972), 345, where the work is listed in the "doubtful" category under "Petrus de Alvernia (de Crocq)," or Peter of Auvergne. For convenience, I shall cite this treatise as "Pseudo-Siger of Brabant, *Physics* (Delhaye ed.)."

5 In commenting on Aristotle's discussion of quantity in the *Metaphysics*, Averroes explained that Aristotle discussed different aspects of quantity in the *Predicaments*, or *Categories*, than in the *Metaphysics*. Thus, in the former, he considered place (*locus*), which is only a passion of quantity, but in the latter he considered the quantity of motion, which is measured by space and time but was omitted from the *Predicaments*. Generally, in the *Predicaments*, Aristotle "intended to enumerate only the well known [or common] species of quantity." See Averroes, *Opera*, vol. 8: *Metaphysics*, bk. 5, comment 18, fol. 125v, cols. 1–2. Although Averroes did not explicitly declare that Aristotle determined the truth in other parts of philosophy while merely presenting common opinions in the *Predicaments*, this became the standard interpretation of the passage in question. For example, John Buridan explained (*Physics*, bk. 4, question 7, fol. 73r, col. 2)

that "The Commentator, in the fifth [book] of the *Metaphysics*, rightly says that in the *Predicaments* Aristotle often spoke with regard to common opinion and not with regard to the determination of truth, namely on those matters which more properly belong to other parts of philosophy."

6 *De caelo* 1.9.279a.14–15. Also see n. 31 of this chapter for the full context, as well as Ch. 5, Sec. 1.

7 *Physics* 4.7.213b.31–34.

8 One of the definitions that Aristotle had rejected was, as Walter Burley explained, that vacuum is something in which there is no heavy or light body, a definition that would have permitted a point to qualify as a void, because a point, which is indivisible, could not receive body, which is divisible. Another rejected definition was that attributed to the Platonists. As Burley put it, "a vacuum is that in which there is nothing determinate, neither an accident nor anything of corporeal substance. And because this description is compatible with prime matter, some, namely the Platonists, say that prime matter and vacuum are the same, just as they thought that prime matter and place are the same." Aristotle rejected this idea because prime matter is inseparable from things, whereas vacuum is assumed separable by those who believe in vacuum. See *Burleus super octo libros Phisicorum* (Venice, 1501, reprinted in facsimile as Walter Burley, *In Physicam Aristotelis expositio et quaestiones* [Hildesheim/New York: Georg Olms Verlag, 1972]), fol. 110r, col. 2, for the first argument and fol. 110v, col. 1, for the second.

9 Thomas Bradwardine, *De causa Dei*, p. 180. The complete reference appears in Ch. 6, n. 90. For similar and contrasting opinions, see below, Ch. 6, Sec. 2a, and n. 41 and 14.

10 Bradwardine said that Aristotle takes it this way in the fourth book of the *Physics*. The association of vacuum with mathematical dimensions was also made in the thirteenth century by the anonymous author of the *Summa philosophiae*, falsely ascribed to Robert Grosseteste (see Ludwig Baur [ed.], *Die philosophischen Werke des Robert Grosseteste. Beiträge zur Geschichte der Philosophie des Mittelalters*, vol. 9 [Münster: Aschendorff 1912], p. 417, lines 10–12). Although the separate vacuum was occasionally identified with mathematical dimensions, it was rarely discussed as a mathematical entity or associated with Euclidean geometry largely because of the impossibilities that would follow from the occupation of such a dimension by a three-dimensional body (see Ch. 1 and Ch. 2, Sec. 2a). The mathematical nature of the volume of a body, which could be conceived as a kind of internal space or vacuum, did come to be identified with internal space (see below, Sec. 2a).

11 As Bradwardine observed, Aristotle discussed the vacuum as motive cause in the fourth book of *De caelo* (see 4.2.309b.16–29).

12 Johannis de Janduno, *Questiones in libros Physicorum Aristotelis. Acced. Heliae Cretensis annotationes* (Venice, 1488), bk. 4, fol. 73r, col. 1. The title of the question is: "Utrum si vacuum esset, esset privatio."

13 Here is the text of the relevant discussion: "Dicendum quod nomine vacui intelligit spacium separatum ab omni corpore naturali non repletum aliquo corpore, naturali aptum tamen repleri. Tunc dico ad questionem quod si esset vacuum posset considerari dupliciter. Uno modo quantum ad ipsum spatium separatum et sic non esset privatio quia privatio est negatio in subiecto apto nato ad habitum, ut patet in 10 *Metaphisice*. Sed ipsum spatium, si esset, non esset talis negatio, ut manifestum est, quare etc. Alio modo posset considerari ipsum vacuum quantum ad negationem habendi corpus et aptitudinem habendi. Et sic esset privatio quia privatio est negatio habitus in subiecto apto nato; et sic esset ipsum vacuum, quare etc.

Sed si queratur absolute utrum vacuum sit privatio, dicendum quod non, quia non est negatio in subiecto apto nato. Dimensio enim separata, que intelligitur nomine vacui, non est negatio alicuius habitus in subiecto apto nato ad illum habitum. Imo huiusmodi dimensio separata ab omni corpore sensibili est omnino non ens extra animam, cum omnis magnitudo sit sensibilis et mobilis." *Ibid*.

Even more explicit than Jandun was Aegidius Romanus (*Egidii Romani in libros De physico auditu Aristotelis commentaria* [Venice, 1502], bk. 4, fol. 84r, col. 2), who argued that if vacuum is a privation, it would be something real (*aliquid reale* or *quid reale*). "For although a negation posits nothing by itself, a privation always determines a certain subject for itself. Indeed a privation is a negation of a subject, as can be had from the fourth book of the *Metaphysics*. Therefore if a vacuum is a space deprived of body, it is necessary that that space and those dimensions be something real in which this privation is based as in a subject. For if such a space, or such dimensions, were nothing, they could not be called a vacuum or a plenum." This argument is hypothetical, however, because Aegidius was convinced of the impossibility of vacuum and therefore denied that we could investigate it in terms of its real existence, or *quid rei*. Vacuum could be analyzed only in terms of its definition or *quid nominis* (see *ibid.*, fol. 77v, col. 2 and above, Ch. 2, Sec. 1). Unless stated otherwise, all translations in this volume are my own.

14 To those who held that void is "what is not full of body perceptible to touch," Aristotle asked what they would say "of an interval that has colour or sound – is it void or not?" Aristotle suggested that their response would be "that if it *could* receive what is tangible it was void, and if not, not." *Physics* 4.7.214a.7–10.

15 P. L. Heath, "Nothing" article in *The Encyclopedia of Philosophy*, vol. 5 (New York: Macmillan and Free Press, 1967), p. 524. Although it was written tongue-in-cheek, Heath enunciated the problems and difficulties of nothing, which he described as "an awe-inspiring yet essentially undigested concept."

16 *Ibid.*

17 "Sciendum quod hoc nomen 'vacuum' est terminus privativus et valet in significando tantum quantum hec oratio: locus non repletus corpore et igitur quando dicitur vacuum est locus non repletus corpore, ly est ponitur pro significat. Et ideo non sequitur: vacuum est locus non repletus corpore, ergo vacuum est." Albert of Saxony, *Physics*, bk. 4, question 8 ("Utrum vacuum esse sit possibile"), fols. 48r, col. 2–48v, col. 1. In the course of his lengthy discussion on vacuum, Walter Burley explained the order of questions that one must confront in treating of the vacuum. He explained that we must first determine "whether the term signifies something intelligible" ("an vox significat aliquid intelligibile"); if it does, we must then inquire what it signifies ("quid significat"); with knowledge of what it signifies, we can then determine whether it exists ("an res est"); and if it does, we must then arrive at its species ("quid sit res in specie"). In sum, Burley explained that "these questions are organized; the second always presupposes the first, since a question about the *quid nominis* always presupposes a question as to whether the name [or term] signifies something; and the question whether it exists presupposes the question about the *quid nominis*; and a question inquiring about the nature of a thing ('quid rei' or 'quid sit res in specie') presupposes the question about whether a thing exists (' an res est')." See Burley, *Physics*, bk. 4, fol. 108v, cols. 1–2.

18 In response to an argument that knowledge of a vacuum presupposes its existence, an argument predicated on the assumption that we lack knowledge only about nonbeing, Albert replied: "Ad nonam, de vacuo est scientia, respondetur quod non; sed bene est scientia de loco et corpore secundum propositiones negativas. Unde sciendo istam: 'nullam vacuum est,' scitur ista 'nullus locus est sine corpore.' Una enim valet aliam, et sic scientia que creditur esse de vacuo est de loco et corpore." Albert of Saxony, *Physics*, fol. 48v, col. 2.

19 William Ockham, *Guillelmus de Occam, Quotibeta septem; Tractatus de sacramento altaris* (Strasbourg, 1491, reprinted in facsimile Louvain: Editions de la Bibliothèque S. J., 1962), Quotlibet primum, question 6 ("Utrum angelus possit moveri per vacuum"), sig. a5v, col. 2 (unfoliated).

20 *Ibid.*

21 *Ibid.* In his *Questions on the Ethics*, John Buridan was prepared to allow that negative propositions that deny the existence of something, such as "a vacuum is nothing," may be included in science as true propositions. As T. K. Scott, Jr., explains ("John Buridan on the Objects of Demonstrative Science, *Speculum* 40 [1965], 667), the inclusion of such negative propositions in science "is based on the rule that a negative categorical proposition asserts only that its terms do not stand for the same things, so that if one of them stands for nothing at all, the proposition is trivially true. And of course, if the things designated by the subject term cannot possibly exist, then such a negative proposition is necessarily true." Descartes also believed it absurd to suppose that nothing could be something, when he denied that nothing could possess extension (see below, sec. 2a and n. 54).

22 Duns Scotus adopted this position. See John Duns Scotus, *God and Creatures: The Quodlibetal Questions*, trans. with an introduction, notes, and glossary by Felix Alluntis, O. F. M,. and Allan B. Wolter, O. F. M. (Princeton, N.J.: Princeton University Press, 1975), question 11, article 2, p. 262.

23 *Questions on De caelo*, bk. 1, question 20, in Ernest Addison Moody (ed.), *Iohannis Buridani Quaestiones super libris quattuor De caelo et mundo* (Cambridge, Mass.: Mediaeval Academy of America, 1942), p. 95. In this passage, Buridan was responding to an argument (p. 92) in which it is assumed that if God annihilated everything within the heaven, a vacuum would exist. And because, under these conditions, nothing more would now exist within the world than outside the world, it follows that beyond the heaven, or world, a vacuum exists. Buridan argued, as we have seen, that such an annihilation would not yield a vacuum of the kind described, and therefore the inference that a vacuum must exist beyond the world is invalid.

24 In his *Physics*, bk. 4, question 7, fol. 73r, col. 2–73v, col. 1., Buridan rejected the attribution of void to the containing surface on the grounds that the concave surface of the lunar sphere is part of a mobile body, namely the lunar sphere itself, from which it follows that a vacuum would be a mobile thing. Marsilius of Inghen also distinguished and rejected the same two types of vacuum, namely a separate space, which he imagined as the space of the world that would be left if the world that occupied it was destroyed, and the surface of a containing body after its material content had been removed. See *Questiones subtilissime Johannis Marcilii Inguen super octo libros Physicorum secundum nominalium viam. Cum tabula in fine libri posita suum in lucem primum sortiuntur effectum* (Lyon, 1518, reprinted in facsimile, Frankfurt: Minerva, 1964), bk. 4, question 13 ("Utrum possibile sit esse vacuum"), fol. 55r, col. 2.

25 *R. P. Gabrielis Vazquez Societatis Iesu Disputationes Metaphysicae desumptae ex variis locis suorum operum* (Antwerp: apud Ioannem Keerbergium, 1618), pp. 369–370.

26 The expression "imaginary space," which usually meant void space, is discussed in Ch. 6, Sec. 1.

27 Vasquez defined *privatio* as "negatio cum aptitudine aliqua, ita ut inane illud sit non repletum, aptum tamen repleri" (*Disputationes Metaphysicae*, p. 369) and then declared (p. 370): "Ad haec de vacuo intra latera superficiei continentis . . . non potest esse privatio sicut neque intra superficiem aliquid esse privatum; illa enim negatio subiectum non habet ac proinde privatio non est."

28 Not only did Vasquez insist that the vacuum conceptualized between the sides or surfaces of a containing body was a pure negation, or absolutely nothing, rather than a privation or simple negation, but he also extended the designation of pure nothing to the imaginary space alleged by many to have existed before the creation of the world. Because no subject could have existed before the creation, imaginary space could not be the negation of anything and hence could only be characterized as a pure negation. ("Verum sive loquamur de spatio imaginario ante mundi creationem sive de eo quod extra caelum esse dicitur; sive de vacuo intra superficiem corporis ambientis, haec solutio futilis est et inanis. Quoniam ut aliqua negatio sit privatio, quae est cum aptitudine, debet esse in aliquo subiecto: aptitudo enim extra subiectum esse nequit quare nulla negatio extra subiectum potest esse privatio, atqui ante mundi creationem nullum erat subiectum illius negationis; ergo nulla erat inanitas, quae esset privatio, sed omnino nihil et pura negatio." *Ibid.*, pp. 369–370.) Where Vasquez described imaginary space as a pure nothing and denied it the status of a privation, Bartholomeus Amicus, using the same terminology to characterize the same imaginary space, would confer a very different status on that elusive entity (see below, Ch. 7, Sec. 2d; for Fonseca's use of *pura negatio*, see Ch. 7, Sec. 2b). Vasquez's distinction between negation and privation and pure negation is virtually identical with the distinction made by John of Jandun, although the latter did not employ the expression "pure negation" (see above, Ch. 2, Sec. 1 and n. 13).

29 "Nam sicut superficies quae intra se corpus continet dicitur repleta, ita etiam cum nullum corpus intra se habet dicitur privata, hoc est, non repleta, apta tamen repleri quam privationem vacuum appellamus." *Ibid.*, p. 370.

30 *Questiones supra libros quatuor Physicorum Aristotelis*, ed. Ferdinand M. Delorme, O. F. M., with the collaboration of Robert Steele, in *Opera hactenus inedita Rogeri Baconi*,

fasc. 8 (Oxford: Clarendon Press, 1928), p. 198. For the English translation and Latin text, see below, Ch. 5, Sec. 1 and n. 6.

31 "It is therefore evident that there is also no place or void or time outside the heaven. For in every place body can be present; and void is said to be that in which the presence of body, though not actual, is possible; and time is the number of movement. But in the absence of natural body, there is no movement, and outside the heaven, as we have shown, body neither exists nor can come to exist. It is clear then that there is neither place, nor void, nor time, outside the heaven." Aristotle, *De caelo*, 1.9.279a.11–19 (Oxford trans. by J. L. Stocks).

32 *Ibid.*, 279a.19–23. See below, Ch. 5, n. 57.

33 William of Ware assumed that prior to the creation of our world, absolutely nothing existed. But he imagined that this nothing is an infinite space with nothing in it. Within that nothing, or infinite space, God created our world and could add as many more as He pleases, even to infinity. See Pierre Duhem, *Le Système du monde, Histoire des doctrines cosmologiques de Platon à Copernic*, 10 vols. (Paris: Hermann, 1913–1959), vol. 9, pp. 381–382. William thus agreed with Bacon that nothing could be equated with vacuum; unlike Bacon, however, William's nothing could receive bodies. Hasdai Crescas also interpreted Aristotle's nothing beyond the world as a vacuum capable of receiving bodies (see below, ch. 5, n. 5). For the reasons why the Stoics, who accepted Aristotle's definition of vacuum, called the nothing beyond the world "void," whereas Aristotle refused to do so, see below, Ch. 5, Sec. 1. It is instructive to compare this approach with the analysis of vacuum by Ockham and Buridan (see above, Ch. 2, Sec. 1 and relevant notes).

34 For the passage, see below, Ch. 8, Sec. 4c. Indeed some four centuries before Bacon, Fridugis the Deacon (d. 834), a student of Alcuin of York, also extolled the virtues and magnificence of nothing. Interpreting the *nihil* ("nothing") in "creation *ex nihilo*" as if it were literally a thing or substance, Fridugis, in his *Letter on Nothing and Darkness* (*Epistola de nihilo et tenebris*) to Charlemagne, declared that "divine power has made out of nothing earth, water, air, and fire, light too and angels, and the soul of man. Our mental capacity, therefore, must be lifted up to the authority of such eminence which can be shaken by no reasoning, refuted by no arguments, impugned by no power. It is this authority which proclaims that those which are the first and foremost among creatures have been created out of nothing. Therefore, nothing is something great and remarkable, and its magnitude from which so many and such noble things have been produced cannot be grasped." Translated from Ernest Dümmler's edition (in *Monumenta Germaniae Historica, Epistolae Karolini Aevi*, tom. II [Berlin, 1895], pp. 552–555) by Hermigild Dressler in John F. Wippel and Allan B. Wolter (eds.), *Medieval Philosophy from St. Augustine to Nicholas of Cusa* (New York: Free Press; London: Collier Macmillan Publishers, 1969), p. 105.

35 In grappling with the problem of whether the world was created out of nothing (*ex nihilo*), or out of some precreation matter, medieval Muslim authors in the Kalam tradition, who were heirs to a Syrian Christian creation formula, confronted the issue of whether the nonexistent (*al-ma'dūm*) was to be interpreted as signifying nothing or something. For an account of this controversy, see Wolfson, *Philosophy of the Kalam*, pp. 359–372 (especially pp. 362–366). Wolfson (p. 361) observes, however, that although "there can be no doubt that the Kalam in its discussion of the problem of the nonexistent has drawn upon the vocabulary used by Democritus and Leucippus, . . . the problem of the nonexistent discussed by the Kalam, as may be judged from the various contexts in which it occurs, has nothing to do with the question whether a vacuum does or does not exist." On the meaning of "Kalam," see ch. 1, pp. 1–58; on the vocabulary of Democritus and Leucippus, see above, Ch. 1 and n. 1.

36 In discussing the opinions of John Philoponus, or Joannes Grammaticus, Averroes observed that although Grammaticus believed that vacuum could not be separated from the body occupying it, there were those who believed that it was separable from body. "Ioannes vero propter hoc obedit huic, scilicet locum esse et dimensionem et vacuum, non finem continentem, ut dicit Aristoteles. Licet apud ipsum non possit inane separari a corpore quoniam dicentes vacuum esse sunt bipartiti: alii enim dicunt ipsum separari a corporibus, et alii non, quorum est Ioannes Grammaticus." Averroes, *Opera*,

vol. 4: *Commentary on the Physics*, bk. 4, comment 43, fol. 141r, col. 2. Although Charles Schmitt rightly observes (*Gianfrancesco Pico della Mirandola [1469–1533] and His Critique of Aristotle* [The Hague: Martinus Nijhoff, 1967], p. 148, n. 64) that Averroes omitted the details of Philoponus' arguments, the mere mention of the distinction was itself of importance. Without mention of Philoponus, Avicenna presented much the same distinction to medieval readers when he declared that the doctors have two opinions about space: "Some of them indeed do not believe that this space could remain void without being filled but they affirm that it [vacuum] is never separated from an occupying body unless another body supervenes. But some of them do not believe this but rather that it is possible that this space is sometimes filled and sometimes indeed void. And these are the opinions of the doctors on emptiness [or void]." Avicenna, *Sufficientia* (Venice, 1508, reprinted in facsimile Frankfurt: Minerva, 1961), bk. 2, ch. 6, fol. 28r, col. 2. Perhaps it was Avicenna whom Albertus Magnus followed when he explained that "some say that this space is always filled, but [nonetheless] separate from any whatever body that is received in it; and some [others] say that sometimes there is a vacuum so that no body is in it." ("Quidam enim dicebant hoc spatium semper esse plenum, sed separatum a quolibet corpore quod suscipitur in ipso; et quidam dicebant aliquando esse vacuum, ita quod nullum corpus est in ipso." *Omnia Opera*, ed. Peter Jammy [Lyon, 1651], vol. 2: *Physicorum lib. VIII*, bk. 4, tract 1, ch. 9, p. 154, col. 1.)

37 Roger Bacon equated vacuum and prime matter as the fifth of five types of void he distinguished. "In a fifth way, vacuum is described as a certain nature or root and foundation which is in potentiality for the reception of many forms . . .; and in this way, prime matter by itself and with respect to its essence is called void; and in this way vacuum is not assumed impossible, but necessary." For the Latin text, see Bacon, *Opera*, fasc. 8: *Physics* (4 books), p. 198. The possible connections between internal space and prime matter will be discussed in the next section.

38 The section on internal space is drawn, almost verbatim, from my article "Principle of the Impenetrability of Bodies," 554–557. The supporting Latin texts, largely omitted from the article, are included here.

39 See Michael Wolff, *Fallgesetz und Massebegriff. Zwei wissenschaftshistorische Untersuchungen zur Kosmologie des Johannes Philoponus. Quellen und Studien zur Philosophie*, ed. by Gunther Patzig, Erhard Scheibe, and Wolfgang Wieland, vol. 2 (Berlin: Walter de Gruyter, 1971), p. 109; for German translations of relevant passages from the Greek text, see pp. 141–143. In his *De opificio mundi*, Philoponus declared that "if one abstracts the forms of all things, there obviously remains the three-dimensional extension only, in which respect there is no difference between any of the celestial and the terrestrial bodies," from which S. Sambursky infers that "anticipating Descartes, Philoponus arrived at the conclusion that all bodies in heaven, as well as on earth, are substances whose common attribute is extension." See Sambursky's article "John Philoponus," in the *Dictionary of Scientific Biography*, vol. 7 (New York: Scribner, 1973), p. 135; the translation is Sambursky's.

40 Giacomo Zabarella (1533–1589) interpreted Philoponus as one who identified prime matter with three-dimensional extension, an interpretation with which he agreed, and which he said had also been adopted by the Stoics. *Jacobi Zabarellae Patavini, De rebus naturalibus libri XXX* (Frankfurt, 1607; first published 1590; reprinted in facsimile, Frankfurt: Minerva, 1966), *De prima rerum materia, liber secundus*, col. 211. Following in the tradition of Plotinus, Simplicus, in his commentary on Aristotle's *Physics*, noted difficulties, and even a contradiction, in Aristotle's conception of matter and then denied extension to prime matter, which acquired tridimensionality only upon receipt of a distinct corporeal form. See Harry A. Wolfson, *Crescas' Critique of Aristotle, Problems of Aristotle's 'Physics' in Jewish and Arabic Philosophy* (Cambridge, Mass.: Harvard University Press, 1929), p. 582. For Wolfson's valuable discussion on the history of the problem of "corporeal form," see part II, n. 18, pp. 579–590.

41 How close one could come to describing internal space without actually proclaiming it is made evident by Zabarella, who, in citing Plotinus with approval, declared that "if we mentally contemplate matter abstracted from all forms, we can conceive nothing other

than a certain extended and indistinct body and a certain void mass, . . ." (*De rebus naturalibus libri XXX*, col. 217.)

42 "Esse autem hoc intrinsecum cuiusque corporis spatium negari non potest, ut videtur, quia invicem sese mutua consequentia inferant. Nam si corpus est, spatium est; et si verum spatium est, in eo corpus est." Toletus, *Commentaria una cum Quaestionibus in octo libros Aristotelis De physica auscultatione* (Venice, 1580), fol. 123r, col. 2.

43 For more of the English translation, see below, Ch. 6, Sec. 2. For Toletus' similar example, see his *Physics*, fol. 123r, col. 2. That adoption of internal space was a means of rejecting the existence of a separate void space was also made clear by Albert Einstein when he explained that "space-time is not necessarily something to which one can ascribe a separate existence, independently of the actual objects of physical reality. Physical objects are not in space, but these objects are *spatially extended*. In this way the concept of 'empty space' loses its meaning." *Relativity, The Special and the General Theory, A Popular Exposition* by Albert Einstein, 15th ed. (New York: Crown Publishers, 1961), p. vi ("Note to the Fifteenth Edition"). The italics are Einstein's. The statement by Einstein and the passage from Hermann Weyl, quoted below in n. 46, were brought to my attention by my colleague, Prof. J. Alberto Coffa, to whom I am indebted.

44 "An vero spatium illud, seu extensio, distinguatur ab ipsa quantitate realiter, vel formaliter, vel ratione, non magnopere refert. . . . Est autem hoc spatium unicuique quantitati suum proprium sicut unicuique motui sua propria temporis successio, et mensura propria et particularis. Et ubicunque sit corpus, locum semper idem habebit suum intrinsecum hoc numero spatium." Toletus, *Physics*, fol. 123r, col. 2. That Toletus believed Philoponus held a doctrine of internal space is made evident a few paragraphs later, when he cited Philoponus as one who believed that internal space is not a separate space but is something that inheres in bodies as a proper accident. ("Et rationes plereque Philoponi hoc ipsum spatium intrinsecum et quantitati proprium ostendunt, sed non separatum, ut ipsa ponebat, sed in rebus ipsis inhaerens tanquam proprium earum accidens." *Ibid.*, fols. 123r, col. 2–123v, col. 1.)

45 Johannes Duns Scotus, *Opera* (Luke Wadding, ed., Lyon, 1639; reprinted in facsimile Hildesheim: Georg Olms Verlag, 1968), vol. 6, part 1: *Quaestiones in Lib. II Sententiarum*, distinction 2, question 6, p. 192.

46 Hermann Weyl recognized that Euclid's geometry was not "a doctrine of *space itself*," but rather, "like almost everything else that has been done under the name of geometry," was "a doctrine of the configurations that are possible in space." *Space-Time-Matter* by Hermann Weyl, trans. from the German by Henry L. Brose (first American printing of the 4th ed. 1922; New York: Dover Publications, n.d.), p. 102. See above, n. 43.

47 Considering the close conformity and agreement between Euclid's *Elements* and Aristotle's conception of definitions, axioms, postulates, and other geometric matters, it is not implausible to suppose that Euclid may have been familiar with the works of Aristotle and perhaps also with the arguments on void space described above. See Sir Thomas L. Heath, *The Thirteen Books of Euclid's Elements*, trans. from the text of Heiberg with introduction and commentary, 2d ed. revised with additions (3 vols.; New York: Dover Publications, 1956; reprinted from the Cambridge edition of 1926), vol. 1, pp. 117–124, 143–151. To my knowledge, Euclid did not use a term for "space" in his geometry. See also below, Ch. 8, n. 31.

48 Shlomo Pines, "Philosophy, Mathematics, and the Concepts of Space in the Middle Ages," in Y. Elkana (ed.), *The Interaction Between Science and Philosophy* (Atlantic Highlands, N.J.: Humanities Press, 1974), p. 84.

49 For more on the role of Euclid's alleged geometric space, see below, Ch. 8, Sec. 4g. As we shall see, the adoption of an infinite space in the seventeenth century resulted primarily from the divinization of space – a process begun in the fourteenth century – and, to a lesser extent, from the needs of physics and cosmology. But it did not arise from any straightforward application of an alleged Euclidean geometric space to the physical world.

50 "Locum autem aliquando consideramus ut rei, quae in loco est, internum, & aliquando ut ipsi externum. Et quidem internus idem plane est quod spatium; . . ." *Principia Philosophiae*, part 2, principle 15 in *Oeuvres de Descartes*, ed. Charles Adam and Paul Tannery, vol. 8, part 1 (Paris: Librairie Philosophique J. Vrin, 1964), p. 48.

51 *Ibid.*, principle 10, p. 45. The translation is from *The Philosophical Works of Descartes*, rendered into English by Elizabeth S. Haldane and G. R. T. Ross (2 vols.; New York: Dover Publications, 1955; 1st ed. 1911; reprinted with corrections 1931 by Cambridge University Press), vol. 1, p. 259. See also principle 11 (*Oeuvres de Descartes*, p. 46), where Descartes reiterated the identification of space and body.

52 *Principles of Philosophy*, part 2, principle 10, p. 259 of the Haldane and Ross translation, which I have altered as follows: (1) I have replaced the words "having removed from a certain space the body which occupied it" with "a body which fills a space has been changed" to render more appropriately the Latin words "adeo ut, mutato corpore quod spatium implet"; and (2) in the phrase "removed the extension of that space," I have altered "removed" to "changed," thus assigning a more accurate sense to the Latin "mutari." For the Latin, see *Oeuvres de Descartes*, vol. 8, part 1, p. 45.

53 *Principles of Philosophy*, part 2, principle 11, p. 259 of the Haldane and Ross translation.

54 *Principles of Philosophy*, part 2, principle 16, p. 262 of the Haldane and Ross translation. Because of its significance, I also cite the full relevant Latin text: "Nam cum ex hoc solo quod corpus sit extensum in longum, latum & profundum, recte concludamus illud esse substantiam, quia omnino repugnat ut nihili sit aliqua extensio, idem etiam de spatio: quod nempe, cum in eo sit extensio, necessario etiam in ipso sit substantia." *Oeuvres de Descartes*, vol. 8, part 1, p. 49.

55 Here is Jandun's discussion: "Prima, si locus esset tale spatium vel dimensio corporalis illius spatii necessario sequitur quod duo corpora sint simul. Sed hoc est impossibile quia, pari ratione, infinita corpora simul essent; etiam totum celum, vel quantitas equalis corporeitati totius celi, posset esse simul in loco unius grani milii, quod est absurdum omnino." *Questions on the Physics*, bk. 4, question 4, in *Questiones Joannis de Janduno De physico auditu noviter emendate . . .* (Venice, 1519), fol. 48r, col. 2. In the context of God's absolute power, Albert of Saxony also imagined a situation in which God could place the whole world within a millet seed and achieve this without any condensation, rarefaction, or penetration of bodies. As a supernatural act, this was not deemed absurd. See my article "Place and Space in Medieval Physical Thought," 154.

56 In response to the question of whether several angels could naturally occupy the same place, Ockham replied: "Ad tertium dico quod sic plura corpora possunt esse in eodem loco per potentiam divinam et etiam naturaliter corpora illa que sunt nata esse partes essentiales alicuius compositi, sicut materia et forma sunt naturaliter in eodem loco adequate. Et per consequens multo magis duo angeli, qui non sunt circumscriptive in loco, possunt esse in eodem loco." *Quotlibeta septem*, Quotlibet primum, question 4, sig. a4v, col. 1. By conceding to God the power to locate two or more bodies in the same place, Ockham probably had in mind article 141 of the Condemnation of 1277, which denounced the opinion that God could not make more than several dimensions exist simultaneously ("Quod Deus non potest facere accidens esse sine subjecto, nec plures dimensiones simul esse." Heinrich Denifle and Emil Chatelain [eds.], *Chartularium Universitatis Parisiensis*, vol. 1 [Paris: Fratrum Delalain, 1889], p. 551.). That Ockham was well aware of the Condemnation of 1277 is evident by his specific citation of an article condemned at Paris (article 219) at the conclusion of the question from which our quotation above was taken and from his citation of yet another article (114) in the second quotlibet, question 10.

57 The simultaneous but natural occupation of a single place by two bodies was mentioned by Ockham, but surely in an equivocal sense. His description would seem to apply to elements in a homogeneous compound in which the elements are related to each other as matter and form in a single body (see Ockham's text in n. 56, above). Thus if the elements lose their respective identities and unite in a new substantial form, which is uniformly distributed over the homogeneous matter of the compound, Ockham's conditions for the simultaneous occupation of a single place by two or more bodies would have been met. But surely the conception with which we are concerned here is one in which the bodies, or dimensions, retain their separate identities even though they occupy the same place.

58 Here is the text of Albert's argument: "Tertio, si corpus naturale indigeret tali spatio separato vel hoc esset ratione sue materie, vel ratione forme, vel ratione suorum accidentium, vel ratione suarum dimensionum. Nec primum, nec secundum, nec tertium quia illis sufficiunt earum proprie dimensiones. Si autem dicatur quartum, ergo ille dimensiones, que

dicuntur esse tale spatium separatum indigent uno alio spatio separato; et illud iterum uno alio, et sic in infinitum." Albert of Saxony, *Physics*, bk. 4, question 1, fol. 43r, col. 2. Albert probably drew the substance of his argument from John Buridan, who gave it in his *Physics*, bk. 4, question 2, fol. 68r, col. 1. In the sixteenth century, Franciscus Toletus offered a similar argument in his commentary on Aristotle's *Physics* (see bk. 4, ch. 5, question 3, fol. 117v, col. 1).

59 "Item ex quo: magnitudo mundi non est aliud quam dimensio et per consequens spacium partium sic et magnitudines partium mundi non sunt nisi spacia et dimensiones. Non apparet ad quid proficeret alia dimensio vel aliud spacium quoniam Deus non indigebat spacio presupposito ad creandum mundum et magnitudinem eius. Quoniam Deus, etiam si esset tale spacium, poterat ipsum annichilare et iterum tale creare. Et si in creando illud spacium non indigebat alio in quo crearet, ita nec in creando magnitudinem mundi indigebat alia magnitudine vel dimensione in qua crearet mundum." Buridan, *Physics*, bk. 4, question 2, fol. 68r, cols. 1–2.

60 The superfluous character of space was manifested in yet another way. John of Jandun argued that as a mere empty, separate, homogeneous dimensionality, space would lack all capacity to influence or affect the bodies that might occupy it. ("Item si locus esset tale spacium separatum nullam virtutem naturalem habens in se, tunc locus nihil conferret locato. Constat enim quod non conferret locato magnitudinem seu dimensionem [corrected from "diminutionem"], quia magnitudinem habet sine tali spatio; nec conferret ei conservationem quia nullam virtutem habet convenientem ipsi locato. Quare, etc." Jandun, *Physics* [1519], bk. 4, question 4, fol. 48v, col. 1.)

61 For Averroes' statement, see above, n. 36. Averroes derived his description ultimately from Philoponus' *Commentary on Aristotle's Physics*, bk. 4, in a section titled "Corollary on Place." Here, after arguing that the place of a body is a void extension distinct from the body occupying it, Philoponus concluded that the void itself is never without an occupying body, just as matter, though it differs from forms, is never without them. For the Greek text, see H. Vitelli (ed.), *Ioannis Philoponi in Aristotelis Physicorum libros quinque posteriores commentaria* in *Commentaria in Aristotelem Graeca*, vol. 17 (Berlin: G. Reimer, 1888), pp. 568–569. The first Greek edition of Philoponus' *Physics Commentary*, published in 1535 by Victor Trincavelli, was followed by Latin translations published at Venice in 1554, 1558, and 1569. In the translation of 1569 by Johannes Rasario (*Ioannes Grammatici, cognomento Philoponi, in Aristotelis Physicorum libros quattuor explanatio, Io. Baptista Rasario Novariensis, interprete*), see cols. 339–340 for the relevant discussion. A German translation of the Greek passage, and much else from the works of Philoponus, appears in *Johannes Philoponos, Grammatikos von Alexandrien (6 Jh. n. Chr.), Christliche Naturwissenschaft im Ausklang der Antike, Vorläufer der modernen Physik, Wissenschaft und Bibel, Ausgewählte Schriften* übersetzt, eingeleitet und kommentiert von Walter Böhm (Munich/Paderborn/Vienna: Verlag Ferdinand Schöningh, 1967), pp. 92–93. For a brief description of Philoponus' theory of space, see S. Sambursky, *The Physical World of Late Antiquity* (New York: Basic Books, 1962), pp. 6–7. Sambursky observes that Philoponus' views are virtually identical with those of Strato of Lampsacus (c. 300 B.C.), whose opinions are described on p. 3 of Samburesky's text. Valuable information on the early history of Greek spatial concepts was preserved by Simplicius in the discussion of place in his commentary on the fourth book of Aristotle's *Physics* (for the Greek edition, see H. Diels [ed.], *In Aristotelis Physicorum libros quattuor priores commentaria* in *Commentaria in Aristotelem Graeca*, vol. 9 [Berlin: Berlin Academy, 1882], pp. 601–645). Because the physics commentaries of Philoponus and Simplicius were not translated into Latin in the Middle Ages, they were not directly influential until the sixteenth century, when they became highly significant following publication of the Greek texts and Latin translations.

62 Pseudo-Siger of Brabant, *Physics* (Delhaye ed.), bk. 4, question 7, pp. 153–154. Although Pseudo-Siger of Brabant (see above, n. 4 of this chapter) named Averroes ("the Commentator") as his source, the latter did not attribute to Philoponus the belief that void is "nothing in reality." Thus Pseudo-Siger of Brabant either learned of this elsewhere or simply assumed it.

63 In his *Examen vanitatis*, first published in 1520, Pico repeated, with approval, Philoponus'

argument by declaring that "place is space, vacant (*vacuum*) assuredly of any body, but still never existing as a vacuum alone of itself. It is like the case of matter, which is something other than form; but, nevertheless, never without form." In the 1520 edition, published at Mirandola, see bk. VI, ch. 4, p. 768. The *Examen vanitatis* also appears in the *Opera Omnia Ioannis Francisci Pici Mirandulae* . . . (Basel: Henricus Petrus, 1573, reprinted in facsimile in *Opera Omnia* [*1557–1573*][*di*] *Giovanni Pico della Mirandola* [*e*] *Gian Francesco Pico*, 2 vols. [Hildesheim: Georg Olms Verlag, 1969]), where the passage appears in vol. 2, p. 1189. The translation cited above is by Charles B. Schmitt, *Gianfrancesco Pico Della Mirandola (1469–1533) and His Critique of Aristotle*, pp. 140–141. Because the Greek edition of Philoponus' *Physics Commentary* was not published until 1535 and Latin translations even later (see above, n. 61), Pico, who knew Greek well (see Schmitt, p. 9), probably consulted Greek manuscripts directly.

64 For Telesio, Patrizi, Bruno, and Campanella, see below, Ch. 8. Like Aristotle, Philoponus denied the existence of an extracosmic void. For the arguments, See H. Vitelli (ed), *Ioannis Philoponis in Aristotelis Physicorum libros quinque posteriores*, pp. 582–583; Rasario's translation of 1569, col. 348; and Böhm's German translation, p. 95.

65 What follows on Philoponus has been drawn from Böhm, *Johannes Philoponos*, pp. 86–89, 92–93, where the page references to Vitelli's Greek edition are also given.

66 See Aristotle, *Physics* 4.4.211b.19–29 and also Aristotle's report of Zeno's argument in 4.1.209a.24–26.

67 Even before Philoponus made this crucial distinction, a version of it was reported by Sextus Empiricus, who flourished around A.D. 200. In relating the opinions of "Dogmatic philosophers," Sextus asserted that "even if, in imagination, we abolish all things, the place wherein all things were will not be abolished, but remains possessing its three dimensions – length, depth, breadth – but without solidity; for this is an attribute peculiar to body." *Sextus Empiricus*, with an English translation by the Rev. R. G. Bury in four volumes (London: William Heinemann; Cambridge, Mass.: Harvard University Press, 1936), vol. 3: *Against the Physicists*, II, 12, p. 217. Although, as a Sceptic, Sextus rejected what he reported, his account of it was known from the sixteenth century on. The same distinction was also made by Simplicius, whose views on space were strikingly similar to those of Philoponus, his contemporary and adversary (see S. Sam, "Place and Space in Late Neoplatonism," *Studies in History and Philosophy of Science 8* [1977], 183).

68 It was first translated into Latin by Nicholas Alfanus (d. 1085), archbishop of Salerno, under the title *Premnon physicon*, and edited by Karl Burkhard, *Nemesii Episcopi Premnon Physicon . . . a N. Alfano Archiepiscopo Salerni in latinum translatus* (Leipzig: Teubner, 1917). Nicholas's quite defective version, with numerous omissions, was far surpassed in 1165, when Burgundio of Pisa (d. 1193) translated it again, this time without omissions. For the edition of Burgundio's translation, see *Némésius d' Emèse, De natura hominis. Traduction de Burgundio de Pise*. Edition critique avec une introduction sur l'anthropologie de Némésius by G. Verbeke and J. R. Moncho (Leiden: E. J. Brill, 1975). On the two medieval translations, as well as three Renaissance translations, see pp. LXXXVI–C. Nemesius was bishop of Emesa, and probably wrote the *De natura hominis* around A.D. 400.

69 Following a denial that soul is three-dimensional because it occupies a three-dimensional body, Nemesius declared: ". . . dicemus quoniam omne quidem corpus tres habet dimensiones, non omne autem quod tres habet dimensiones corpus est; etenim locus et qualitas, incorporea existentia, secundum accidens in tumore quantitantur." *De natura hominis*, ed. Verbeke and Moncho, pp. 25–26.

70 "Locus enim in spatio est, quod longitudine et latitudine et altitudine corporis occupatur; . . ." *Magistri Petri Lombardi, Parisiensis Episcopi, Sententiae in IV libris distinctae*. Edito tertia ad fidem codicum antquiorum restituta. Tom I, pars II, liber I et II (Grottaferrata [Rome]: Editiones Collegii S. Bonaventurae ad Claras Aquas, 1971), bk. 1, distinction 37, ch. 9, par. 2, p. 274. As the source of his quotation, Peter cited St. Augustine's *De diversis quaestionibus 83*. Obviously, the latter treatise would have been yet another possible source for a three-dimensional concept of place. I have characterized Peter's definition as "ambiguous" because it is compatible with a three-dimensional sense of empty space as well as with Aristotle's definition of place as a two-dimensional containing

surface. For a further discussion of Peter's concept of place, see below, Ch. 6, Sec. 2b and notes 64 to 66.

71 The Hebrew text with accompanying translation, introduction, and notes has been published by Wolfson, *Crescas' Critique of Aristotle*. With an infinite void surrounding our finite world, which is a plenum, Crescas' cosmology is more akin to that of the Stoics than the atomists, although in contrast to the Stoics, Crescas not only allowed for the possibility of the existence of other worlds beyond ours but eventually conceded that "we are unable by means of mere speculation to ascertain the true nature of what is outside the world . . ." (p. 217). For an explicit statement by Wolfson that for Crescas the vacuum does not exist within the universe, but only outside, see his *The Philosophy of Spinoza* (2 vols.; Cambridge, Mass: Harvard University Press, 1934), p. 275.

72 As Crescas put it, "the impenetrability of bodies is due not to dimensions existing apart from matter, but rather to dimensions in so far far as they are possessed of matter." Wolfson, *Crescas' Critique of Aristotle*, p. 187. Although discussion of extracosmic void has been reserved for the second part of this study, it is appropriate to include Crescas at this point, because the crucial distinction between material and immaterial dimensions is applicable to void within and beyond the world.

73 Wolfson, *Crescas' Critique of Aristotle*, p. 187.

74 Milič Čapek (ed.), *The Concepts of Space and Time, Their Structure and Their Development* (Dordrecht/Boston: Reidel Publishing, 1976), p. 91. The section from Gassendi in Čapek's volume (pp. 91–95) is drawn from Gassendi's *Syntagma philosophicum*, physica, sectio I, liber 2: "De Loco et Duratione Rerum," in *Opera Omnia* (Florence, 1727), vol. I, pp. 162–163, 170. The translation is by Čapek and Walter Emge. In the recent facsimile reprint of Gassendi's *Opera Omnia* (Lyon, 1658) by Friedrich Fromann Verlag (Stuttgart-Bad Cannstatt, 1964), see vol. I, p. 182. For the Latin text of this passage from Burgundio of Pisa's twelfth-century translation of Nemesius' *De natura hominis*, see above, Ch. 2, Sec. 4 and n. 69.

75 Čapek, *Concepts of Space and Time*, p. 93.

76 John Locke, *An Essay Concerning Human Understanding*, ed. with an introduction, critical apparatus, and glossary by Peter H. Nidditch (Oxford: Clarendon Press, 1975), bk. 2, ch. 13, par. 23, p. 178. Earlier in the same treatise, Locke had argued that "those who dispute for or against a *Vacuum*, do thereby confess, that they have distinct *Ideas* of Vacuum and *Plenum*, i.e. that they have an *Idea* of Extension void of Solidity, though they deny its existence; or else they dispute about nothing at all" (bk. 2, ch. 13, par. 21 [bis], p. 177).

77 In effecting this result, God is first assumed to have caused all motion of the material universe to cease. "Whoever then will allow, that God can, during such a general rest, annihilate either this Book, or the Body of him that reads it, must necessarily admit the possibility of a *Vacuum*. For it is evident, that the Space, that was filled by the parts of the annihilated Body, will still remain, and be a Space without Body. For the circumambient Bodies being in perfect rest, are a Wall of Adamant, and in that state make it a perfect impossibility for any other Body to get into that Space." *Essay Concerning Human Understanding*, bk. 2, ch. 13, par. 21 [bis], pp. 176–177. To demonstrate the existence of vacuum, Locke also appealed to our experience of motion (see par. 22, pp. 177–178).

78 *Ibid.*, par. 21 [bis], p. 177.

79 *Ibid.*, par. 26, p. 180.

CHAPTER 3
The possibility of motion in void space

1 Although Aristotle did not describe instantaneous motion as the simultaneous occupation of all the points to be traversed, it is clearly implied. During the Middle Ages, it was conceived as the negation of a successive motion, which was characterized as the acquisition of successive parts of a distance in contrast to the simultaneous acquisition of the whole of it. Buridan (*Physics*, bk. 4, question 9, fol. 74r, col. 2) defined succes-

sion as follows: ". . . Hoc nomen successio manifeste connotat quod continue pars post partem acquiratur, dispositio secundum quam est motus et non tota simul."

2 See Roger Bacon, *Opera hactenus inedita Rogeri Baconi*, fasc. 13: *Questiones supra libros octo Physicorum Aristotelis*, ed. Ferdinand M. Delorme, O. F. M., with the collaboration of Robert Steele (Oxford: Clarendon Press, 1935), p. 235.

3 "Queritur consequenter utrum si vacuum esset grave moveretur in eo. Arguitur quod sic quia Aristoteles probat quod in instanti moveretur, ergo moveretur in vacuo." Buridan, *Physics*, bk. 4, question 10, fol. 76v, col. 2. In his first conclusion, Buridan explained that because Aristotle believed that vacuum is absolutely impossible, and because anything whatever can be derived from the impossible, it follows that "if a vacuum exists, a heavy body could be moved in it," and it also follows that "if a vacuum exists, no heavy body could be moved in it." ("Prima conclusio, scilicet quod Aristoteles concessisset istam consequentiam tanquam bonam et similiter istam vacuum est, ergo nullum grave movetur in eo quia ipse credidit quod simpliciter esset impossibile vacuum esse et ad impossibile sequitur quodlibet. Ideo concessisset istas: si vacuum esset grave moveretur in eo; et si vacuum esset nullum grave moveretur in eo." *Ibid.*, fol. 77r, col. 1.) Buridan was of the opinion that Aristotle did not wish to assert that motion in a void would be instantaneous in the sense that an instantaneous motion is a specific kind of actual motion to be contrasted to finite motion. On the contrary, "Aristotle wished to conclude from this that it would be impossible for a heavy body to be moved by its own gravity in a void . . . " (" . . . ideo volebat ex hoc Aristoteles concludere quod impossibile esset grave per suam gravitatem moveri in vacuo . . ." *Ibid.*, fol. 77v, col. 1; see also my article "Motion in the Void," 279).

4 "Questio III. *De motu in vacuo*. Duo possent quaeri in hac quaestione: primum, an si daretur vacuum posset in eo fieri motus aliquis; secundum, admisso quod efficiatur motus, an talis motus in instanti fieret, an in tempore et successive." *In Arist. Stag. libros Physicorum quibus ab adversantibus tum veterum tum recentiorum iaculis Scoti philosophia vindicatur a P.P. Magistris Bartholomeo Mastrio de Meldula, . . . et Bonaventura Belluto de Catana* (2d ed.; Venice: typis Marci Ginammi, 1644), p. 898, col. 1. Among those who denied motion in a vacuum, our authors included Albertus Magnus, Aegidius Romanus, Walter Burley, John of Jandun, Albert of Saxony (who is presumably intended by the abbreviation "Saxo"), and others. Averroes was listed among authors who generally denied motion in a vacuum but allowed at least one exception; in Averroes' case, that exception is the motion of things projected into a vacuum. With respect to the second position and instantaneous motion, our authors declared: "Quo ad secundum, data hypothesi quod res aliqua in vacuo moveretur, dicunt aliqui motum in instanti absolui debere, non in tempore, ex quo absurdo postea deducunt impossibilitatem motus in vacuo; ita Averroes, Burlaeus, Aegidius, Jandunus, Venetus, et Pereira." *Ibid.* Virtually the same distinctions were made by Bartholomeus Amicus, *In Aristotelis libros De physico auditu dilucida textus explicatio et disputationes . . .* (2 vols.; Naples, 1626–1629), vol. 2, p. 753, col. 1, who gives as the first opinion the denial of the possibility of motion in a vacuum ("Prima opinio est dicentium in eo non posse fieri motum") and cites Aquinas as one of its supporters. The possibility of such motion constitutes the second opinion, with some authors believing it to be instantaneous and others temporal ("Secunda opinio est dicentium in vacuo, si daretur, posse fieri motum; ita Simplicius, Philoponus, Averroes, D. Thomas, Burlaeus, Niphus, Vicomercatus, Soto, Villapand.; licet differant quod alii velint fieri in instanti, alii in tempore, fundamentum quia nulla est repugnantia cur ibi non possit fieri motus.") We observe, with some bewilderment, that Aquinas was said to both deny and affirm the possibility of motion in a vacuum. For yet another similar approach, see the Complutensian disputations on the *Physics* of Aristotle in *Collegii Complutensis Discalceatorum Fratrum Ordinis B. Mariae de Monte Carmeli Disputationes in octo libros Physicorum Aristotelis iuxtam miram Angelici Doctoris D. Thomae et scholae eius doctrinam . . . nunc primum in Galliis excusae* (Paris, 1628), pp. 438–439.

As we shall see below, Burley's discussion of motion in a vacuum was given a variety of interpretations. If Bartholomeus Mastrius and Bonaventura Bellutus assigned contradictory opinions to him, others would say that he believed motion in a void would be instantaneous (for example, Bartholomeus Amicus, *op. cit.*, vol. 2, p. 754, col. 2, and the Complutensian *Disputationes in octo libros Physicorum*, p. 439, col. 2). In fact, Burley's opinion is complicated by the fact that he distinguished between pure elemental and mixed, or compound, bodies. Although his position with respect to pure elemental bodies is confused and uncertain, there is little doubt that, under certain conditions, he thought mixed bodies, as well as animals, could move temporally in a void (see below, Ch. 3, secs. 4, 7b, d). Curiously, in discussions on Burley, the crucial distinction between elemental and mixed bodies was ignored.

5 For my discussion of this, see Edward Grant, "Aristotle, Philoponus, Avempace, and Galileo's Pisan Dynamics," *Centaurus 11* (1965), 79–95. The article is concerned primarily with demonstrating that Aristotle, Philoponus, and Avempace did not use a concept of specific weight.

6 This opinion is apparently embedded in Avempace's commentary on the seventh book of Aristotle's *Physics*. An Arabic manuscript of it in the Bodleian Library, Oxford, has been reported by Shlomo Pines, "La dynamique d'Ibn Bājja," in *Mélanges Alexandre Koyré: L'Aventure de la science* (Paris: Hermann, 1964), pp. 442–468, especially pp. 460–461; see also Grant, "Aristotle, Philoponus, Avempace, and Galileo's Pisan Dynamics," 95, n. 38. For a general sketch of Avempace's life and works, see Shlomo Pines, "Ibn Bājja," *Dictionary of Scientific Biography*, vol. 1 (New York: Scribner, 1970), pp. 408–410.

7 The whole of Averroes' commentary on bk. 4, text 71, has been translated by John E. Murdoch and Edward Grant in E. Grant (ed.) *A Source Book in Medieval Science* (Cambridge, Mass.: Harvard University Press, 1974), pp. 253–262. In the Bekker numbers of the modern Greek edition of Aristotle's *Physics*, text 71 is equivalent to *Physics* 4.8. 215a.24–215b.20. In vol. 4 of the Junctas edition of Averroes' commentary on the works of Aristotle, the Latin text appears on fols. 158v, col. 1–162r, col. 1. For a discussion of Avempace's views and the problem of motion in a vacuum, see the excellent article by Anneliese Maier, "Der freie Fall im Vakuum," in Anneliese Maier, *An der Grenze von Scholastik und Naturwissenschaft* (2d ed.; Rome: Edizioni di Storia e Letteratura, 1952), pp. 219–254. See also my articles "Bradwardine and Galileo: Equality of Velocities in the Void," *Archive for History of Exact Sciences 2*, 4 (1965), 344–364, and "Motion in the Void and the Principle of Inertia in the Middle Ages," *Isis 55* (1964), 265–292, on which I shall draw in what follows.

8 What Avempace actually denied was that "the ratio of the motion of one and the same stone in water to its motion in air is as the ratio of the density of water to the density of air, unless we assume that the motion of the stone takes time only because it is moved in a medium. Averroes, *Opera*, vol. 4: *Physics*, bk. 4, comment 71, fol. 160r, col. 1. In effect, Avempace rejected the Aristotelian formulation that $V/V' = R/R'$, where V is velocity, or speed, and R is the density, or resistance, of the medium.

9 Grant, *Source Book in Medieval Science*, pp. 256–257, and Averroes, *Opera*, vol. 4: *Physics*, bk. 4, comment 71, fol. 160r, cols. 1–2. The relatively brief section of Averroes' comment 71 devoted to Avempace's criticism and Averroes' reply were previously translated by Ernest A. Moody, "Galileo and Avempace: The Dynamics of the Leaning Tower Experiment," *Journal of the History of Ideas 12* (1951), 184–186. For the most part, the translation by Murdoch and Grant in the *Source Book in Medieval Science* follows Moody's version.

10 Grant, *Source Book in Medieval Science*, p. 257; Averroes, *Opera*, vol. 4: *Physics*, bk. 4, comment 71, fol. 160r, col. 2.

11 *Source Book in Medieval Science*, p. 257. (Thus Averroes explained that according to Avempace, "non sequitur ut illud quod movetur in vacuo, moveatur in instanti quoniam tunc non aufertur ab eo nisi tarditas que accidit ei propter medium, et remanet ei motus naturalis; et omnis motus est in tempore ergo illud quod movetur in vacuo movetur in tempore necessario et motu divisibili. Et nullum sequitur impossibile." Averroes, *Opera*, vol. 4: *Physics*, bk. 4, comment 71, fol. 160v, col. 1.) It is worth noting that in Averroes' direct quotation from Avempace, the latter did not mention void or motion in a void. It is Averroes who rightly drew the inference in the passage just quoted that Avempace's opinion implied finite motion in a void. For a detailed analysis of Avempace's position and Averroes' critique, see the notes to the

translation by Murdoch and Grant, which are cited above in n. 7 of this chapter.

12 The anonymous author of a fourteenth-century compendium of natural philosophy expressed the commonly held doubt that successive motion could occur in a vacuum when he explained that "it is commonly held that local motion could not occur without resistance; and because resistance arises from the medium, it follows that if a vacuum existed, no succession would occur in it because there is no medium in a vacuum." ("Ad hec dubium communiter dicitur quod motus localis fieri non possit sine resistentia; et quia resistentia provenit ex parte medii, relinquitur ergo quod si vacuum esset, nulla esset successio eo quod in vacuo nullum esset medium." MS Bibliothèque Nationale, fonds latin, 6752, fol. 153r. For further information on this work, which has thus far been identified only in the single manuscript mentioned here, see below, Ch. 6, n. 37.)

13 See Grant, Source Book in Medieval Science, p. 257. After explaining that we observe the greatest slowness and the greatest speed in celestial motions (see above, Ch. 3, Sec. 2 and n. 10), Avempace declared: "Et hoc non est nisi propter distantiam motoris in nobilitate a moto; cum igitur fuerit nobilior, tunc illud, quod movetur ab eo, erit velocius; et cum motor fuerit minoris nobilitatis, erit propinquior moto, et tunc motus erit tardior." Averroes, Opera, vol. 4: Physics, bk. 4, comment 71, fols. 160r, col. 2 – 160v, col. 1. Avempace's vagueness stands in contrast to John Philoponus' discussion of the same problem in the sixth century A.D. Philoponus argued clearly that velocity of fall in a void was directly, and time of fall inversely, proportional to the weight of the falling body. See Grant, Source Book in Medieval Science, p. 257, n. 14.

14 For these two expressions see Ernest A. Moody, "Ockham and Aegidius of Rome," Franciscan Studies 9 (1949), 424–425. At least two additional variants may be noted, one by Marsilius of Inghen, who referred to it as incompositas terminorum (see Marsilius of Inghen, Physics, bk. 4, question 9, fol. 51v, col. 2), and another by Pseudo-Siger of Brabant, who cited it as repugnantia terminorum (Pseudo-Siger of Brabant Physics [Delhaye ed.], bk. 4, question 25, p. 181).

15 Bacon, Opera, fasc. 13: Physics (8 books), p. 234.

16 After posing the general problem of whether any kind of motion could occur in a vacuum, Bacon continued: "si esset ibi possibile translationem fieri, queritur utrum esset in tempore vel instanti. Quod in tempore, videtur. Aristoteles, in .8. Physicorum: prius et posterius in spatio sunt causa in translatione facta super spatium; set prius et posterius in translatione faciunt prius et posterius in tempore; ergo in illa translatione esset prius et posterius in tempore, et ita motus." Ibid. For a translation of the last part of this passage, see below, n. 22 of this chapter. Although Bacon cited the eighth book of Aristotle's Physics as a justification for relating space and time (perhaps 8.1.251b.10–12 is intended), a more relevant passage in the Physics from which the distantia terminorum argument may have been derived is 4.11.219a.14–19, which is cited below, Ch. 3, Sec. 3b and n. 22.

17 Although in the fourteenth century Albert of Saxony rejected the distantia terminorum (he used the expression incompossibilitas terminorum), he furnished an excellent description of its essential features and the manner in which it was interpreted. "There is an imagination [or conception]," he declared, "that, if there were no resistance of a medium but a vacuum could be imagined, a simple heavy body would nevertheless descend successively and with a finite velocity because it is impossible that it be simultaneously in the terminus from which, and the terminus to which, [it moves]. And because this is impossible, it is necessary that it be moved successively from the terminus from which [it moves] to the terminus to which [it moves], so that the incompossibility [or incompatibility] of being simultaneously in different places is the cause of the finite velocity of motion of simple heavy and light bodies when surrounded by the resistance of the medium. And those who are of this opinion say that it is the cause of the finite velocity of celestial motion which is without any of the aforementioned resistances." From Albert of Saxony's Physics, bk. 4, question 10, fol. 49v, col. 2, as translated in Grant, Source Book in Medieval Science, p. 273, n. 6. I have added the italics here. According to Albert, then, "those who are of this opinion" assign the cause of all motion, whether through resistant or nonresistant media or in vacua, to the fact that the termini of any motion are distinct

and separated by a distance. Because a body cannot be in any two termini simultaneously, it had to be assumed that an interval of time was required to move from one terminus to the other. Thus the separation of the termini – i.e., "the incompossibility of the termini" (*incompossibilitas terminorum*) – was said "to cause" the successive and finite motion solely because the body in motion could not possibly occupy both termini simultaneously. Because the role of an external resistance was to prevent instantaneous motion, and because the incompossibility of the termini caused finite, successive, temporal motion, Albert included it as one of seven different kinds of external resistance (*ibid.*).

Making the same point somewhat differently, an anonymous fourteenth-century commentator, omitting all mention of the expressions *distantia* or *incompossibilitas terminorum*, argued that successive temporal motion could occur in a hypothetical vacuum because the instants of time that would measure it cannot be immediate: "therefore if some thing were moved from one extremity of a void place to another, either such a thing is in each extremity of the place in the same instant, or not. If it is [in each extremity in the same instant], then it follows that the extremities of the place are immediate [that is, in contact] and, consequently, there is no place. However, if they are not [immediate], then it follows that the thing is in one instant in one extremity, and in another instant in the other extremity. It must be concluded, therefore, that in the intervening [or middle] time that thing is moved in the void place. Therefore a motion can be made in a vacuum." ("Secundo ad idem arguitur quia instantia non sunt immediata. Si ergo res aliqua ab una extremitate loci vacui moveatur ad aliam aut ergo talis res in eodem instanti est in utraque extremitate loci, aut non. Si sic, tunc sequitur quod loci extremitates sunt immediata et consequenter nullus est locus. Si autem non, tunc sequitur quod in uno instanti est in una extremitate loci et in altero in altera. Relinquitur ergo quod in tempore medio movetur in tali loco vacuo. Ergo in vacuo fieri potest motus." MS Bibliothèque Nationale, fonds latin, 6752, fols. 153r–153v. For additional information on this manuscript, see below, Ch. 6, n. 37.) Although our anonymous author conceded that motion could occur in a hypothetical vacuum, he emphatically denied that vacua can occur naturally. In considering the possibility of motion in a vacuum, it is therefore essential to assume what is false and impossible. Nevertheless, "it is sometimes appropriate to assume impossible things for the investigation of truth." ("Querendum est tamen an in vacuo possit fieri motus. Et, licet tale dubium presupponat falsum et impossibile, quia tamen impossibilia presupponere aliquando decet pro veritatis inquisitione. Ideo supposito ut pro impossibili quod vacuum esset, queritur utrum in eo fieri possit motus." *Ibid.*, fol. 153r.)

18 Moody correctly declared ("Ockham and Aegidius of Rome," 425) that "St. Thomas was the recognized advocate, or even originator, of the thesis that *distantia terminorum* is the essential and sufficient cause of the temporal character of motion; . . ."

19 Aquinas' argument is embodied in the following text:
   "1016. Sed *contra hanc rationem Aristotelis* insurgunt plures difficultates. Quarum quidem *prima* est quod non videtur sequi, si fiat motus per vacuum quod non habeat proportionem in velocitate ad motum qui fit per plenum. Quilibet enim motus habet determinatam velocitatem ex proportione potentiae motoris ad mobile, etiam si nullum sit impedimentum. Et hoc patet *per exemplum et per rationem.*

   1017. *Per exemplum* quidem in corporibus caelestibus; quorum motus a nullo impeditur, et tamen eorum est determinata velocitas secundum determinatum tempus.

   1018. *Per rationem* autem, quia ex hoc ipso quod in magnitudine, per quam sit motus, est accipere prius et posterius in motu; prius autem et posterius est in motu ex temporis ratione; sequitur motum esse in determinato tempore." *S. Thomae Aquinatis in octo libros De Physico auditu sive Physicorum Aristotelis commentaria*, editio novissima, ed P. Fr. Angeli-M. Pirotta, O. P. (Naples: M. D'Auria Pontificus Editor, 1953), bk. 4, lectio 12, p. 224. See also my translation of this passage in Grant, "Motion in the Void," 269. The sections cited here have been slightly altered. Already in 1256, when he was about thirty years old, Aquinas had concluded that even if all media and resistance were removed, motion could still occur in measurable time, because the body itself would offer sufficient resistance to cause the motion to take a certain time. The relevant discussion, enunciated in a context of a denial that glorified bodies would be moved from place to place instantaneously, is from Aquinas' *Sentence Commentary*, bk. 4, distinction 44, question 2, article 3, ql. 3 ad 2, and has been translated by James A. Weisheipl, O. P., "Motion in a Void: Aquinas and Averroes," in *St. Thomas Aquinas 1274–1974, Commemorative Studies*, foreword by Etienne Gilson (Toronto: Pontifical Institute of Medieval Studies, 1974), pp. 481–483 (for further discussion, see below, Ch. 3, Sec. 5). To this early dynamic argument, Aquinas added the kinematic argument described here from his *Commentary on the Physics*.

Despite Aquinas' unambiguous position, C. de Waard, *L'expérience barométrique, ses antécédents et ses explications: Etude critique* (Thouars: Imprimerie Nouvelle, 1936), p. 13, has mistakenly included him among those who assumed instantaneous motion in a void.

20 "Resistentia autem talis potest causari ex repugnantia terminorum motus quia non potest transiri ab extremo in extremum nisi per medium; omnis autem motus localis fit super aliquam magnitudinem. Nunc autem termini isti possunt simul esse, et ideo remoto impedimento potest fieri motus remanente sola ratione magnitudinis." Pseudo-Siger of Brabant, *Physics* (Delhaye ed.), bk. 4, question 25, p. 181. Obviously, the expression *repugnantia terminorum* is here equivalent to *distantia terminorum* (see above, n. 14 of this chapter).

21 On Albert, see above, n. 17 of this chapter. Among those who conceived void space as a resistance by virtue of the *distantia terminorum*, A. Maier also identifies Peter John Olivi, William of Ware, and William of Ockham (*An der Grenze von Scholastik und Naturwissenschaft*, pp. 228–234; see also Moody, "Galileo and Avempace," 385–388).

22 *Physics* 4.11.219a.14–19. The translation is by Hardie and Gaye. In his discussion of the *distantia terminorum* argument, Walter Burley cited this very passage to justify motion in a vacuum, declaring that as "the Philosopher says in his treatise on time that before and after in time are taken from before and after in magnitude...." ("... Philosophus tractatu de tempore dicit quod prius et posterius in motu accipiuntur a priori et posteriori in magnitudine...." Burley, *Physics*, fol. 116r, col. 2). It is curious that Roger Bacon chose to ignore this clear passage in favor of a rather vague statement in the eighth book of the *Physics* (8.1.251b.10–12), where Aristotle said: "Further, how can there be any 'before' and 'after' without the existence of time? Or how can there be any time without the existence of motion?" Bacon, however, interpreted this passage as if it were actually the *distantia terminorum* argument, explaining that "in the eighth of the *Physics*, Aristotle [declares] that before and after in space are the cause of the translation made over a space; but before and after in translation [or motion] make before and after in time; therefore, in this translation [or motion] there would be before and after in time, and so also in the motion" (*Opera*, fasc. 13: *Physics* [8 books], p. 234; for the Latin text, see above, n. 16 of this chapter).

23 If it did so serve, it would have been an ironic twist, because the section on time in which it occurs follows immediately after the section on vacuum in which Aristotle had argued that motion in a void would of necessity be instantaneous. In the *De rerum natura*, Lucretius used the concept of the successive occupation of the space between two separated surfaces to argue for the existence of vacuum (see below, Ch. 4, Sec. 4e). Although his argument obviously used a basic idea in common with that of the *distantia terminorum*, and both arguments were intended to show that the concept of a void was at least intelligible, Lucretius seems to have had no connection with the history of the *distantia terminorum* argument.

24 Aristotle, *De sensu* 446a.26–446b.3, in the Oxford translation. The argument is cited in Grant, *Source Book in Medieval Science*, p. 334, n. 4. Aristotle rejected the finite transmission of light on the grounds that "any given time is divisible into parts; so that we should assume a time when the sun's ray was not as yet seen, but was still traveling in the middle space." Because we cannot distinguish the sun's rays in the prior parts of their path from the posterior parts, Aristotle assumed the speed of light to be instantaneous. In *De anima* 2.7.418b.20–27, Aristotle repeated Empedocles' argument and again rejected it.

25 "Ergo cum movetur aliquid in inani, necesse est ut interrumpat spacium vacuum motu

in tempore, aut non in tempore. Sed impossibile est ut sit non in tempore quia prius incidit partem spacii quam totum spacium. Ergo oportet ut sit in tempore." *Sufficientia*, bk. 2, ch. 8, in *Avicenne perhypatetici philosophi ac medicorum facile primi opera in lucem redacta ac nuper quantum ars niti potuit per canonicos emendata: Logyca; Sufficientia; De celo et mundo; De anima; De animalibus; De intelligentiis; Alpharabius de intelligentiis; Philosophia prima* (Venice: Octavianus Scotus, 1508, reprinted in facsimile, Frankfurt; Minerva, 1961), fol. 30r, col. 2.

26 "Haec autem et his similia induxerunt Avicennam quod dixit quod si motus est in vacuo, in rei veritate quod ille esset in ipso tempore; et consentit ei Avempace." Albertus Magnus, *Opera*, vol. 2: *Physics*, bk. 4, tract 2, ch. 7, p. 173, col. 2. But in his discussion of vacuum in the *Sufficientia*, Avicenna emphatically rejected natural and violent motion in a void (see *Sufficientia*, bk. 2, ch. 8, *ed. cit.*, fol. 30v) and mustered numerous arguments against it. In addition to her observation that Avicenna rejected motion in a vacuum, A. Maier describes the corrupt status of the medieval Latin translation of Avicenna's treatise (see Anneliese Maier, *Zwei Grundprobleme der scholastischen Naturphilosophie* (2d ed.: Rome: Edizioni di Storia e Letteratura, 1951), Part II: Die Impetustheorie, pp. 129–132. In light of the poor quality of the available text of Avicenna's *Sufficientia*, it is likely that Albert misinterpreted Avicenna's true opinion.

27 See A. I. Sabra, "Ibn al-Haytham," in *Dictionary of Scientific Biography*, vol. 6 (New York: Scribner, 1972), pp. 196, col. 2–197, col. 1. The translator is unknown.

28 See Frederick Risner, *Opticae Thesaurus Alhazeni Arabis libri septem nunc primum editi. Eiusdem liber De crepusculis et nubium ascensionibus. Item Vitellonis Thuringopolonis libri X* (Basle, 1572), pp. 37–38 and Lindberg's n. 17 in Grant, *Source Book in Medieval Science*, p. 395.

29 *De multiplicatione specierum*, bk. II, ch. 3 (in *The Opus Majus of Roger Bacon*, ed. J. H. Bridges [London, 1900], vol. 2), as translated by David C. Lindberg in Grant, *Source Book in Medieval Science*, p. 431.

30 Bacon, *Opera*, fasc. 13: *Physics* (8 books), p. 235. In much the same manner, Albertus Magnus argued that if a separate vacuum has three dimensions, it must be a body and therefore unable to receive any other body because two bodies cannot occupy the same place (*Physics*, bk. 4, p. 177, col. 2).

31 One who ignored it was Thomas Aquinas.

32 In what follows on the problem of yielding, I have relied largely on my article "Principle of the Impenetrability of Bodies," 564–569.

33 Johannes Duns Scotus, *Opera*, vol. 6, part 1: *Quaestiones in Lib. II Sententiarum*, distinction 2, question 9 ("Utrum Angelus possit moveri de loco ad locum motu continuo"), p. 300.

34 "Ad primam rationem, cum dicitur in illo spacio posset fieri motus, et cetera, verum est si esset tale prius et posterius quod posset cedere mobili. Sed si non esset tale prius et posterius quod posset cedere mobili, non esset verum. Modo quamvis una pars in spacio vacuo esset prior alia, cum una pars esset propinquior mobili quam alia. Tamen illa pars propinquior non posset cedere ipsi mobili; et ideo si moveretur ad eam oporteret quod duo corpora simul essent in uno et eodem loco et unum corpus penetraret aliud, quod est impossibile." Jandun, *Physics* (1519), fol. 57r, col. 1.

35 "Ulterius forte dubitaret aliquis si corpus poneretur in vacuum, utrum vacuum ei cederet. Dicendum quod nunquam corpus cedit alteri corpori nisi sit rarefactibile vel condensabile. Vacuum autem quia dicit dimensiones separatas a talibus passionibus, hic ei competere non possent, nec habet unde cedat . . . . Si enim vacuum cedere non potest corpori posito in ipso, ponentes corpus recipi in vacuo oportet hoc impossibile concedere quod duo corpora simul esse possent. Predicti ergo philosophi ponendo vacuum propter motum ne cogerentur ponere duo corpora simul esse incidebant in illud inconveniens: ut oporteret eos dicere duo corpora simul esse posse. Si enim corpus recipitur in pleno, quia plenum credit, non oportet quod dimensiones simul cum dimensionibus existant: sed si recipitur in vacuo, quia vacuum non habet unde cedat, oportebit dimensiones cum dimensionibus simul esse." Aegidius Romanus, *Physics*, fol. 83r, col. 2.

36 Burley probably had in mind the Greek atomists, and perhaps the Pythagoreans, whose

opinions had been transmitted by Aristotle and Averroes. Although the atomists were generally conceived to have held a concept of void akin to Burley's first type, it is by no means certain that they actually did (see below, n. 38 of this chapter, for another interpretation). However, what he described as the first kind of vacuum is compatible with the conception of Philoponus and Crescas described earlier (see above, Ch. 2, Secs, 3a and 4).

37 Burley, *Physics*, fol. 116v, col. 2.

38 What has been described here appears to have much in common with the ancient atomists, who, according to G. S. Kirk and J. E. Raven (*The Presocratic Philosophers, A Critical History with a Selection of Texts* [Cambridge: Cambridge University Press, 1957], p. 408), "had no conception of bodies occupying space" but "for them the void only exists where atoms are not, that is, it forms gaps between them." A consequence of this interpretation is that void space would be discontinuous. The concept of void space yielding to body and being displaced by it was described and rejected by William of Auvergne. In his *De Universo*, written between 1231 and 1236, William declared that a vacuum must be either divisible or not. If not, motion through it would be impossible, just as it would be if air and water were assumed indivisible, "since every body which is moved through another [body] makes a path for itself by division in it or through it." Thus if a void were truly indivisible, it would be impenetrable and be the strongest and most solid of all things, in which event, we might well ask: If it has such a nature, in what sense is it a vacuum?

But if void is divisible, and a body moves through it, the body will displace, or expel, parts of the vacuum equal to itself. At this point, William departed from those who supported the opinion described by Burley. The displacement of vacuum by a body moving through it signified for William that the displaced void was itself in a void space now occupied by the body that displaced it. William used this argument in support of Aristotle's claim that if a dimensional void exists, it would need a three-dimensional place, and so on, *ad infinitum*. See *Guilelmi Alverni . . . Opera Omnia* (Paris, 1674, reprinted in facsimile, Frankfurt: Minerva, 1963), vol. 1, p. 607.

39 Burley, *Physics*, fol. 117r, col. 1.

40 "Sed contra hoc arguitur per rationem Philosophi sic: quantitas separata nullam proportionem habet in subtilitate et spissitudine ad medium plenum corpore naturali quia quantitas separata ab omni qualitate non habet raritatem vel densitatem, quia raritas et densitas sunt qualitates. Ergo nulla est proportio inter motum factum in medio pleno corpore naturali et motum factum in medio vacuo quantitate; et per consequens motus factus in medio pleno quantitate separata sit subito et in instanti." *Ibid.*, fol. 116v, col. 2.

41 "Dicendum quod raritas et densitas, seu spissitudo et subtilitas, uno modo sunt qualitates consequentes calidum et frigidum. Alio modo accipiuntur pro approximatione partium quantitatis adinvicem, vel pro elongatione partium adinvicem." *Ibid.*

42 " . . . sic dico quod in quantitate separata non sunt raritas et densitas et quod quantitas bene potest separari a raritate et densitate, que sunt qualitates tangibiles." *Ibid.*, fols. 116v, col. 2–117r, col. 1.

43 The complete relevant discussion reads: "Si autem accipiatur raritas et densitas ut sunt remotio vel propinquitas partium quantitatis inter se. . . . Unde dico quod loquendo de raritate et densitate vel de spissitudine et subtilitate secundo modo, scilicet secundum quod sunt relationes necessario consequentes quantitatem sub disiunctione, quoniam omnis quantitas est necessario rara vel densa, sic est verum quod qualis est proportio medii ad medium in subtilitate talis est proportio motus ad motum in velocitate." *Ibid.*, fol. 117r, col. 1.

44 "Et ideo dico quod quantitas est sic activa quia est resistiva, quamvis non sit activa sic quod sit alterativa." *Ibid.*

45 After declaring that some believe that a heavy body could fall in a vacuum because of the *distantia terminorum*, Burley explained that these same individuals then reply to the charge that such a motion would be impossible because of a lack of resistance in a void by distinguishing two kinds of resistance, "scilicet quedam positiva, et illa est semper cum resistentia et reactione. Et talis resistentia nunquam est sine violentia

resistentis, et hoc si resistens vincatur sicut patet quando aliquod corpus penetrat terram, vel movetur contra aquam currentem, vel contra motum aeris." *Ibid.*, fol. 117r, col. 2.

46 "Resistentia privativa est incompossibilitas aliquorum ad tertium vel alicuius unius ad aliqua diversa. Verbi gratia, posito quod lapis posset esse simul cum aere, sic quod lapis non pelleret aerem extra locum suum secundum quod aer et lapis possent esse simul in eodem loco adequato, tunc aer non resistit lapidi positive quia non oportet aerem violenter moveri ad motum lapidis. Tamen aer secundum diversas partes eius resisteret lapidi privative et illud non est aliud dicere quam quod lapis non posset esse simul in diversis partibus aeris. Et ista resistentia requiritur necessario in omni motu locali recto; et talis resistentia sufficit ad motum localem." *Ibid.*

47 Aristotle's purpose, of course, had been to show the impossibility of the existence of void, whereas those who sought to render intelligible the concept of finite motion in a void adopted Aristotle's identification of void and body as a means of achieving this end.

48 Burley made this explicit with respect to the first kind of void, namely that which was assumed capable of receiving bodies. "Sed primum modum ponendi dimensionem separatam, si grave vel leve esset positum in tali spatio quanto quod spatium non cederet corpori naturali, sed compareretur corpus naturale secum; sic dicendum quod non movetur motu divisibili in tali spatio quia ex quo permittit secum corpus grave vel leve nullo modo resisteret grave vel levi et per consequens non esset resistentia ibi ex parte medii, nec est aliqua resistentia ex parte mobilis, quia idem non resistit sibi ipsi. Ideo dico quod corpus grave vel leve fieret subito ab uno extremo talis spatii ad aliquod extremum, nec transiret medium, nec aliqua parte ipsius spatii. . . ." *Ibid.*, fol. 116v, col. 2.

49 If motion could be successive in a vacuum, Burley explained that this would vitiate Aristotle's arguments, which he obviously accepted. "Item illa propositio Philosophi esset falsa, videlicet que est proportio medii ad medium in subtilitate, talis est proportio motus ad motum in velocitate, quia certum est quod vacui ad plenum nulla est proportio in subtilitate. Item secundum istud sequitur quod in eodem tempore pertransiretur duo spatia equalia ab eodem mobili quorum unum est plenum et aliud vacuum, quod est impossibile. Consequentia patet, quia omnis motus gravis et levis quantumcumque velox posset fieri in tempore quantumcumque tardo, ut patet quia medium potest subtiliari in infinitum." *Ibid.*, fol. 117r, col. 2. In addition to the inadequacy of the *distantia terminorum*, the arguments cited here led Burley to reject the vacuum as the cause of finite motion within itself. Albertus Magnus had earlier argued against the *distantia terminorum* on similar grounds. He insisted that if a body fell in a vacuum with a finite speed and finite time, one could imagine a material medium with just the proper density so that it would cause that same body to fall with the same speed and time over the same distance as in the void, a patently absurd consequence of finite motion in a void. Albertus Magnus, *Opera*, vol. 2: *Physics*, bk. 4, tract 2, ch. 6, p. 172, col. 1. Aquinas, who accepted the possibility of finite motion in a void, may have had Albertus Magnus in mind when he argued that "Nothing absurd follows if a body were to move at the same rate of time in a plenum as in a void, if we imagine an extremely subtle body, because the greater the determined subtlety of the medium, the greater is the tendency to diminish the slowness in motion. Hence one can imagine a subtlety so great that it would tend to make less slowness than that slowness which the resistance of the body would have. And so the resistance of the medium would add no slowness to the motion." Translated from Aquinas' *Sentence Commentary*, bk. 4, distinction 44, q.2, a.3, ql.3, ad 2, by James A. Weisheipl, "Motion in a Void: Aquinas and Averroes," p. 483. According to Aquinas, one can imagine a material medium so subtle and rare that it no longer affects the velocity of fall, and the body will fall in that medium as if it were in a void.

50 In an otherwise interesting article, "Walter Burley and Text 71," *Traditio 16* (1960), 395–404, Herman Shapiro analyzed the texts we have discussed and erroneously attributed to Burley the very opinion that he had rejected. That is, Shapiro concluded that Burley justified finite motion in an extended vacuum by virtue of the *distantia terminorum* and by the resistance offered by the rarity and density of the vacuum. We

are told (p. 403) that "for Burley the resistant value encountered in a void is always expressible by a positive integer," which is assumed constant, thus permitting ratios of motion to be formulated and comparisons to be made. Shapiro associated the resistance of the vacuum with the divisibility of the constituent parts, which Burley expressed in terms of rarity and density. However, the resistance is a "pure mathematical resistance not subject to such variations of density as arise from the physical 'elongation' or 'propinquity' of substantial and/or qualitative parts" (ibid.). Shapiro concludes that "Burley has projected a view of the void as being a magnitude of space both empty and full; substantially featureless, and yet quantitatively endowed; real, and yet ideal – all at the same time" (ibid.). Shapiro's serious misinterpretations and embellishments stem from his mistaken belief that Burley assigned a resistive function to an extended vacuum. In the admittedly difficult arguments, this is precisely what Burley rejected. It was because of the lack of resistance that Burley sought to justify motion in a vacuum *ex parte mobilis*, an aspect of Burley's discussion not even mentioned by Shapiro.

51 "Dicimus autem quod resistentia mobilis ad motorem est non talis quod movens non possit movere mobile vel quod mobile inclinetur ad oppositum, quia sic est precise in motu violento. Sed talis resistentia intelligitur quod mobile est sub aliquo ubi cui non potest immediate succedere terminus motus intentu a primo motore infinite virtutis quoniam mobile non est in plena obedientia motoris infiniti cum vigore. Per resistentiam autem medii ad mobile et ad motorem intelligo omne illud quod necessario precedit inductionem termini producendi que sunt mediata (media?) naturaliter ordinata inter formam quam habet mobile et terminum ad quem. Talia inquam mediata (media?) resistunt mobili et motori ut mobile non statim possit poni virtute finita motoris in termino ad quem." *Habes Nicholai Bonetti viri perspicacissimi quattuor volumina: Metaphysicam, videlicet naturalem phylosophiam, predicamenta, necnon theologiam naturalem . . .* (Venice, 1505), fol. 62v, col. 2. A Franciscan and student of Duns Scotus, Nicholas Bonetus (ca. 1280–1343) taught at Paris in the 1320s, during which time he probably wrote his commentary on Aristotle's *Physics*.

A similar opinion was held by Johannes Canonicus, or Johannes Marbres, a follower of Franciscus de Marchia, who explained that the "succession of a motion arises not only from the resistance of a medium, or of a mover to the thing moved or that to which it is moved, but it can arise from the order of the forms that are acquired and arranged naturally, just as appears in the motion of the supercelestial bodies where, in the absence of a resistant medium, there is no resistance of a medium to a mobile, or of a mover to the body that is moved." (". . . successio motus non solum est ex resistentia medii ad mobile, vel motoris ad motum vel ad quod movetur, immo potest esse ex ordine formarum acquisitarum naturaliter ordinatarum sicut apparet in motu corporum supercelestium ubi nulla est resistentia medii ad mobile vel motoris ad ipsum corpus quod movetur, cum nullum medium ibi fuerit." *Joannis Canonici questiones super VIII lib. Physicorum Aristotelis perutiles; nuperrime correcte et emendate additis textibus. Commentorum in margine una cum utili reportorio cunctorum auctoris notabilium indice* [Venice, 1520], fol. 43r, cols. 1–2.) Shortly after, Canonicus declared explicitly that succession is caused solely by the order of the positions, or locations (*ubi*), that are acquired by a body in motion (". . . dico quod successio que causatur ex solo ordine ipsorum ubi acquiredorum . . ." *Ibid.*, fol. 43r, col. 2). Canonicus probably completed his *Questions on the Physics* between 1321, the year Franciscus Mayronis wrote his *Sentence Commentary*, which Canonicus cited, and 1323, the year St. Thomas Aquinas was canonized, because the latter is identified by Canonicus as *frater Thomas* rather than *sanctus Thomas* (see Maier, *Zwei Grundprobleme*, p. 199, n. 56). For a description of Canonicus' treatise, see Anneliese Maier, *Ausgehendes Mittelalter: gesammelte Aufsätze zur Geistesgeschichte des 14. Jahrhunderts* (3 vols.; Rome: Edizioni di Storia e Letteratura, 1964–1977), vol. 1, ch. 10 ("Verschollene Aristoteleskommentare des 14. Jahrhunderts"), pp. 239–250. The numerous manuscripts and editions of Canonicus' *Physics* are listed by Charles H. Lohr, "Medieval Latin Aristotle Commentaries, Authors: Jacobus-Johannes Juff," *Traditio* 26 (1970), 183–184. Lohr mentions that Canonicus also cited Gerard of Odo, Peter Aureoli, and Thomas Anglicus. For more on Canonicus and Bonetus, see below, Ch. 3, Secs. 6 and 8.

52 Many in the Middle Ages assumed that distance, rather than velocity, is proportional to time. See Marshall Clagett, *The Science of Mechanics in the Middle Ages* (Madison, Wis.: University of Wisconsin Press, 1959), p. 678.

53 Aegidius Romanus, *Physics*, bk. 4, fols. 80v, col. 2-81r, col. 1 and 97v, col. 2-98r, col. 1. Because of its length, I have omitted the Latin text.

54 In truth, Aegidius argued that the descent of a body in a vacuum is not actually a motion because it does not occur in time, which, as Aristotle insisted, is the measure of motion. Motions and changes can only occur in time when, as we have seen, there is a contrary disposition, or resistance, a condition that cannot obtain in a vacuum. ("Dicendum quod, ut in capitulo de vacuo diximus, sicut tota causa quare requiritur tempus in introductione forme est contraria dispositio in materia, sic tota causa quare requiritur tempus in motu locali sumitur ex resistentia medii vel mobilis. Propter quod si grave descenderet in vacuo, quia ibi non esset resistentia medii cum vacuum sit, in eo esset motus non haberet unde resisteret nec etiam esse ibi resistentia mobilis, cum in tali motu mobile non sit actu distinctum a motore quia gravia non moventur ex se; nec sunt divisibilia in talia duo quorum unum est per se movens et aliud per se motum, sequeretur quod in vacuo gravia descenderent in non tempore. Ideo concludebatur supra quod cum de ratione motus sit quod mensuretur tempore, quod in vacuo non est motus." *Ibid.*, bk. 4, fol. 98r, col. 1.) Although the descent of a body through a vacuum is not categorized as a motion, the vacuum is conceived as an extended magnitude, and any body imagined to fall through it must occupy different positions (*mutata esse*) within that extension. Different instants (*instantia*), or moments, must correspond to each of these different positions. Thus a plurality of instants must exist when a body is imagined to fall in a vacuum. Each instant cannot, however be measured by a corresponding moment in the heavens where actual continuous time is measured. For if such correspondences could be made, the moments in a vacuum would be measurable by regular time and, as a consequence, regular temporal motion could occur in a vacuum, which Aegidius emphatically denied. And yet Aegidius held that between these instants that measure the fall of a body in a vacuum, there is no temporal interval. Indeed the totality of instants that might measure that fall do not even constitute a magnitude. In their totality, they do not equal more than one instant of celestial time. Aegidius concluded that the plurality of instants that measure the fall of a body in a vacuum correspond to a single instant of celestial time. Thus not only are the instants in the vacuum not linked by time, but the different positions occupied by the descending body are not linked by any motion, because time is the measure of motion. Aegidius, however, insisted that there is a succession between the instants because the descending body could not be simultaneously in the different parts of void space. It must occupy those different and distinct parts of space successively, even though the totality of instants corresponding to each punctual location of the body equals only one instant of real celestial time. ("Dicamus ergo quod si grave descenderet in vacuo esset ibi dare plura instantia absque tempore medio, sicut esset ibi dare plura mutata esse absque moveri medio. Et inter illa plura instantia esset successio; nec esset simul corpus grave in diversis partibus spatii. Sed secundum illa plura instantia successive esset in pluribus partibus spatii, non obstante quod omnia illa instantia non essent plus quam unum instans." *Ibid.*, fol. 98v, col. 1.) For a brief discussion of Aegidius' views, see Maier, *An der Grenze von Scholastik und Naturwissenschaft*, pp. 225-227. I have also benefiited from a paper by John M. Quinn, O. S. A. (Villanova University), "The Problem of the Void in Giles of Rome," delivered on May 6, 1978 at the thirteenth annual Medieval Conference at Kalamazoo, Michigan. Father Quinn generously supplied me with a copy of that paper. Where Father Quinn has rendered *mutatum esse* as "moment" I have conceived it as position or location in order to contrast it to *instans*, which must be translated as "instant" or "moment."

55 In Buridan's words (*Physics*, bk. 4 question 9, fol. 76v, col. 2), "the incompossibility of the termini is inadequate to explain that a mutation [or instantaneous change] might be successive because there could be an incompossibility of termini in an instantaneous change, even as God could make a body in the heavens reach the earth instantaneously." See also my article "Motion in the Void," 277.

56 See Maier, "Der freie Fall im Vakuum," in her *An der Grenze von Scholastik und*

Naturwissenschaft, pp. 249–250, and Grant, "Bradwardine and Galileo: Equality of Velocities in the Void," 347–348. For Albert of Saxony's discussion of infinite velocity in a vacuum, see below, Sec. 7a.

57 Marsilius of Inghen, *Physics,* bk. 4, question 9, fol. 51v, cols. 1–2. In this argument, Marsilius, who regularly used the expression *incompositas terminorum,* is clearly relying on Aristotle, *Physics* 4.8.216a.11–20.

58 *Ibid.,* fol. 51v, col. 2.

59 *Ibid.* In the text of Marsilius' *Physics* attributed to Duns Scotus and published in the latter's *Opera Omnia* (Lyons, 1639), vol. 2, p. 258, col. 2, *incompositas terminorum* has been consistently presented as *incompossibilitas terminorum.* Albert of Saxony used the same concept of ever-increasing velocities to justify motion in a vacuum; see below, Sec. 7a.

60 *Ibid.*

61 In rejecting the *distantia terminorum* as the exclusive cause of successive motion, Burley insisted that it had to be supplemented "by resistance on the part of the medium, or on the part of the mobile, or from both." ("Ad primam rationem, cum dicitur quod sola distantia inter terminos sufficit ad motum successivum. Dico quod non est verum. Immo simul cum distantia terminorum requiritur resistentia ex parte medii vel ex parte mobilis, vel ex parte utriusque." Burley, *Physics,* fol. 117r, col. 1; see also fol. 116v, col. 1.) Aegidius Romanus (Giles of Rome) said much the same thing. ("Dicamus ergo cum Commentatore quod tempus in motu requiri potest vel propter resistentiam mobilis vel propter impedimentum medii, vel propter utrumque." Aegidius, *Physics,* bk. 4, fol. 81r, col. 2.)

62 It is obvious that no conception of resistance, whether *ex parte medii* or *ex parte mobilis,* would have provided an effective response to Buridan's example (above, n. 55 of this chapter). If God decided to move a body instantaneously over an intervening distance, whether void or full, no resistance inside or outside of the mobile could reduce that infinite speed to a finite one. The value of Buridan's illustration, and others like it, may lie in the encouragement it provided for scholastics to explain succession in a vacuum by means that were either supplemental to the *distantia terminorum* or wholly opposed to it.

63 "Deinde, quia in gravibus et levibus remota forma, quam dat generans, remanet per intellectum corpus quantum; et ex hoc ipso quod quantum est in opposito situ existens, habet resistentiam ad motorem." Thomas Aquinas, *Physics,* bk. 4, lecture 12, par. 1024, p. 225. See my translation of this passage in "Motion in the Void," 270 and n. 16. In his *Physics,* written around 1269–1270 (see James A. Weisheipl, O. P., *Friar Thomas d'Acquino* [Garden City, N.Y.: Doubleday, 1974], p. 375), Aquinas elaborated an earlier version of this opinion enunciated around 1256 in his *Commentary on the Sentences* (see Weisheipl, *Friar Thomas d'Acquino,* pp. 358–359 and above, n. 19 of this chapter).

64 In the judgment of John of Jandun, Aquinas and his followers identified the *corpus quantum,* not the *distantia terminorum,* as a sufficient condition for motion in a void space. "Addunt autem aliqui, scilicet sanctus Thomas et sui sequaces, quod si esset huiusmodi spatium vacuum, mobile in ipso existens ex eo ipso quod esset corporeum in opposito sui existens resistentiam sufficientem haberet ad motorem. Et hec resistentia sufficeret ad motum localem in illo spatio, ut dicunt." Jandun, *Questions on the Physics,* bk. 4, question 11, fol. 56v, col. 2.

65 In *Physics,* 8.4.256a.1–2, Aristotle suggested that in natural motion, light and heavy things "are moved either by that which brought the thing into existence as such and made it light and heavy [the medieval term for this causal agent was *generans*], or by that which released what was hindering and preventing it." I have added the bracketed material. It was essentially this view of the *generans* that Aquinas adopted as the cause of natural motion. For a further discussion, see Grant, *Source Book in Medieval Science,* p. 265; for Averroes' description, see *ibid.,* p. 263.

66 Aquinas' contemporary, Pseudo-Siger of Brabant, also identified a body's *corpus quantum* as a resistance. Offering much the same argument as Aquinas, he declared: "Quod autem dicimus quod, cum removeamus formam gravium et levium, non remanet nisi materia, dicendum quod remanet corpus actu quantum; remanente autem corpore quanto remanent

aliqua resistentia, et ideo quamvis separetur forma gravis et levis, adhuc remanebit quod possit facere resistentiam, et ideo dicendum quod, remanente prioritate et posterioritate in medio, potest manere motus." Pseudo-Siger of Brabant, *Physics* (Delhaye ed.), bk. 4, question 25 ("Utrum in vacuo possit fieri motus"), p. 182. Because Pseudo-Siger identified the *distantia terminorum* as resistance in the void (see above, n. 20 of this chapter, where it is called *repugnantia terminorum*), he included in his physics two distinct, and apparently ontologically equal, resistances in empty space.

67 Aegidius declared that "in motu autem gravium et levium solum requiritur tempus propter medii impedimentum, non autem propter resistentiam mobilis; nam cum ibi ratio movendi sit forma, quia forma abstracta ab huiusmodi corporibus gravibus vel levibus non remanet nisi materia. Quia materia non habet per quod resistat in motu talium non potest requiri tempus propter resistentiam mobilis, sed solum propter impedimentum medii, quod si dicatur quod abstracta forma, si non remanet materia cum qualitate, quia qualitas se tenet ex parte forme, remanet tamen materia cum quantitate, quia quantitas se tenet ex parte materiae, ut patet per habita. Propter hoc non requiretur ibi tempus, nam, ut ostensum est, ratione quantitatis non requiritur tempus in motu nisi ipsam quantitatem comitetur aliqua contraria dispositio vel aliqua resistentia vel aliquod aliud impedimentum. Cum ergo queritur utrum in vacuo esset motus in non tempore, dicendum quod motus ille qui requirit tempus solum propter resistentiam medii, cuiusmodi est motus gravium deorsum vel levium sursum, fieret in non tempore." Aegidius Romanus, *Physics*, fol. 81r, col. 2. For Aegidius' views on instantaneous motion, see above, Ch. 3, Sec. 4.

68 By "mover" in this context, the reader should understand an efficient, not a final, cause.

69 Throughout the discussion described here, Averroes spoke of "simple" bodies, by which was ordinarily meant a body composed of one of the four pure elements. But the argument applies as well to inorganic compounds composed of two or more of those elements. By the fourteenth century, however, as we shall see below, the difference between pure elemental bodies and compounds, or mixed bodies, was of great significance.

70 For the relevant texts, see my article "Motion in the Void," 270, n. 13 and 14. The arguments from Averroes' comment 71 have been translated by Murdoch and Grant in Grant, *Source Book in Medieval Science*, pp. 261–262. For the difficulties in Averroes' account of the relationship between form, body, and external resistance, see my translation of Averroes' comment 28 on bk. 3 of Aristotle's *De caelo* in *Source Book*, pp. 263–264, especially n. 12, where there are references to other related selections. See also Weisheipl, "Motion in a Void: Aquinas and Averroes," pp. 478–480.

Walter Burley revealed how scholastics interpreted Averroes' explanation of succession in motion by means of resistance. The three kinds of resistance arise by virtue of the relationship between a mover and the thing that it moves. The first way is restricted to the mobile alone, which offers resistance to its mover, as is the case in celestial motions, "since in these motions there is no resistance except that between mobile and mover," by which Burley meant that in the incorruptible ether of the heavenly region there is no external resistance. Hence the resistance that causes the finite motion of the celestial spheres arises from a direct relationship between the celestial mover, or intelligence, and the physical sphere. Temporal motion arises in yet a second way by virtue of the resistance of a medium only, "as in [natural] motions of simple [i.e., elemental] heavy and light bodies, since in these the thing that is moved is not actually divided into a mover and thing moved." The debate on temporal motion in a void was largely concerned with this case. The third way consists of a combination of the first two ways, "namely from the thing moved and from the media, as is the disposition of animals that are moved in water." The soul, which is the mover in an animal, must overcome not only the resistance of its own body but also the resistance of the external medium in which the animal finds itself. ("Et dicit Commentator quod ista resistentia quam ponit inter motorem et rem motam contingit tripliciter: vel ex parte moti tantum, sicut est in motu corporum celestium, quoniam in motibus illorum non est aliqua resistentia nisi mobilis ad motorem. Alio modo contingit ista resistentia ex parte medii tantum, sicut in motibus corporum simplicium levium et gravium, quoniam in illis res mota non dividitur in motorem et rem motam in actu. Tertio modo contingit ista resistentia ex utroque, scilicet ex re mota et ex mediis, sicut est dispositio in animalibus que moventur in aqua. Et sic patet ex dictis Commentatoris quod causae successionis in motu sunt tres, videlicet resistentia solius mobilis ad motorem, sicut accidit in motibus corporum celestium; alia causa est resistentia solius medii ad motorem, sicut accidit in motu naturali gravium et levium; tertia causa est resistentia ex parte utriusque, scilicet ex parte mobilis et ex parte medii, sicut accidit in motu animalis in aere vel in aqua, quoniam in tali motu corpus resistit anime et etiam medium resistit maxime si medium econtra moveat." Burley, *Physics*, bk. 4, fol. 116r, col. 1.

71 After explaining the conditions under which motion in a void would occur in no time (see below, n. 74 of this chapter), Aegidius declared: "Sed motus ille qui requirit tempus propter resistentiam mobilis cuiusmodi est motus violentus, et motus animalium, vel motus eorum que moventur ex se; huiusmodi motus fieret in tempore." Aegidius Romanus, *Physics*, bk. 4, fol. 81r, col. 2.

72 Like Aegidius Romanus, Walter Burley drew this inference from Averroes. But Burley made such self-motion in a vacuum dependent on the animal having a fixed support on which to walk, as for example, the earth itself. If a vacuum existed above the earth, an animal could walk on the earth while surrounded by a vacuum. A bird, however, could not fly in a vacuum because there would be no body, presumably air, by which it could support its flight. ("Et si dicatur ponamus quod super terram sit vacuum ita quod totus locus aque et aeris esset vacuus, posset ne animal moveri in illo vacuo figendo pedes suos super terram. Dico ut mihi videtur quod animal bene posset ambulare in vacuo existente super terram figendo pedes ad terram quia haberet fixionem et resistentiam intrinsecam sufficientem. Posset etiam animal saltare in vacuo. Et tamen non esset possibile animal volare in vacuo quia ibi non esset corpus cui inniteretur in volando." Burley, *Physics*, fols. 117v, col. 2–118r, col. 1.) In his *Questions on the Physics*, Albert of Saxony expressed an opinion similar to Burley's (see Grant, *Source Book in Medieval Science*, p. 339). The same opinions were repeated in the seventeenth century by Bartholomeus Amicus (*Physics*, vol. 2, p. 753, col. 2) and, in a joint work, by Bartholomeus Mastrius and Bonaventura Bellutus (*In Arist. Stag. libros Physicorum*, p. 899, col. 2).

73 "Cum ergo queritur utrum animalia moverentur in tempore in vacuo, dici potest quod si ille motus non esset mediante appetitu, sed mediante gravitate, ut puta si animal caderet de alto deorsum, in tali motu idem iudicium esset de animali quod de aliis gravibus et levibus. Esset enim huius motus in non tempore quia non haberet per quid resisteret mobile moventi. Sed si fieret talis motus mediante appetitu fieret in tempore quia in tali motu gravitas se teneret ex parte mobilis per quam resisteret moventi. Cum ergo motus animalis, ut animal est, sit mediante appetitu, quia in tali motu requiritur tempus, bene dictum est quod in vacuo gravia et levia moverentur in instanti, animalia autem in tempore." Aegidius Romanus, *Physics*, bk. 4, fol. 82v, col. 1.

74 "Cum ergo queritur utrum in vacuo esset motus in non tempore, dicendum quod motus ille qui requirit tempus solum propter resistentiam medii, cuiusmodi est motus gravium deorsum, vel levium sursum, fieret in non tempore." *Ibid.*, fol. 81r, col. 2.

75 For Aristotle's explanation of how the air can cause continuous motion, see *Physics*, 8.10.266b.25–267a.21.

76 On the development of medieval theory, the best account is Clagett, *Science of Mechanics in the Middle Ages*, pp. 505–540. See also the important study by Maier, "Die Impetustheorie," *Zwei Grundprobleme*, pp. 113–314.

77 Here is the full Latin text, from which I have translated only a few sentences: "Motus etiam violentus potest esse in vacuo. Et quando dicitur de motu proiectorum, dico quod melius fieret in vacuo quam in pleno propter resistentie carentiam. Et quando dicitur quod res proiecta in vacuo non haberet a quo moveretur sicut in pleno, dico quod res proiecta movetur naturaliter virtute proiicientis et non a partibus medii per quod fertur, quod apparet de sagitta proiecta per aerem. Unde non potest intelligi vel dici quod moveatur propter impulsum partium aeris. Sensibiliter enim apparet quod aer non facit illum motum, sed magis prestat ipse impedimentum, ergo, etc." Johannes Canonicus, *Physics*, fol. 43r, col. 1. This was Canonicus' response to the objection made at the beginning of the question that because void lacks a medium, such as air, which could

serve to push the arrow along, violent motion could not occur in a vacuum. "Nec violentus, probatur de proiectis, quoniam cum proiectum a proiieciente dimissum fuerit non habet unde moveatur in vacuo, cum proiectio fiat in aere, ut sagitta, proiecta movetur a partibus aeris." *Ibid.*, fol. 42v, col. 2.

In her citation of the first passage above (from MS Vat. lat. 3013, fol. 47r), Maier (*Zwei Grundprobleme*, p. 200. n. 58) has substituted *materialiter* for *naturaliter*, on grounds that the latter makes no sense. In the context, however, *materialiter* seems to clarify little or nothing. Perhaps we ought to accept *naturaliter*, with the clear understanding that it does not signify natural motion, but rather that an impressed force is communicated to the body by a natural process. which enables it to move with violent motion in a void.

78 The use of impressed forces as explanatory mechanisms for motion in a vacuum by Johannes Canonicus and Nicholas Bonetus (see below) is treated by Maier as little more than a repetition of the ideas on impressed force developed at some length by Franciscus de Marchia in the latter's *Sentence Commentary*, which was read at Paris in 1319–1320 but has been left to us in the form of a *Reportatio* dated 1323 (see Maier, *Zwei Grundprobleme*, pp. 197, 198–200). But in de Marchia's relevant Latin text, which Maier reproduced (*ibid.*, pp. 166–180; for a translation of part of that text, see Clagett, *Science of Mechanics in the Middle Ages*, pp. 526–530), there is no discussion of motion in a vacuum. Hence Canonicus and Bonetus extended the application of impressed forces to render violent motion in a void intelligible, no small achievement. If they were not the first, they must surely be counted among the very earliest scholastics to consider the problem. Because Bonetus and Canonicus were contemporaries and both wrote on our subject in the 1320s (see Maier, *Zwei Grundprobleme*, pp. 198 and 199, n. 56, and n. 51 above, this chapter), priority between them has yet to be determined. In the seventeenth century, Bartholomeus Amicus also thought that the violent motion of a projectile could be effected in a vacuum by means of an impressed force (*impetus impressus*). Whether the force is permanent or transient is left unmentioned. After explaining that animals could move progressively in a vacuum, Amicus stated that things can be moved in it by an impressed force: "Hinc etiam patet idem dicendum esse de motu violento proiectorum, si hic motus fiat ab impetu impresso; si autem fiat a medio non erit." Bartholomeus Amicus, *Physics*, vol. 2, p. 753, col. 2(B). In a later clarification, Amicus described the impressed force, rather than the external medium, as the principal cause of motion. ("Ad 4 respondetur negando motum proiectorum oriri medio ut causa adaequata, vel principali, sed ut causa adiuvante impetum impressum, qui est causa principalis." *Ibid.*, p. 754, col. 2[B].)

79 Bonetus, *Habes Nicholai Bonetti*, fol. 63r, cols. 1–2. For the Latin text, see my article "Motion in the Void," 273, n. 26.

80 Bonetus, *Habes Nicholai Bonetti*, fol. 63r, col. 2. The Latin text is cited in my article "Motion in the Void," 274, n. 27.

81 Although the passage is cited in "Motion in the Void," 274, n. 28, its importance warrants repetition here: "Fertur autem aliquibus quod motus violentus posset esse in vacuo absque hoc quod movens primum sive proiiciens coniungatur mobili vel realiter vel virtualiter. Et ratio huius dicti est ista: quia in motu violento mobili imprimitur aliqua forma non diu permanens sed quasi transiens, et quamdiu illa forma durat posset esse motus in vacuo; illa autem deficiente cessat motus." Bonetus, *Habes Nicholai Bonetti*, fol. 63, col. 2.

82 *Ibid.*, fol. 63v, col. 1; see Grant, "Motion in the Void," 274, n. 29, for the Latin text.

83 A possible source for the concept of self-expending impetus is Simplicius' commentary on Aristotle's *De caelo*, which was translated from Greek to Latin in 1271 by William of Moerbeke. Here Simplicius attributed such a concept to Hipparchus. For an English translation, see Morris R. Cohen and Israel E. Drabkin, *Source Book in Greek Science*, p. 209. The translation is reproduced by Clagett, *Science of Mechanics in the Middle Ages*, p. 543; for Abu'l-Barakat's use of the idea among the Arabs, see pp. 513–514.

84 On Buridan's impetus theory, see Clagett, *Science of Mechanics in the Middle Ages*, pp. 532–540, 557–564, and Maier, *Zwei Grundprobleme*, pp. 201–235.

85 For further discussion of the problem of inertial motion in a vacuum, see Grant, "Motion in the Void," 265–292.
86 See Aristotle, *De caelo* 1.2.268b.27–30.
87 For Aristotle, all mixed, or compound, bodies were composed of all four elements (see *De generatione et corruptione* 2.8.334b.31–335a.24). In the Middle Ages not all mixed bodies were thought to be constituted of all four elements. Those that had only two or three were characterized as "imperfect" (*mixtum imperfectum*) (see Maier, *An der Grenze von Scholastik und Naturwissenschaft*, p. 244. For Walter Burley's quite different interpretation of perfect and imperfect mixed bodies, see below, Ch. 3, Sec. 7b.
88 Perceptible fire, for example, was not the pure element fire but a compound of two or more elements. It was the same for earth, water, and air.
89 *De caelo* 1.2.268b.30–269a.6.
90 See his *Physics*, bk. 4, question 10, fol. 77r, col. 2, where these two arguments constitute the eighth conclusion of a question on whether a heavy body could be moved in a vacuum. For Aristotle's description, see above, Ch. 1, for the first and *Physics*, 4.8.215b. 23–216a.7 for the second.
91 *Ibid.*, fol. 77r, col. 2.
92 *Ibid.*, fols. 77r, col. 2–77v, col. 1.
93 See Albert of Saxony, *Physics*, bk. 4, question 9, fol. 49r, col. 2, and Grant, *Source Book in Medieval Science*, p. 273, for the translation.
94 Earlier in the fourteenth century, Johannes Canonicus had made speed in a vacuum directly proportional to the heaviness (*gravitas*), or weight, of a body. (In response to a claim that no natural motion would be possible in a vacuum, Canonicus declared: "dico quod differentie tarditatis vel velocitatis provenire possunt sicut ratio supponit ex differentia gravitatis maioris vel minoris, supposita indeterminatione medii. Et tunc dico quod successio, que causatur ex solo ordine ipsorum ubi acquirendorum, potest velocitari secundum magis et minus ex sola differentia gravitatis maioris vel minoris in moventibus." Johannes Canonicus, *Physics*, bk. 4, question 4, fol. 43r, col. 2.) The general conviction that the speed of a falling body is directly proportional to its weight was widely adopted in the fourteenth century. Although Aristotle had hinted at it, John Philoponus made it explicit (see Grant, "Aristotle, Philoponus, Avempace, and Galileo's Pisan Dynamics," 84–85).
   In the early seventeenth century, Tommaso Campanella would invoke both the *limitatio agentis* (which he attributed to Avempace and Aquinas) and the incompossibility of the termini arguments as evidence that if artificial vacua were created, motion in them would not be instantaneous, as Aristotle believed (see pp. 27–28 of the Italian edition of Campanella's *Del senso delle cose e della magia* and p. 364 of the English translation cited below in Ch. 8, n. 72).
95 See Grant, *Source Book in Medieval Science*, p. 273. In a question inquiring whether heavy elemental bodies could have an internal resistance by means of which successive motion could be caused in the absence of an external medium ("Utrum grave simplex habeat resistentiam intrinsecam per quam possit fieri successio in eius motu circumscripta resistentia extrinseca"), Marsilius of Inghen described and rejected a variety of internal resistances. Among seven arguments against the "limitation of the agent" (*limitatio agentis*), the third is noteworthy because it denies to elemental bodies one of the essential properties of a resistance, namely that it must have a tendency to move the body in a direction contrary to that in which the agent pushes it. But if an element had within it a resistance inclining it in a contrary direction, it would no longer be a simple body. ("Tertio, quia omnis resistentia in motu est aliquo modo contraria inclinationi motoris. Sed in elemento simplici non sunt inclinationes contrarie quia iam non esset elementum simplex." Marsilius of Inghen, *Physics*, bk. 4, question 8, fol. 51r, col. 2. The enunciation of the question cited above appears on fol. 50v, col. 2.)
96 See the seven types of internal resistance described by Marsilius of Inghen, *Physics*, fols. 50v, col. 2–51r, col. 1.
97 *Ibid.*, fol. 51r, col. 2.
98 "Item omne mobile est continue in loco sibi equali, ut patet octavo huius. Si igitur grave

subito descenderet in vacuo, in eodem instanti esset sursum et deorsum et in toto medio. Ergo si minimum grave, ut terra minima, moveretur in maximo vacuo, illa terra in uno instanti occuparet totum vacuum et per consequens esset equalis toto vacuo, quod est impossibile, quia ponamus quod vacuum sit maius in centuplo quam ista terra minima." Burley, *Physics*, fol. 116r, col. 2.

99 In the next and ninth question of the fourth book of his *Questions on the Physics*, Marsilius of Inghen inquired "whether an external medium is necessarily required in the motions of heavy and light simple [elemental bodies]" ("Utrum in motibus gravium et levium simplicium necessario requiratur medium extrinsecum." *Physics*, fol. 51v, col. 1). Based on all the arguments, he concluded that an external medium is essential ("Ex quibus omnibus sequitur conclusio principalis, scilicet quod in motibus gravium et levium simpliciter necessario requiritur medium extrinsecum." *Ibid.*, fol. 52r, col. 1). Also see above, Ch. 3. n. 12.

100 What follows is drawn from my translation in Grant, *Source Book in Medieval Science*, pp. 337–338. In *Questions on the Physics*, see bk. 4, question 11, fol. 50v, col. 2.

101 For the technical details of the distinction, which hinges ultimately on differences between the expressions (1) *descenderet in infinitum velociter*, or *descenderet infinite velociter*, and (2) *in infinitum velociter descenderet*, or *infinite velociter descenderet*, and the means of distinguishing between them in terms of the distinction between *determinate* and *merely confused suppositio*, see Grant, *Source Book in Medieval Science*, p. 338, n. 23, 24.

102 Marsilius of Inghen used a similar procedure to argue that all velocities in a void would be equal because infinite. Because any body in a void can be imagined to move with yet a greater velocity, and this can proceed ad infinitum, Marsilius insisted that all motions in a vacuum would be infinite, though whether syncategorematic or categorematic is left unspecified. Perhaps Marsilius and Albert should be included among those scholastics who conceived instantaneous motion in a vacuum as a separate category of motion, contrasting it to finite and successive motion in a vacuum. See above, Ch. 3, Sec. 1.

103 See the fourth conclusion in Grant, *Source Book in Medieval Science*, p. 341.

104 See the fifth conclusion, *ibid*. Marsilius of Inghen repeated this idea (*Physics*, bk. 4, question 12, fol. 54v, col. 2), observing that in a vacuum any excess of motive force over resistance could move an elemental body, whereas in a plenum this might not occur because of the external resistance.

105 In her article, "Die Struktur der materiellen Substanz," Anneliese Maier discussed the nature of *mixta* and the various medieval theories on how elements subsisted in a compound, or *mixtum*. See her *An der Grenze von Scholastik und Naturwissenschaft*, pp. 3–140, especially pp. 18–22. For selections from Thomas Aquinas and Albert of Saxony on the problem of "How Elements Persist in a Compound," see Grant, *Source Book in Medieval Science*, pp. 603–614. Our concern here will be primarily with the motion of such bodies and but little with the nature of the relationships between the forms of the constituent elements and the compound body itself, a problem that took its origin from Aristotle's *De generatione et corruptione*, bk. 1, ch. 10.

106 Although writing somewhat after the 1320s, John Buridan explicitly rejected Aristotle's theory of a predominant element as the cause of upward or downward motion (". . . videtur mihi dicendum quod auctoritas Aristotelis debet sic corrigi, quod mixtum movetur secundum naturam elementi dominantis – id est, secundum qualitatem vel qualitates elementi vel elementorum habentium vel habentes dominium in trahendo superius vel inferius." Buridan, *De caelo* [Moody ed.], bk. 1, question 7, p. 35, lines 20–24) and adopted the theory we have just described (*ibid.*, pp. 34–35).

107 "Ad illud dubium dico quod mixta sunt duplicia, scilicet quedam imperfecta in quibus elementa sunt in actu completo distincta secundum situm et quasi colligata ut sunt corpora porosa in quibus est aer in actu. Alia sunt mixta perfecta, ut illa in quibus nullum elementum existit in actu completo, sed in tota materia est forma mixti uniformis." Burley, *Physics*, fol. 117v, col. 1.

108 "Et de talibus mixtis est duplex opinio, videlicet una que ponit quod in tali mixto perfecto non manent elementa in actu, sed quod simpliciter corrumpuntur, sed remanet

quedam forma subtantialis media perficiens totam materiam; et eodem modo est de gravitatibus et levitatibus elementorum, scilicet quod non manent actu in mixto perfecto, sed quadam forma media. Idem enim est iudicium de formis substantialibus elementorum et gravibus et levibus elementorum sive gravitas et levitas sint forme substantiales sive non. Et secundum istam opinionem consequenter est dicendum quod huiusmodi mixtum perfectum non posset moveri in vacuo plusquam elementum purum, quia in eo non est nisi una forma que est principium motus et materia prima, sicut vero [altered from "nec"] in elemento. Et ideo in eo non est aliquid resistens motui et sic non habet resistentiam intrinsecam, nec in vacuo est aliqua resistentia intrinseca; sed omnis motus requirit aliquam resistentiam, ut dictum est. Ideo secundum istam opinionem dico quod mixtum perfectum nullo modo posset moveri in vacuo." *Ibid.* For Burley's arguments that pure elemental bodies would fall in a vacuum instantaneously, see fol. 118v, col. 1, which is a comment on Aristotle, *Physics*, 4.8.216a.12–20.

109 Burley assumed that mixed bodies that are composed of contraries have internal resistances. ("Mixta sunt composita ex contrariis. Ergo in eis est resistentia intrinseca sufficiens ad motum." *Physics*, fol. 117v, col. 1.)

110 "Alia est opinio que ponit quod in mixto perfecto sunt elementa aliquo modo in actu et secundum illam opinionem uniformiter est dicendum de mixtis perfectis et imperfectis; et esset consequenter dicendum quod omnia mixta possent ex se moveri in vacuo propter resistentiam intrinsecam, ut superius est argutum." *Ibid.*

Marsilius of Inghen described mixed bodies in terms of intension and remission of qualities, where the motive qualities are remitted by their contraries, "as heaviness is remitted by a certain degree of lightness, and contrarily; or, at least, it [i.e., a mixed body] is understood as an aggregate of heavy and light, as if within the heavy parts there is contained a certain light body." Under these conditions, a heavy mixed body consisting of all four elements could move successively in a vacuum. ("Quantum ad secundum de motu mixti in vacuo dico quod in proposito intelligimus per mixtum illud quod determinat sibi omnes qualitates citra summum gradum et cum hoc determinat sibi qualitates motivas remissas per contrarium, ut gravitas remissa per aliquos gradus levitatis, et econtra; vel saltem per mixtum intelligitur aggregatum ex gravi et levi, ut si infra partes gravis contineatur aliquod corpus leve." Marsilius of Inghen, *Physics*, bk. 4, question 12, fol. 54v, col. 2.)

111 As an anonymous fourteenth-century author expressed it in his discussion of the fall of mixed bodies through a void, "it seems that resistance not only arises with respect to the [external] medium, but also from the participation of the degrees of lightness or heaviness of elements [in a mixed body]. Therefore, if a mixed body were moved in a vacuum, it could have a resistance and succession in its motion quite apart from an [external] medium. A motion could, therefore, be made [in a vacuum.]" ("Videtur ergo quod resistentia non solum proveniat ex parte medii, quin imo etaim ex parte participationis graduum levitatis aut gravitatis elementorum. Si ergo in vacuo mixtum grave moveretur, adhuc preter medium posset habere resistentiam et successionem in suo motu. In vacuo ergo fieri potest motus." MS Bibliothèque Nationale, fonds latin, 6752, fol. 153r.) For more on this treatise, see below, Ch. 6, n. 37.

112 "Intelligendum tamen quod licet omne mixtum posset ex se moveri in vacuo, non tamen in quocumque loco positum, ut si esset aliquod mixtum imperfectum in quo non esset ignis omnino, illud non posset ex se moveri in vacuo posito in loco ignis quia ibi omnia elementa ex quibus componitur essent gravia et non resisterent in motu descensus. Et tale mixtum positum in vacuo existente in loco ignis non haberet resistentiam intrinsecam." Burley, *Physics*, fol. 117v, col. 1. Marsilius of Inghen presented virtually the same illustration when he declared that "if there were a mixed body containing only certain [elements], then it would be moved instantly in a certain place, as [for example] a mixed body of earth and air would be moved instantly in the place of fire." ("Secundo, quia si esset mixtum ex aliquibus tantum, quia tunc in aliquo loco moveretur subito, ut mixtum ex terra et aere moveretur subito in loco ignis." *Physics*, bk. 4, question 12, fol. 54v, col. 2.) Under the same conditions, John Buridan, in apparent opposition to Burley and Marsilius, would appear to have opted for the possibility of finite motion when he declared "quod non solum moveretur mixtum grave si esset in igne secundum naturam

terrae vel secundum naturam aquae vel etiam secundum naturam aeris; imo ista tria elementa simul, vel qualitates eorum, traherent illud mixtum ad inferius donec esset in aere, et tunc amplius aer non traheret sed aqua et terra simul donec esset in aqua, et tunc amplius aqua non traheret ipsum deorsum sed sola terra." Buridan, *De caelo* (Moody ed.), bk. 1, question 7, p. 35.

113 "Et sic est possibile [altered from "impossibile"] quod tale mixtum imperfectum moveatur in vacuo in uno loco et non in alio propter hoc quod unum elementum intrinsecum potest resistere aliis in uno loco et non in alio." Burley, *Physics*, fol. 117v, col. 1.

114 "Illud tamen mixtum posset moveri ex se in vacuo posito in loco terre vel aque et forte in loco aeris, quia aer resistit motui descensus in loco suo secundum quod patet in motu gravis simplicis facti in medio aeris." *Ibid.*, fol. 117v, col. 1. Because air resists the descent of a heavy elemental body, Burley assumed that it will also actively resist as a constituent of a mixed body even when the latter is in a void that was formerly the natural place of air. (Albert of Saxony adopted the same position some years later; see below, Ch. 3, Sec. 7b and n. 117, where Buridan opposed the opinion that an element in a mixed body could actively resist the motion of that body when falling through its own natural place.) Although Maier (*An der Grenze von Scholastik und Naturwissenschaft*, p. 244) was of the opinion that Burley was confused in allowing the air of the mixed body to function as a resistance, there can be no doubt that it was his intention. Indeed in the earlier example (see above, Ch. 3, Sec. 7c and n. 112), in which all the elements but fire were in a mixed body located in the void space that was formerly the natural place of fire, we might inquire as to the consequence of adding fire to the mixed body. In this event, the fire of the mixed body would function as an internal resistance in opposition to the downward thrust of the combined heaviness of the other three elements. It seems obvious that the body would move downward with a finite, instead of instantaneous, speed.

115 Cited here from my translation in Grant, *Source Book in Medieval Science*, pp. 336–337. It appears in Albert of Saxony's *Physics*, bk. 4, question 11, fol. 50v, col. 1. In the seventeenth century, Bartholomeus Amicus repeated Albert's example even to the extent of assigning the very same numbers to the different elements. See Amicus, *Physics*, vol. 2, p. 755 (B–D).

Not all authors who discussed the rise and fall of mixed bodies were concerned with their motion in a void. The most notable exception was John Buridan, who, in his *Questions on De caelo* and *Questions on the Physics*, confined himself to the fall of mixed bodies in material media. See Buridan, *De caelo* (Moody ed.), bk. 1, question 7 ("utrum mixtum movetur secundum naturam elementi dominantis"), pp. 31–35, and *Physics*, bk. 4, question 9, fols. 75r, col. 2–75v, col. 1. See also Maier's discussion on Buridan in her *An der Grenze von Scholastik und Naturwissenschaft*, pp. 239–240.

116 It was apparently customary to assign a value of 1 to the resistance of an internal medium through which a mixed body fell. See Grant, "Bradwardine and Galileo: Equality of Velocities in the Void," 354, n. 31.

117 In the medieval theory of the motion of mixed bodies, an element in that body was held to behave differently in its natural place when the latter was a plenum than when it was void. Albert's example illustrates the latter condition, where, in the natural place of air, the gross weight of air in the mixed body actively resists the downward thrust of the heavier gross weight of the earth and thus operates as an active resistance in its own natural place when the latter is void. John Buridan (*Physics*, bk. 4, question 9, fols. 75r, col. 2–75v, col. 1) disagreed with this interpretation in an example involving the descent of a heavy mixed body through successive natural places that are filled with resisting matter. Because the external fire functions as an external resistance to obstruct the passage of the mixed body in the sphere of fire, the fire within the mixed body is assumed inactive and therefore incapable of functioning as an internal resistance. We see this in Buridan's comparison of the fall of a mixed body with that of an equally heavy pure elemental body. On the assumption that both are of equal magnitude and shape, Buridan assigned the pure elemental body 8 degrees of heaviness and the mixed body 8 degrees distributed equally among its four elements, such that the earth and water were each assigned 2 degrees of heaviness and air and fire 2 degrees each of lightness. Now if both bodies fall through the plenum of the sphere of fire, the 8 degrees of heaviness in the

pure elemental body will act simultaneously against the external resistance of the fire; but in the mixed body, the 2 degrees of fire "would neither move nor resist," so that only the 6 degrees are active in the downward motion of the mixed body through the fire. Under these circumstances, the heavy pure body will move more quickly through the sphere of fire than will the mixed body, even though their respective resistances are equal and each body was assigned the same number of initial degrees. In light of this example, Maier's statement (*An der Grenze von Scholastik und Naturwissenschaft*, p. 239) that Buridan gave no examples employing numbers, or degrees, is obviously mistaken. Indeed he also gave numerical examples in his *Questions on De caelo*. See also above, n. 114).

118 Grant, *Source Book in Medieval Science*, p. 339; Albert of Saxony, *Physics*, bk. 4, question 12, fol. 51r, col. 1.

119 In general agreement with Albert of Saxony, Marsilius of Inghen emphasized our lack of experience with the vacuum and stressed that his conclusions about the behavior of bodies in it were only probable and conjectural. ("Et notandum quod quia nunquam expertum est quid fieret in vacuo, ideo posito quod si vacuum esset nullus sciret quid sequeretur. Ideo predicte conclusiones posite sunt verisimiliter et coniecturando." Marsilius of Inghen, *Physics*, bk. 4, question 12, fol. 54v, col. 1.)

120 Here is the full text of Burley's example: "Et sic est possibile tale mixtum esse quod propter magnitudinem resistentie intrinsece tardius movetur in vacuo posito in uno loco quam in pleno in alio loco in quibus ista elementa sua movent que in priori vacuo resistebant. Verbi gratia, ut si esset aliquod mixtum compositum solum ex terra et aqua ita proportionatum quod aqua predominetur super terram et quod ille excessus non sit tantus quantum eadem aqua, si esset per se, excederet aerem in loco aeris ad dividendum ipsum et movendum per ipsum deorsum. Et tale mixtum positum in vacuo sub loco aque, id est, in loco terre, moveretur sursum aliqua velocitate, supposito quod aqua naturaliter moveretur a loco terre, sicut ad presens gratia exempli suppono et illud mixtum si poneretur superius in loco aeris descenderet velocius. Probo, quoniam si esset ibi sola aqua, que est in illo mixto, ipsa descenderet velocius quam ascenderet huiusmodi mixtum in vacuo supradicto. Ergo cum in hoc mixto sit ista aqua cum terra que est gravis in isto loco aeris et inclinat versus deorsum, sequitur quod ipsum totum compositum precise ex ista aqua et terra multo velocius descendit in illo aere quam ascendit in predicto vacuo. Eo tota causa huius est quia tale mixtum positum in vacuo in tali loco habet resistentiam intrinsecam quam non habet in alio medio pleno, scilicet in loco aeris. Et sic patet quod aliquod mixtum tardius movetur in vacuo posito in uno loco quam in pleno posito in alio loco." Burley, *Physics*, fol. 117v, cols. 1–2.

121 Following upon the text of the preceding note, Burley said: "Impossibile est tamen quod aliquod mixtum eque velociter moveatur in pleno sicut in vacuo, si tantam resistentiam habeat intrinsecam utrobique et cum hoc habet resistentiam extrinsecam in pleno quia ubicumque est maior resistentia ibi est motus tardior." *Ibid.*, fol. 117v, col. 2.

122 As Burley expressed it, "some mixed body assumed in one place in a vacuum could be moved slower than when assumed in another place in a plenum." ("Et sic patet quod aliquod mixtum tardius movetur in vacuo posito in uno loco quam in pleno in alio loco." *Ibid.*, fol. 117v, cols. 1–2.)

123 At this point, a significant problem about the internal relationships of mixed bodies arises. In Albert of Saxony's example, fire and air were each assigned a degree of 1. When the mixed body is in fire, the degree of 1 for air is added to the total heaviness of the body; but when the body is in air (as well as in water or earth), air is added to the total lightness of the body. The one degree assigned to water operates similarly. When the body is in fire and air, the one degree of water is added to the total heaviness of the body, but is added to the total degree of lightness when the body is in water or earth. From this, it seems plausible to infer that mixed bodies composed of the four elements do not possess two, or even three, separate lightnesses or heavinesses. This problem was actually discussed by the anonymous author of a fourteenth-century compendium of natural philosophy. Although a mixed body has four primary qualities (hot, cold, wet, and dry), it does not possess one heaviness corresponding to earth and another to water. Rather, we must imagine that "a body's heaviness is nothing other than its density and its lightness nothing

other than its rarity. Therefore, since different densities and rarities are not assumed in a mixed body, nor are different heavinesses and lightnesses to be assumed. It is, however, true that in so far as a mixed body participates more in the quality of fire, namely hotness, so much the rarer it is; but if the same mixed body [also] participates in the nature of air, it does not become [even] rarer because of this, since the rarity of fire contains the rarity of air. The same must be said about heaviness [that is, the heaviness of earth contains the heaviness of water]." ("Sed restat dubium an successio proveniat aut provenire possit ex diversitate graduum levitatis aut gravitatis participatorum. Ad cuius inquisitionem videre oportet an in mixto alia sit levitas correspondens igni et alia correspondens aeri; et sic de gravitate terre vel aque similiter queri potest.

Ad hoc dubium dicendum apparet quod in corpore gravi non sunt ponende distincte gravitates nec distincte levitates. Unde licet mixtum in se habeat quatuor primarias qualitates ad sensum alias declaratum, non tamen habet gravitates vel levitates correspondentes que adinvicem differant sic quod in mixto alia sit gravitas correspondens terre et alia ipsi aque. Unde imaginandum est quod gravitas corporis nihil aliud est nisi eius densitas; et levitas nihil aliud est nisi eius raritas. Quia ergo in mixto non ponuntur diverse densitates nec raritates, consequenter nec diverse gravitates vel levitates ponende sunt. Verum est tamen quod quanto mixtum magis participat qualitatem ignis, scilicet caliditatem, est tanto magis rarum. Et si idem mixtum participat naturam aeris, tunc propter hoc non est magis rarum quia raritas ignis continet raritatem aeris; et de gravitate conformiter est dicendum." MS Bibliothèque Nationale, fonds latin, 6752, fol. 153v. For more on this codex, see below, Ch. 6, n. 37).

Albert of Saxony's expressions "vacuum of air" or "vacuum of fire" were repeated in the early sixteenth century by Johannes Dullaert, *Questiones super octo libros Physicorum Aristotelis* . . . (Paris, 1506), bk. 4, question 2 (unfoliated).

124 Although Aristotle held that natural motion is accelerated, and therefore quicker at the end than at the beginning, it is obvious that Albert was not concerned here with accelerated or decelerated motion. In his examples, the motion through each natural place is uniform and alters by a quantum jump in the next natural place.

125 Drawn from my translation in Grant, *Source Book in Medieval Science*, p. 337. I have added "[elemental]" to the translation.

126 Scholastics differed as to the function of natural place. Some thought each natural place attracted a particular element and caused its natural motion, just as a magnet attracted iron. Buridan rejected this opinion and argued that the function of a natural place was to conserve certain bodies appropriate to it, which it did by virtue of an elementary power it possessed with respect to a single quality, as well as by celestial influence. Thus a natural place could be conceived as a final and formal cause acting by reason of its power to conserve the elemental body appropriate to it. A heavy body tends naturally downward because it is naturally preserved there. Although celestial influence is assumed to be continually diffused through the sublunar region, it has a different power near the heaven than farther from it, for which reason heavy and light bodies arrange themselves differently. To explain how each natural place functions as a final and formal cause and affects the bodies appropriate to it, Buridan used an example of a man moving toward a fire, which he does not only to be near it (final cause) but also to receive the heat inherent in the fire (formal cause). See Buridan, *De caelo* (Moody ed.), bk. 4, question 2 ("Utrum loca naturalia gravium et levium sunt causae motuum ipsorum"), pp. 249–250. In his *Questions on De caelo*, bk. 3, question 7, fol. 124r, col. 1, Albert of Saxony repeated Buridan's arguments almost verbatim. For scholastics, such as Albert of Saxony, who discussed the motion of mixed bodies in a sublunar void, the properties described here for natural place would have been fully operative.

127 "Sed illud magnum mixtum equevelociter movebitur in vacuo sic parvum mixtum quia ex quo sunt equalis commixtionis; eadem est proportio coniuncta ad levitatem in utroque et in vacuo non est resistentia extrinseca. Ergo illa duo mixta equaliter moventur in vacuo." Burley, *Physics*, fol. 117r, col. 2. The context of Burley's statement is a general argument that a mixed body will move with equal speed in air and in a vacuum, a consequence that he rejected (see above, Ch. 3, Sec. 7b and n. 121).

128 The translation is by H. Lamar Crosby, Jr. (ed. and trans.), *Thomas of Bradwardine, His*

"Tractatus de proportionibus" (Madison, Wis.: University of Wisconsin Press, 1955), p. 117. Whether Burley anticipated Bradwardine on the idea of equality of fall in a vacuum for unequal, homogeneous, mixed bodies cannot be determined because the date for the composition of the fourth book of Burley's *Physics* commentary can only be fixed as completed by 1327 in Paris (see Murdoch and Sylla, "Burley, Walter," in *Dictionary of Scientific Biography*, vol. 2 [New York: Scribner, 1970], p. 609), whereas Bradwardine's treatise was completed in 1328 in Oxford. With such close completion dates, it would be difficult to assign priority. Indeed both may have drawn on some as yet unknown earlier source.

129   "Octava conclusio. Mixta consimilis compositionis equaliter moventur in vacuo, sed non equaliter in pleno. Primum patet ex quo: essent consimilis mixtionis consimilis esset proportio motive potentie super totam resistentiam in uno sicut in alio, ex quo non haberent resistentiam nisi intrinsecam." Albert of Saxony, *Physics*, fol. 51r, col. 2. (Johannes Dullaert assumed the validity of the same proposition in his early-sixteenth-century *Questions on the Physics* [Paris, 1506], bk. 4, question 2 [unfoliated].) In a curious subsequent argument, Albert then denied that two unequal, homogeneous, mixed bodies would fall with equal speeds in a plenum. He assumed that mixed body A has a ratio of heaviness (motive force) to lightness (internal resistance) of 8 to 4, and mixed body B a ratio of 4 to 2. Because both bodies will fall in the same external medium, Albert assigned that medium a resistive capacity of 1 (see above, n. 116). The absolute values of the numbers now come to dominate Albert's procedure. If we add the common external resistance to the internal resistances, the ratios will be altered to 8/5 and 4/3. Because 8/5 > 4/3, it follows that body A will move more quickly in the plenum than body B. If the bodies were truly homogeneous, Albert's data suggest that body A is twice the size, or magnitude, of body B. The addition of a uniform external medium should, therefore, leave unaltered the equal ratios of internal motive force and resistance of the two bodies, which ought to fall with equal speeds in the plenum as well as in the vacuum. But here Albert brought his conclusion into conformity with Aristotle's physics, which demanded that a heavier body fall more quickly in a plenum than a lighter body. The Latin text, which will not be reproduced here, follows immediately after the passage quoted above, and appears with translation in Grant, "Bradwardine and Galileo: Equality of Velocities in the Void," 358, n. 42; an English translation of the entire eighth conclusion may be consulted in Grant, *Source Book in Medieval Science*, p. 341. Not all who discussed the motion of mixed bodies considered whether unequal, homogeneous, mixed bodies would fall with equal velocities in void and plenum. Two notable exceptions were Buridan and Marsilius of Inghen.

130   See the edition and translation by H. Lamar Crosby, Jr., cited above, n. 128. For an analysis of the concept of "ratio of ratios" (*proportio proportionum*), see *Nicole Oresme: De proportionibus proportionum and Ad pauca respicientes*, edited with introductions, English translations, and critical notes by Edward Grant (Madison, Wis.: University of Wisconsin Press, 1966), especially pp. 14–65; a brief summary appears in Edward Grant (ed. and trans.), *Nicole Oresme and the Kinematics of Circular Motion: Tractatus de commensurabilitate vel incommensurabilitate motuum celi*, ed. with an introduction, English translation, and commentary (Madison, Wis.: University of Wisconsin Press, 1971), pp. 73–76, n. 113.

131   See Clagett, *Science of Mechanics in the Middle Ages*, p. 212.

132   Trans. by Clagett, *Science of Mechanics in the Middle Ages*, p. 213, who also added the italics. In Crosby's edition, the Latin text appears on p. 118.

133   Trans. by Clagett, *Science of Mechanics in the Middle Ages*; Crosby's text, p. 118.

134   Duhem, (*Le Système du monde*, vol. 8, pp. 108–109), who only knew Albert of Saxony's opinion on equality of fall of unequal, homogeneous, mixed bodies, offered a quite different explanation of its origin. He saw it as a necessary consequence of the direct application of a basic rule from the seventh book of Aristotle's *Physics* (250a.4–6 and 25–28). In fact, not only is the rule Duhem cites never mentioned in discussions of fall in the vacuum in the fourth book of the *Physics*, but that alleged rule was grossly violated by Bradwardine. All evidence points to the distinction we have stressed above as the fundamental source of Bradwardine's opinion. For the detailed arguments, see my articles

"Bradwardine and Galileo: Equality of Velocities in the Void," 350–351, n. 22, and "On the Origin of the Medieval Version of Equality of Fall for Unequal Bodies in the Void: A Critique of Duhem's Explanation," *Actes du XIᵉ Congrès International d'Histoire des Sciences, Varsovie-Cracovie 24–31 Août 1965* (Warsaw/Cracow, 1967), vol. 3, pp. 19–23.

135 Aegidius Romanus was one who faithfully adhered to Aristotle's argument and considered it absurd to suppose that lead and chaff (which are mixed bodies but not identified as such by Aegidius), for example, could fall in a vacuum with equal velocities. But this was an inevitable consequence in a vacuum where no material medium existed to resist the fall of bodies and produce the differences in speed that are ordinarily observed. ("Cum ergo in vacuo hanc differentiam salvare non possimus in eo omnia mobilia erunt eque velocia, quod est impossibile. Notandum autem totam vim rationis in hoc consistere quod cum de ratione mobilium sit quod differentia mobilia sint differenter velocia quia differentes dividunt medium. Cum hoc in vacuo salvare non possumus, quia ibi nulla divisio esse potest, impossibile est vacuum esse. Si enim esset vacuum, non citius descenderet per ipsum plumbum quam palea, quod est omnino absurdum." Aegidius Romanus, *Physics*, fol. 82v, col. 1.)

136 Although animals were actually mixed bodies composed of varying proportions of the four elements, their fall in a vacuum was not considered from this standpoint. We have already seen that Aegidius Romanus and Walter Burley had allowed that animals could move in a vacuum by a progressive motion as self-moved entities (see above, Ch. 3, Sec. 5, and n. 71–73). Self-propulsion of this kind was the result of desire and will acting against the resistance of bodies. But Aegidius, who did not accept the doctrine of mixed bodies, insisted that all inorganic bodies would fall in a vacuum with instantaneous motion and applied this judgment to all animals that fell through space from a height. For Aegidius, all animals in this situation were exactly like inorganic bodies (see above, n. 73). Burley, however, was committed to another position. Because he accepted the finite fall of mixed bodies in vacua, it would appear that he would also have been committed to the belief that an animal, *qua* material mixed body, would also fall with temporal motion in a void. But Burley considered only the willful progressive self-motion of animals in a vacuum and ignored the problem of their possible accidental free fall. Whether he would have characterized them as mixed bodies is moot.

137 The frequent citation of medieval authors – especially Aquinas, Scotus, John of Jandun, Johannes Canonicus, Burley, and Albert of Saxony – bears witness to this reliance.

138 From Philoponus' *Commentary on Aristotle's Physics*, as translated by Drabkin and Cohen, *Source Book in Greek Science*, p. 218. See also Grant, "Aristotle, Philoponus, Avempace, and Galileo's Pisan Dynamics," 86.

139 In the Middle Ages, Johannes Canonicus adopted a position similar to that of Philoponus. See above, n. 94 of this chapter.

140 *Galileo Galilei On Motion and On Mechanics*, comprising *De motu* (c. 1590), translated with introduction and notes by Israel E. Drabkin, and *Le Meccaniche* (c. 1600), trans. with introduction and notes by Stillman Drake (Madison, Wis.: University of Wisconsin Press, 1960). The Latin text appears in *Le Opere di Galileo Galilei, Edizione Nazionale*, ed. by Antonio Favaro (23 vols.; Florence: 1891–1909), vol. 1 (1890), pp. 251–419. In the following description of Galileo's interpretation of motion in a void, I draw heavily on my articles "Motion in the Void," 287–288, and "Bradwardine and Galileo: Equality of Velocities in the Void," 355–364.

141 *De motu* (Drabkin trans.), p. 41.

142 *Ibid.*, p. 47.

143 *Ibid.*

144 *Ibid.*, p. 30. Drabkin (p. 30, n. 9) observed that "though it is widely held . . . that Galileo was here influenced by Benedetti, it should be noted that Benedetti's proof involves the severing of connected weights (cf. Aristotle, *De caelo* 301), while Galileo's involves the joining of separate weights."

145 Recognizing that one of the two homogeneous bodies falling in a plenum might be so small that its fall would actually be impeded by the medium, Galileo explained that

"our conclusion must therefore be understood to apply to [two] bodies when the weight and volume of the smaller of them are large enough not to be impeded by the small viscosity of the medium. . . ." *Ibid.*, p. 30.

146 *De motu* (Drabkin trans.), pp. 48–49. Galileo's final and correct version of this important proposition appeared in the First Day of the *Discourses*, where *all* bodies of whatever composition and specific weight fall with equal, uniformly accelerated speeds in the void, but not in a plenum. See *Discorsi e dimostrazione matematiche intorno a due Nuove Scienze* in *Le Opere di Galileo Galilei*, vol. 8 (Florence, 1898), pp. 117, 119. For English translations, see *Dialogues Concerning Two New Sciences by Galileo Galilei*, trans. from the Italian and Latin into English by Henry Crew and Alfonso de Salvio, with an introduction by Antonio Favaro (New York: Macmillan, 1914), pp. 72, 74, and *Galileo Galilei Two New Sciences, Including Centers of Gravity & Force of Percussion*, trans. with introduction and notes by Stillman Drake (Madison, Wis.: University of Wisconsin Press, 1974), pp. 76, 78.

147 *De motu* (Drabkin, trans.), p. 46.

148 As Galileo explained, "Therefore, the body will move in a void in the same way as in a plenum. For in a plenum the speed of motion of a body depends on the difference between its weight and the weight of the medium through which it moves. And likewise in a void [the speed of] its motion will depend on the difference between its own weight and that of the medium. But since the latter is zero, the difference between the weight of the body and the weight of the void will be the whole weight of the body. And therefore the speed of its motion [in the void] will depend on its own total weight. But in no plenum will it be able to move so quickly, since the excess of the weight of the body over the weight of the medium is less than the whole weight of the body. Therefore its speed will be less than if it moved according to its own total weight." *Ibid.*, pp. 45–46. Throughout this passage, "weight" is to be taken as "specific weight." Even in the later *Discourses*, Galileo retained the law that the velocity of a body is determined by the difference in its specific weight and that of the medium. Although he fully recognized that natural downward motion is uniformly accelerated, and took this into account when he introduced the concept of terminal velocity (*Opere*, vol. 8, p. 119; pp. 74 and 78 of the translations by Crew and Drake, respectively), all of Galileo's numerical examples in which velocities are compared by differences in specific weights of body and medium assume uniform downward speed and are almost identical in form with those given many years earlier in *De motu*.

149 For Galileo's explicit rejection of Aristotle's claim that heavy and light bodies will fall with equal speeds in a void because of a lack of material resistance, see *De motu* (Dabkin trans.), p. 48 and pp. 61–63. Galileo argued that only in a vacuum can differences in weights be truly determined. He acknowledged (*De motu*, p. 49) that Duns Scotus, Thomas Aquinas, and Philoponus had disagreed with Aristotle on the possibility of motion in a void, but they failed to refute Aristotle's basic view "that the speed in one medium is to the speed in the other, as the rareness of the first medium is to the rareness of the second. And no one up to now has ventured to deny this relation." In a note (p. 49, n. 23), Drabkin observes that both Philoponus and Benedetti had indeed denied this relation. We may add that in the Middle Ages, Aquinas, and almost all who adopted the position of Avempace as reported by Averroes, denied, if only implicitly, that same relation. Acceptance of temporal motion in a hypothetical vacuum was simply incompatible with the relation attributed by Galileo to Aristotle. Although medieval scholastics violated Aristotle's rule, they did not formulate a coherent replacement for it. In this sense, Galileo may have developed the first systematic "law" for the fall of bodies in void and plenum.

150 In refuting Aristotle's claim that it is the air that moves a projectile after it has lost contact with its motive force, Galileo argued in favor of a self-expending impetus as the source of that continuous motion. He presented a *reductio ad absurdum* to demonstrate that the impressed motive force, which he assumed to be the operative cause of projectile motion, must continually diminish. For on the assumption that the impressed motive force is the same at two different points of time, "it will be shown that the forced motion is never diminished, but continues always and without end at the same speed, with the

motive force always remaining the same. But this is surely most absurd" (*De motu*, Drabkin trans., p. 85). A body propelled through a vacuum by a self-expending impetus would gradually lose velocity and come to rest (see above, Ch. 3, Sec. 6, for Bonetus' arguments).

151 In chapter 12 of *De motu*, Galileo argued "in opposition to Aristotle . . . that the absolutely light and the absolutely heavy should not be posited; and that even if they existed, they would not be earth and fire as he believed" (*De motu*, Drabkin trans., p. 55). If there could be a "heaviest" thing, it would not be called heaviest "except in comparison with other things which are less heavy, since the heaviest cannot be defined or conceived except insofar as it lies below the less heeavy; and in the same way, . . . it is impossible for anything to be called lightest, except in comparison with things which are less light and above which it rises; and . . . the lightest substance is not that which lacks all weight – for this is void, not some substance – but that which is less heavy than all other substances that have weight." *Ibid.*, p. 60. In Galileo's opinion, "it is clearly absurd to posit an [absolutely] greatest lightness or heaviness. For just as, when any speed, however great, is assumed, another speed greater than it can be assigned; so, when any heaviness or lightness, however great, is assumed, another greater than it can be assigned." *Ibid.*, pp. 117–118.

152 The "limitation of the agent" may have been identified with a body's weight. See above, Ch. 3, Sec. 7a and n. 94, 95. Thomas Bradwardine reported and rejected what may have been a mathematical version of Avempace's opinion (see Crosby [ed. and trans.], *Thomas of Bradwardine, His "Tractatus de proportionibus,"* pp. 86–93; see also Grant, *Source Book in Medieval Science*, pp. 292–296).

153 For the different ways in which ratios could produce velocities, see Grant (ed. and trans.), *Nicole Oresme: De proportionibus proportionum and Ad pauca respicientes*, pp. 14–24, and Grant, *Source Book in Medieval Science*, pp. 292–305).

154 Although Ernest Moody ("Galileo and Avempace," 186, 417) argued that Avempace determined velocities by a subtraction of specific gravities, there is no support for this interpretation in Averroes' report of Avempace's opinion in the former's *Commentary on the Physics*, bk. 4, comment 71; nor is there any evidence that Philoponus employed the concept of specific gravity or specific weight in his discussions of natural fall in void and plenum. For discussions on both Avempace and Philoponus, see Grant, "Aristotle, Philoponus, Avempace, and Galileo's Pisan Dynamics," 84–91; for the full textual account of Avempace's ideas as reported by Averroes, see Grant, *Source Book in Medieval Science*, pp. 256–262. The medieval Latin authors who followed Avempace's opinion did not interpret that "law" in terms of specific gravities.

155 On Benedetti, see Drabkin (trans.), *De motu*, p. 29, n. 9.

156 For these, see above, n. 129 and Sec. 7b. In comparing the fall of a heavy mixed body and a pure elemental body in the same medium, the density of which is adjusted in various ways, Thomas Bradwardine was able to demonstrate, in his *Tractatus de proportionibus*, that the same heavy mixed body could fall with speeds greater than, equal to, or less than that of the same elemental body. See Grant, "Bradwardine and Galileo: Equality of Velocities in the Void," 352–353.

CHAPTER 4

Nature's abhorrence of a vacuum

1 Pierre Duhem (*Le Système du monde. Histoire des doctrines cosmologiques de Platon à Copernic* [10 vols.; Paris: Hermann, 1913–1959], vol. 8, p. 158) mentions William of Auvergne as one who used such expressions. In the Greek text of the *Problemata*, falsely ascribed to Alexander of Aphrodisias and probably written in late antiquity, we find, in a discussion of the siphon, the expression ἀπειλή τοῦ κενοῦ ("horror of the vacuum") (see the edition of this text by Julius Ludwig Ideler, *Physici et Medici Graeci Minores* [2 vols.; Amsterdam: Adolf M. Hakkert, 1963; reprint of 1841–1842 edition], vol. 1, bk. 2, problem 59, p. 69). Although the Pseudo-Alexander *Problemata* was translated into Latin in 1302 by Peter of Abano, I was unable to locate the expression in the two extant manuscripts of that translation (Madrid: Escorial, Real Biblioteca, f.1.11, and

Città del Vaticano: Biblioteca Vaticana, MS Reg. 747; see below, n. 10 of this chapter. In what follows, Duhem's chapter, "L'horreur du vide," in *Le Système du monde*, vol. 8, pp. 121–168, plays a significant role (an almost identical version of this chapter was published by Duhem in A. G. Little [ed.], *Roger Bacon Essays* [Oxford: Clarendon Press, 1914], pp. 241–248 [Chapter X: "Roger Bacon et l'horreur du vide"]). I have also drawn heavily on my article "Medieval Explanations and Interpretations of the Dictum that 'Nature Abhors a Vacuum'," *Traditio* 29 (1973), 327–355.

2 Translated by Hermann Gollancz in *Dodi Venechdi (Uncle and Nephew), the Work of Berachya Hanakdan, . . . to which is added the first English Translation from the Latin of Adelard of Bath's 'Quaestiones Naturales'* (London: Oxford University Press, 1920), pp. 143–144. For the Latin text, see Martin Müller (ed.), "Die *Quaestiones Naturales* des Adelardus von Bath," in *Beiträge zur Geschichte der Philosophie und Theologie des Mittelalters*, vol. 31, heft 2 (Münster: Aschendorff, 1934), pp. 53–54. A summary of Adelard's clepsydra discussion appears in Marshall Clagett, "Adelard of Bath," *Dictionary of Scientific Biography*, vol. 1 (New York: Scribner, 1970), pp. 61–62.

3 The basis for the date is the existence of a thirteenth-century manuscript version, British Museum, Sloane MS 2030, fols. 110r–114r, which is mistakenly titled *Liber Aristotelis de conductibus aquarum*. The *Pneumatics*, written originally in Greek but no longer extant, was the fifth book of Philo's *Mechanics*. An edition of the medieval Latin text, with a German translation, was published by Wilhelm Schmidt (ed. and trans.), *Heronis Alexandrini Opera*, vol. 1 (Leipzig: Teubner, 1899), pp. 459–489. The much longer Arabic text was published with a French translation by Baron Carra de Vaux, "Le Livre des appareils pneumatiques et des machines hydrauliques par Philon de Byzance," in *Notices et Extraits des Manuscrits de la Bibliothèque Nationale et Autres Bibliothèques* (Paris: L'Académie des Inscriptions et Belles-Lettres, 1903). For a further discussion of Philo's *Liber de ingeniis spiritualibus*, see my article "Henricus Aristippus, William of Moerbeke and Two Alleged Mediaeval Translations of Hero's *Pneumatica*," *Speculum* 46 (1971), 656–669.

4 Schmidt ed., p. 464; also pp. 468, 472.

5 ". . . quia si evacuaretur aer, statim succedit aliquid corporum que ipsi aeri commiscentur, quia sui pro natura inpelluntur." Philo added that he agreed with the professors of natural science who think this way ("et hoc quidem asserunt professores sciencie naturalis, quibus similiter opinamur"). *Ibid.*, p. 476.

6 *Conciliator differentiarum philosophorum et precipue medicorum* (Mantua, 1472), fol. 30v, col. 2. The italics are mine. For the texts and translations, see my article "'Nature Abhors a Vacuum'," 329, n. 4.

7 The discussions of Philo, Adelard, and Peter of Abano are best understood within the framework of Aristotle's rejection of void.

8 The translation is by W. K. C. Guthrie in the Loeb Classical Library (London: William Heinemann; Cambridge, Mass.: Harvard University Press, 1939). The Greek terms cited above were translated by forms of *trahere* in the Latin translation made from the Greek text of *De caelo* by William of Moerbeke in the thirteenth century. The sense of suction was thus properly preserved. In the earlier translation from the Arabic by Michael Scot, no counterpart for the Greek terms appears, leaving the impression that water rises to replace the displaced air by means of a violent pushing motion rather than by some kind of suction power. Both translations accompany Averroes' commentary on *De caelo* in *Aristotelis opera cum Averrois commentariis* (9 vols; Venice, Junctas ed., 1562–1574, reprinted in facsimile, Frankfurt: Minerva, 1962), vol. 5 (see fol. 264, col. 1 for Moerbeke's version and col. 2 for Michael Scot's).

9 Averroes, *Opera*, vol. 5: *De caelo*, bk. 4, comment 39, fols. 264v, col. 1–266v, col. 1.

10 In rejecting a position formulated by Themistius, Alexander, after declaring that air is warm, said: "ergo infrigidabitur ex aqua inspissabitur et veniet in quantitatem minorem et propter necessitatem vacui sequitur aqua aerem et implet loca vacua evacuata ab aere." *Ibid.*, fol. 265v, col. 1(I). The work in which Alexander is alleged to have made this statement is not mentioned. However, in the second book of the *Problemata*, falsely ascribed to Alexander during the Middle Ages and Renaissance (see F. Edward Cranz, "Alexander Aphrodisiensis," *Catalogus translationum et commentariorum: Medieval and*

Renaissance Latin Translations and Commentaries, ed. by P. O. Kristeller [Washington, D.C.: Catholic University of America Press, 1960], vol. 1, p. 126, col. 2), there is a discussion on vacuum in which the author attributed the rise of water in a tube, or siphon, to the force of a vacuum rather than to the sucking action of the mouth. He also observed that, in order to prevent formation of a vacuum, cupping glasses, after being heated, draw blood into the place previously occupied by the fire. Other actions are similarly ascribed directly to the force of the vacuum.

The *Problemata* attributed to Alexander of Aphrodisias circulated in two recensions, one in two books, the other in four. In 1302, the short recension, which presumably contained the section on vacuum, was translated from Greek to Latin by Peter of Abano. Because only two manuscripts of Abano's translation have thus far been identified (Cranz, p. 127; see above, n. 1 of this chapter) and both seem to lack this section, it probably exerted little influence during the fourteenth and fifteenth centuries. The work was again translated from Greek to Latin by Giorgio Valla sometime around 1465–1467. In the 1490 edition of Valla's translation, the relevant passage appears on signature d_{iii} but omits the expression "horror vacui" (see above, n. 1 of this chapter). Two other translations of Pseudo-Alexander were also made during the fifteenth and sixteenth centuries by Theodore of Gaza and Angelo Poliziano (see Brian Lawn, *The Salernitan Questions. An Introduction to the History of Medieval and Renaissance Problem Literature* [Oxford: Clarendon Press, 1963], p. 97). A summary of Pseudo-Alexander's arguments appears in Giovanni Battista della Porta's *Pneumaticorum libri tres* (Naples, 1601), pp. 9–10. Della Porta rejected the idea that fear, or dread, of a vacuum (*metus vacui*) could constitute a force. As a nonexistent entity, the vacuum, or fear of it, was incapable of acting on anything.

11 *Liber primus Communium naturalium Fratris Rogeri*, partes tertia et quarta, ed. by Robert Steele, in *Opera hactenus inedita Rogeri Baconi*, ed. by Robert Steele (fasc. III (Oxford: Clarendon Press, 1911), pp. 219–224.

12 *Jean de Paris (Quidort) O. P., Commentaire sur les Sentences, Reportation, Livre I*, ed. by Jean-Pierre Muller, O. S. B., in *Studia Anselmiana, Philosophica, Theologica*, edita a professoribus Instituti Pontificii S. Anselmi de Urbe, fasc. 47 (Rome: Orbis Catholicus, Herder, 1961), quaestio 154 (distinctio 44, quaestio 3), pp. 468–470. Quidort probably wrote his commentary sometime around 1294–1295 (see Charles H. Lohr, S. J., "Medieval Latin Aristotle Commentaries, Authors: Johannes de Kanthi–Myngodus," *Traditio* 27 [1971], 273).

13 The *Summa philosophiae* has been published by Ludwig Baur in "Die philosophischen Werke des Robert Grosseteste," *Beiträge der Geschichte der Philosophie des Mittelalters*, vol. 9 (Münster: Aschendorff, 1912), pp. 275–643; for particular and universal natures, see pp. 590–594.

14 Johannes Quidort distinguished between a "particular nature" (*natura particularis*) and a "special nature" (*natura specialis*). The former is in the individual of a species; for example, Socrates and Plato are individuals in the species man. As the nature common to the species as a whole, the special nature was superior to the particular nature. The two expressions were frequently used interchangeably. Quidort distinguished four basic natures – particular, special, universal, and "most universal" (*natura universalissima*) – which he identified with God. See Muller's edition of Quidort's *Commentaire sur les Sentences*, p. 468, lines 10–18.

15 Walter Burley used these and other expressions in his *Questions on the Physics*, bk. IV, question 6 ("Utrum vacuum possit esse") in MS Basel Universitäts-bibliothek F. V. 12 (see below, n. 107, 110, and 145 of this chapter). Because no printed edition of Burley's *Questiones on the Physics* exists, a word about the manuscript is in order. It probably dates from the late fourteenth century and contains only 43 of the 108 questions listed in a table of *questiones* on folios 108r–108v (for a description of the manuscript by its discoverer, see S. Harrison Thomson, "Unnoticed *Questiones* of Walter Burley on the Physics," *Mitteilungen des Instituts für österreichische Geschichtsforschung* 62 [1954], 390–405; another manuscript, containing Burley's questions on books I, V, VI, and VII, but not the question listed above, from which we shall quote below, has been identified by Vladimir Richter in Cambridge, Gonville and Caius 512, fols. 109r–

234v [I], 220r–234v [V–VIII]; cf. Albert Zimmermann, *Verzeichnis ungedruckter Kommentare zur Metaphysik und Physik des Aristoteles aus der Zeit von etwa 1250–1320*, vol. 1 [Leiden/Cologne: Brill, 1971], 226–228).

16 As Quidort put it: "Natura autem universalis, quae est virtus caelestis cum suo motore sic diffusa, quia est universalior, intendit illud quod sibi convenit magis." *Commentaire sur les Sentences*, p. 469, lines 39–40.

17 Among the descriptions that Bacon and Burley offered for the universal nature we find *virtus regitiva universi*. See *Opera*, fasc. 3, p. 222, lines 25–26, for Bacon, and fol. 65v, col. 2 of the Basel manuscript, for Burley. For Amicus' use of the same expression, see below, n. 75 of this chapter.

18 Bacon, *Opera*, fasc. 3, p. 220, lines 11–15.

19 Dumbleton, *Summa logicae et philosophicae naturalis*, pars sexta, ch. III. See Duhem, *Le Système du monde*, vol. 8, pp. 162–163, where the source of Dumbleton's passage is mistakenly given as Bibliothèque Nationale, fonds latin, 16621 instead of Bibliothèque Nationale, fonds latin, 16146. For a brief discussion of this error, see Grant, " 'Nature Abhors a Vacuum'," 338, n. 22. The ultimate source of Dumbleton's opinion may have been Avicenna, who, according to Peter of Abano (in his commentary on Aristotle's *Problemata*, completed at Padua in 1310), believed that nature's abhorrence of a vacuum was so powerful that the heavens themselves would, if necessary, descend to fill a potentially void space (see Duhem, *Le Système du monde*, vol. 8, p. 164). For a discussion of universal nature and mention of Dumbleton and other scholastics who accepted it, see C. de Waard, *L'Expérience barométrique*, pp. 57–59.
    Not all scholastics were of the opinion that the sides of the heaven would meet to prevent a vacuum. In bk. 4, question 26 of the *Questions on the Physics*, Pseudo-Siger of Brabant argued that the existence of a vacuum in the concavity of the heavens is less impossible than a contactual meeting of the sides of the heavens to prevent the occurrence of that vacuum. Although both are impossible, the existence of an actual vacuum in the heavens is more likely than a collapse of the sides of the celestial spheres. ("Maius autem impossibile esset latera caeli concurrere, et ideo magis ponendum esset vacuum quam latera caeli concurrere." Pseudo-Siger of Brabant, *Physics* [Delhaye ed.], p. 184.

20 "But a negation is not the cause of an affirmative [action]; 'that a vacuum not occur' is a negation." ("Set negacio non est causa affirmacionis: 'ne fiat vacuum' est negacio." Bacon, *Opera*, fasc. 3, p. 219, lines 28–29.) In discussing the failure of water to descend from a clepsydra, Walter Burley offered the same interpretation when he declared that "it does not suffice to say that the water is prevented from descending 'that a vacuum not occur,' for this is not a cause." ("Similiter ista questio querit causam positivam prohibentem descensum aque. Non sufficit ergo dicere quod prohibetur descendere 'ne fiat vacuum,' hec enim non est causa." Burley, *Questiones circa libros Physicorum*, fol. 65v, col. 2; for more on the positive cause, see below, n. 110 of this chapter.)

21 It is virtually certain that bodies were thought to possess their particular natures as inherent properties. The status of the universal nature is, however, vague. Is it also an intrinsic property of all matter? Or is it a separately existing agent that acts on bodies from outside when need arises? Burley, who spoke of a "celestial agent" (see above, n. 15 of this chapter), and probably Bacon, seems to have conceived it as an independent entity acting on bodies externally. However, in a work falsely ascribed to Aegidius Romanus, the author spoke as if both natures are inherent to bodies. Identifying God with the universal agent, Pseudo-Aegidius explained that "although every particular agent confers particular tendencies on [each] generated thing – for example, lightness or heaviness so that it can be moved to its place – God, nevertheless, as universal agent, not only confers on the parts of each thing a universal tendency to be joined and united when its better preservation demands it, but He also confers on things, insofar as they are parts of the universe, a tendency to be joined [or connected] when necessity demands it." ("Unde quamvis quodlibet agens particulare tribuat rei genitae particulares appetitus, verbi gratia, levitatem aut gravitatem ut ad suum locum moveatur [corrected from *moventur*], Deus, tamen, tanquam agens universale, sicut tribuit partibus cuiuslibet rei appetitum universalem se coniungendi atque copulandi quando melior sui conservatio id postulat, ita etiam tribuit rebus, quatenus sunt partes

universi, appetitum sese coniungendi quando id postulat necessitas." *Commentationes Physicae et Metaphysicae: in Physicorum libros octo, De coelo libros quatuor . . . tradita a . . . Fratre Aegidio Romani. . . .* [Ursellis, 1604], Physicorum Aristotelis liber quartus, p. 182.)

22 The anonymous author of the *Liber de causis* distinguished a remote first cause (*causa prima longinqua*) from a proximate second cause (*causa propinqua*). In the operations of nature, the remote cause dominates and can act when the proximate cause is absent. For these and other similarities, see my translation of the relevant passage in Grant, " 'Nature Abhors a Vacuum'," 330–331, n. 6. The Arabic version was probably an "original" treatise based on Proclus' Στοιχείωσις Θεολογική (it was Thomas Aquinas who made this discovery after a comparison of the *Liber de causis* with William of Moerbeke's recently completed Latin translation of the Greek text of Proclus' work that Moerbeke titled *Elementatio theologica*). The Arabic and Latin versions were edited by Otto Bardenhewer, *Die pseudo-aristotelische Schrift Ueber das reine Gute bekannt unter dem Namen "Liber de causis"* (Freiburg: Herderhandlung, 1882); a recent Latin edition based on ninety manuscripts has been published by Adriaan Pattin, *Le Liber de causis* (Louvain, 1966).

23 For example, see Roger Bacon, *Opera*, fasc. 13: *Physics* (8 books), pp. 225–230; John of Jandun, *Physics* (Venice, 1519), bk. 4, question 10, fols. 55v, col. 2–56v, col. 1; John Buridan, *Physics*, bk. 4, question 7, fols. 72v, col. 2–73v, col. 2; Marsilius of Inghen, *Physics*, bk. 4, question 13, fol. 55v, col. 1; and Friedrich Sunczel, *Collecta et exercitata Friderici Sunczel Mosellani liberalium studiorum Magistri in octo libros Phisicorum Arestotelis* [sic] *in almo studio Ingolstadiensi* (Hagenau, 1499), bk. 4, question 7 (unfoliated). The sections in Buridan and Marsilius have been translated in my *Source Book in Medieval Science*, pp. 326–328. As we shall see, some of these experiments were probably derived from Averroes' lengthy discussion in his commentary on *De caelo*, comment 39, in Averroes, *Opera*, vol. 5, fols. 264v, col. 1–266v, col. 1.

24 See Albert of Saxony, *Physics*, bk. 4, question 13 ("Utrum condensatio et rarefactio sint possibiles"), fols. 51v, col. 2–52r, col. 1. Although Albert made no mention of the vacuum in these experiments, he seemed to imply it. Apart from rejecting atomist arguments that explained condensation and rarefaction by appeal to the alleged existence of interstitial vacua, the lengthy medieval analyses of condensation and rarefaction played little role in the history of the concept of vacuum, because, as Chrysostom Javelli put it in the sixteenth century, "rarefaction and condensation occur without a vacuum" ("rarefactio et condensatio fiunt sine vacuo"; Chrysostomus Javelli, O. P., *Totius rationalis, naturalis, divinae ac moralis philosophiae compendium* . . . [2 vols.; Lyon, 1568], vol. 1: *Quaestiones acutissimae super octo libros Physices Aristotelis*, bk. 4, p. 566, col. 1.). For Aristotelians, who assumed a material plenum in which the occurrence of void was naturally impossible, condensation and rarefaction could not produce vacua. "From the Aristotelian point of view, although there is rarefaction and condensation, the last bit of air, which marks the boundary between full space and an empty one, can never be extracted by natural means." (Charles B. Schmitt, "Experimental Evidence For and Against a Void: The Sixteenth-Century Arguments," *Isis* 58 [1967], 362.) Medieval discussants of condensation and rarefaction sought to explain the expansion and contraction of bodies in terms of a continuous material plenum. These attempts will be described below. Not until the sixteenth century did anti-Aristotelian vacuists argue that air could be sufficiently exhausted from closed vessels to produce a partial or complete vacuum.

25 As will be seen below (Sec. 5), during the sixteenth century some of these experiments would once again be invoked as evidence for the existence of vacua.

26 It was so designated by Albertus Magnus, *Opera*, vol. 2: *Physics*, bk. 4, tract 2, ch. 9, p. 179, col. 2; Roger Bacon, *Opera*, fasc. 8: *Physics* (4 books), p. 198, lines 7–8; John of Jandun, *Physics* (Venice, 1519), fol. 55r, col. 2; Aegidius Romanus, *Physics*, fols. 84r, col. 2 and 84v, col. 2; Walter Burley, *Physics*, bk. 4, fols. 109v, col. 1; 119r, col. 2; 119v, cols. 1–2; 120r, cols. 1–2; and by Nicholas of Autrecourt in his *Exigit ordo executionis*, the text of which was edited from a single known manuscript by J. Reginald O'Donnell, C. S. B., in "Nicholas of Autrecourt," *Mediaeval Studies 1* (1939), 218, lines 43–44. Autrecourt's section on vacuum, which will be discussed below, appears on

217–222. For a translation of the whole of it, see Grant, *Source Book in Medieval Science*, pp. 352–359. The entire *Exigit* has been translated in *The Universal Treatise of Nicholas of Autrecourt* by Leonard A. Kennedy, Richard E. Arnold, and Arthur E. Millward, with an introduction by Leonard A. Kennedy (Milwaukee, Wis.: Marquette University Press, 1971); for the section on vacuum, see pp. 87–95.

27 See John of Jandun, *Physics* (Venice, 1519), bk. 4, question 10, fol. 55r, col. 2. It was also used by Nicholas of Autrecourt (see O'Donnell, "Nicholas of Autrecourt," 220, line 31; Grant, *Source Book in Medieval Science*, p. 357).

28 Roger Bacon, *Opera*, fasc. 13: *Physics* (8 books), p. 236.

29 *Ibid.*, fasc. 8: *Physics* (4 books), p. 198, lines 7–8.

30 Albertus Magnus, *Opera*, vol. 2: *Physics*, p. 180, col. 1, and Conimbricenses, *Physics*, bk. 4, ch. 9, article 2, col. 90.

31 Albertus Magnus, *Opera*, vol. 2; *Physics*, p. 179, col. 2; John of Jandun, *Physics* (Venice, 1519), fol. 55r, col. 2, who used the expression "vacuum esse spatium separatum"; Nicholas of Autrecourt, *Exigit ordo executionis*, in O'Donnell, "Nicholas of Autrecourt," 219, lines 9–10, and 220, line 6 (see my translation, *Source Book in Medieval Science*, pp. 355 and 357); and Walter Burley, *Physics*, fol. 119r, col. 2.

32 John of Jandun described both as follows: "Intelligendum quod quidam antiquorum philosophorum posuerunt esse vacuum in rebus propter rationes predictas et dixerunt *vacuum esse spatium separatum* ab omnibus corporibus et virtutibus naturalibus et a motu. Istud autem vacuum quidam dixerunt esse extra corpora naturalia et distinguens abinvicem ipsa corpora naturalia et sensibilia et infinitum esse tale spatium extra celum. Alii dixerunt *vacuum imbibitum* vel inexistens corporibus naturalibus inter partes ipsorum et per hoc fieri condensationem et rarefactionem, cum enim partes corporis per vacuum distantes appropinquant se invicem replentes illud vacuum vel etiam cum de foris ingrediatur aliquod corpus in illa vacua, tunc dicebant corpus condensari et spissari." *Physics* (Venice, 1519), bk. 4, question 10, fol. 55r, col. 2.

33 As Aristotle reported it (*Physics* 4.9.216b.30–217a.4), vacuists distinguished two kinds of interstitial vacua. In the first and more significant type, the vacua exist separately and are of sufficient size to allow the movement of particles. The second type, however, have no separate existence and are too minute for material particles to move within them. Their sole purpose is to contribute lightness, or buoyancy, to the rare bodies that possess them. Although Aristotle believed the second type "less impossible" than the first, it was the first, and more customary, sense of interparticulate vacua that became the major focus in subsequent discussions of condensation and rarefaction.

34 Aristotle, *Physics* 4.6.213b.19–21.

35 *Physics* 4.7.214a.29–214b.9.

36 Averroes, *Opera*, vol. 4: *Physics*, bk. 4, comment 55, fol. 150r, col. 1.

37 Buridan, *Physics*, bk. 4, question 7, fol. 73r, col. 1.

38 John of Jandun made the transformation of water to air explicit when he declared ". . . ita est causa huius potest esse quia cum aqua imponitur vasi habenti cineres calidos, tunc aqua illa corrumpitur et resolvitur in aerem et exit evaporando." *Physics* (Venice, 1519), bk. 4, question 10, fol. 56r, col. 2.

39 "Ad rationem de cineribus, Commentator bene dat causam propter quam ita est, scilicet enim cineres, si sint novi et maxime sunt calidi et sicci et activi virtualiter, ideo agunt in aquam infusam evaporando magnam partem sive quantitatem ex ea; et etiam virtute aque subintrantis plures partes subtiles cineris vel etiam inter cineres incluse exeunt aqua intrante, non enim erant partes cineris continue adinvicem sed erat multus aer inclusus. Et sic tandem possibile esset quod iste pottus plus receiperet de aqua quam si non essent ibi cineres, sicut etiam si in illo potto essent frustra ferri igniti et candentis." Buridan, *Physics*, fol. 73v, col. 2. In addition to John of Jandun, the same position was adopted by Albert of Saxony, *Physics*, bk. 4, question 8, fol. 48r, col. 2 (for the statement of the problem) and fol. 48v, col. 2 (for the response). Although agreeing that some parts of the water are corrupted, Walter Burley (*Physics*, fol. 110r, col. 1) also spoke of the corruption of air and ashes. He deduced the corruption of the ashes from the observation that when the ashes are subsequently dried, their quantity is diminished. For Burley,

then, all the constituent ingredients in this experiment are somewhat diminished at the end (see also fol. 111v, col. 1, for Burley's further discussion).

40 See also John of Jandun, *Physics* (Venice, 1519), fol. 56r, col. 2.

41 "Nam quamcumque spatii partem ponerentur occupare aut aqua non reciperetur in illa aut duo corpora simul erunt." *Ibid.*, fol. 56r, col. 2. The proponents of interstitial vacua would, of course, deny these consequences.

42 Aristotle, *Physics* 4.6.213b.15–18.

43 "De doliis autem habentibus utres quasi nihil respondit Aristoteles quia forte supponit illud argumentum unum manifeste falsum." Jandun, *Physics* (Venice, 1519), fol. 56r, col. 2.

44 "Cum enim uter sit quoddam corpus naturale necesse est ut occupet et obstruat et impleat aliquam partem spatii existentis in dolio et in illa simul non recipietur vinum nisi duo corpora sint simul in uno eodem loco. Et sic non tantum de vino omnino recipiet quantum faceret sine utribus, sed forte quia parum minus est vinum quod recipitur cum utribus in dolio et sine utribus. Ideo quasi equale videtur, cum tamen non sit. Et forte diceret aliquis quod uter habet quamdam virtutem consumptivam vini, et ideo consumit aliquas partes vini propter quod tantum ibi recipitur sicut sine utre. Sed si removeatur et mensuretur diligenter non invenitur equale." *Ibid.*

45 *Physics* 4.9.217a.11–19.

46 *Physics* 4.9.217a.20–217b.19.

47 See Aristotle, *De Generatione et Corruptione*, bk. 1, ch. 5, especially 321a.10–29.

48 On this, see Harold H. Joachim's description in his *Aristotle On Coming-To-Be and Passing-Away (De Generatione et Corruptione), a revised text with introduction and commentary* (Oxford: Clarendon Press, 1922), p. 124.

49 *Categories* ch. 8. 10a.20–22.

50 *Physics* 4.9.217b.12–19.

51 Each of these categories was suggested by Aristotle in one place or another without any definitive resolution. In his commentaries, Averroes took little note of these difficulties and rested content to explicate each discussion as he came upon it. Marc Antonio Zimara collected the conflicting statements of the Philosopher and his Commentator and published them as the eighth contradiction in his "Solutions to the Contradictory Statements of Aristotle and Averroes" (*Solutiones contradictionum in dictis Aristotelis et Averrois*), which appears in *Aristotelis opera cum Averrois commentariis*, vol. 4, fol. 488r, col. 1. Zimara concluded that Averroes truly believed that dense and rare are contrary qualities and therefore in the category of quality. For more on this problem, see Anneliese Maier, "Das Problem der quantitas materiae," in her *Die Vorläufer Galileis im 14. Jahrhundert. Studien zur Naturphilosophie der Spätscholastik* (Rome: Edizioni di Storia e Letteratura, 1949), pp. 28–29, n. 3.

52 For Marsilius of Inghen's discussion of the problem, see Grant, *Source Book in Medieval Science*, pp. 350–352. The relationship between quantity of matter and volume as a special aspect of the problem of rarefaction and condensation is discussed by Maier, "Das Problem der quantitas materiae," *Die Vorläufer Galileis im 14. Jahrhundert*, pp. 26–52.

53 For the full titles of the edition and translations, see above, n. 26 of this chapter. A major study of his thought appears in Julius R. Weinberg, *Nicholas of Autrecourt. A Study in 14th-Century Thought* (Princeton, N.J.: Princeton University Press, 1948); a good brief summary is provided by Frederick Copleston, S. J., *A History of Philosophy*, vol. 3 (Westminster, Md.: Newman Press, 1953), pp. 135–152.

54 Armand Maurer, *Medieval Philosophy* (New York: Random House, 1962), p. 289.

55 See O'Donnell's edition of the *Exigit ordo executionis*, 219, lines 8–10; also Grant, *Source Book in Medieval Science*, p. 355.

56 For typical defenses of Aristotle and arguments against the possibility of interstitial vacua, see, for example, Roger Bacon, *Opera*, fasc. 13: *Physics* (8 books), pp. 236–238; Aegidius Romanus, *Physics*, fols. 84r, col. 2, 84v, col. 2–85v, col. 2; and Paul of Venice, *Expositio Pauli Veneti super octo libros Phisicorum Aristotelis super commento Averrois cum dubiis eiusdem* (Venice, 1499), bk. 4 (unfoliated), beginning with Paul's

comment on text 79 and continuing to the end of his comment on text 82. For scholastic arguments rejecting interstitial vacua in the sixteenth and early seventeenth centuries, see Domingo Soto, *Super octo libros Physicorum Aristotelis commentaria* . . . (Salamanca, 1555), bk. 4, fols. 67v, col. 1–68r, col. 2, and Federicus Pendasius, . . . *Physicae auditionis texturae libri octo nunc primum in lucem editi* . . . (Venice, 1604), p. 524.

57 O'Donnell edition of *Exigit ordo executionis*, 218, lines 43–46; Grant, *Source Book in Medieval Science*, p. 355.

58 O'Donnell (ed.), *Exigit ordo executionis*, p. 217, lines 33–37; Grant, *Source Book in Medieval Science*, p. 353.

59 O'Donnell (ed.), *Exigit ordo executionis*, p. 217, lines 11–15; Grant, *Source Book in Medieval Science*, p. 354.

60 O'Donnell (ed.), *Exigit ordo executionis*, p. 221, lines 25–33; Grant, *Source Book in Medieval Science*, p. 358.

61 What follows is largely drawn from my article "The Arguments of Nicholas of Autrecourt for the Existence of Interparticulate Vacua," *Actes de XIIe Congrès International d'Histoire des Sciences*, Paris, 1968 (Paris: Albert Blanchard, 1971), tome IIIA, pp. 65–68. The supporting texts can be found in O'Donnell's edition and my translation.

62 The process that Autrecourt described and attacked was frequently called "mutual replacement" or *antiperistasis*, the Greek term used by Aristotle in his brief mention of it in *Physics* 4.8.215a.15 and 8.10.267a.15–21. Although Aristotle envisioned mutual replacement to occur simultaneously, it is not obvious that Autrecourt's description of a simultaneous exchange of places between contiguous bodies is exactly what Aristotle had in mind. And if it was, it is again questionable whether it applied to violent as well as natural motions (see Grant, *Source Book in Medieval Science*, pp. 353–354, n. 18). But the kinds of problems Autrecourt raised are those that should have been of concern to supporters of a material plenum.

63 See O'Donnell (ed.), *Exigit ordo executionis*, 218, line 46–219, line 7; Grant, *Source Book in Medieval Science*, p. 355.

64 O'Donnell (ed.), *Exigit ordo executionis*, 207 (section titled *De indivisibilibus*) and 219, lines 29–44.

65 See Maimonides' *Guide of the Perplexed*, bk. 1, ch. 73 in Moses Maimonides, *The Guide of the Perplexed*, trans. with an introduction and notes by Shlomo Pines, with an introductory essay by Leo Strauss (Chicago: University of Chicago Press, 1963), pp. 194–214.

66 See Schmidt's edition, pp. 476–479, for the Latin text and German translation.

67 It extends over bk. 4, comment 39 of Averroes' commentary on *De caelo*, which is a discussion of Aristotle 312b.2–312b.19. See Averroes, *Opera*, vol. 5: *De caelo*, fols. 264v, col. 1–266v, col. 1. A possible reference to the candle experiment appears in St. Ambrose's *Hexameron*, where, in explaining the power of heat, he declared that "when physicians burn a small candle and attach it to the inside of certain types of vases, narrow in the spout, rather flat on top and hollow within, how does it happen that this heat attracts to itself all the moisture?" St. Ambrose, *Hexameron, Paradise and Cain and Abel*, trans. by John J. Savage (New York: Fathers of the Church, 1961), p. 58. Although the instrument described here may be a cupping glass, the fact that the candle is said to be fastened in the vessel makes this seem implausible.

68 Averroes, *Opera*, vol. 5: *De caelo*, fol. 265v, col. 1(H).

69 For the Arabic version and French translation, see Baron Carra de Vaux, "Le livre des appareils pneumatiques . . . par Philon de Byzance," *Notices et Extraits des Manuscrits de la Bibliothèque Nationale et Autres Bibliothèques*. For the French translation of the candle experiment, see pp. 127–128.

70 Duhem (*Le Système du monde*, vol. 8, p. 125) was convinced that Averroes had fashioned his version from the Arabic translation of Philo's treatise. But the differing versions, and the failure to mention the name of Philo although including those of Themistius and Alexander, seem to tell against Duhem's contention. Indeed if Averroes was influenced by the Greek pneumatic tradition, it was through Hero of Alexandria rather than Philo of Byzantium. In discussing the manner in which water is drawn up into a narrow-mouthed, egg-shaped cup used by physicians, Hero explained that "when

they wish to fill these with liquid, after sucking out the contained air, they place the finger on the vessel's mouth and invert them into the liquid; then, the finger being withdrawn, the water is drawn up into the exhausted space, though the upward motion is against its nature." *The Pneumatics of Hero of Alexandria from the Original Greek*, trans. for and ed. by Bennet Woodcroft (London: Taylor, Walton and Maberly, 1851), pp. 3–4. The translation was made by J. G. Greenwood. Reprinted in facsimile with introduction by Marie Boas Hall (London: Macdonald; New York: American Elsevier, 1971). It was possible, then, for Averroes to have derived from Hero that part of the candle experiment involving the inversion of the vessel over water. According to A. G. Drachmann, *Ktesibios, Philon and Heron, A Study in Ancient Pneumatics, Acta Historica Scientiarum Naturalium et Medicinalium*, vol. 4 (Copenhagen: Ejnar Munksgaard, 1948), p. 126, Hero of Alexandria was probably familiar with Philo's *Pneumatica* and made use of the work, although taking pains to alter the details of some of the devices and experiments.

71 For the Latin text, see Averroes, *Opera*, vol. 5, fol. 265v, col. 2. At the conclusion of his discussion, Averroes said that, following such an experiment, he once broke the vessel and found water on the sides of it (". . . et iam fregi quoddam vas post statum aquae in eo, et inveni eam applicatam in lateribus vasis").

72 *Ibid.*, fol. 266r, col. 1.

73 I now cite a summary of all three causes that Averroes furnished to explain the ascent of the water: "Iste igitur motus componitur ex tribus causis. Quarum prima est motus partis igneae ad superius. Secunda est transmutatio partis igneae ad minorem quantitatem. Tertia est motus aquae per se propter fortitudinem et velocitatem motus, quas acquisivit ex istis motibus duobus." *Ibid.*

74 In Marsilius of Inghen's description of the candle experiment (*Physics*, fol. 55v, col. 1), the candle's flame was also extinguished by the rising water. For a translation, see Grant, *Source Book in Medieval Science*, p. 327. Peter of Abano gave much the same description in his *Conciliator differentiarum philosophorum et precipue medicorum* (Mantua, 1472), differentia XIV, fol. 30v, col. 2. Also included are the opinions of Themistius and Alexander, which Peter drew from Averroes. For Duhem's discussion of Peter of Abano's description of the candle experiment, see *Le Système du monde*, vol. 8, p. 129.

75 For the description of Canonicus' candle experiment, see his *Physics*, bk. 4 question 4, fol. 42v, col. 1; for his explanation, see fol. 43r, col. 2. Duhem, *Le Système du monde*, vol. 8, p. 157, includes a French translation of part of this discussion. Canonicus specifically denied that any new air or water is generated to prevent formation of the vacuum. In the seventeenth century, Bartholomeus Amicus gave a similar overall description of the candle experiment. The problem that required a response was the manner in which the volume of the destroyed fire could be filled, because it occupied twice the volume of an equal quantity of the matter of air and 100 times the volume of an equal quantity of the matter of water. Formation of a vacuum is prevented, however, because the air will rarefy and, in obedience to "a universal regulative power of the whole universe" (*virtus regitivae totius universi*), the water will rise from the vessel below sufficiently to fill the potentially void space. Bartholomeus Amicus, *Physics*, vol. 2, tractatus 21 (*De vacuo*), p. 745, col. 2; see also above, n. 17 of this chapter, for use of a similar expression by Bacon and Burley.

76 "Quarta experientia: si vitrum habens longum rostrum, ut phiala, calefiat circa ignem ut aer in ea rarefiat et postea rostrum ponatur ad superficiem aque frigide, ipsa aqua ascendit quia aer in phiala condensatur propter contrariam frigiditatem aque. Ergo nisi ascenderet aqua fieret vacuum." Sunczel, *Physics*, bk. 4, question 7 (unfoliated). Cf. Hero of Alexandria's experiment above, n. 70 of this chapter. In the fourteenth century, Albert of Saxony included virtually the same example in a question on "whether condensation and rarefaction are possible" ("Utrum condensatio et rarefactio sint possibiles"). See his *Physics*, bk. 4, question 13, fol. 52r, col. 1. For John Buridan's version, see his *Physics*, bk. 4 question 11 ("Utrum rarefactio et condensatio sunt possibiles"), fol. 77v, col. 2.

77 A significant device, which employed heat and was relevant to discussions on vacuum, was the cupping-glass used by physicians and surgeons since Greek antiquity. As André

Kenny explains ("Vacuum Theory and Technique in Greek Science," *Transactions, Newcomen Society 37* [1964/1965], 52), the cupping-glass was "used for drawing matter through the skin of a patient or from scratches or incisions made for this purpose in the skin. A partial vacuum, as we should call it, was produced by burning some in-flammable substance, often charcoal or chaff, inside the glass while it was held against the patient's skin. After a while the glass would support its own weight and begin to act on the skin." Hero of Alexandria provided two alternative explanations in the intro-duction to his *Pneumatics.* In one, the fire in the cupping-glass consumes and rarefies the air, thus reducing the volume of air within the glass and causing the contiguous matters, presumably the flesh and blood of the patient, to fill the reduced space (see Woodcroft, *Pneumatics of Hero of Alexandria,* p. 4). Hero's other explanation (*ibid.,* p. 6; Kenny, "Vacuum Theory and Technique in Greek Science," 52) is quite similar except that the quantity of air within the cupping-glass is now said to be reduced by the escape of the rarefied air through the pores of the cupping-glass itself. The cupping-glass was undoubtedly mentioned in the medieval medical tradition (for Francis Adams's brief description of Greek and Arabic usages of this device, see *The Seven Books of Paulus Aegineta,* trans. from the Greek by Francis Adams [3 vols.; London: printed for the Sydenham Society, 1846], vol. 2, pp. 325–328), and occasionally in treatises that could have been known to medieval natural philosophers, and, for example, in William of Moerbeke's 1271 Latin translation of Simplicius' commentary on *De caelo,* where, in commenting on Aristotle's remarks in 4.5.312b.4–12 (for the passage, see above, Ch. 4, Sec. 1), Simplicius mentioned syphons and cupping-glasses (for the Latin text from Moerbeke's translation, see below, n. 79 of this chapter); in Averroes' brief mention of it in his commentary on *De caelo* (Averroes, *Opera,* vol. 5, fol. 265v, col. 2[K]); and perhaps in Peter of Abano's Latin translation of the *Problemata* of Pseudo-Alexander of Aphrodisias (such a discussion appears in Giorgio Valla's translation, cited above in no. 10 of this chapter; I failed to locate such a passage in Escorial La Real Biblioteca, Manuscritos F.I.11, which is one of the two extant manuscripts of Abano's translation).

78 Philo, *Liber de ingeniis spiritualibus,* p. 464 of Schmidt's edition. See also Drachmann, *Ktesibios, Philon and Heron,* pp. 47–48. For an English translation of the passage, see Kenny, "Vacuum Theory and Technique in Greek Science," 49.

79 "Vi enim nihil prohibet aerem sursum ferri locum ignis, sicut et aqua in locum aeris trahitur vi ab aere quando ait una sit superficies trahentis aeris et aque tracte. Etenim in syphonibus et in medicinalibus ventosis per que trahitur aqua et sanguis, et aer trahens et aqua que trahitur appropinquantia invicem superficiebus propriis determinata sunt corpora existentia. Et quamdiu utique discrete superficies ipsorum et solum tangunt invicem, manet secundum regionem utrumque. Quando vero due superficies in unum convenerint spiritu aut calefactione confundente ipsas et veluti commixtionem quandam faciente, tunc alterum ab altero tanquam pars ipsius factum iam trahitur quando velocior qui ad sursum aeris trahentis fit quam propria aque reptio ad deorsum. Nam antequam dissolata et separata sit unio superficierum prevenit sursum trahi tanquam connexum." *Simplicii philosophi acutissimi commentaria in quatuor libros De caelo Aristotelis. Guil-lermo Morbeto interprete* (Venice, 1540), fol. 117r, col. 1. For a French translation from the Greek text of part of this passage, see C. de Waard, *L'Expérience barométrique,* p. 55.

80 The approximate and tentative date is assigned by Lynn Thorndike, *The "Sphere" of Sacrobosco and Its Commentators,* (Chicago: University of Chicago Press, 1949), pp. 21–23.

81 *Ibid.,* p. 287, for the Latin text. In the fourteenth century, Buridan, using wine instead of water, declared that "by drawing up the air standing in the reed, you [also] draw up the wine by moving it above [even] though it is heavy. This happens because it is neces-sary that some body always follow immediately after the air which you draw upward in order to prevent formation of a vacuum." Buridan, *Physics,* bk. 4, question 8, fol. 73v, col. 1; the translation is taken from Grant, *Source Book in Medieval Science,* p. 326.

82 My translation as it appears in Grant, *Source Book in Medieval Science,* p. 327. The passage appears in Marsilius' *Physics,* bk. 4, question 13 "Utrum possibile sit esse vacuum"), fol. 55v, col. 1. The curved siphon is also described in Philo's *Liber de*

ingeniis spiritualibus, sec. 9 (Schmidt ed., pp. 478–480). The *Tractatus de inani et vacuo* may have been a medieval compilation based upon Philo, Palladius Rutilius (fourth century A.D.), and perhaps Hero of Alexandria; or it may have been a compilation from Arabic sources. The three experiments that Marsilius described in the question cited above (see Grant, *Source Book in Medieval Science*, p. 327) appear in Philo's work, but not in Hero's *Pneumatica*. For a detailed discussion of the anonymous *Tractatus de inani et vacuo*, see Grant, "Henricus Aristippus, William of Moerbeke and Two Alleged Mediaeval Translations of Hero's *Pneumatica*," 660–662, n. 21.

83 "Propterea est [i.e., vacuum] indecor et quedam deordinatio in toto universo. Ideo natura, scilicet communis, abhorret vacuum iiii huius, ut videlicet res naturales in esse conserventur. Ideo etiam multis potest probari hoc experientiis quo res naturales, etiam contra naturam propriam, conantur quodammodo impedire vacuum. Et prima, quia grave ascendit ne committatur vacuum quando invenes calamis aquam sursum trahunt et bibunt. Et leve descendit, ut aer et exalatio in concavitatibus terre; et sic leve quiescit sub gravi." Sunczel, *Physics*, bk. 4, question 7 (unfoliated).

84 *Physice perscrutationes Magistri Ludovici Coronel Hispani Segoviensis* (Paris, after 1511), bk. 4, fol. 86r, cols. 1–2. The discussion is too lengthy for citation. Roger Bacon proposed a similar pair of options that nature might exercise (see below, Ch. 4, Sec. 4d and n. 106).

85 For similar versions of, and responses to, this experiment, see John of Jandun, *Physics* (Venice, 1519), bk. 4, question 10, fol. 56r, col. 1 (for description) and 56v, col. 1 (for response); Chrysostomus Javelli, *Totius rationalis, naturalis, divinae ac moralis philosophiae compendium*, vol. 1: *Quaestiones acutissimae super octo libros Physices Aristotelis*, bk. 4, p. 566, col. 2; Johannes Canonicus, *Physics*, bk. 4, question 4, fol. 42v, col. 1; Friedrich Sunczel, *Physics*, bk. 4, question 7 (Quinta experientia) (unfoliated). Although numerous other instances could be cited, two others will be mentioned as variations on the theme. Marsilius of Inghen (*Physics*, bk. 4, question 13, fol. 55v, col. 1; translated in Grant, *Source Book in Medieval Science*, pp. 327–328) replaced the water with air when he described a closed body the inner concavity of which is "filled with air and placed in water or in something that is made intensely cold so that the air contained in the vessel will be condensed, since condensation follows upon coldness. Hence the air will occupy a smaller place than before, and, as a consequence, there will be a vacuum." He then explained further that "it is impossible that the air in this body be condensed by coldness, however cold it may be, unless the vessel is broken or there is an opening in it through which another body enters." In contrast to Marsilius, Franciscus Toletus (*Physics*, fols. 130v, col. 2–131r, col. 1) described the experiment in its usual terms but explained the prevention of the vacuum quite differently (fol. 131v, col. 1) by assuming that as the water froze and contracted, it gave off subtle vapors that filled the place left vacant (the passage is translated by Schmitt, "Experimental Evidence for and Against a Void," 357).

86 Because of its intrinsic interest, I present somewhat more of the Latin text than I have translated: "Unde cum aqua per corpus fistulare ascendit dicendum est minorem esse eius resistentiam quo ad ascensum quam resistentia corporis fistularis quo ad divisionem. Et cum natura minus difficile semper amplectatur potius per ascensum aque quam per fractionem corporis proceditur." Coronel, *Physice perscrutationes*, bk. 4, fol. 86r, col. 1.

87 Bacon, *Opera*, fasc. 13: *Physics* (8 books), p. 228.

88 *Ibid.*, p. 229.

89 *Physics*, bk. 4, question 7, fol. 73v, col. 1; for a translation, see Grant, *Source Book in Medieval Science*, p. 326.

90 For similar descriptions, see *Commentariorum Collegii Conimbricensis Societatis Iesu in octo libros Physicorum Aristotelis Stagiritae* (Lyon: sumptibus Horatii Cardon, 1602), secunda pars, col. 90. See also Schmitt, "Evidence for and against a Void," 355, where the passage from the Jesuits at the University of Coimbra is translated and a reference to the *Physics* of Domingo Soto is given.

91 With teams of horses on each side, Buridan's description bears an interesting resemblance to Otto von Guericke's famous experiment in 1654 with the "Magdeburg Hemispheres." Of course, where Buridan's horses labored mightily to demonstrate nature's

abhorrence of a vacuum, von Guericke's pulled in opposite directions to demonstrate creation of an artificial vacuum and the powerful pressure of the atmosphere. For a fuller discussion and references, see Grant, *Source Book in Medieval Science*, p. 326, n. 17.

92 "Item ponatur quod in aliquo vase compactissimo et fortissimo sit unum solum foraminem et in illo ponatur os follis; et ipsa follis nullum habet foraminem in aliquo laterum suorum. Tunc eleventur et separentur latera follis; tunc queritur utrum aer existens in illo vase ingrediatur follem aut non: si sic, ergo vas remanet vacuum; si non, ergo inter latera follis ad se invicem distantia erit vacuum, quia pono quod follis sit omniquaque sic obturata et conclusa et continua quod nullum corpus possit in ea ingredi per partes exteriores; et quod sic applicetur illud os follis ad ipsum vas quod nihil possit ingredi in ipsum vas per suum foraminem." John of Jandun, *Physics* (Venice, 1519), bk. 4, question 10, fols. 55v, col. 1–56r, col. 1. The same experiment is described by Franciscus Toletus (*Physics*, bk. 4, question 10, fol. 130v, col. 2; for a translation, see Schmitt, "Experimental Evidence for and against a Void," 355) and Bartholomeus Amicus (*Physics*, vol. 2, bk. 4, tractatus 21 [De vacuo], p. 744, col. 1 [D–E]).

93 "Ad aliud dico quod illis suppositis numquam aliqua virtus elevaret latera follis nec abinvicem separaret, sed quod prohiberet natura universalis propter fugam vacui." Jandun, *Physics*, fol. 56v, col. 1. Without appeal to universal nature, Toletus (*Physics*, fols. 131r, col. 2–131v, col. 1) and Amicus (*Physics*, p. 744 col. 2[C–D]) agreed that the sides of the bellows could not be separated.

94 Because they formed the basis of my article " 'Nature Abhors a Vacuum',", it will be convenient to repeat much that was included there, although rearrangements, alterations, and additions have been freely made.

95 As Duhem observed (*Le Système du monde*, vol. 8, p. 136), the clepsydra embraced a wide variety of decanting vessels, including spigot, funnel, pipette, and chantepleure (or *cantaplora* in Latin). Writing around 1220, Alexander of Neckham, in a treatise titled *De utensilibus*, defined a number of these vessels, which he classified generically as *clessedra* (see Duhem, *Le Système du monde*, vol. 8, p. 136; in his *De naturis rerum*, cap. XIX ["Quod nullus locus sit vacuum, vel diu vacuus"], Neckham described the clepsydra, though he did not name it, as an "urceus habens fundum multis foraminibus distinctum et orificium superius." See Thomas Wright [ed.], *Alexandri Neckam De naturis rerum libri duo; with the poem of the same author, De laudibus divinae sapientiae* [London, 1863, reprinted 1967], p. 64).

96 In the dialogue between Adelard and his nephew, the latter recalled a visit to a witch in whose house "there was a vessel of remarkable powers which was brought out at meal times. Both at top and bottom it was pierced with many holes; and when water for washing the hands had been put into it, so long as the servant kept the upper holes closed by putting his thumb over them, no water came out of the lower holes; but as soon as he removed his thumb, there was at once an abundant flow of water for the benefit of us who were standing round. This seemed to me the effect of magic. . . ." *Quaestiones naturales*, ch. 58 in *Dodi Venechdi (Uncle and Nephew)*, p. 143.

97 Adelard went on to explain that "when . . . the entrance is closed to that which is to come in, in vain will the exit be open for the departing element: thanks to this loving waiting, it will be all in vain that you open an exit for the water, unless you give an entrance to the air. . . . Hence it happens that if there be no opening in the upper part of the vessel, and an opening be made at the lower end, it is only after an interval, and with a sort of murmuring, that the liquid comes forth." *Ibid.*, pp. 143–144.

98 Among the translations, perhaps the earliest interpretation of the water's contrary behavior as an action intended to prevent formation of a vacuum was to be found in Philo of Byzantium's *Liber de ingeniis spiritualibus*, ch. XI (Schmidt ed., pp. 480, 482) and in Averroes' *Commentary on the Physics*, bk. 4, comment 51, *Opera*, vol. 4, fol. 148r, col. 1, which was a rather meager account. Although Aristotle mentioned the clepsydra in *Physics* 4.6.213a ff., *De caelo* 2.13.294b.21, *De respiratione* 473a.15–473b.10, and *Problems* 16.8 (the last two works were not by Aristotle but derive from his school), he did not interpret the behavior of the water as an indication of the non-existence of vacuum, but as evidence of the corporeality of air (in ch. II of his work [Schmidt ed., pp. 460, 462], Philo also demonstrated the corporeality of air by means

of a vessel with only one hole at the bottom). A later thirteenth-century source is Simplicius' commentary on *De caelo*, translated from Greek to Latin in 1271 by William of Moerbeke. Simplicius provided a substantial description of the clepsydra and also explained that it is to prevent a vacuum that the water does not fall when the upper orifice is closed (see *Simplicii . . . commentaria in quatuor libros De celo Aristotelis*, fol. 84r, col. 2; for an English translation of the Greek text, see W. K. C. Guthrie, *A History of Greek Philosophy*, vol. 2 [Cambridge: Cambridge University Press, 1965], p. 221). By the end of the thirteenth century, mention of the clepsydra and the behavior of the liquid to prevent formation of a vacuum became commonplace.

A considerable literature now exists on the meaning and significance of ancient Greek discussions of the clepsydra; for a recent article, which also includes additional bibliography, see Thomas D. Worthen, "Pneumatic Action in the Klepsydra and Empedocles' Account of Breathing," *Isis 61* (1970), 520–530. A useful brief account is furnished by W. K. C. Guthrie in a note to his translation of Aristotle's *On the Heavens* (Loeb Classical Library; London: William Heinemann; Cambridge, Mass.: Harvard University Press, 1939), pp. 226–229,

99 Marsilius of Inghen, *Physics*, fol. 55v, col. 1; for a translation, see Grant, *Source Book in Medieval Science*, p. 327.

100 After explaining that air is more easily rarefiable or condensible than water, Buridan illustrated with the clepsydra by explaining that "si dolium plenum vino perfectissime bene ligatum et obstructum et perforetur inferius ad trahendum vinum, vinum non exibit vel valde modicum exibit quia non potest multum tali rarefactione rarefieri, nec potest aer subintrare ad replendum pro eo quod exiret." *Physics*, bk. 4, question 11, fol. 78r, col. 1.

101 Instead of a multiplicity of holes in the bottom of the clepsydra, Buridan spoke of only one.

102 Buridan, *Physics*, fol. 78r, col. 1.

103 *Ibid.*, fols. 77v, col. 2–78r, col. 1. The consequence about the heavens becoming involved in the condensations and rarefactions unless offset by compensatory actions elsewhere in the universe was probably drawn from Aristotle's discussion in *Physics* 4.9.216b.21–217a.19. Buridan's position implied that any rectilinear motion involves the motion of everything else in the world. By the assumption of interstitial vacua, Nicholas of Autrecourt believed that he could avoid such difficulties.

104 O'Donnell, "Nicholas of Autrecourt," 219, line 45–220, line 7; for the translation, see Grant, *Source Book in Medieval Science*, p. 356. Autrecourt also reported an alternative explanation that bears a resemblance to Adelard of Bath's interpretation (see above, Ch. 4, Sec. 1). In this account, the failure of the water to flow from the clepsydra is not explained by appeal to "nature's abhorrence of a vacuum" but by imagining "that the whole universe has its mode of fullness (*modum plenitudinis*), so that to the extent that something exists it is able to enter into the universe of existent beings, which is considered to be good. Now, if this body should flow forth, another could not enter, and the whole universe would not be a total plenum and two bodies would be [in the same place] simultaneously." Grant, *Source Book in Medieval Science*, p. 356; the Latin text appears in O'Donnell, "Nicholas of Autrecourt," 220, lines 8–13. Thus if the water left the clepsydra, not only could another body not enter to fill the void left by the departing water, but the water that entered the material plenum outside the clepsydra would have to occupy the same space as the hypothetical volume of the body that was unable to enter the clepsydra. Thus not only would a vacuum be left in the clepsydra, but two bodies outside would occupy the same place simultaneously. To avoid this situation, it was deemed impossible for the water to depart and disrupt the material plenitude of the world. To conform to this interpretation of the argument, I have here altered the last few lines of my translation in the *Source Book in Medieval Science*. Autrecourt rejected this explanation and opted for nature's abhorrence of a vacuum as the cause of the water's failure to fall from the clepsydra.

105 *Opera*, fasc. 13: *Physics* (8 books), bk. 4 ("Quaestio: utrum sit ponere vacuum infra celum"), p. 230.

106 Bacon characterized the collapse of the vessel's sides as "essentially unnatural" (*innaturale essentialiter*), whereas water at rest above and outside of its natural place is merely

"accidentally unnatural" (*accidens unnaturale*). *Ibid.* The basis for this distinction is obviously the greater disruption of nature that follows upon the collapse of a vessel. In the sixteenth century, Luis Coronel considered a similar pair of options open to nature. See above, Ch. 4, Sec. 4b.

107 "Ymmo videtur quod oportet dicere quod virtus regitiva universi vel aliquod agens superceleste regens ordinem universi faciat [text: *faciens*] aquam moveri sursum ne fiat vacuum; et similiter impediat aliquando aquam moveri deorsum ne fiat vacuum. Contra probo quod agens universale non prohibeat aquam descendere in cantaplora quia si orificium cantaplori obturetur, tunc non descendet; si non obturetur, tunc descendet. Sed agens celeste eiusdem virtutis est et uniformiter agit sive orificium illius vasis obturetur sive non. Si ergo agens celeste impediret descensum aque quando orificium obturetur, ipsum similiter impediret descensum aque quando orificium non obturatur. Similiter ex quo tale agens est agens universale indifferens est ad hunc effectum et ad eius oppositum. Ergo oportet ibi ponere aliam causam particularem determinantem illud agens celeste ut nunc prohibeat descensum huius aque et alias faciat aliam aquam ascendere. Et tunc redit difficultas de illa causa particulari ibi ponenda." Burley, *Questiones circa libros Physicorum*, fols. 65v, col. 2–66r, col. 1.

108 "Moreover, the Commentator shows that the celestial agent has control over lower [i.e., sublunar] things where he says, in the seventh [book] of the *Metaphysics*, comment 32, that it is impossible that a corporeal agent should move through [or across] matter except with the mediation of supercelestial bodies. Similarly, in the first [book] of the *De celo*, comment 24, the Commentator says that the celestial bodies conserve the elements." ("Quod autem agens celeste habet regere hec inferiora paret Commentatorem [sic] septimo *Metaphysice*, commento 32, ubi dicit quod impossibile est ut agens corporeum transmutet materiam nisi mediantibus corporibus supercelestibus. Similiter primo *De celo*, commento 24, dicit quod corpora celestia conservant elementa." *Ibid.*, fol. 66v, col. 2.)

109 Burley gave no indication as to whether the downward motion of the air was to be understood as a natural or violent motion as it replaced the departing water. In his discussion of the vessel filled half with wine and half with air (see above, Ch. 4, Sec. 4d and n. 100–102), Buridan characterized as violent the downward motion of the air as it replaced the departing wine ("ut michi videtur quod talis condensatio vel rarefactio est quasi violenta;" *Physics*, fol. 78r, col. 1). On the assumption that water and earth are both withdrawn from below air, Peter of Abano considered the downward motion of air as natural by virtue of its relative heaviness (as an intermediate element, it possessed both heaviness and lightness).

110 Here is Burley's text in support of the ideas described in this paragraph: "Ad rationes. Ad primam dico, sicud ultimo dicebatur, quod aliquod agens celeste prohibet aquam in cantaplora ne descendat et idem agens facit aliquando aquam ascendere ne sit vacuum, natura enim abhorret vacuum. Ideo illud agens, quod regit naturam et ordinem naturalem in universo, illud idem salvat plenitudinem in universo quia si est vacuum in aliqua parte universi, tunc deficeret aliqua pars requisita ad perfectionem universi et tunc universum non esset perfectum. Ideo illud agens celeste, quod regit ordinem universi et salvat perfectionem eius, illud prohibet aquam descendere ne fiat vacuum ibi et deficiat aliqua pars requisita ad perfectionem universi. Unde quando arguitur quod oportet ibi ponere causam positivam, iam posita est causa positiva, scilicet agens celeste, salvans ordinem universi et ei correspondens effectus positivus, scilicet salvare perfectionem et plenitudinem universi. Unde iste effectus privatus 'ne sit vacuum' non est proprius effectus, sed effectus suus est positivus, scilicet ut universum sit plenum et perfectum. Universum autem non esset perfectum si esset vacuum quia tunc in aliqua parte universi, vel 'unum nichil' esset pars universi. . . . Ad primum, in contrarium, quando accipitur quod agens celeste uniformiter agit super orificium cantaplore sive obturetur sive [non], dicendum quod uniformiter agit quantum ad effectum primum. Semper enim uniformiter intendit hunc effectum, scilicet salvare plenitudinem et perfectionem universi. Tamen realis effectus secundarii intenti difformiter agit. Aliquando enim facit aquam moveri, aliquando quiescere et semper propter effectum eundem, primo [sic] scilicet propter plenitudinem et perfectionem universi. Nec est inconveniens

agens celeste habere contrarios effectus in istis inferioribus quia sol aliquando est causa generationis aliquando causa corruptionis." Burley, *Questiones circa libros Physicorum*, fol. 66v, col. 2.

111 Burley argued as follows: "to the second [principal argument], it should be said that the universal agent is adequately determined to this effect, namely to make the water rest in the clepsydra, for the reason that if in this effect the plenitude and perfection of the whole could not be saved [or preserved], then, since an end [or goal] imposes the necessity [for achieving that end] and determines those things that are [necessary] for achieving it, it is necessary that the celestial agent seeking this end, namely to preserve the plenitude [or fullness] of the universe, be made (*necessitatur*) to make the water rest there. Therefore, when you seek what determines this agent toward this effect, I say that it is determined by the end [or goal] of the universe and is necessitated by that end. It is unnecessary to assume another determinant than the end [or goal], for [it is] the end [or goal] that determines the agent to make those things that exist for the sake of producing the end [or goal]." For the translation, which I have altered slightly, and the Latin text, see Grant, " 'Nature Abhors a Vacuum'," 337–338, n. 20.

112 Pseudo-Aegidius Romanus, *Commentationes physicae et metaphysicae Physicorum libros octo*, p. 182. My translation is drawn from Grant, " 'Nature Abhors a Vacuum'," 338, n. 21, where the Latin text is also cited. A similar judgment was expressed by Themon Judaeus, who insisted that the heaven, or celestial region, could not act on the inferior terrestrial world through a vacuum "because an agent must be immediate to the patient; the heaven is an agent and inferior things are under its action, . . ." (". . . quia agens debet esse immediatum passo et celum est agens et illa inferiora patiuntur ab illo, . . ." in Themon's *Questions on the Four Books of the Meteorology*, bk. 1, question 1 in *Questiones et decisiones physicales insignium virorum: Alberti de Saxonia in octo libros Physicorum*, . . . *Thimonis in quatuor libros Meteororum*, . . . *Recognitae rursus et emendatae summa accuratione et iudicio Magistri Georgii Lokert Scotia quo sunt Tractatus proportionum additi* [Paris, 1518], fol. 156r, col. 1).

113 My translation from Duhem's French translation of John Dumbleton's *Summa* (Duhem omitted the Latin text) in *Le Système du monde*, vol. 8, p. 162. The manuscript of Dumbleton's *Summa naturalium*, which Duhem cited, is erroneous. For the correct reference, see above, n. 19 of this chapter.

114 R. E. Latham, trans. *On the Nature of the Universe* (Harmondsworth, Middlesex: Penguin Books, 1951), pp. 38–39. On Lucretius and the *distantia terminorum*, see above, Ch. 3, n. 23. For the Latin text, see *De rerum natura libri sex*, ed. Cyril Bailey, editio altera (Oxford: Clarendon Press, 1922). C. de Waard describes and discusses Lucretius' argument in *L'Expérience barométrique*, p. 14.

115 Despite the existence of at least two manuscripts of the *De rerum natura* during the Middle Ages from the ninth century on, which eventually served as the basis of the modern edition, it seems to have been unread and unknown. For further information, see Grant, " 'Nature Abhors a Vacuum'," 339, n. 24.

116 Despite a suggestive statement by Blasius of Parma, it is not likely that Averroes was the source. In his *De tactu corporum durorum (On the Contact of Hard Bodies)*, Blasius described the experiment in question and, after noting the possibility that a vacuum might be left in the center before air rushed to fill it, declared that the Commentator [Averroes] gives this as a reason why many philosophers believe that two hard bodies could not mutually touch. ("Et hec ratio Commentator facit multos philosophos inclinare ad credendum quod duo corpora dura non possint se tangere." *Questio Blasij de Parma De tactu corporum durorum*, the title of Blasius' treatise, is not mentioned on the title page of the edition I have used but simply appears as the last treatise in *Questio de modalibus Bassani Politi; Tractatus proportionum introductorius ad calculationes Suisset; Tractatus proportionum Thoma Bradwardini; Tractatus proportionum Nicholai Oren; Tracatus de latitudinibus formarum Blasij de Parma; Auctor sex inconvenientum* [Venice, 1505]. The folios of Blasius' treatise are unnumbered, but the quotation appears in the first column of the second page of his brief work.) It is likely that Blasius drew this opinion from Averroes' *Commentary on De anima (Aristotelis opera cum Averrois commentariis*, suppl. II, bk. 2, comment 113, fols. 109v–110r), where Averroes said

that it is impossible that two dry bodies touch in air or water without air or water intervening. But nowhere did Averroes mention anything remotely resembling the argument of Lucretius. Aristotle's passage from *De anima* does, however, play a role in our story and is discussed below, in this section.

117 Among those who adopted this position, we may include, Roger Bacon, *Opera*, fasc. 13: *Physics* (8 books), pp. 225–226; Pseudo-Siger of Brabant, *Physics*, p. 178; John of Jandun (Venice, 1519), *Physics*, fol. 55v, col. 2; Complutenses, *Physics*, p. 436; and Conimbricenses, *Physics*, secunda pars, col. 89.

118 Albertus Magnus insisted that upon separation air would instantly fill the intervening space. No momentary vacuum is mentioned or considered. See his *Physics*, bk. 4, ch. 7, p. 176, col. 2.

119 Translated from Bacon, *Opera*, fasc. 13: *Physics* (8 books), p. 226. John of Jandun also adopted this position when he declared: "Indeed if they [i.e., two plane bodies] were to be separated, it would be necessary that first one part then another be separated successively. And as much space as there is between those parts, just so much air would enter so that the air would enter successively as the planes are successively separated. It is impossible, however, that all parts [of the planes] be separated simultaneously and uniformly because of the fear of a vacuum." *Physics*, fol. 56v, col. 1. Essentially the same solutions were offered by Pseudo-Siger of Brabant, *Physics*, p. 180; Conimbricenses, *Physics*, secunda pars, cols. 95–96; and Complutenses, *Physics*, p. 436, which also suggested, as did Bacon, that if this solution fails, the two planes will be unable to separate.

120 In referring to Bacon's remarks quoted above, Duhem (*Le Système du monde*, vol. 8, p. 142) observes: "Chacun sait qu'un verre plongé dans l'eau se laisse soulever sans grand effort; au contraire, l'adhérence de deux disques plans se peut observer sans aucune difficulté. Il est clair que notre auteur n'avait tenté ni l'une ni l'autre des deux épreuves." Duhem seems to have missed the point. Whether or not Bacon had tried either of these experiences himself, or had ever observed them personally, is irrelevant. All instances of separation between two plane surfaces were for Bacon instances in which the surfaces were inclined; otherwise no separation could occur. Under the conditions described, no observation could refute Bacon's claim.

121 My interpretation is based on the following text: "Contra sive sit sive non, non minus stat argumentum quia posito quod inequaliter elevetur, ita quod magis fiat elevatio ex una parte quam ex alia, nichilominus aer circumdans ineque cito replebit partes interiores sicud exteriores et per consequens reliquitur ibi vacuum. Ymno dicitur quod pars tabule elevabitur ante partem in infinitum." Burley, *Questiones circa libros Physicorum*, bk. 4, question 6, fol. 66r, col. 1. Burley's remarks are perhaps susceptible of another plausible interpretation. Instead of conceiving the intervening space as a series of minute parts, perhaps Burley wished only to consider that part of the space between the points of the surface still in contact and those no longer in contact. The consequences, however, are identical with those of the previous interpretation. Following the angular separation of the surfaces, and assuming that a finite time is required for the incoming air to sweep from the open end to the point where the surfaces are still in contact, a momentary vacuum would occur prior to the arrival of the air at the point of contact. Such an opinion had already been reported in the thirteenth century by Pseudo-Siger of Brabant. ("Si elevetur prius secundum alteram partem, adhuc erit ponere vacuum, quia aer non potest pervenire a parte prius elevata usque ad aliam circumferentiam in instanti." *Physics*, p. 178.) Pseudo-Siger eventually adopted the position that separation could occur only when one surface was inclined toward another. He therefore concluded (Sec. 4e) that the existence of a momentary vacuum was undemonstrated by the experiment of the plane surfaces.

122 "Contra per Philosophum 10 *Metaphysice* verum continuum est cuius motus est unus. Volo ergo quod illa tabula sit mota (?) continua vel aliqua pars eius. Ergo quantumcumque aliqua pars eius movetur totum movetur. Si ergo tale corpus elevetur erit ibi vacuum antequam aer potest replere partes interiores spatii." Burley, *Questiones circa libros Physicorum*, fol. 66r, col. 1.

123 After describing how a vacuum is conceived to form upon the separation of two plane surfaces, Burley explained that some deny the possibility of the formation of such a

vacuum "impugnando casui uno modo quod non est possibile aliqua corpora esse suma-tur plana." He refuted this argument as false: "Item hec responsio, licet sit falsa, non vitat difficultatem quia argumentum est satis efficax duobus corporibus non planis. Si enim coniungantur, duo lapides non plani antequam aer circumstans posset replere partes centrales oportet quod repleat partes exteriores illius spacii. Et sic in illo priori erit ibi vacuum." *Ibid.*, fol. 66r, col. 1.

124 That the third article probably represented Blasius' own opinion may be inferred from his declaration that "In the third article, I shall determine the proposed question for the affirmative side according to what seems to me must be said." ("In tertio articulo determinabo propositam questionem pro parte affirmativa secundum quod mihi vide-bitur fore dicendum." *Questio De tactu corporum durorum*, p. 2, col. 2 of the edition cited above in n. 116.) In the absence of foliation and pagination, I have numbered the pages for convenience.

125 Because Blasius spoke of a circumference, radius, and center, I have assumed that the surfaces are circular.

126 The second of five suppositions, which says that "whatever the velocity by which some air is moved, other air can be moved infinitely more quickly. . . ." ("Secunda suppositio: quacumque velocitate data qua aliquis aer movetur in infinitum potest velocius alius aer moveri . . ." Blasius of Parma, *Questio De tactu corporum durorum*, p. 4, col. 2.)

127 "Et modus est iste: quia sit *b* aer qui est iuxta centrum, et sit *a* aer qui est iuxta cir-cumferentiam, modo dico sic ad hoc quod *a* aer iuxta circumferentiam exiens moveatur ad extra requiritur tempus mensurans istum motum. Sed quacumque velocitate data qua moveatur iste aer *a* ad extra, potest aer centralis *b* moveri in centecuplo velocius per unam suppositionem et in millecuplo. Quare in eodem tempore erunt isti aeres *a* et *b* moti ad extra." *Ibid.*, p. 5, col. 1.

128 "Tertia difficultas de eo quod dictum est in declarando unam conclusionem [i.e., the first conclusion under discussion here]. Dictum fuit enim quod in approximando duo plana adinvicem, aer centralis erat eque cito extra circumferentiam sicut erat aer qui erat iuxta circumferentiam quod non videtur esse verum cum unus et idem sit motor propellens exterius aerem centralem et aerem qui erat iuxta circumferentiam et cum una et eadem esset applicatio propellentis ad istos aeres." *Ibid.*, p. 5, col. 1.

129 ". . . licet iste aer centralis et iste circumferentialis moveantur ab eodem motore, ut scilicet a lapide plane qui deorsum movetur versus alium, non tamen moventur secun-dum eandem proportionem eo quod aer situatus in circumferentia habet in motu suo aerem extrinsecum sibi continue magis et magis resistentem propter maiorem et maiorem condensationem eius; sed aer centralis habet continue minorem et minorem resistentiam quia aer sibi immediatus continue est minor et minor et continue est rarior et rarior. Et hoc faciliter potest intelligenti patere. Et ideo licet iste aer et ille moveantur ab eodem motore non tamen secundum eandem proportionem, ut patet." *Ibid.*, p. 6, col. 1.

130 The loci of such discussions were separate treatises usually titled *De instanti*, or com-mentaries and *questiones* on Aristotle's *Physics*, bk. 6, ch. 5 (236a.3–4, 12–27; 236b.3–8, 33–34) and bk. 8, ch. 8, 263b.10–15. See Curtis Wilson's description of the problems in his *William Heytesbury, Medieval Logic and the Rise of Mathematical Physics* (Madi-son, Wis.: University of Wisconsin Press, 1956), pp. 29–31. Wilson distinguishes two phases in the approach of the Schoolmen (pp. 31–32). The first "in which, *natu-raliter loquendo*, the Schoolmen sought to apply the distinction [i.e., of intrinsic and extrinsic boundaries] directly to the physical magnitudes of the Aristotelian world – lengths, velocities, weights, and indeed whatever physical quantity may be conceived as continuous; the other in which, *logice loquendo* or *sophistice loquendo*, and applying the distinction to problems which were imaginable but presumably not capable of physical realization, they arrived at results which are of interest for the mathematical analysis of the continuum and of infinite aggregates."

131 How a quantity of air could rarefy indefinitely without a break in continuity and for-mation of a vacuum is not explained, but simply assumed. See above, Ch. 4, Sec. 3a.

132 A reference to the fourth supposition, which declared: "No minimum distance can be assigned by which distant bodies are separated, so that I wish to say that if any bodies

are newly separated, they will previously have been separated by half that distance. And this is true unless we should say that any bodies are first separated now when they begin to exist, since two bodies could now be generated, one in the east, one in the west. However, the supposition [excludes the latter case and] comprehends [only] bodies which begin their separation by local motion [rather than by distinct generations in widely separated places]." ("Quarta suppositio. Non est dare minimam distantiam per quam corpora distantia distant, et velim dicere quod si aliqua corpora nunc noviter distant prius per medietatem distantie distabant. Et hoc est verum nisi diceremus aliqua corpora nunc primo distare cum inceperunt esse quia nunc possent generari duo corpora, unum in oriente et reliquum in occidente. Suppositio autem intelligit de corporibus que inceperunt distare per motum localem." *Questio De tactu corporum durorum,* p. 5, col. 1.

133 "Secunda conclusio. Possibile est duo corpora dura et plana abinvicem equidistanter elevari. Probatur conclusio quia si non, hoc esset: vel eo quod daretur vacuum vel motus in instanti, ut communiter dicitur. Sed quod hoc non sequatur declaro quia cum tu queris de aere existente in circumferentia *vel* eque cito movetur ad centrum sicut ad punctum medium semidiametri *vel* prius ad punctum medium semidiametri quam ad centrum, dico quod prius movetur ad punctum medium semidiametri quam ad centrum. Et cum tu concludis ergo est tunc vacuum in centro, nego adhuc consequentiam quia quandocumque fuit aer in puncto medio semidiametri prius fuit alius aer in centro. Volo tamen quod omnis aer qui est in centro prius fuit in puncto medio semidiametri. Unde nullus est primus aer qui primo fuerit in centro quia sicut nulla est prima distantia qua ista corpora plana nunc distant, ut dicit una suppositio, sic non est aliquis aer qui primo subingressus est. Et ideo quicumque aer est nunc in centro prius fuit in puncto medio semidiametri. Et sic non sequitur aliquid inconveniens." *Ibid.*

134 2.11.423a.22–423b.1. On Averroes' comment on this passage, see above, n. 116 of this chapter.

135 Bacon insisted that nothing could intervene between a plane surface and air in air or a plane surface and water in water. Therefore, "if my palm touches the Seine, then, if the palm is raised from the water, it is necessary to assume that in the central point [of my palm] there would be a dimension [or space] without air and water. Hence there is a void, or at least a disposition toward a void, which is false." *Opera,* fasc. 13: *Physics* (8 books), p. 226. Even more to the point was John of Jandun's criticism: ". . . someone might say that two bodies could never be so perfectly joined and in mutual contact in air or in water without there being air or water between them, as Aristotle seems to intend in the second [book] of *De anima,* in the chapter on touch. But he [Aristotle] assumes what is false, for this might not suffice because although two solid bodies cannot be brought into mutual contiguity and contact in air unless air intervenes, nevertheless a solid and plane body could be applied to a surface of water so that nothing could lie between. And then if this body were raised simultaneously and separated from the water, the previous doubt would return because air cannot immediately [i.e., instantaneously] arrive at the middle of the space between them [i.e. the surfaces]. Therefore it seems that a vacuum will remain." *Physics* (Venice, 1519), bk. 4, question 10, fols. 56r, col. 2–56v, col. 1.

136 "Ymmo dicitur aliter quod impossibile est aliqua corpora solida tangere se in aere vel in aqua nisi aer vel aqua intercipiatur. Hoc enim dicit Philosophus secundo *De anima.* Unde si duo corpora solida coniungantur et deinde separentur abinvicem aer interceptus rarefit et replet locum quam potest quousque adveniat aer circumstans." Burley, *Questiones circa libros Physicorum,* fol. 66r, col. 1.

137 "Contra, possibile est quod aer interceptus sit ita rarus quod non possit iterum rarefieri ut repleat tantum locum quantus intercipitur [text: *incipitur*] inter illa corpora post separationem. Possibile enim est quod ille aer interceptus sit aer rarissimus. Volo enim quod duo corpora tangant se in aere rarissimo vel in igne rarissimo et redit argumentum." *Ibid.,* fol. 66r, cols. 1–2.

138 How bending or folding one of the two surfaces, which implies a part-after-part separation, would avoid either instantaneous motion of the inrushing medium or the dreaded

vacuum is left unexplained. We have already seen that Burley rejected part-by-part separation of surfaces. Nor does it seem plausible that Burley intended that one of the original surfaces be conceived as bent or folded prior to contact, as the words "nisi altera tabularum flecteretur sive plicaretur" might suggest (for the full text, see the following note). Whatever merit might attach to this interpretation derives from the fact that in what immediately follows, Burley argued that, whether the original surfaces are plane or not, separation cannot occur.

139 The text on which the interpretations in this paragraph are based follows: "Ad primum: in contrarium quando arguitur de igne rarissimo intercepto dicendum quod si ignis rarissimus interciperetur inter duas tabulas, tunc una tabula non posset elevari nisi altera tabularum flecteretur sive plicaretur. Tamen licet tabula non posset elevari a tabula potest tamen separari ab alia tabula trahendo a latere tabulam a tabula. Nec est cura quantum sive sint tabule plane sive non plane et sive equaliter fiat elevatio in partibus tabule sive inequaliter quia nullum corpus interceptum posset rarefieri ad replendum locum. Nullo modo potest tabula elevari a tabula nec equaliter nec inequaliter." *Ibid.*, fol. 67r, col. 1.

140 "Item si aliquis magnus saxus [sic] supponitur alteri posito, tunc quod ibi est aer medius, magnus lapis superpositus posset moveri per illum aerem. Lapis enim potest esse tante gravitatis quod ille aer non sufficit ad resistendum lapidi. Ergo lapis superpositus permoveri <potest> per illum aerem interceptum et per consequens potest ita approximari alteri lapidi quod nullus aer intercipitur quia possibile est corpus esse tante gravitatis quod aer non resistet motui suo." *Ibid.*, fol. 66r, col. 2.

141 "Si enim lapis moveatur et debeat in fine motus coniungi cum aere vel cum aqua eque cito coniungitur medium superficiei lapidis ad superficiem aque sicud partes exteriores illius superficiei quia totus lapis simul movetur et aliqua superficies lapidis tota simul coniungitur aque. Sed corpus prius interceptum inter lapidem et aquam non eque cito recedit a medio sicud a partibus exterioribus illius spacii. Ergo ibi derelinquitur vacuum." *Ibid.*

142 Because Burley omitted discussion of these alleged arguments, we cannot determine whether they were similar to those employed by Blasius of Parma described above.

143 "Propter istas rationes dicitur quod si duo corpora debeant adinvicem coniungi tunc partes exteriores aeris intercepti et similiter partes interiores simul recesserunt ab illo loco qui erat inter illa corpora. Licet enim partes aeris successive recedant, tamen ille partes aeris simul recesse sunt et ideo possunt tam partes exteriores quam interiores illorum corporum simul et semel coniungi." *Ibid.*

144 Here is the text for the summary that follows in this paragraph: "Ad aliud quando dicitur quod magnus lapis alteri superpositus moveri posset per illum aerem interceptum dicendum quod quantumcumque sit lapis magnus et gravis non movebitur propter hoc per illum aerem interceptum: quia infinita pars lapidis equaliter tendit deorsum versus aliam lapidem in partibus exterioribus et interioribus. Ymmo si lapis superior deorsum moveri per aerem interceptum, eque cito tangeret alium lapidem secundum suas partes exteriores sicud secundum partes interiores centrales. Et hoc non posset esse nisi aer interceptus recederet eque cito secundum onmes partes suas extra circumferentias. Et sic recederet sine motu, quod est impossibile. Ideo quantumcumque sit magnus lapis superpositus non descendet per illum aerem interceptum quia tunc est possibile quod totus aer interceptus cedat subito." *Ibid.*, fol. 67r, col. 1.

145 "Et quando accipitur quod lapis potest esse tante gravitatis quod aer non sufficiet ad resistendum, dicendum quod licet aer in propria virtute non sufficeret ad resistendum lapidi, tamen, in virtute agentis superioris salvantis perfectionem et plenitudinem universi, sufficit aer motus ad resistendum lapidi quantumcumque gravi. Unde aer interceptus ideo non cedit lapidi quia detinetur ab agente superiori prohibente vacuum forte." *Ibid.* On the universal celestial agent, see above, Ch. 4, Sec. 2.

146 "Tamen corpus multum grave propinquius tangit aliud quam corpus minus grave, licet enim semper sit aer medius inter corpora solida, aliquando est spissior, aliquando magis tenuis." *Ibid.*

147 "Item contra istam responsionem [i.e., Burley's theory] videtur sequi derisoria, puta quod animal semper ambulet super aerem vel aquam quia animal in ambulando num-

quam attingeret terram, sed semper ambularet super aerem qui, scilicet, intercipitur inter pedem et terram." *Ibid.*, fol. 66r, col. 2.

148 "Ad aliud quando accipitur quod animal numquam ambularet super terram sed semper super aerem vel aquam dicendum quod nullum est inconveniens. Hoc concedere nec est hoc deridendum ex quo est verum. Verumptamen secundum communem usum loquendi animal dicitur ambulare super illud quod supportat: huiusmodi est terra et non aer. Ideo dicitur animal ambulare super terram et non super aerem quia terra sustinet et supportat et non aer." *Ibid.*, fol. 67r, cols. 1–2.

149 Where Lucretius believed in extended and interstitial vacua, Hero accepted the latter but thought that separate, extended vacua could only be produced artificially. On Hero, see his *Pneumatica* (Greenwood trans.), pp. 6–10; this section is reprinted in Cohen and Drabkin, *Source Book in Greek Science*, pp. 251–254.

150 Except for Galileo's discussion of the separation of surfaces, what follows is drawn largely from Schmitt, "Experimental Evidence for and Against a Void," 352–366. For a lengthy seventeenth-century description and discussion of numerous traditional experiments on vacuum (including the freezing of water in a vessel, the bellows, the separation of polished plane surfaces, the candle experiment, and the clepsydra), all of which are interpreted in the medieval manner as denying the existence of vacuum, see Bartholomeus Amicus, *Physics*, vol. 2, pp. 743, col. 1–745, col. 2.

151 *Ibid.*, 356. Schmitt quotes the text from *Bernardini Telesii De rerum natura*, ed. Vincenzo Spampanato (3 vols.; Modena/Genoa/Rome: A. F. Formiggini, 1910–1923), vol. 1, p. 88. For Blaise Pascal's full and proper explanation of the bellows phenomenon, see the *Physical Treatises of Pascal: The Equilibrium of Liquids and the Weight of the Mass of Air*, trans. by I. H. B. Spiers and A. G. H. Spiers, with introduction and notes by Frederick Barry (New York: Columbia University Press, 1937), pp. 68–69; the passage is also included in Grant, *Source Book in Medieval Science*, pp. 330–331.

152 *Ibid.* Schmitt cites the Venice edition of 1593 (in this volume, I use the Ferrara edition of 1587) and Benjamin Brickman's translation, "On Physical Space, Francesco Patrizi," *Journal of the History of Ideas* 4 (1943), 234.

153 *Ibid.*, 358.

154 *Ibid.*, 360.

155 *Ibid.*, 361. The references to Hero's *Pneumatica*, which are given by Schmitt, are to the Greenwood translation, pp. 3–4, and the Greek text in Schmidt, *Heronis Alexandrini Opera*, vol. 1, pp. 8–10. For references to Pierre Gassendi and Patrizi, who also repeated this experiment and interpreted it much as did Turnèbe, see Schmitt, *ibid.*, 361–362 and n. 23. Patrizi suggested that only a partial vacuum could be created in this way.

156 Galileo Galilei, *Two New Sciences, Including Centers of Gravity & Force of Percussion*, trans. with introduction and notes by Stillman Drake (Madison, Wis.: University of Wisconsin Press, 1974), pp. 19–20 [59].

157 On this, of course, Sagredo and Galileo were following a longstanding medieval tradition in rejecting Aristotle's opinion that motion in a void would be instantaneous. See above, Ch. 3.

158 *Two New Sciences* (Drake trans.), pp. 20–21 [60].

159 Galileo challenged his own opinion only to the extent that Sagredo was permitted to raise a perplexing question. If, as Salviati argued, void functions as a resistance to material separation, void itself must be the cause of the initial resistance to separation of the slabs or plates. But if a vacuum does not form until the slabs are separated, how can resistance to separation be attributable to an as yet unformed void? The effect would precede the cause! Galileo's explanation of this paradox was left vague and uncertain. Somewhat later in the First Day, Galileo, through Salviati, declared "that a repugnance to the void is undoubtedly what prevents the separation of two slabs except by great force." (*Two New Sciences* [Drake trans.], p. 26 [66].) But does the vacuum occur before or after separation?

The solution may lie in Galileo's important claim that solid bodies are held together by the attractive force of minute, indivisible, dimensionless, interparticulate vacua whose collective cohesive force must be overcome before a material body can be broken or fractured. For, as Salviati said in response to an objection by Simplicio, "one may reply that although such voids are very tiny, and as a result each one is easily overpowered, still the innumerable multitude of them multiplies the resistances innumerably, so to speak" (*ibid.*, pp. 27–28 [67]). Salviati went on to demonstrate "that in a finite continuous extension it is not impossible for infinitely many voids to be found" (*ibid.*, p. 28 [68]). Ought we to understand from this argument that the total resistance to separation of two perfectly polished slabs or plates in direct contact arises from the infinite number of interparticulate vacua disseminated not only between the particles of each slab but also between the corresponding particles of each surface at every point of contact? May we then infer that when sufficient force is applied to cause separation of the slabs, an *actually extended*, though momentary, vacuum will result before the inrushing air fills it? On this interpretation, the proper cause of resistance to separation is attributable to the infinity of indivisible, interparticular vacua, and the actually extended, though momentary, vacuum between the slabs is but the effect or consequence of separation.

Evidence that Galileo also believed in the existence of minute, but actually extended vacua, in contrast to his more widely discussed dimensionless vacua with which we have been concerned in the preceding paragraph, may be inferred from Salviati's remark on p. 64 [105]. That Galileo rejected the existence in nature of void spaces of perceptible size, or *vacua separata*, to use medieval terminology, is evident by Salviati's statement on p. 71 [112], where it was also conceded that they could be produced artificially. For a discussion of the problem posed by Salviati "that in a finite continuous extension it is not impossible for infinitely many voids to be found," see A. Mark Smith, "Galileo's Theory of Indivisibles: Revolution or Compromise?" *Journal of the History of Ideas* 37, no. 4 (1976), 571–588.

160 See *Physical Treatises of Pascal*, trans. Spiers and Spiers, pp. 67–75, which is also reprinted in Grant, *Source Book in Medieval Science*, pp. 329–333.

CHAPTER 5
The historical roots of the medieval concept of an infinite, extracosmic void space

1 In the Latin West, medieval discussions on vacuum are separable into three quite distinct subtopics, all of which arose from Aristotle's intense attacks against the possible existence of void in any manner or place, namely (a) the separate, or extended, void (*Physics* 4, chs. 6–8), (b) the interparticulate void (see Aristotle's major discussion in *Physics* 4, ch. 9), and (c) the extracosmic void. The first two concerned intracosmic void and were considered in Chs. 1–4.

2 *De caelo* 1.9.279a.12–13, 17–18 (Oxford trans.; J. L. Stocks, trans.).

3 *Ibid.* 13–14.

4 This definition of void occurs at least twice in substantially the same form; see *De caelo* 1.9.279a.14–15 and *Physics* 4.7.214a.8–19. See above, Ch. 1 and Ch. 2, Sec. 1. The entire relevant passage from *De caelo* is cited in Ch. 2, n. 31.

5 Although Aristotle insisted that this nothing beyond the world was not to be interpreted as a vacuum, it was occasionally so construed, as when the Spanish Jew, Hasdai Crescas (ca. 1340–1412), in his *Or Adonai* (*Light of the Lord*), held that one could properly infer the existence of extracosmic vacuum from Aristotle's own arguments. Crescas's position as summarized by Wolfson, was that "according to Aristotle the world is finite, and beyond the outermost sphere there is no body. The absence of a body beyond the universe naturally means the absence of a plenum. The absence of a plenum must inevitably imply the presence of a non-plenum. Now, a non-plenum necessarily means some kind of potential space, actually devoid of any bulk, which, however, it is capable of receiving. Such a potential space is what is called a vacuum, for by definition a vacuum is nothing but incorporeal intervals or extensions. Thus, beyond the universe there must be a vacuum." Harry Austryn Wolfson, *Crescas' Critique of Aristotle, Problems of Aristotle's 'Physics' in Jewish and Arabic Philosophy* (Cambridge, Mass.: Harvard University Press, 1929), pp. 60–61. On interpretations of vacuum as nothing or something, see above, Ch. 2, Sec. 1, especially the section on Roger Bacon, who concluded from

Aristotle's arguments that a vacuum could exist beyond the world but, in contrast to Crescas, denied it the capacity to receive bodies.

6 "Tertio modo dicuntur vacuum spatium in quo omnino nullum corpus est, nec est aptum natum ad aliquod corpus recipiendum; et hoc modo est ponere, scilicet extra celum." Bacon, *Opera*, fasc. 8: *Physics* (4 books), bk. 4, p. 198. For a further discussion of these lines in a different context, see above, Ch. 2, Sec. 1.

7 The fact that Bacon's extracosmic void lacked the capacity to receive bodies suggests that he did not attribute tridimensionality to it.

8 *Physics* 3.4.203a.4–7.

9 *Physics* 4.6.213b.23–25.

10 Trans. by F. M. Cornford, "The Invention of Space," in *Essays in Honour of Gilbert Murray* (London: Allen & Unwin, 1936), p. 233. Because the works of Archytas have not survived, the translation was made from fragment 30 of Eudemus, presumably as quoted by Simplicius in the latter's commentary on Aristotle's *Physics*, 108a. Cornford ("The Invention of Space," pp. 233–234) identifies Lucretius (*De rerum natura*, 1, 968) and John Locke (*Essay*, ii, 13, 21) as among those who repeated the substance of Archytas' argument. Cornford's article has been reprinted in Milič Čapek (ed.), *The Concepts of Space and Time, Their Structure and Their Development* (Dordrecht/Boston: D. Reidel, 1976), pp. 3–16. On Archytas, see also Max Jammer, *Concepts of Space, The History of Theories of Space in Physics*, 2d ed. (Cambridge, Mass.: Harvard University Press, 1969), pp. 9–10.

11 Simplicius' *Physics* commentary was not translated from Greek to Latin until the sixteenth century.

12 ". . . the Stoics, however, thinking that there is a vacuum beyond the sky, prove it by this kind of assumption: let it be assumed that someone standing motionless at the extremity [of the world] extends his hand upward. Now if his hand does extend, they take it that there is something beyond the sky to which the hand extends. But if the arm could not be extended, then something will exist outside that prevents the extension of the hand; but if he then stands at the extremity of this [obstacle that prevents the extension of his hand] and extends his hand, the same question as before [is asked], since something could be shown to exist beyond that being." My translation as it appears in my article, "Medieval and Seventeenth-Century Conceptions of an Infinite Void Space Beyond the Cosmos," *Isis* 60 (1969), 41 (cited hereafter as Grant, "Infinite Void Space"). The translation is from *Simplicii . . . Commentaria in quatuor libros De celo*, fol. 44v, col. 2.

An interesting argument for the existence of extracosmic void was presented by al-Ghazali, as quoted in Averroes' *Tahafut al-Tahafut (The Incoherence of the Incoherence)*, a lengthy attack on al-Ghazali. On the assumption that God's power is absolute, al-Ghazali held that one might argue that God could continually increase the height (that is, size) of the world by one cubit, then two cubits, and so on, ad infinitum. "Now we affirm," he declared, "that this amounts to admitting behind the world a spatial extension which has measure and quantity, as a thing which is bigger by two or three cubits than another occupies a space bigger by two or three cubits, and by reason of this there is behind the world a quantity which demands a substratum and this is a body or empty space. Therefore, there is behind the world empty or occupied space." Al-Ghazali repudiated this argument as one that relies on possibilities that are an "illusion of imagination." Simon van den Bergh (trans.), *Averroes' Tahafut al-Tahafut (The Incoherence of the Incoherence)* (Printed Oxford: Oxford University Press; London: Luzac, 1954), vol. 1, p. 51; for further discussion, see below, n. 33 of this chapter. Al-Ghazali's argument was perhaps related to, or even derived from, Aristotle's argument in *De caelo* 2.4.287a.13–23, where Aristotle argued that if the world were any shape other than spherical, "there will at one time be no body where there was body before, and there will be body again where now there is none" (Loeb trans. by W. K. C. Guthrie). Under these circumstances, void and place would necessarily exist beyond the world. Al-Ghazali's argument played little role in the Latin tradition.

13 See above, Ch. 5, Sec. 1 and n. 4. For a Stoic who adopted essentially this definition, see Cleomedes (first century A.D.), *On the Circular Motion of the Heavenly Bodies*, bk.

1, ch. 1; for the Greek text, see H. Ziegler (trans.), *De motu circulari corporum caelestium* (Leipzig: Teubner, 1891), p. 8, and in German translation, see Arthur Czwalina (trans.), *Kleomedes Die Kreisbewegung der Gestirne* in *Ostwald's Klassiker der exakten Wissenschaften* (Leipzig: Engelmann, 1927), p. 3. For my translation of the relevant passage, see Grant, "Infinite Void Space," 40. An excellent discussion of the Stoic concept of extracosmic void and its relationship to Aristotle appears in David E. Hahm, *The Origins of Stoic Cosmology* (Columbus, Ohio: Ohio State University Press, 1977), pp. 103–107.

14 Hahm, *Origins of Stoic Cosmology*, p. 106.

15 For the means by which Stoics kept the cosmos from disintegrating and dissipating into the void, see *ibid.*, pp. 107–126; see also Grant, "Infinite Void Space," 40–41, n. 11.

16 Ziegler, *De motu circulari*, pp. 14, 16; Czwalina, *Kleomedes Die Kreisbewegung*, pp. 5–6; see also Grant, "Infinite Void Space," 41, n. 12.

17 Hahm, *Origins of Stoic Cosmology*, p. 106.

18 Aristotle himself may have been the source of this conception when he declared (*Physics* 4.6.213b.2–3) that "it might perhaps be maintained that, though material existence is continuous throughout the cosmos, the void is something existing outside it." The translation is that of P. H. Wicksteed and F. M. Cornford in the Loeb Classical Library.

19 See Gassendi's remarks in Čapek, *Concepts of Space and Time*, p. 91.

20 Among those who specifically cite, or allude to, the Stoic argument, we may mention Thomas Aquinas, John Buridan, Albert of Saxony, Nicole Oresme, and Richard of Middleton (see below, Ch. 6, n. 29). For the references, see Grant, "Infinite Void Space," 41–42 and n. 13–15. In n. 14, where a summary of Aquinas' rejection of the Stoic argument is given, Aquinas agreed with Aristotle (*De caelo* 1.9.279a.19–24) that not void, but only immutable separate substances exist beyond the world. See below, n. 57 of this chapter. Although Arabs also discussed, and even accepted, the existence of extracosmic void space, as did al-Rāzī (ca. 854–925 or 935), these ideas do not appear to have reached medieval Europe. For a description of al-Rāzī's spatial theory, see Salomon Pines, *Beiträge zur Islamischen Atomenlehre* (Berlin: A. Heine, 1936), pp. 45–49, and Jammer, *Concepts of Space*, pp. 91–92.

21 For the impact of the condemnation on natural philosophy, see Edward Grant, "The Condemnation of 1277, God's Absolute Power, and Physical Thought in the Late Middle Ages," *Viator 10* (1979), 211–244, and a much shortened version of it in Norman Kretzmann, Anthony Kenny, and Jan Pinborg (eds), *Cambridge History of Later Medieval Philosophy* (Cambridge: Cambridge University Press, in press). For the most part, I have relied here on the shortened version. The Latin text of the 219 articles has been published in its original order by H. Denifle and E. Chatelain, *Chartularium*, vol. 1 (1889), pp. 543–555. To facilitate their use, the articles were grouped by subject matter and published by Pierre F. Mandonnet, O. P., *Siger de Brabant et l'Averroïsme Latin au XIIIᵐᵉ siècle . . .* IIᵐᵉ Partie: Textes inédits (Louvain: Institut Supérieur de Philosophie de l'Université, 1908), pp. 175–191. Using Mandonnet's reorganized version, Ernest L. Fortin and Peter D. O'Neill translated the condemned articles into English, in *Medieval Political Philosophy: A Sourcebook*, ed. Ralph Lerner and Muhsin Mahdi (New York: Free Press of Glencoe, 1963), pp. 337–354. Their translation has been reprinted in Arthur Hyman and James J. Walsh (eds.), *Philosophy in the Middle Ages: The Christian, Islamic, and Jewish Traditions* (Indianapolis: Hackett Publishing, 1973), pp. 540–549. Selected articles relevant to medieval science have been translated by Grant, *Source Book in Medieval Science*, pp. 45–50.

22 Descriptions of the events leading up to, and including, the Condemnation of 1277 appear in: Pierre F. Mandonnet, *Siger de Brabant et l'Averroïsme Latin au XIIIᵐᵉ siècle . . .*, Iʳᵉ Partie: Étude Critique (Louvain: Institut Supérieur de Philosophie de l'Université, 1911), ch. 9 ("Condemnation du Péripatétisme 1277"), pp. 214–251; Fernand Van Steenberghen, *Siger de Brabant d'après ses oeuvres inédites*, vol. 2 (*Les Philosophes Belges*, vol. 13 [Louvain: Editions de L'Institut Supérieur de Philosophie, 1942]), ch. 2 ("La Philosophie à l'Université de Paris avant Siger de Brabant"), pp. 357–497; Pierre Duhem, *Le Système du monde*, vol. 6, ch. 1, pp. 3–69, and vol. 8, ch. 8, pp. 7–120; Gordon Leff, *Paris and Oxford Universities in the Thirteenth and Fourteenth Centuries* (New York: Wiley, 1968), part 2, ch. 4, pp. 187–238; and, finally John W. Wippel, "The Condemnations

of 1270 and 1277 at Paris," *The Journal of Medieval and Renaissance Studies* 7 (1977), 169–201. For the possible sources of each of the 219 articles and the reasons for its condemnation, see Roland Hisette, *Enquête sur les 219 articles condamnés à Paris le 7 mars 1277, Philosophes Médiévaux*, vol. 22 (Louvain: Publications Universitaires; Paris: Vander-Oyez, 1977).

23 It had already been enunciated in St. Peter Damian's *De divina omnipotentia* in the eleventh century (see the translation of the relevant section in John F. Wippel and Allan Wolter, O. F. M. [eds.], *Medieval Philosophy: From St. Augustine to Nicholas of Cusa* [New York: Free Press, 1969], pp. 143–152, especially 148–149) and in Peter Lombard's *Sentences* in the twelfth century (see *Magistri Petri Lombardi . . . Sententiae . . . ,* tom I, pars II, liber I et II, bk. I, distinctio 42, ch. 3, par. 6, pp. 297–298; for my translation of the passage, see my article "The Condemnation of 1277," 214, n. 10.

24 On the difference between God's absolute power (*potentia absoluta*) and ordained power (*potentia ordinata*), see Grant, "The Condemnation of 1277," 215; see also Amos Funkenstein, "Descartes, Eternal Truths, and the Divine Omnipotence," *Studies in History and Philosophy of Science* 6 (1975), 185, n. 1.

25 "Quod impossibile simpliciter non potest fieri a Deo, vel ab agente alio. – Error, si de impossibili secundum naturam intelligatur." Denifle and Chatelain, *Chartularium*, vol. I, p. 552.

26 Although the propositions relevant to Thomas Aquinas were annulled on February 14, 1325, the condemned articles were otherwise in effect during the fourteenth century. Indeed no specific cancellation seems to have been promulgated, and their effectiveness may simply have waned and faded with time.

27 In this group are included Richard of Middleton, Duns Scotus, Thomas Bradwardine, Walter Burley, William of Ockham, Peter Aureoli, Jean de Ripa, John Buridan, and Nicole Oresme.

28 "Quod prima causa non posset plures mundos facere." Denifle and Chatelain, *Chartularium*, vol. I, p. 545.

29 "Quod Deus non possit movere celum motu recto. Et ratio est, quia tunc relinqueret vacuum." *Ibid.*, p. 546. Cleomedes the Stoic declared that if we imagine the world moved from its place, a vacuum would be left behind (p. 6 of Ziegler's Greek edition, cited above in n. 13 of this chapter). The work of Cleomedes was unknown in the Middle Ages, but even if known, the medieval debate would have differed by virtue of the theological context of the problem.

30 See above, Ch. 2, Secs. 1 (and n. 23), 2b (and n. 59), and 4 (and n. 77, Locke); Ch. 3, Secs. 7a, b; and below, Ch. 6, Sec. 2a (and n. 37), and notes 22, 34 (Descartes), and 35; and Ch. 7, Sec. 2d (and n. 18, John Major). For the use of imaginary annihilations in spatial discussions of the seventeenth century, see below Ch. 8, n. 169. On the general use of the method of annihilation in fourteenth-century terminist philosophy and its subsequent role in the seventeenth century, see the excellent analyses by Funkenstein, "Descartes, Eternal Truths, and the Divine Omnipotence," 191, and "The Dialectical Preparation for Scientific Revolutions: On the Role of Hypothetical Reasoning in the Emergence of Copernican Astronomy and Galilean Mechanics," in Robert Westman (ed.), *The Copernican Achievement* (Berkeley: University of California Press, 1975), pp. 182–187.

31 It is true that Thomas Aquinas, and others, believed that no proper demonstration had been offered for or against the eternity of the world. Thomas had even argued that creation of the world need not necessarily occur in time and that the idea of creation does not require a beginning. To Aquinas, the idea of creation involved dependence, the dependence of the mutable world as effect on an immutable God as first cause. Nevertheless, Aquinas believed on faith that the world had a beginning in time. See *St. Thomas Aquinas, Siger of Brabant, St. Bonaventure On the Eternity of the World (De Aeternitate Mundi)*, trans. from the Latin with an introduction by Cyril Vollert, Lottie H. Kendzierski, and Paul M. Byrne (Milwaukee, Wis.: Marquette University Press, 1964), pp. 14–16. In the minds of some hostile critics, Aquinas may have been identified as one who allowed the possible eternity of the world. Because of this, one or more of the numerous articles condemning the eternity of the world in 1277 may have had Aquinas in mind.

32 *De caelo* 3.2.301b.31–302a.9 in the translation by W. K. C. Guthrie (Loeb Classical Library).

33 Averroes, *Opera*, vol. 5: *De caelo*, bk. 3, comment 29, fol. 199v, col. 1 (H–I).

Averroes also enunciated much the same argument in his *Tahafut al-Tahafut* (*The Incoherence of the Incoherence*). In this significant treatise, which is Averroes' section by section refutation of al-Ghazali's *Tahafut al Falasifa* (*The Incoherence of Philosophy*), Averroes declared that "he who believes in the temporal creation of the world and affirms that all body is in space, is bound to admit that before the creation of the world there was space, either occupied by body, in which the production of the world could occur, or empty, for it is necessary that space should precede what is produced." *Averroes' Tahafut al-Tahafut* (*The Incoherence of the Incoherence*), vol. 1, "The First Discussion," p. 52. Al-Ghazali, whose arguments were included verbatim by Averroes, had sought to uphold creation and also to deny the necessity of a pre-creation void space by insisting that it was mere imagination to assume possible spatial existence prior to the creation of the world ("And our answer is . . . 'It belongs to the illusion of imagination to suppose possibilities in space behind the existence of the world'." *Ibid.*, p. 51). Averroes' *Tahafut al-Tahafut* was translated from Arabic into Latin by Calonymos ben Calonymos in 1328 at Arles. Although this version, which bore the title *Destructio Destructionum*, included the first sixteen discussions, and presumably the passage included here, it omitted the last four physical discussions. The impact of this translation on fourteenth-century Latin thought is largely unknown, although Simon van den Bergh believes it influenced Nicholas of Autrecourt (*Tahafut al-Tahafut*, vol. 1, pp. xxx–xxxi). A Latin edition of it was published at Venice in 1497 by Bonetus Locatellus, along with a lengthy commentary by Agostino Nifo (it was reprinted at Lyons in 1517, 1529, and 1542). Another Latin translation, this time from an earlier Hebrew translation, was made in the sixteenth century by Calo Calonymos, or Kalonymos ben David the Younger, and published at Venice in 1527. The latter translation has been published by Beatrice Zedler, *Averroes' "Destructio Destructionum Philosophiae Algazelis" in the Latin Version of Calo Colonymos* (Milwaukee, Wis.: Marquette University Press, 1961). For a discussion of the various translations and the significant impact that the *Destructio* had on sixteenth-century Aristotelian thought, see Zedler's introduction; for a "Chart of Latin Editions of the *Destructio Destructionum*," see pp. 55–56.

34 Averroes, *Opera*, vol. 5: *De caelo*, fol. 200r, col. 1(A). In his *Commentary on the Physics*, bk. 4, comment 6 (*Opera*, vol. 4, fols. 123v, col. 2–124r, col. 1), Averroes, in commenting on Aristotle, *Physics* 4.1.208b.25–33, considered what he called "the third common argument" about place, namely that it is a vacuum. He explained that "dicere enim vacuum esse et quod est locus sine corpore est famosum propter existimationem et est necessarium in existimatione eis qui generant mundum secundum totum, scilicet existimare vacuum precedere necesse est enim ut locus precedat corpus; et cum totum est generatum, sequitur ut vacuum precedat generationem totius." The context of the discussion was Aristotle's mention of Hesiod's (in the Junctas edition of Averroes' *Physics* commentary, Hesiod becomes Homer) belief that chaos preceded the world, which implied that space precedes the things in it. Averroes went on to mention that the Muslim Mutakallimun, who in the Latin translation are called *Loquentes nostrae legis*, are among those who believe that a vacuum preceded the generated world. For a discussion of the Mutakallimun see Wolfson, *The Philosophy of the Kalam*, pp. 1–2.

35 Albertus Magnus presented and rejected Averroes' dilemma as follows (*Opera*, vol. 2: *Physics*, p. 146, col. 1): "Quod autem dicit Averroes, quod omnes qui generant mundum per creationem incidunt in dictum Hesiodi quia necesse est esse vacuum ante mundum in quo creetur mundus; et cum falsum sit esse vacuum, falsum est mundum esse creatum, non est recipiendum." For Albertus' solution, see below, note 42. In his *Summa Theologiae* (part I, question 46, article 1, argument 4), Thomas Aquinas repeated Averroes' argument that a pre-creation void would necessarily exist if the world were created (see St. Thomas Aquinas, *Summa Theologiae*, Latin text and English translation, introductions, notes, appendices, and glossaries [New York and London: Blackfriars in conjunction with McGraw-Hill and Eyre & Spottiswoode], 1967, vol. 8, pp. 66 [Latin] and 67 [English]) and rejected it later (p. 73) when he concluded that "before the world existed there

was no place or space." The author of the thirteenth century *Summa Philosophiae*, falsely ascribed to Robert Grosseteste, also cited Averroes' argument that a pre-creation void must precede a created world but went on to explain that those who believe this (apparently he did not) conceive this pre-creation place not as a natural being but as nonbeing. They interpret the relationship of being and nonbeing in the same manner in which actuality is opposed to privation. Thus it is that they suspect that a vacuum, equated with privation, exists beyond the heaven, in contrast to the plenum, which lies within the heaven and is equated with natural being. ("Supponentes itaque mundum novum esse, id est creatum, necessario iuxta Averroëm ponunt in totali mundi spatio fuisse, non tanquam aliquod ens naturale, sed tanquam non-ens ubi, et aliquid esse possit, quod est mundus, licet autem non-ens a cognitione nostra ex relatione ad ens fingitur sicut ex respectu actus privatio opposita, ac sic supra caelum vacuum suspicimus, quia infra caelum plenum intuemur." Ludwig Baur (ed.), *Die philosophischen Werke des Robert Grosseteste* in *Beiträge der Geschichte der Philosophie des Mittelalters*, vol. 9 [Münster: Aschendorff, 1912], p. 417, lines 14–20). For Thomas Bradwardine's repudiation of Averroes' argument, see below, Ch. 6, Secs. 3b, e.

36 For example, see articles 9, 52, 87, 90, 91, 93, 94, 95, 98, 99, 185, and 202 in Grant, *Source Book in Medieval Science*, pp. 48–50.

37 In article 201, we learn that anyone who believes that the world has been generated "assumes a vacuum because a place necessarily precedes what is generated in that place; therefore, before the generation of the world there existed something without a thing located in it, which is a vacuum." ("Quod, qui generat mundum secundum totum, ponit vacuum quia locus necessario precedit generatum in loco; et tunc ante mundi generationem fuisset sine locato, quod est vacuum." Denifle and Chatelain, *Chartularium*, vol. 1, p. 554. I have altered the translation of this article as it appears in my *Source Book in Medieval Science*, p. 50. The objectionable aspect of article 201 lay in the clear implication that a void place *necessarily* preceded the world if the latter was actually created. If so, then God required that a void exist in order to create the world, a requirement that would have been a restriction on His absolute power to have created the world without the need of an empty space in which to locate it.

38 ". . . cum ille situs imaginarius nullam naturam habeat positivam, sicut nullus ignorat; alias etenim esset aliqua natura positiva que nec esset Deus, nec a Deo, cuius oppositum secundum et tertium huius docent; essetque coaeterna Deo, quod nullus recipit Christianus." Bradwardine, *De causa Dei*, p. 177. The translation is mine and is cited from Grant, *Source Book in Medieval Science*, p. 558. For the full context of the discussion, see below, Ch. 6, n. 98.

39 Following immediately after the text in the preceding note, Bradwardine explained that in order to prevent belief in a pre-creation eternal void space that was not from God, "Bishop Stephen of Paris condemned an article asserting that 'many things are eternal,' and that even if He wished, God could not destroy them; therefore He is deprived of omnipotence." ("Quare et Stephanus Parisiensis episcopus damnavit articulum asserentem quod multa sunt aeterna, nec Deus posset illa destruere etsi vellet quare et omnipotentia privaretur." Bradwardine, *De causa Dei*, pp. 177–178. For the full passage, see below, Ch. 6, n. 98. The article cited by Bradwardine is probably 52, which denounced the claim "That that which is self-determined, as God, either always acts or never acts; and that many things are eternal." See Grant, *Source Book in Medieval Science*, p. 48. Bradwardine might, of course, have invoked article 201 (see above, n. 37 of this chapter), which specifically rejected the necessity of a pre-creation void space. But he seems to have preferred an article that condemned the existence of all things that might be independent of, and coeternal with, God. Such a general article would also have included a coeternal, pre-creation void.

In his Hebrew treatise *Milhamot Adonai*, or *Milhamot Hashem* (*The Wars of the Lord*), written in southern France, Levi ben Gerson (Gersonides) (1288–1344) conceded that creation *ex nihilo* committed one not only to a pre-creation void space but also to the existence of an extramundane void. Because he believed vacuum impossible, Levi concluded with Plato, that the world was created from primordial matter, some of which now lies outside our world. See Seymour Feldman, "Platonic Themes in Gersonides' Cosmology," in *Salo Wittmayer Baron Jubilee Volume* (Jerusalem: American

Academy for Jewish Research, 1975), pp. 388, 390–391. Levi's position would have been condemned in Latin Christendom.

40 Without mention of Bradwardine, whose 1618 edition of *De causa Dei* he may have known, Bartholomeus Amicus (*Physics*, vol. 2, p. 760, col. 2[A–B]), citing Pedro Fonseca (see below, this note) with approval, also argued against those who believed that imaginary space could have a positive existence or essence. Although Amicus held that imaginary space is eternal and exists outside the world, he declared nevertheless that "before the creation of the world there was no positive being except God – [for such a positive being] would either be by itself [i.e., independent] and would [therefore] be a God; or it would be [produced or created] by another, [namely,] either from this God – but since it is assumed God produced nothing before this world, this matter is settled – or it would be [produced or created] by another God. But it is repugnant to natural reason that there be many gods." ("Prima conclusio: spatium imaginarium formaliter non est quid positivum sive secundum essentiam, sive secundum existentiam, accipiatur. Probatur cum Fonseca 5, *Met.* c.13, q.7, sect. 1, contra Philoponum, 4 *Phys.*, digressione de loco, et alios. Primo, ex generali ratione, quia spatium fuit in aeternitate et est extra hoc universum, ut infra. Sed ante mundi creationem nullum fuit ens positivum existens praeter Deum, quia vel esset a se et esset Deus; vel ab alio: vel ab hoc Deo, ut supponitur, quia hic Deus nihil produxit ante hunc mundum, ut ea de re constat; vel ab alio Deo et repugnare etiam secundum rationem naturalem dari plures Deos." Bartholomeus Amicus, *Physics*, vol. 2, p. 760, col. 2[A–B].) In arguing that space is not a true species of quantity, Pedro Fonseca (1528–1599), whom Amicus cited, had earlier argued that "space is not a true being, for if it were, it would be *either* an uncreated being, which is appropriate only to God; *or* it would be a created being, which it cannot be because space could not begin to exist, for wherever it is now, there, necessarily, it always was and always will be." ("Nobis tamen . . . hoc tantum ostendendum est hoc loco fieri non posse ut spatium sit vera species quantitatis. Quod primum ex eo constat: quia non est verum ens, si enim esset, aut increatum ens esset, quod in solum Deum convenit; aut creatum, quod fieri nequit quia spatium non potuit incipere esse, ubicunque enim nunc est, ibi necessario semper fuit, semperque erit." *Commentariorum Petri Fonsecae Lusitani, Doctoris Theologi Societatis Iesu, in libros Metaphysicorum Aristotelis Stagiritae* [4 vols.; Cologne: Zetzneri Bibliopolae, 1615]; reprinted in facsimile with the title *Petri Fonsecae Commentariorum in Metaphysicorum Aristotelis Stagiritae Libros* [2 vols.; Hildesheim: George Olms Verlag, 1964], vol. 2, bk. 5, ch. 13, question 7, sec. 1, col. 701.) How Amicus, and long before him, Bradwardine, explained the relationship between an infinite, uncreated space and an infinite, eternal, omnipresent God will be described below. On the meaning of the expression "imaginary space" (*spatium imaginarium*), see below, Ch. 6, Sec. 1.

Although theological reasons were the usual grounds for denying a pre-creation void, al-Ghazali had used another argument. According to Averroes's report in the *Destructio Destructionum*, al-Ghazali had denied a pre-creation void space by designating such a conceptual possibility as an "illusion of imagination" (see above, n. 33 of this chapter).

41 For a description and analysis of the various ways in which medieval Aristotelian commentators assigned or denied a place to the last sphere and to the world itself, see my forthcoming article, "The Medieval Doctrine of Place: Some Fundamental Problems and Solutions," in *Filosofia e scienza nella tarda scolastica: Studi in memoria di Anneliese Maier* (Rome: Edizioni di Storia e Letteratura; forthcoming, 1981).

42 ". . . quando Deus fecit mundum, fecit locum vel dimensionem recipientem mundum; et ideo neque erat locus, neque dimensiones, nec etiam sunt nunc extra ipsum caelum." Pseudo-Siger of Brabant, *Physics* (Delhaye ed.), bk. 4, question 24, p. 179 (on the false attribution of this work to Siger of Brabant, see above, Ch. 2, n. 4). Both Albertus Magnus and Roger Bacon agreed with Pseudo-Siger that God created the world and its place simultaneously. See Albertus Magnus, *Opera*, vol. 2: *Physics*, p. 146, col. 1, and Roger Bacon, *Opera*, fasc. 8: *Physics* (4 books), pp. 193–194.

43 See S. Sambursky, *The Physical World of Late Antiquity* (New York: Basic Books, 1962), p. 4.

44 *De somniis*, I, 63–64, in *Philo with an English translation*, ed. and trans. by F. H.

Colson, G. H. Whitaker, and R. Marcus (12 vols.; Loeb Classical Library. Cambridge, Mass.: Harvard University Press; London: William Heinemann, 1929–1962), vol. 5 (1934), p. 329. The italics are mine. Also see Sambursky, *Physical World of Late Antiquity*, p. 4. Although medieval Jewish philosophers knew nothing of Philo's work, he was of great significance to early Christians, especially Ambrose (see A. H. Armstrong [ed.], *The Cambridge History of Later Greek and Early Medieval Philosophy* [Cambridge: Cambridge University Press, 1970], p. 157).

45 Wolfson, *Crescas' Critique of Aristotle*, p. 123. See also below, Ch. 6, n. 20.

46 "Ergo unus est et ubique diffusus est." Cited from *Quod idola Dii non sint* in *S. Thasci Caeceli Cypriani Opera Omnia*, ed. William Hartel (Vienna, 1868), p. 26 (Cyprian's *Opera* constitutes vol. 3, part 1 of the *Corpus Scriptorum Ecclesiasticorum Latinorum*).

47 Arnobius of Sicca, *The Case Against the Pagans*, newly translated and annotated by George E. McCracken (2 vols.; Westminster, Md.: Newman Press, 1949), vol. 1, p. 180. In his *De spatio reali* (1702), pp. 24, 87, Joseph Raphson cited the Latin text of this passage twice (for the full title of Raphson's work, see below, Ch. 8, n. 277).

48 The full passage reads: "He is in every place, for He cannot be in any place at all – but that every place is present to Him for Him to occupy, although He Himself can be received by no place, and therefore He cannot anywhere be in a place, since He is everywhere but in no place." *The Theological Tractates*, with an English translation by H. F. Stewart and E. K. Rand; *The Consolation of Philosophy* with an English translation by "I. T." (1609), revised by H. F. Stewart (Cambridge, Mass.: Harvard University Press, 1953), p. 21.

49 *De spatio reali*, pp. 24, 85, 87.

50 The Old and New Testaments were, of course, a prime source for such citations, as were also works by Hermeticists, Neoplatonists, Stoics, Cabbalists, and Christians. For a list of frequently cited biblical sources drawn from Newton's *Principia*, see Jammer, *Concepts of Space*, p. 113. Numerous citations from, or references to, these varied treatises appear throughout this volume.

51 Augustine's question ("Ubi erat Deus antequam esset caelum et terra?") appears in his *Evodii de fide contra Manichaeos*, ch. (or paragraph) 37 (see *Corpus Scriptorum Ecclesiasticorum Latinorum*, editum consilio et impensis Academiae Litterarum Caesareae Vindobenensis, vol. 25, sec. 6, part 2: *Contra felicem de natura boni; Epistula Secundini contra Secundinum accedunt; Evodii de fide contra Manichaeos; et Commonitorium Augustini quod fertur*, ed. by Joseph Zycha [Prague/Vienna/Leipzig: F. Tempsky/G. Freytag, 1892], p. 967). Augustine's reply ("In se habitabat Deus, apud se habitabat, et apud se est Deus") appears in *Enarrationes in Psalmos*, psalm 122, 4, in *Corpus Christianorum, series Latina: Aurelii Augustini Opera*, vol. 40, part 10, 3: *Enarrationes in Psalmos CI-CL* (Turnholt: Brepols, 1956), p. 1817. Augustine's description bears a resemblance to that given by Philo Judaeus.

52 The title of the chapter is: "We ought not to think of infinite extents of times (*de infinitis temporum spatiis*) before the [creation of the] world, nor of infinite extents of places [or spaces] (*de infinitis locorum spatiis*) beyond the world, since just as there are no times before the [creation of the] world, so there are no places beyond it." My translation is from *Corpus Scriptorum Ecclesiasticorum Latinorum*, vol. 40 (Vienna Academy, 1899), p. 517. The quotations in the remainder of this paragraph are from the translation by Marcus Dods, *The City of God* (2 vols.; New York: Hafner Publishing, 1948; first published Edinburgh: T & T Clark, 1872), pp. 441–442.

53 Among those who saw fit to cite this argument in the context of a discussion of extracosmic space were the Coimbra Jesuits (*Conimbricenses*), *Commentariorum Collegii Conimbricensis in octo libros Physicorum Aristotelis Stagiritae* (Cologne, 1602), bk. 8, ch. 10, question 2, col. 518; John Major, *In Primum Sententiarum ex recognitione Io. Badii* (Paris: apud Badium, 1519), fol. 93r, col. 2; Petrus Barbay, *Commentarius in Aristotelis Metaphysicam*, 4th ed. (Paris, 1684), p. 408; Phillip Faber, *Disputationes Theologicae Librum Primum Sententiarum Complectentes . . . secundum seriem distinctionum Magistri Sententiarum et questionum Scoti paucis exceptis ordinate . . .* (Venice, 1613), bk. 1, distinctio 37, pp. 338, col. 1, and 344, col. 2; Gabriel Vasquez, *Disputationes Metaphysicae*, p. 378; Bartholomeus Amicus, *Physics*, vol. 2, p. 763, col.

2(C); and Franciscus Bona Spes, *Commentarii Tres in Universam Aristotelis Philosophiam* (Brussels: apud Franciscum Vivienum, 1652), p. 177, col. 1. On the basis of this passage, some (Major, the Coimbra Jesuits, and Amicus) mistakenly concluded that Augustine really believed in the existence of infinite extracosmic space with an omnipresent God, whereas others (Petrus Barbay, Gabriel Vasquez, Phillip Faber, and Franciscus Bona Spes) rightly understood that Augustine was actually denying spaces beyond the world. For John Major, see below, Ch. 7, Sec. 1; for Barbay, Ch. 7, n. 17; and for Bona Spes, Ch. 7, Sec. 2f and n. 157.

54 For Augustine's specific use of the *Asclepius*, see A. D. Nock and A. J. Festugière, *Corpus Hermeticum* (4 vols.; Paris: Société d'Edition "Les Belles Lettres," 1945 and 1954), vol. 2, pp. 264–266. The passage from *Asclepius* quoted below was not, however, actually cited by Augustine. The *Asclepius* was known to Augustine in a third- or fourth-century Latin translation from the original Greek. For an account of the translation into Latin of the other Greek treatises that made up the Hermetic corpus, see Frances A. Yates, *Giordano Bruno and the Hermetic Tradition* (New York: Vintage Books, 1969; first published by the University of Chicago Press, 1964), pp. 12–17. The translations were made by Marsilio Ficino in the fifteenth century.

55 The profound impact of the Hermetic literature on Renaissance thinkers and Giordano Bruno in particular was described by Frances A. Yates in her *Giordano Bruno and the Hermetic Tradition*. She explained (p. 245) that although "Bruno would not have found in the Hermetic writings the conception of an infinite universe and innumerable worlds, the spirit in which he formulates such a conception is to be found in them." Along with a brief passage from the *Corpus Hermeticum*, Yates cited in *Giordano Bruno* the last few lines of this paragraph as potentially significant, because "Bruno had but to add to this that there *is* an infinite space outside the world and it *is* full of divine beings and he would have his extended Hermetic gnosis of the infinite and the innumerable worlds." As we shall see below, Thomas Bradwardine did something similar and equally dramatic some two centuries before Bruno.

56 Translation by Walter Scott (ed. and trans.), *Hermetica, the Ancient Greek and Latin Writings which Contain Religious or Philosophic Teachings Ascribed to Hermes Trismegistus* (4 vols.; Oxford: Clarendon Press, 1924–1936), vol. 1, pp. 319–321. Although the edition of *Asclepius* by A. D. Nock and A.-J. Festugière in vol. 2 of their *Corpus Hermeticum* is better than Scott's, the latter's translation seems to agree well with both versions. I have also discussed both St. Augustine and the *Asclepius* in Grant, "Infinite Void Space," 42–43. For another significant passage from the *Asclepius*, see below, Ch. 7, Sec. 2b and n. 69.

57 One notes with interest that after denying the existence of matter, place, void, or time beyond the heaven, Aristotle allowed (*De caelo* 1.9.279a.19–23) that there are, nonetheless, entities there that are neither born in place nor affected in any way by change. Indeed, "changeless and impassive, they have uninterrupted enjoyment of the best and most independent life for the whole aeon of their existence." *Aristotle on the Heavens* (Guthrie trans.). This passage was occasionally cited as evidence that God exists beyond the heavens, "from which it was inferred that God was in an imaginary space or void." Thus the Franciscan Phillip Faber (1564–1630) observed that there were those who cited "the authority of Aristotle in *De caelo*, I, text 100 [which includes 1.9.279a.19–23], where he clearly said that God is beyond the heaven" (*Disputationes theologicae*, p. 338, col. 1, sec. 23; the bracketed addition is mine). In refuting this interpretation of Aristotle, Faber declared that "many things could be said about what was in his [Aristotle's] mind, but, briefly, it must be especially noted that, according to Aristotle, there is no vacuum beyond the heaven, and, consequently, there is no imaginary space; but there is nothing [out there]. Thus he [Aristotle] could not say that intelligences and God, which are according to him outside the heaven in the same place, are in a vacuum or in an imaginary space. . . . Besides, when he [Aristotle] posits intelligences outside the heaven, he does not assume them in any great imaginary space, as the adversaries would have it, but [he posits them] beyond the farthest motion and there deduces that they are not aged, or changed, or altered, nor is it natural that they be in a place. They live the best life because outside the heaven there is no place, or time, or motion. for these are

passions of natural body. But outside the heaven there is no body, but only intelligences which are incorporeal and thus lack passions according to Aristotle. Therefore, we hold that God is outside the heaven, but not in a vacuum or imaginary space. . . . And if you should say that while He is outside the heaven, it is necessary that He be there in some manner, therefore how is He to be there if He is not there as in some place? I reply that He is there in Himself, not as in a place, as we shall discuss below against the fifth argument" (*Disputationes Theologicae*, p. 345, col. 1, sec. 53). See above, n. 20 of this chapter, where Aquinas appears in essential agreement with Faber, and below, Ch. 7, n. 67, where the Coimbra Jesuits interpreted Aristotle's statement as evidence that God existed beyond the world in an imaginary space.

58 Although the passage from *Asclepius* quoted above concludes with the assertion that void "cannot possibly be empty of spirit and of air," our anonymous author is here referring to the cosmos itself and not to the extramundane. It is only the concluding lines of the first paragraph that are relevant to our investigation and they clearly imply that the region outside the cosmos would not be void because it is filled with spiritual beings – i.e., "filled with things apprehensible by thought alone." It is of interest that Aristotle mentioned (*Physics* 4.6.213a.23–31) that there were those who mistakenly believed that things full of air were empty, an opinion that the anonymous author of *Asclepius* apparently accepted.

59 See *Magistri Petri Lombardi Parisiensis Episcopi, Sententiae in IV libris distinctae* Grottaferrata [Rome]: Editiones Collegii S. Bonaventurae Ad Claras Aquas, 1971), vol. 1, part 2, books 1 and 2, pp. 263–275. Wherever appropriate, a commentary on Peter Lombard's *Sentences* will be cited in the customary manner as a *Sentence Commentary*, rather than a *Sentences Commentary*, which, though technically correct, is rarely used.

60 As we shall see below in the course of our discussion, it could also be taken up in Commentaries on Aristotle's *Physics* (for example, Bartholomeus Amicus), *De caelo* (as with Nicole Oresme), and *Metaphysics* (as with Petrus Barbay and Pedro Fonseca, and perhaps also Francisco Suarez, who considered the problem in his *Disputationes Metaphysicae*, which, however, is not a direct commentary on Aristotle's *Metaphysics*).

CHAPTER 6
Late medieval conceptions of extracosmic ("imaginary") void space

1 By comparison with what followed, Roger Bacon's few relevant lines described in Ch. 2, Sec 1, qualify as a "minor exception." Because Bacon lectured on the *Physics* at the University of Paris in 1255, it is likely that he wrote these lines before 1277.

2 See *Aristoteles Latinus I 6–7, Categoriarum Supplementa: Porphyrii Isagoge translatio Boethii et Anonymi Fragmentum vulgo vocatum 'Liber sex principiorum,'* ed. by Laurence Minio-Paluello with the assistance of Bernard G. Dod (Bruges and Paris: Desclée de Brouwer, 1966), p. 47, lines 8–9.

3 Sacrobosco merely repeated Aristotle's argument (*De caelo* 2.4.287a.12–13) that if the world were many-sided, rather than spherical, its rotation would entail the occupation of successively vacant places. The argument was dismissed as false. See Thorndike, *The "Sphere" of Sacrobosco and its Commentators*, pp. 80–81 (Latin text) and 120 (English translation). When Nicole Oresme repeated this argument in 1377, it was under the direct influence of the doctrine of God's absolute power, for he allowed that God might indeed have created the world in a nonspherical shape with protruding parts. The rotation of such a world would indeed leave behind successive vacua. See *Nicole Oresme: Le Livre du ciel et du monde*, ed. by Albert D. Menut and Alexander J. Denomy, C. S. B.; trans. with an introduction by Albert D. Menut (Madison, Wis.: University of Wisconsin Press, 1968), p. 177.

4 The extent of Grosseteste's discussion of extracosmic void is the following: "Quinta: quod ymaginacio ponit spacium infinitum, et si spacium infinitum sit extra celum, sicud dicit ymaginacio, si est repletum corpore, infinitum est. Si est vacuum sequitur, ut post patebit, quod sit plenum, quia locum esse vacuum est impossibile, et ita quod infinitum sit." Richard C. Dales (ed.), *Roberti Grosseteste Episcopi Lincolniensis Com-*

mentarius in VIII libros Physicorum Aristotelis (Boulder, Colo.: University of Colorado Press, 1963), pp. 58-59.

5 See above, Ch. 5, Sec. 2. We have seen (Ch. 5, Sec. 3) that Christians were compelled to reject the possibility of an *uncreated eternal void*. The latter would be coeternal with God and therefore repugnant to faith.

6 Beginning in 1272, masters of arts had to swear that they would not "presume to determine or even to dispute any purely theological question." For Lynn Thorndike's translation of the statute and other references, see Grant, *Source Book in Medieval Science*, pp. 44-45.

7 Aristotle, *Physics* 3.4.203b.22-24. For a Latin version, see Aristotle's *Physics*, bk. 3, text 32, in *Aristotelis Opera cum Averrois commentariis*, vol. 4, fol. 99v, cols. 1-2.

8 Aristotle, *Physics* 3.4.203b.25-29. Although Aristotle produced a rather detailed discussion of imagination in *De anima*, bk. 3, ch. 3 (427a.16-429a.9), it is irrelevant to the concept of imaginary space.

9 Aristotle, *Physics*, bk. 3, text 32, and Averroes' *Commentary on the Physics*, bk. 3, comment 32, in *Aristotelis Opera cum Averrois commentariis*, vol. 4, fol. 99v, cols. 1 and 2, respectively. The Arabs appear to have discussed imaginary space long before Averroes. According to Shlomo Pines ("Philosophy, Mathematics, and the Concepts of Space in the Middle Ages," Y. Elkana [ed.], *The Interaction Between Science and Philosophy* [Atlantic Highlands, N.J.: Humanities Press, 1974], p. 85), Ibn al-Haytham (Alhazen) spoke to an "imaginary vacuum" (khalā mutakhayyal), which he identified with three-dimensional void space. Apparently he failed to indicate whether it was to be interpreted as a physical reality or a mathematical concept.

10 Aristotle, *De caelo*, 1.9.278a.18-21, in the Loeb Library translation by W. K. C. Guthrie. The summary in this paragraph is of the following passage by Averroes: "Deinde dixit 'Et iste sermo sequitur sermonem, etc.,' id est, et testatur huic sermoni, scilicet mundos esse plures, qui ponit in creatore mundi exemplar, secundum quod fecit mundum dicens enim exemplar esse contingit ei necessario dicere ut mundus habeat plura individua." Averroes, *Aristotelis Opera cum Averrois commentariis*, vol. 5: *De caelo*, bk. 1, comment 92, fol. 61v, col. 2. Averroes went on to mention that Plato assumed an eternal exemplar and that because Plato's creator made this world, he could also make others.

11 Although Aristotle regarded the argument just described as sound, he rejected a plurality of worlds because he believed that all the matter in existence is actually contained in our world, with none left over to produce additional cosmic instantiations. Thus our world is unique. Had sufficient surplus matter been available, however, one or more other worlds would certainly have existed. For Averroes' discussion, see *De caelo*, bk. 1, comment 95 (*Opera*, vol. 5, fols. 63v, col. 2-64r, col. 2).

12 What follows in the rest of this paragraph is drawn from Averroes, *Opera*, vol. 5: *De caelo*, bk. 1, comment 92, fol. 62r, col. 1.

13 At the risk of being obvious, I remind the reader that Averroes, like Aristotle, rejected extracosmic void as well as a plurality of worlds. In what we have presented here, his objective was to describe the reasoning used by those (for example, the *Loquentes*) who came to believe in extracosmic vacuum.

14 Trans. by Thomas Gilby, O. P., in Aquinas, *Summa Theologiae*, vol. 8: Creation, Variety and Evil (1a. 44-49), part I, question 46, article 1, reply to argument 8, p. 75. In his interpretation of this passage, John of St. Thomas (John Poinsot), the eminent seventeenth-century Thomistic commentator, observed that Aquinas designated what lies beyond the heaven, or world, as an "imaginary place" (*locus imaginarius*). John distinguished between a *vacuum privativum* within the world, which is defined "as a place that is not filled with body" but "nevertheless, capable of being filled," and a *vacuum negativum*, which lies outside the world. Because John, like St. Thomas, denied that bodies can exist beyond the world, he also inferred that no containing places can exist there, from which it follows that the *vacuum negativum* cannot receive bodies. Indeed John of St. Thomas assumed the actual existence of the *vacuum negativum* when he declared: "Suppono secundo vacuum negativum *de facto dari*, cum supra

coelum nullum sit corpus, cum ibi finiatur hoc universum." (John of St. Thomas, *Cursus Philosophicus Thomisticus*, new ed. by P. Beatus Reiser, O. S. B. [3 vols.; Turin: Marietti, 1933; it appears that the various parts were first published between 1633 and 1635], vol. 2, p. 361, col. 1. The italics are mine.) Although John implied that Aquinas also believed in the extracosmic *vacuum negativum*, there is no clear evidence of this. It is obvious, however, that Roger Bacon had assumed the existence of a vacuum of this kind (see above, Ch. 2, Sec. 1).

15 For the Latin text, see Pseudo-Siger of Brabant, *Physics* (Delhaye ed.), p. 179.

16 One could also be led to extracosmic void by assuming that a vacuum, equated with nonbeing or privation, must exist beyond the heaven to serve as contrast to the plenum of our world that lies within the heaven and is equated with being. This is indeed the position reported by the author of the *Summa philosophiae*, falsely ascribed to Grosseteste. For the text, see above, Ch. 5, n. 35. Whether the author of the *Summa* accepted this is left unclear. The argument he described bears a resemblance to one proposed by Roger Bacon cited above, Ch. 2, Sec. 1.

17 My translation from Menut's edition of Oresme's *Le Livre du ciel et du monde*, bk. 1, ch. 24, p. 176. See also my article "Infinite Void Space," 48 and n. 45, for the French text.

18 Long before Oresme, al-Rāzī (ca. 854–925 or 935), or Rhazes, argued that reason told even simple folk that empty space exists outside the world. See Pines, "Philosophy, Mathematics, and Concepts of Space in the Middle Ages," in *Interaction Between Science and Philosophy*, p. 84. For Abu'l-Barakāt al-Baghdadi (ca. 1080–d. after 1164/1165), reason or the estimative faculty was responsible for the innate knowledge in the human mind that empty three-dimensional space filled with bodies is prior to the notion of a plenum (*ibid.*, p. 85).

19 *The Universe Around Us* (New York: Macmillan, 1929), p. 68. A similar judgment was made by John Locke when he declared: "I would fain meet with that thinking Man that can in his Thoughts, set any bounds to Space, more than he can to Duration; or by thinking, hope to arrive at the end of either; . . ." *An Essay Concerning Human Understanding*, bk. 2, ch. 13, par. 21, ed. with an introduction, critical apparatus and glossary by Peter H. Nidditch (Oxford: Clarendon Press, 1975), p. 176.

20 In a lengthy refutation of Aristotle's doctrine of place, Crescas assumed the existence of an infinite three-dimensional void space. See Wolfson, *Crescas' Critique of Aristotle*, pp. 60–61 (for Wolfson's summary and analysis) and 186–189 (for the Hebrew and English texts of Crescas' arguments). Although Spinoza seems to have been influenced by Crescas (Wolfson, *Philosophy of Spinoza*, vol. 1, pp. 224, 275–281), the latter's role in the history of the concept of extracosmic space is uncertain. Pico della Mirandola, for example, was aware of Crescas' ideas but rejected an actually existent separate void space when, in his *Examen vanitatis*, bk. 6, ch. 4, p. 768, he declared that "place is space, vacant (*vacuum*) assuredly of any body, but still never existing as a vacuum alone of itself." Trans. by Charles Schmitt, *Gianfrancesco Pico della Mirandola (1469–1533) and His Critique of Aristotle* (The Hague: Martinus Nijhoff, 1967), pp. 140–141. Crescas may have been the first scholar in Western Europe since Greek antiquity to have adopted unequivocally the existence of an infinite thre-dimensional void space. But despite his statement that God "is the Place of the world" (*Crescas' Critique of Aristotle*, p. 201) and omnipresent, Wolfson is of the opinion (p. 123) that Crescas stopped short of identifying God with that infinite space. For this reason, and the uncertainty of his influence, Crescas plays a minor role in this volume. It would appear, however, that his anti-Aristotelian position was constructed without direct access to Greek authorities, but solely by reacting to Aristotle's arguments within a context of traditional Arabic and Hebrew commentary literature. For more on Crescas, see above, Ch. 2, Sec. 4.

21 In agreeing with Averroes, Henry declared: "Et ideo, ut dicit [i.e., Averroes] videmus istos non credere demonstrationibus nisi imaginatio concomitetur eos. Non enim possunt credere plenum non esse, aut vacuum, aut tempus extra mundum. Neque possunt credere hic esse entia non corporea, neque in loco, neque in tempore. Primum non possunt credere quod imaginatio eorum non stat in quantitate finita, et ideo mathematicae

imaginationes et quod est extra coelum videntur eis infinita." *M. Henrici Goethals a Gandavo, Doctoris Solemnis, Socii Sorbonici, Ordinis Servorum B. M. V. et Archid. Tornacensis, Aurea Quodlibeta . . . Hac postrema editione commentariis doctissimis illustrata M. Vitalis Zuccolii Patavini Ordinis Camaldulensis Theologi Clarissimi* (2 vols.; Venice, 1613), quodlibet 2, question 9 ("Utrum Angelus secundum substantiam suam sine operatione sit in loco"), fol. 60r, col. 2. I am grateful to Dr. George Molland, who cited this passage (from the Paris 1518 edition) in his doctoral dissertation and brought it to my attention.

22 After declaring that the kinds of arguments used in support of the existence of extra-cosmic void are invalid, Buridan assumed that all matter within the lunar orb is anni-hilated and asked what would happen if a stone were moved varying distances from the lunar pole. Buridan believed we would assume the existence of a space in which that stone moves, just as we wrongly infer the existence of an extracosmic space: "Dico quod in hoc est difficile satisfacere ymaginationi quia semper apparet imaginationi quod ibi esset spacium, sicut semper sensui apparet quod sol non sit maior equo et quod valde minor terra. Tamen in talibus intellectus debet corrigere illas apparentias sensus et ymaginationes." *Physics*, bk. 3, question 15 ("Utrum est aliqua magnitudo infinita"), fol. 57v, col. 2. In discussing the question "whether another earth of the same species as the earth of this world could be made beyond the world" ("Utrum posset extra mundum istum fieri altera terra eiusdem speciei cum terra huius mundi"), Godfrey of Fontaines (ca. 1250–d. after 1306) denied that a vacuum would exist between them, although he conceded that we could imagine such a vacuum. ("Nec etiam oporteret poni vacuum; vacuum enim est locus inanis, id est superficies corporis, apta nata con-tinere corpus, non tamen continens. Sed si mundus alius esset, sic haberet locum suum sicut iste; nec esset vacuum inter istum et illum, quia nihil natum continere, non tamen continens nisi imaginatum, sicut modo imaginamur extra caelum vacuum." M. de Wulf and A. Pelzer (eds.), *Les quatre premiers Quodlibets de Godefroid de Fontaines, Les Philosophes Belges*, Textes et Etudes, vol. 2 (Louvain: Institut Supérieur de Philosophie de l'Université, 1904), quartum quodlibet, quaestio VI, p. 255.

23 "Quidam dixerunt vacuum extra mundum separatum a mundo. Contra quia si extra mundum esset vacuum illud esset nobis manifestum in hoc mundo et probatur dupli-citer. Primo quia aer videtur aliquod vacuum vel nihil cum non sit visibilis. Et tamen exprimitur ipsum aerem non esse vacuum ut patet in utre inflato a quo sensibiliter senti-mus exire aerem. Secundo quia si pisces maris essent se rei experiremur hoc et si non visu saltem tactu." Franciscus de Mayronis, *Expositio super octo libros Phisicorum* in *Tria principia Antonii Andree Ordinis Minorum; Expositio Fran. Mayronis Ordinis Minorum super octo libros Phisicorum; Formalitates eiusdem Francisci Mayronis . . .* (Ferrara, 1490; folios unnumbered), bk. 4, Capitulum: *De vacuo*, sig. kiiii, col. 2. Although Aris-totle (*Physics* 4.216b.17–19) is the ultimate source of Mayronis' statement, the latter probably derived the substance of his remarks directly from Averroes (*Opera*, vol. 4: *Physics*, bk. 4, comment 78, fol. 167r, col. 1), who, in commenting on Aristotle's text, declared that "if there were [void], as claimed by those who say it can be conceived, namely that there is a separate vacuum beyond the world in which the world exists, it would be necessary that some trace [or evidence] (*residuum*) of it be apparent to us. But we do not sense it within the world. Indeed air is sensed to be a body by touch; but if it were judged according to vision [or sight] it [i.e., air] would be [thought to be] nothing since it does not have color, for vision grasps colored things. Then he [Aristotle] says and 'if there were thickness' ('et si spissitudo esset, etc.'), and similarly if it were thought that the thickness of it [i.e., water] had the hardness of iron, its tangibility could not be judged by the sense of sight but [only] by the sense of touch." (The strangeness of some of this discussion derives from a corrupt text that reads "et similiter existi-matur de aqua et si spissitudo esset frustum ferri . . ." when Aristotle actually says, "For air is something, though it does not *seem* to be so – nor, for that matter would water, if fishes were made of iron; . . ." [Oxford trans.].) For appeals by Albert of Saxony and Marsilius of Inghen based on our lack of experience with a real existent vacuum, see above, Ch. 3, Sec. 7b and n. 118 and 119.

24 This seems to be the import of Nicole Oresme's statement that "pour ce que la

congnoissance de nostre entendement depent de noz senz qui sont corporelz, nous ne povons comprendre ne proprement entendre quelle est ceste espasce incorporelle qui est hors le ciel. Et toutevoies rayson et verité nous fait congnoistre que elle est." *Le Livre du ciel et du monde* (Menut ed.), bk. 1, ch. 24, p. 176, lines 320–322. Thus where Franciscus de Mayronis (see n. 23) and Oresme agreed that our inability to perceive extracosmic void should lead us to reject it, Oresme appealed to reason to override experience.

25 Assuming that actual bodies are potentially receivable into imaginary space, the Coimbra Jesuits insisted that the dimensions of such a space would surely not be imaginary in the sense of a fiction, nor would they depend on the mind, but they are imaginary only in the sense described above, which translates the following passage: ". . . sed quia imaginamur illas in spatio proportione quadam respondentes realibus ac positivis corporum dimensionibus." Conimbricenses, *Physics* (Cologne, 1602), bk. 8, ch. 10, quaestio 2, articulus 4 ("Quidnam sit imaginarium spatium quo pacto Deus in eo existat. Solutio argumentorum primi articuli."), col. 519. See also Grant, "Infinite Void Space," 52.

26 The descriptions mentioned here are taken from Otto von Guericke, *Experimenta nova [ut vocantur] Magdeburgica de vacuo spatio*, p. 52, cols. 1–2. The whole of this section is translated by Grant, *Source Book in Medieval Science*, pp. 563–564; see also Grant, "Infinite Void Space," 54, n. 72. The passage in which imaginary space is identified with "God Himself" was drawn by von Guericke from the *De perfectione divino* by Lessius. For a more detailed set of meanings of the term "imaginary" in the expression "imaginary space," see Bartholomeus Amicus, *Physics*, vol. 2, p. 759, col. 1–760, col. 2. Despite the numerous meanings assigned to imaginary space, A. Koyré's insistence ("Le vide et l'espace infini au xiv⁰ siècle," *Archives d'histoire doctrinale et littéraire du moyen âge* 24 [1949], 52) that the designation "imaginary" by scholastic Aristotelians was intended as a denial of *independent* existence to extracosmic space appears reasonable (for an apparent exception, however, see below, Ch. 7, Sec. 2d). As we shall see below, the status accorded imaginary space was one of *dependent* existence based on an intimate association, or relationship, to God.

27 See my article "Place and Space in Medieval Physical Thought," in Machamer and Turnbull (eds.), *Motion and Time, Space and Matter*, p. 150, where this passage is translated from Buridan's *Questions on De caelo* (Moody ed.), bk. 1, question 15, p. 79, lines 1–9.

28 *Ibid.* The creation of an infinite void posed special problems, as will be seen below. Although Buridan denied the existence of extracosmic void, he was convinced that if such a space did exist, it would have to be infinite.

29 Buridan, *Physics*, bk. 4, question 10, fol. 77v, col. 1. The italics are mine. For a complete translation, which has been slightly corrected here, see E. Grant, "Jean Buridan: A Fourteenth Century Cartesian," *Archives internationale d'histoire des sciences* 16 (1963), 252; the Latin text appears on 251–252, n. 4. See also above, Chap. 2, Sec. 2a and n. 43. In his version of the Stoic argument, Albert of Saxony (*Physics*, bk. 1, question 9, fol. 93v, col. 2) denied Buridan's interpretation by insisting that even without obstacles, no one could extend an arm beyond the extremity of the world because no receptacle – that is, no place or space – exists beyond the world to receive it. For the argument of Archytas and the Stoics on extracosmic void, see above, Chap. 5, Sec. 1. Other medieval scholastics who mentioned the Stoic argument discussed here are Thomas Aquinas, Richard of Middleton, and Nicole Oresme (for references, see Grant, "Infinite Void Space," 41–42, n. 14 and 15); among later scholastics who described some version of the Stoic argument, see Phillip Faber, *Disputationes Theologicae*, p. 338, col. 2, sec. 24, and Gabriel Vasquez, *Disputationes Metaphysicae*, p. 362; Bartholomeus Amicus (*Physics*, vol. 2, p. 742, col. 2 [E]) was content merely to mention the Stoic belief in an infinite extracosmic void.

30 In arguing against those who claimed that if God could create a bean beyond the world, this very act would in and of itself imply a space because the terms "within" (*intra*) and "without" (*extra*) signify a place or space ("Item si extra celum crearentur due fabe ille essent extra invicem et extra celum; sed 'intra' et 'extra' significant locum vel spacium. Ergo ibi est locus et spacium." Buridan, *Physics*, fol. 57r, col. 2).

Buridan replied that an extracosmic space would indeed exist under the conditions described, but the space in question would be the magnitude of the bean itself, not a separate void dimension ("Tunc ergo ad rationem que arguit quod ibi sit spacium quia supra vel ultra posset Deus creare fabam, dico quod verum est et tunc extra esset spacium et illud spacium non esset nisi magnitudo fabe, que ante non erat." *Ibid.*, fol. 58r, col. 2).

31 The supporting text follows immediately after the Latin passage in n. 30: "Ad aliam dico quod illa faba nec crearetur in indivisibili nec in spacio divisibili quia non esset in loco, nec in aliquo spacio, sicut nec totalis mundus est modo in loco vel in aliquo spacio, sed collective loquendo omnes magnitudines partium mundi sunt omnia spacia." *Ibid.*, fol. 58r, col.2. For Buridan's more considered argument against the concept of a separate space for the world, see above, Ch. 2, Sec. 2b and n. 59 for the Latin text. Buridan there reiterated, at greater length, that the magnitudes and dimensions of the world and its parts are actually spaces.

32 Buridan, *Questions on De caelo* (Moody ed.), bk. 1, question 17, p. 79, lines 21–23. The English translation is quoted from my article "Place and Space in Medieval Physical Thought," *Motion and Time, Space and Matter* (Machamer and Turnbull eds.), p. 151.

33 Although the analogy presented here is mine, not Marsilius', the latter conceived the relationship between an extracosmic body and the world from which it is separated by a vacuum in the same manner as most scholastics would have interpreted the relationship between the sides of a clepsydra emptied of all body. On the clepsydra and its role in medieval discussions of nature's abhorrence of a vacuum, see above, Ch. 4, Sec. 4d. Descartes fully accepted the dominant medieval interpretation when he declared that "if it is asked what would happen if God removed all the body contained in a vessel without permitting its place being occupied by another body, we shall answer that the sides of the vessel will thereby come into immediate contiguity with one another. For two bodies must touch when there is nothing between them, because it is manifestly contradictory for these two bodies to be apart from one another, or that there should be a distance between them, and yet that this distance should be nothing; for distance is a mode of extension, and without extended substance it cannot therefore exist." *Principia Philosophiae*, part 2, principle 18, as translated in *The Philosophical Works of Descartes*, rendered into English by Elizabeth S. Haldane and G. R. T. Ross, vol. 1, p. 263. See also Koyré, *From the Closed World*, pp. 101–102.

34 The argument to which Marsilius responded declares: "Quia Deus potest creare unum lapidem vel unam fabbam supra convexum ultime sphere et illam movere motu recto elongando a celo, tunc ille lapis plus distabit a celo quam prius quia prius tetigit celum et nunc non tangit et non distat nisi per spacium separatum intermedium. Igitur de facto est ibi aliquod spacium separatum." Marsilius rejected this argument in the following reply: "Ad primam [rationem] dico quod propter motum localem rectum non oportet ponere aliquod spacium circumdans mobile quod movetur, et ideo admisso casu quod Deus moveret fabbam, etc., dico quod fabba non tangeret celum nec distaret a celo, sed potest distare, posito quod Deus crearet aliquod corpus intermedium." Marsilius of Inghen, *Physics*, bk. 4, question 2, fol. 46v, col. 2. I have added the bracketed word.

Albert of Saxony offered a similar argument when, after conceding that God could create a stone beyond the last heaven, or world, he explained further that if God moved this stone, its motion would have the nature (*ratio*) of a local motion except that it could move neither closer to nor farther from the convex surface of the last sphere. ("Consequenter concedo quod Deus potest talem lapidem creatum extra celum movere, non motu locali, sed motu eiusdem rationis cum motu locali absque hoc quod fieret propinquior vel remotior lateribus celi." *De celo et mundo*, bk. 1, question 9, fol. 93v, col. 2.) For Albert, distance alteration is possible only when a material medium, functioning as a space, exists. Because such a medium does not exist beyond the world, Albert agreed with Marsilius that wherever a stone might be placed beyond the world, it would actually be in contact with the convex surface of the outermost sphere. Only if God also created a space for it, that is, created a corporeal interval for it, could the stone vary its distance with respect to the last sphere. ("Deus potest illum lapidem creatum extra mundum movere elongando ipsum a celo. Dico quod verum est; sed hoc non potest esse nisi etiam crearet spatium in quo moveretur et per quod haberet certum situm ad centrum vel celum et ad partes celi. Sed post quam Deus absolveret illum lapidem a determinato situ ad celum ipse nec esset prope celum nec longe a celo et de hoc dicebatur in corpore questionis." *Ibid.*, fol. 94r, col. 1.)

Descartes adopted virtually the same attitude as did Marsilius and Albert of Saxony. As Koyré explained (*From the Closed World*, p. 101), for Descartes "*nothing* can have no properties and therefore no dimensions. To speak of ten feet of void space separating two bodies is meaningless: if there were a void, there would be no separation, and bodies separated *by nothing* would be in contact." For Descartes's discussion, see *Principia Philosophiae*, part 2, principle 18 in *Oeuvres de Descartes* (Adam and Tannery, eds.), vol. 8, p. 50, and p. 263 of the Haldane and Ross translation, where Descartes argued that if God removed all the matter in a vessel, the sides of the vessel would immediately come into contact. The collapse of the sides of a vessel following the creation of a vacuum within was also commonly assumed in the Middle Ages (see above, Ch. 4, Secs. 4b, d).

35 Marsilius of Inghen first presented the argument he would reject: "Tertio, creet Deus extra mundum tres mundos tangentes se invicem secundum puncta, eo modo quo sphere tangunt se, tunc latera istorum mundorum abinvicem distant, ex non nisi per spacium, igitur [etc.]," and then his rebuttal: "Ad tertiam dico quod illi mundi sic se tangentes non distarent nisi secundum lineas circulares secundum quas est corpus intermedium." *Physics*, fol. 46v, col. 2. John Buridan would offer a similar argument by assuming that if everything were annihilated within the orb of the moon, the poles would not be separated from each other by a measurable rectilinear distance, but their separation could be measured only by a circular or curvilinear distance. ("Dico ergo quod in predicto casu annichilationis eorum que sunt infra orbem lune unus polus non tangeret alterum polum, nec distaret ab altero polo secundum rectitudinem quia non esset spacium rectum medium per quod distant. Sed posset concedi distare secundum distantiam circularem vel curvam." *Physics*, bk. 3, question 15, fol. 58r, col. 1.) Buridan also mentioned three worlds in contact with intervening void spaces, but used that illustration differently than Marsilius (*ibid.*, bk. 3, question 15, fol. 57r, col. 2).

In his *Opus Majus* Roger Bacon formulated a version of this argument in an attack against a plurality of worlds. He argued that "if there were another universe, it would be of spherical figure, like this one, and there cannot be distance between them, because in that case there would be a vacant space without a body between them, which is false. Therefore they must touch; but they cannot touch each other except in one point by the twelfth proposition of the third book of the *Elements*, as has already been shown by circles. Hence elsewhere than in that point there will be vacant space between them." Robert B. Burke (trans.), *The Opus Majus of Roger Bacon* (2 vols.; Philadelphia: University of Pennsylvania Press, 1928, reprinted New York: Russell and Russell, 1962), vol. 1, p. 186. Bacon thus rejected the existence of other spherical worlds because of the impossibility of void space existing between them. Under the conditions he described, even if the worlds touched, void space would necessarily lie between the surfaces, which can touch only at a point. Because the *Opus Majus* was composed around 1267 (see the article "Roger Bacon" by A. C. Crombie and J. D. North in *Dictionary of Scientific Biography*, vol. 1 [New York: Scribner, 1970], p. 378), Bacon's remarks antedate the Condemnation of 1277. As a good Aristotelian, Bacon believed that the void space that would necessarily lie between spherical worlds in contact was sufficient to reject as absurd the concept of a plurality of worlds. In the aftermath of the Condemnation of 1277, however, the void space that would lie between the spherical worlds was not sufficient to condemn the possibility of other worlds. True, Marsilius of Inghen, as well as Bacon, denied that distances could be measured in the intervening void. But Marsilius allowed that distances could at least be measured along the circumferences of the different worlds and did not use the existence of intervening void space as grounds for dismissing the possibility of other worlds. The difference between their approaches was largely determined by the Condemnation of 1277. Bacon's summary dismissal of a plurality of worlds because of the intervening vacua would have been unacceptable after 1277, when it would have been necessary to concede the possibility that God could create not only such worlds but also the intervening vacua. Indeed it is possible that Averroistic opinions of this kind may have been responsible for Bacon's condemnation and imprisonment by his order between 1277 and 1292 (see Crombie and North, "Roger Bacon," p. 378).

36 I have discussed the issue of distance measurement and its application to vacua in my article "Place and Space in Medieval Physical Thought," 147–148, 151–152. Some of my earlier thoughts are repeated here.

37 Richard's statement was in response to a claim that not even God could create a vacuum within which no distances can exist or be measured and simultaneously cause things to be separated by that vacuum. In his definition of vacuum, Richard described it as "a capacity for distance between [the termini of] which there is absolutely no distance." As a resolution of this dilemma, Richard declared that God could create a dimension in that vacuum after He had annihilated everything between earth and heaven. However, that dimension would exist independently of everything else (see *Clarissimi Theologi Magistri Ricardi de Media Villa . . . Super quatuor libros Sententiarum Petri Lombardi quaestiones subtilissimae* [4 vols.; Brixiae (Brescia), 1591]; reprinted under the title *Richardus de Mediavilla* [Frankfurt: Minerva, 1963], vol. 2, bk. 2, distinction 14, article 3, question 3, p. 186).

In an anonymous fourteenth-century compendium of natural philosophy, the author replied to a series of arguments in which vacua were imagined as created (what follows is drawn from MS Bibliothèque Nationale, fonds Latin, 6752, fols. 152r–153r). After the enunciation of five assumptions about quantity and distance, he argued that if God destroyed the earth, a vacuum would not remain because the sides of the bodies containing and surrounding the earth (air and water) would immediately (that is, instantaneously) come into contact. Belief in the inevitability of such contact derived from the conviction that vacuum is not a distance and cannot therefore preserve the separation of the surrounding sides after earth has been annihilated.

Our author included two other similar arguments to be subsequently rejected. If, with all other things remaining constant, God were to remove a portion of the sky or part of an element and ttransfer it beyond the heaven, He could create a vacuum within the world in the place where the matter was removed. Indeed God could also produce a vacuum in the world by condensing or contracting something without simultaneously causing a compensating rarefaction elsewhere. Because of the basic similarity of the two arguments, our author provided a single response, which was to deny that a vacuum can be formed in either of these ways. The matter surrounding the potentially empty spaces thus produced would come into immediate contact, because it would no longer be separated by a distance but only by a vacuum, which is not a distance and cannot, therefore, serve to separate things. "For just as it is impossible that Sortes [that is, Socrates] be white without whiteness, so it is impossible that things be really distant [or separated] without a real distance." By a maxim of theology, however, God can conserve any distinct things separately and can therefore preserve Sortes without whiteness; and, similarly, He can conserve a distance without things that are distant, and conversely. Nevertheless, it is impossible that God can make a thing be distant without a distance, because this implies a contradiction.

Because of its length, the Latin text is omitted. For a discussion of MS BN 6752 and a description of its contents, see Lynn Thorndike, *A History of Magic and Experimental Science* (8 vols.; New York: Columbia University Press, 1923–1958), vol. 3, ch. 33, pp. 568–584.

38 Roger Bacon, it will be recalled (above, Ch. 2, Sec. 1), denied *all* properties to extra-cosmic void, even the ability to receive bodies.

39 "Dico quod aliqua contingit distare dupliciter. Uno modo per se, scilicet quia est aliqua distantia positiva dimensionis corporalis in medio eorum." Quoted from Henry of Ghent's *Quodlibeta*, XIII, question 3 (Venice, 1613), vol. 1, fol. 293v, col. 2. Koyré, "Le vide et l'espace infini," 63, n. 2, has cited the same passage from the edition of Paris, 1518. Although omitting the Latin texts, Pierre Duhem summarized Henry of Ghent's arguments and partially translated them into French (*Etudes sur Léonard de Vinci* [3 vols.; Paris: Hermann, 1906–1913], seconde série, pp. 447–451).

40 Henry of Ghent, *Quodlibeta*, XIII, question 3, vol. 1, fol. 293v, col. 2; Koyré, "Le vide et l'espace infini," 63 n. 2. I have constructed the figure and assigned letters to represent the bodies.

In the anonymous fourteenth-century treatise on natural philosophy cited earlier (see above n. 37), the author may have had in mind something akin to Henry of Ghent's position when he described and rejected two opinions (MS Bibliothèque Nationale, fonds latin, 6752, fols. 151v–152r) that sought to attribute quantity or distance to void space. The two opinions are a response to the following question: Could God bring the sides of a void space closer together? If He could not, it follows that the sides were not separated and no void space existed. But if it is said that God could bring them closer, it then follows that the sides of that void were indeed separated and that the intervening void in that place is a quantity. Because a vacuum cannot be a quantity or body, two opinions were apparently formulated to explain how such a void place can have sides that are separated and yet not be a real quantity or distance.

The first of these opinions held that although the distance separating the sides of the void place is not directly measurable through the vacuum because of the lack of a material medium, the void place is measurable by comparison to something external to it. Thus one side surrounding the vacuum is nearer to things closer to it than is another side of the same vacuum. The anonymous author rejected this opinion on the grounds that when related to something outside the vacuum, we must inquire "by what" (*per quid*) is one side more distant than another? If it is farther by nothing, then one side of a vacuum would be no more distant than another, because nothing is equivalent to no more distant. But if it is argued that one side of the void place is "more distant" (*magis distant*) by something (*per aliquid*), "then it follows that there is something in this place, namely distance and quantity and, consequently, the place is not void, the opposite of which they assumed."

Those who formulated the second opinion, characterized as "more subtle," maintained that the sides of such a void place are not separated by "real distance" (*distantia realis*) but by a distance that is "produceable" (*producibilis*) between the sides, by which is apparently meant "produceable" by God. For we are told that proponents of this opinion imagine that if the sides of such a void place were "immediate" (*immediata*), that is, in contact, "then God could not produce an intervening quantity unless those sides were mutually separated. Therefore they say that if some place were made void, the sides would be related such that a distance or quantity can be produced between them without a new separation [or elongation]. Therefore they conclude that the sides are separated by a possible quantity [or distance]." The anonymous author objected because this conception presupposes what it must prove. "For they presuppose that a distance can be produced between the sides of a vacuum without a new separation [or elongation] of those sides. But they have to prove this because it follows from this that such sides would not be immediate [that is, in contact], the opposite of which was deduced." Our author was here arguing that if a place were made void, its sides would instantly come into contact. Therefore, if a void were to be produced, those sides would have to be separated – that is, a new separation [or elongation] would have to be made. But the advocates of this opinion would argue that after the matter of a place is removed and it is left void, a possible distance remains, and the sides of the place would not come into immediate contact, which is what they have not proved.

Our anonymous author was further disturbed by the fact that proponents of this opinion have, in effect, equated "possible distance" (*possibilis distantia*) with real, quantitative distance. For a possible distance can also be equated with two sides in immediate contact, because those sides are potentially capable of being separated to produce an extended void space. Therefore, bodies actually in contact would be actually distant! The lengthy Latin text has been omitted.

41 "Et propterea si extra caelum extremum esset a Deo aliquod corpus factum vel alius mundus sine contactu ad caelum, tunc inter utrunque dicendum esset esse vacuum secundum mensuram determinatam corporis, quia natum esset inter illa recipi: sed alibi non, sicut nec modo extra dicendum est esse vacuum, sicut nec plenum, sed purum nihil, . . ." Henry of Ghent, *Quodlibeta*, XIII, question 3, vol. 1, fol. 294r, col. 1; Koyré, p.

64, n. 2. Bradwardine's twofold conception of vacuum seems similar to Henry's. The privative sense, which Bradwardine described as "the pure lack of a plenum" (*carentia pura pleni*), corresponds to Henry's pure nothing; the positive sense, which could be taken as a corporeal dimension, is akin to Henry's dimensional vacuum (see above, Ch. 2, Sec. 1, and n. 9). Buridan would later adopt a position almost opposite to that of Henry of Ghent. For Buridan the void between bodies created beyond the world would not serve to separate the bodies, which would come into contact; but a body in contact with the convex surface of the last sphere would be in its own internal space and not in a pure nothing. See above, Ch. 6, Sec. 2.

42 "Jean de Ripa I Sent. Dist. XXXVII: De modo inexistendi divine essentie in omnibus creaturis," édition critique par André Combes et Francis Ruello, présentation par Paul Vignaux, *Traditio 23* (1967), 232, lines 68–78. Cited hereafter as Combes, "De Ripa."

43 The paradoxes derived here from Aristotle's concept of place were widely discussed in the Middle Ages, beginning with Robert Grosseteste and continuing all the way to the seventeenth century. A variety of responses were developed, including a distinction between the material and formal aspects of place and the concept of the "identity of place by equivalence." On these and related problems on place, see my forthcoming article cited in Ch. 5, n. 41 above, and Duhem, *Le Système du monde*, vol. 7, pp. 158–302 ("Le Lieu").

44 Using a similar argument, John Buridan (*Physics*, bk. 3, question 15, fol. 58r, col. 1) arrived at an opposite conclusion from De Ripa, namely that a distance could not be measured when the matter between two bodies was annihilated. Buridan imagined that the only existent things in the world are three one-foot-square stones arranged in a row and in direct contact. If God should then destroy the middle stone, the two end stones would not come into contact, nor would they draw apart or approach each other. And yet the void "distance" between them is not measurable because no material medium exists through which measurements could be made. Duns Scotus, however, arrived at the same judgment as De Ripa. On the assumption that space is a privation, Scotus insisted that, at the very least, a potentially positive distance would exist in a void (see John Duns Scotus, *God and Creatures, The Quodlibetal Questions,* trans. with an introduction, notes, and glossary by Felix Alluntis, O. F. M., and Allan B. Wolter, O. F. M. [Princeton, N.J.: Princeton University Press, 1975], pp. 262, 263). Although it does not occur in a discussion on void, William Ockham seems to have shared this opinion in his consideration of the doctrine of place, where he argued that distance measurements between any two points must be conceived in the abstract, as if intervening bodies do not exist (see *The "Tractatus successivis" Attributed to William Ockham,* ed. with a study on the life and works of Ockham by Philotheus Boehner, O. F. M. [St. Bonaventure, N.Y.: Franciscan Institute, 1944], pp. 89–90).

45 The translations and the Latin texts of these three condemned articles follow:
"139. That an accident existing without a subject is not an accident, except equivocally; and that it is impossible that a quantity or dimension exist by itself because that would make it a substance. ("Quod accidens existens sine subjecto non est accidens, nisi equivoce; et quod impossibile est quantitatem sive dimensionem esse per se; hoc enim esset ipsam esse substantiam.")

140. That to make an accident exist without a subject is an impossible argument implying a contradiction. ("Quod facere accidens esse sine subjecto, habet rationem impossibilis, implicantis contradictionem.")

141. That God cannot make an accident exist without a subject, nor make several dimensions exist simultaneously [in the same place]." ("Quod Deus non potest facere accidens esse sine subjecto, nec plures dimensiones simul esse.")"
For the Latin texts, see Denifle and Chatelain, *Chartularium,* vol. 1, p. 551; for the translations, see E. Grant, "The Condemnation of 1277," *Viator 10* (1979), 232, n. 76. Because these articles were condemned, it had to be conceded that God was capable of performing the feats they denied.

46 After assuming that God can make an accident without a subject and conserve accidents after separating them from their subjects, Buridan declared: "Secondly, it seems to me that a penetration of dimensions is not impossible for God. Indeed He can make

several bodies exist simultaneously in the same subject or in the same place without their differing in position, namely that one should lie outside another positionally. Therefore God can make an absolute dimension or space separated from every natural substance [and He can also make this space] such that it can receive natural bodies without yielding [to them]." Buridan, *Physics*, bk. 4, question 8, fol. 74r, col. 1.

Naturally speaking, however Buridan denied that a void dimensional space could exist because it would require a penetration of dimensions when occupied by body and because it would be an accident without a subject (*Physics*, bk. 4, question 7, fol. 73r, col. 2). Earlier, in bk. 3, question 15 ("Whether an infinite magnitude exists"), Buridan had already denied that an accident or dimension could exist naturally without a subject (*Physics*, fol. 67v, col. 1).

Years before Buridan, Walter Burley (*Physics*, fol. 116v, col. 1) had also invoked God's absolute power to create a three-dimensional space without substance or qualities and likened it to the sacrament of the altar, in which a quantity exists without inhering in a subject. For further discussion of the impact of articles 139, 140, and 141 on Burley and Buridan, see Grant, "The Condemnation of 1277," 233–235.

47 After proposing the question, "Utrum est aliqua magnitudo infinita," Buridan presented the opinion that he would reject: "Arguitur quod sic quia extra celum est aliquod spacium et illud est infinitum quia non est ratio quare esset spacium alicuius certe quantitatis potius quam maioris nisi poneretur infinitum; et etiam quia quocunque spacio finito extra celum posito et concesso remaneret questio et omnis difficultas ut prius, videlicet, utrum ultra illud finitum datum esset spacium aliud. Ergo ficticium est dicere quod ultra celum sit spacium nisi concedatur esse infinitum. . . ." Buridan, *Physics*, bk. 3, question 15, fol. 57r, col. 2. The problem of an actually existing infinite space should be seen against the larger background of the general medieval concern with the possible existence of an actual infinite – i.e., whether God could create an actual infinite. It was a problem that emerged from a conflict between the peripatetic doctrine of the infinitely large and the absolute power of the Christian God. Pierre Duhem has described aspects of the problem in *Le Système du monde*, vol. 7 (chs. 1 and 2, pp. 3–157, which treat the infinitely large and infinitely small), and in his earlier *Etudes sur Léonard de Vinci*, seconde série, Notes, E: "Sur les deux infinis," pp. 368–407. Anneliese Maier also considered the medieval problem of the actual infinite in *Die Vorläufer Galileis im 14. Jahrhundert*, pp. 196–215. Nowhere in these discussions do Duhem or Maier discuss infinite void space as a possible instance of a divinely created actual infinite. In the appendix of *Le Traité 'De l'infini' de Jean Mair*, ed. Hubert Elie (Paris: Librairie Philosophique J. Vrin, 1938), Elie provided a series of French translations on the actual infinite drawn from Duns Scotus, Albert of Saxony, Robert Holkot, Marsilius of Inghen, and Luis Coronel. The possibility of an actual infinite space is not mentioned among these extracts, although John Major briefly considered it in Elie's edition of his *Propositum de infinito* (see below, Ch. 7, Sec. 1).

48 Only too well aware of the Condemnation of 1277 (see above, Ch. 5, Sec. 2, and n. 45, 46 of this chapter), Buridan subsequently denied the necessity of this conclusion because one must concede that God could create a *finite* space beyond the world of any size whatever, and no reason for this need be sought other than "the simple will of God." In fact, he concluded this brief rebuttal by doubting that any space lies beyond the world. ("Tunc igitur ad rationes: quando dicitur si extra celum sit spacium ipsum est infinitum dico quod Aristoteles concessisset istam consequentiam sed tamen propter potentiam supernaturalem dico quod ipsa non est necessaria quia Deus posset ibi creare spacium finitum quantum placeret sibi de cuius quantitate. Non esset querenda ratio nisi simplex voluntas Dei. Sed tamen opinor etiam quod ibi non sit aliquid spacium, scilicet ultra corpora nobis apparentia et ea que ex sacra scriptura tenemur credere." *Physics*, bk. 3, question 15, fol. 58r, col. 2.) Albert of Saxony adopted a similar position when he declared that God could create a finite space beyond the world of whatever size He pleased without making it infinite. (". . . dico quod extra mundum Deus posset creare bene spatium finitum tantum quantum sibi placeret absque hoc quod illud spatium esset infinitum." Albert of Saxony, *De celo*, bk. 1, question 9, fol. 93v, col. 2.)

49 Buridan, *Physics*, fol. 57r, col. 2. Although Descartes would characterize space as "in-

definitely extended" rather than actually infinite (see Koyré, *From the Closed World*, pp. 104, 118; the two descriptions are really the same, though for theological reasons Descartes preferred the former), the equation of space and body enunciated here is substantially identical with Descartes's description (see above, Ch. 2, Sec. 2a). Although Buridan was prepared to accept the equation of space and bodily dimension for finite bodies (see above, Ch. 2, Sec. 2a), he would reject it for infinite extension, the existence of which he denied.

50 Buridan, *Physics*, fol. 57v, col. 1. As will be seen below, Jean de Ripa would disagree with Buridan and claim that God exceeds an actual infinite. It should be noted that although Buridan conceded that no demonstration can be formulated to show that God did not create a magnitude or space beyond our world, it is unlikely that He did so. For if God had wished to make other mundane creatures, He could have done so by simply increasing the size of our world rather than creating other worlds. And if He did not make other worlds, why should He have made a space beyond the world that would have served no purpose (*ibid.*, fol. 57v, col. 2)?

51 See Buridan, *De caelo* (Moody ed.), bk. 1, question 15, p. 71, lines 4–19. The order in which Buridan wrote his *Questions on De caelo* and *Questions on the Physics* is presently unknown.

52 That the creation of an actual infinite was a theological question exclusively reserved for consideration by the theologians is indicated by Buridan when he abandoned ultimate determination of the question to the "lord theologians" and declared his intention to abide by their resolution of the question. ("Tamen de isto et de pluribus immo de omnibus que dicam in ista questione ego dimitto determinationem dominis theologis et acquiescere volo determinationi eorum." *Physics*, fol. 57v, cols. 1–2.) On Buridan's difficulties with theologians and theological problems, see E. Grant, "Scientific Thought in Fourteenth-Century Paris: Jean Buridan and Nicole Oresme," *Machaut's World: Science and Art in the Fourteenth Century*, ed. Madeleine Pelner Cosman and Bruce Chandler, in *Annals of the New York Academy of Sciences*, vol. 314 (New York: New York Academy of Sciences, 1978), pp. 107 and 118, n. 12, 13.

53 Besides Buridan, whose final opinion is uncertain, we may include Richard of Middleton, Duns Scotus, Walter Burley, Thomas Bradwardine, and Albert of Saxony among the major scholastic figures who denied that God could create an actual infinite entity, or categoric infinite, as it was often called. In his *Questions on De caelo* (bk. 1, question 15, p. 71, lines 6–7 of the Moody ed.), Buridan declared that "many deny that God could make an infinite" ("multi negarent quod deus posset facere infinitam").

54 Pierre Duhem mentions some of these authors on p. 126 of his lengthy discussion of the infinitely large in *Le Système du monde*, vol. 7, Ch. 2, pp. 89–157. Anneliese Maier reconsiders the problem in *Die Vorläufer Galileis im 14. Jahrhundert*, pp. 196–215 ("Das Problem des aktuell Unendlichen"). In her *Metaphysische Hintergründe der spätscholastischen Naturphilosophie, Studien zur Naturphilosophie der Spätscholastik*, vol. 4 (Rome: Edizioni di Storia e Letteratura, 1955), p. 381, n. 9, Maier briefly mentions the views of Gerard of Bologna, Johannes Baconthorpe, and Paul of Perugia. To this list, we may add John Major in the sixteenth century (see below, Ch. 7, Secs. 1, 2).

55 *Metaphysische Hintergründe*, p. 381, n. 9.

56 See Maier, *Die Vorläufer Galileis im 14. Jahrhundert*, pp. 203–204. Not every infinitist would have accepted actual infinites in all of the categories mentioned here (see Maier, *Metaphysische Hintergründe*, p. 381, n. 9).

57 My discussion is based on the edition of distinction 37 in Combes, "De Ripa," 191–267. Because the text is readily accessible, I shall cite the Latin text only where it seems especially appropriate and helpful.

58 "*Prima conclusio.* Prima sit ista: *Deus est realiter presens infinito vacuo ymaginario extra celum.*" Combes, "De Ripa," 233, lines 1–2.

59 Combes, "De Ripa," 228, lines 64–68, and 230, lines 33–35. Thomas Aquinas had argued that there was no logical contradiction in supposing that something made by God might yet have never been without existence. As a cause that acts instantaneously, God need not precede His effect in duration. Therefore, God may have created the world

without temporal beginning. As Cyril Vollert explains Aquinas' position, "the idea of creation does not logically require beginning. The creature exists as an effect of God's creative act; if the Creator wills the creature to exist without inception of its duration, it so exists; if the Creator wills it to have a finite, limited duration, it has a beginning of its existence. Therefore creation is essentially dependence in being. . . . A universe without initial instant would still be a created universe dependent on the first Cause, and its successive unfolding in time would leave it infinitely inferior to God's eternity," because "God alone is without beginning, end, and succession; He alone is changeless, intemporal." *De Aeternitate Mundi*, p. 15; for the list of works in which Aquinas considered the possible eternity of the world, see p. 14; for Aquinas' discussion in the *De Aeternitate Mundi*, see pp. 19–25.

60 "Secunda conclusio: *Totius vacui possibilis seu situs ymaginarii infinitas immense exceditur ab immensitate reali et presentiali divina.*" Combes, "De Ripa," 235, lines 26–28.

61 "Deus spera intelligibilis infinita." Combes, "De Ripa," 237, lines 76–77. For further information on this expression, see below, Ch. 6, Sec. 3c and n. 108.

62 In a totally different context, Robert Grosseteste had argued for unequal infinites that could be proportionally related. See A. C. Crombie, *Robert Grosseteste and the Origins of Experimental Science 1100–1700* (Oxford: Clarendon Press, 1953), pp. 101–102.

63 Although each species of being has its own essential perfection that is not subject to variation of intensity, De Ripa believed that God could assign to individuals of that species specific latitudes that could vary in intensity from some particular minimum all the way to an actually infinite intensive latitude of being. "Thus de Ripa concludes," as Janet Coleman explains ("Jean de Ripa O. F. M. and the Oxford Calculators," *Mediaeval Studies* 37 [1975], 163), "that in some species . . . there can be an infinite intensive latitude of being that terminates exclusively in God. Implied here is a great chain of perfected being, a continuous infinite series of perfected species, indivisible in their perfection and yet infinitely intense at the *summum* of each species. The picture is one of infinite subsets within an infinite set, all within an infinite 'immense' set, the latter serving as an exclusive term beyond all infinite created intensities." See Coleman's article (130–189) for an extended consideration of De Ripa's analysis of the intension and remission of latitudes of form as found in the latter's *Sentence Commentary*, bk. 1, distinction 17. For more on De Ripa's concept of the perfection of species, see John E. Murdoch, "*Mathesis in Philosophiam Scholasticam Introducta*, The Rise and Development of the Application of Mathematics in Fourteenth Century Philosophy and Theology." *Arts Libéraux et Philosophie au Moyen Age. Actes du Quatrième Congrès International de Philosophie Médiévale*, Montréal 27 août – 2 septembre 1967 (Montreal: Institut d'Etudes Médiévales; Paris: Librairie Philosophique J. Vrin, 1969), pp. 238–241.

64 "1. Quibus modis aliquid dicatur locale vel circumscriptibile. Duobus namque his modis dicitur in Scriptura aliquid locale vel circumscriptibile, et e converso, scilicet vel quia dimensionem capiens longitudinis, altitudinis et latitudinis, distantiam facit in loco, ut corpus; vel quia loco definitur ac determinatur, quoniam cum sit alicubi non ubique invenitur: quod non solum corpori, sed etiam omni creato spiritui congruit." Petrus Lombardus, *Sententiae in IV Libris Distinctae*, tom. I, pars II, p. 270 (bk. 1, distinction 37, ch. 6, par. 1).

65 It appears that Peter gave no thought to the problem of the relationship of the three-dimensional place to the three-dimensional body that occupied it. Did he conceive the place as separate but coextensive with the occupying body? Or did he perhaps conceive the place as internal to the body? For a discussion of these problems, see above, Ch. 2, Sec. 4 and n. 70 for a repetition of Peter's three-dimensional definition of place.

66 A body thus completely occupies and fills its place (*ubi circumscriptivum*), and, because it is also somewhere, is fully delimited and circumscribed by the termini of that place (*ubi definitivum*). A created spiritual entity, by contrast, is delimited and defined locally only by the terminus of a place – for it cannot be everywhere at once – but does not occupy its place as a dimensional entity. Peter elaborated this in the next paragraph: "2. De spiritu creato: quod secundum alterum modum nec localis nec circumscriptibilis est, secundum alterum vero est; corpus vero omnino locale et circumscriptibile; Deus vero

omnino inlocalis et incircumscriptibile. Omne igitur corpus omni modo locale est; spiritus vero creatus quodam modo localis est, et quodam modo non est localis. Localis quidem dicitur, quia definitione loci terminatur, quoniam cum alicubi praesens sit totus, alibi non invenitur; non autem ita localis est, ut dimensionem capiens, distantiam in loco faciat." *Ibid.*, Tom. I, p. 270, par. 2. For more on the manner in which a spiritual entity occupies a place, see the next note.

67 Although Peter Lombard did not specifically employ this formal terminology, he used the terms "circumscriptibile" and "definitur" in identical ways (see above, n. 64 and 66) and probably influenced the choice of terms. For Gregory of Rimini's (d. 1358) brief history of the distinction, see Combes, "De Ripa," 215–216, n. 12. The *ubi definitivum* also came to be characterized by the assumption that a spiritual substance could fill not only the whole of the place that delimited it but the whole of that spiritual substance, for example, an angel or soul, was in every part of its place or *ubi definitivum* (for uses of this concept, see below, Ch. 6, Sec. 3f [Fishacre], Ch. 7, Sec. 2a [Suarez], and Ch. 7, Sec. 2e [Maignan]). For a discussion of the various conceptions of *ubi definitivum* and *ubi circumscriptivum*, see E. Grant, "The Concept of *Ubi* in Medieval and Renaissance Discussions of Place," *Manuscripta* 20 (1976), 71–80. In his *Nouveaux Essais sur l'Entendement Humain*, Leibniz briefly described three kinds of *ubi*, the circumscriptive, definitive, and repletive, after which he admitted, "Je ne sais si cette doctrine des écoles mérite d'être tournée en ridicule, comme il semble qu'on s'efforce de faire." *Nouveaux Essais sur l'Entendement Humain.* Chronologie et introduction par Jacques Brunschwig (Paris: Garnier-Flammarion, 1966), pp. 189–190.

68 In the first article ("Utrum divina immensitas per vacuum ymaginarium infinitum sit possibilis adequari?"), De Ripa declared: "Dico autem hic de ubi diffinitivo, quoniam secus est de ubi circumscriptivo, ad quod res determinatur ex modo quantitativo." Combes, "De Ripa," 215, lines 13–14.

69 *Ibid.*, 221, lines 1–3; see also 216, lines 15–25, and 195–196 for Vignaux's remarks. Because a spiritual substance need not be coextensive with its place, it could occupy a place smaller than its intensive perfection is capable of occupying.

70 *Ibid.*, 217, lines 31–35.

71 According to Janet Coleman ("Jean de Ripa," 163), although De Ripa concluded "that no identity exists between intension and extension," he did allow that "a parallel remains between quality and quantity. This is because de Ripa sees a degree of intensity as a divisible quantitative mode of the indivisible essence."

72 *Ibid.*, 224–225, lines 62–66. According to Koyré (*Newtonian Studies*, p. 195), the denial to God of the power to create an actual infinite was not considered a limitation on His infinite creative power because "the impossible is not a limit." De Ripa, however, did assume it was a limitation, as did John Buridan (see above, Ch. 6, Sec. 2b), who even considered the creation of an actual infinite as a limitation of divine power because of the clear implication that God could then create nothing greater.

73 This was made apparent when De Ripa explained that if God wished, He could create an infinitely powerful secondary cause that could then produce an infinite body. But if this infinitely powerful secondary cause could produce an infinite body, why could it not also occupy an infinite place? Combes, "De Ripa," 224, lines 58–61.

74 *Ibid.*, 238, lines 101–105.

75 *Ibid.*, 231–232, lines 57–60. Because De Ripa contrasted imaginary and positive places, with the former being void, it seems reasonable to assume that positive places are to be associated only with the Aristotelian plenum, where a place is defined as the innermost surface of the separable, material body that surrounds what is contained. This interpretation gains support from De Ripa's attack against the Aristotelian concept of place in the first two paradoxes cited above in Ch. 6, Section 2a, where he sought to show that restricting the concept of place to positive places only would lead to absurdities with respect to motion and measurement.

76 *Ibid.*, 232, lines 66–68, and 233, line 4; 234, line 9; also 232, n. 35. The condemned article is number 49 and is cited above, in n. 29 of Ch. 5. For its use by Bradwardine, see below, Ch. 6, Sec. 3b. Bradwardine, along with Richard of Middleton, Nicole Oresme,

Marsilius of Inghen, Walter Burley, John Buridan, and Gaietanus de Thienis from the Middle Ages are discussed in my article "The Condemnation of 1277, God's Absolute Power, and Physical Thought in the Late Middle Ages," 226-232; for Amicus, Gassendi, and Clarke, who also made use of this illustration of the world's rectilinear motion, see p. 243 of my article. For further consideration of the world's motion in this volume, see below, Ch. 7, n. 73 (for Gabriel Vasquez and Phillip Faber), Ch. 8, Sec. 4b (Gassendi), Ch. 8, Sec. 4m (Clarke), and Ch. 8, n. 397 (Oresme).

77 Combes, "De Ripa," 232, lines 60-63.
78 *Ibid.*, 232, lines 63-66.
79 *Ibid.*, 234, lines 20-22.
80 *Ibid.*, 234, lines 16-20.
81 For De Ripa's judgment that God would alter locally if He were not also in imaginary as well as positive places, see Combes, "De Ripa," 230, lines 33-36, and 234, lines 12-14.
82 See *ibid.*, 234, n. 38. On Augustine's views, see above, Ch. 5, Sec. 4; for further discussion of his idea that God is in Himself, see below, Ch. 7, n. 110. Although De Ripa conceded that "God is not properly in such a place [i.e., imaginary place or space] but in Himself" (Combes, "De Ripa," 230, line 33), and thus seemed to agree with Augustine and Peter Lombard, he did diverge from them by interpreting "in Himself" spatially. God does not actually need imaginary places, but He could have created an infinite imaginary void and does appear to occupy it. Its creation was conceived neoplatonically, because it "flows intelligibly from the divine existence, as from a cause, in a certain a priori way by reason of His immensity and immense power for [producing] the existence of a creature" (my translation from *ibid.*, 228, lines 64-66).
83 The distinction may be linked with Peter Lombard's definition of positive place as derived from St. Augustine, namely "a place in a space is what is occupied by the length, width, and depth of a body" (see above, Ch. 2, Sec. 4 and n. 70). Without the body in that place, we have an imaginary place. Whether that imaginary empty space is three-dimensional or not is usually left unanswered or ambiguous.
84 Combes, "De Ripa," 222, lines 33-35; 235, lines 26-28 and 35-37. Elsewhere De Ripa used the expression "and should such a place [i.e., imaginary place] exist" ("et huiusmodi situs esset"; 238, line 99).
85 *Ibid.*, 235, lines 29-34.
86 *Ibid.*, 235, lines 35-40.
87 See *ibid.*, 235, lines 41-48, for the arguments embracing both alternatives.
88 *Ibid.*, 238, lines 101-104.
89 *Ibid.*, 238, line 105, 240, line 122.
90 The work was first published by Henry Savile in London, 1618 under the title *De causa Dei contra Pelagium et de virtute causarum ad suos Mertonenses libri tres . . . opera et studio Dr. Henrici Savilii, Collegii Mertonensis in Academia Oxoniensis custodis, ex scriptis codicibus nunc primum editi* (London, 1618, reprinted in facsimile Frankfurt: Minerva, 1964). The quotations below from the *De causa Dei*, pp. 177-180, are drawn from my translation in Grant, *Source Book in Medieval Science*, selection 73: "On a God-filled Extramundane Infinite Void Space," pp. 555-568. The section on Bradwardine appears on pp. 556-560. For a French translation, with accompanying Latin text, see Alexander Koyré's excellent article "Le vide et l'espace infini," with the section on Bradwardine appearing on 80-91. An important summary article on Bradwardine, with a detailed bibliography, has been published by John E. Murdoch, "Bradwardine, Thomas," in the *Dictionary of Scientific Biography*, vol. 2 (New York: Scribner, 1970), pp. 390-397.
91 As with Spinoza's more famous *Ethics*, Bradwardine employed a quasi-mathematical form, enunciating propositions and deriving numerous corollaries therefrom.
92 Cited from my translation in Grant, *Source Book in Medieval Science*, pp. 556-557; see also Grant, "Infinite Void Space," 44. Because of its importance I also offer the Latin text, with punctuation altered where necessary:

"1. Prima, quod Deus essentialiter et praesentialiter necessario est ubique, nedum in mundo et in eius partibus universis;

2. Verumetiam extra mundum in situ seu vacuo imaginario infinito.

3. Unde et immensus et incircumscriptus veraciter dici potest.

4. Unde et videtur patere responsio ad Gentilium et Haereticorum veteres quaestiones: Ubi est Deus tuus? Et ubi Deus fuerat ante mundum?

5. Unde et similiter clare patet quod vacuum a corpore potest esse, vacuum vero a Deo nequaquam."

De causa Dei, p. 177. In the edition, the Arabic numerals appear to the right of the corollaries.

93 *De causa Dei*, p. 177; See Koyré, "Le vide et l'espace infini," 83–84, for the French translation and Latin texts; also my translation, p. 557 of Grant, *Source Book in Medieval Science*.

94 This is article 49 of the Condemnation of 1277. For the Latin text, see above, Ch. 5, n. 29. For De Ripa's mention of it, see Ch. 6, Sec. 2b and n. 76.

95 My translation cited from Grant, *Source Book in Medieval Science*, p. 557; for the Latin text, see *De causa Dei*, p. 177, and Koyré, "Le vide et l'espace infini," 84, n. 4, and 85, n. 1.

96 To deny to God the ability to move the world was also considered a restriction on His absolute power by Jean de Ripa (see above, Ch. 6, Sec. 2b). Bradwardine, however, centered his argument on the size of the world, whereas De Ripa couched it in terms of positive versus imaginary places.

97 Grant, *Source Book in Medieval Science*, p. 558; *De causa Dei*, p. 177; and Koyré, "Le vide et l'espace infini," 85, n. 1.

98 Because of its importance, I give the entire relevant passage: "Moreover, as no one can ignore, this imaginary place (*situs imaginarius*) could have no positive nature, for otherwise there would be a certain positive nature which is not God, nor from God, the opposite of which was shown in the second and third [chapters] of this [book]; such a nature would be coeternal with God, something no Christian can accept. It was for this reason that Bishop Stephen of Paris condemned an article asserting that 'many things are eternal,' and that even if He wished, God could not destroy them; therefore, He is deprived of omnipotence. The argument of these [Aristotelians] tells against them. For if, according to the assumption of the Philosopher and his followers, there could be no void, nor any imaginary place not filled with body, [then] the world is eternal, which is heretical, and which these people [themselves] deny; or, prior to the creation of the world, there was an imaginary void place, unoccupied by any body. The previously cited article [of Bishop Stephen of Paris] condemned this reply and its irrational argument." My translation, Grant, *Source Book in Medieval Science*, p. 558; Latin text in *De causa Dei*, pp. 177–178, and Koyré, "Le vide et l'espace infini," 86, n. 3. For a similar statement by Bartholomeus Amicus, see below, Ch. 7, Sec. 2 and n. 92.

99 "Est ergo perfectionis et potentiae infinitae quod Deus necessario sit ubique." *De causa Dei*, p. 180(E). In view of the lack of modern translations for some of the following sections, and because of the overall importance of Bradwardine's discussion, the relevant Latin texts will be cited in full.

100 "Rursum Deus in principio creans mundum et quamlibet partem eius, simul fuit cum mundo et qualibet parte eius." *De causa Dei*, p. 178(A) and Koyré "Le vide et l'espace infini," 87, n. 2.

101 This opinion was shared by Jean de Ripa (Combes, "De Ripa," pp. 230–231, lines 30–47, and 234, lines 12–14). Nicole Oresme made much the same point when he declared that "if God made another world or several outside of this world of ours, it would be impossible that He not be in these worlds, and without moving Himself, because God cannot possibly be moved in any way whatsoever" (*Le Livre du ciel et monde*, bk. II, Ch. 2, p. 279 of Menut's edition and translation). Although Aristotle's Unmoved Mover of the *Metaphysics* (bk. 12, ch. 7) corresponds to the Christian God described here, he held in *De caelo* (2.3.286a.10–13) that the immortal life of a god necessarily demands that it be in eternal, circular motion.

102 "Sic ergo erat [i.e., God] in mundo quomodo per quem mundus factus est. Vel ergo prius natura fuit ibi quam creavit ibi et quam creatura fuit ibi, vel e contra. Si prius natura fuit ibi, hoc non fuit per creaturam, nec per creare, sed per seipsum et non noviter quia tunc per seipsum noviter mutaretur; fuit ergo ibi aeternaliter per seipsum.

Quare, et eadem ratione, fuit aeternaliter per seipsum ubique in vacuo, seu situ imaginario infinito, et adhuc est ubique similiter extra mundum. . . . Ad idem, creato mundo in hoc situ, Deus fuit hic. Vel ergo noviter vel aeternaliter. Si noviter, hoc fuit per aliquam mutationem vel motum in Deo, vel in alio, puta mundo; sed non per mutationem in Deo, sicut quintum ostendit; nec per mutationem in mundo quia Deus prius natura fuit ibi quam mundus crearetur, vel quomodolibet mutaretur, sicut praecedentia manifestant. Non est ergo Deus noviter factus ibi, sed aeternaliter fuit ibi. Quare, et eadem ratione, aeternaliter fuit, est, et erit ubique in situ imaginario infinito." *De causa Dei*, p. 178(B–C); Koyré, "Le vide et l'espace infini," 87, n. 2, and 88, n. 1.

103 Grant, *Source Book in Medieval Science*, p. 559. "Praeterea perfectius est esse in aliquo situ ubique, et in sitibus multis simul, quam in unico situ tantum; . . . *De causa Dei*, p. 178(E) and Koyré, "Le vide et l'espace infini," 89, n. 1.

104 Grant, *Source Book in Medieval Science*, p. 559. "Potest igitur Deus esse in situ quo voluerit sine indigentia creaturae, et non noviter, quia non per motum sui, sicut quintum ostendit; ergo aeternaliter quiescendo. Deus ergo essentialiter per seipsum in omni situ ubique aeternaliter et immobiliter perseverat. . . . Est ergo Deus necessario, aeternaliter, infinite ubique in situ imaginario infinito; unde et veraciter omnipraesens sicut et omnipotens dici potest." *De causa Dei*, pp. 178(E)–179(A); Koyré, "Le vide et l'espace infini," 89, n. 1, and 90, n. 2.

105 See Sidney Greenberg, *The infinite in Giordano Bruno with a translation of his Dialogue "Concerning the Cause Principle, and One"* (New York: King's Crown Press, 1950), p. 49.

106 The same derivation is noted in Combes, "De Ripa," 211, n. 6. In an article on "Omnipresence" in the *New Catholic Encyclopedia*, (17 vols; New York: McGraw-Hill, 1967–1979), vol. 10 (1967), p. 689, M. F. Morry has declared that "God's omnipresence has a relationship to divine immensity of actuality to aptitude. For immensity is the infinite plenitude of subsistent being that is free from all spatial limitations and, thus, is able to be present in all things. Immensity implies the power to be everywhere. Omnipresence is the actual exercise of the power to be everywhere. Whereas immensity is an essential, and eternal attribute in God, omnipresence is relative to created being." Although Bradwardine may well have accepted this description of the relationship between omnipresence and immensity, it is clear that he has derived God's immensity from his "demonstration" that God is omnipresent in an infinite void space. On the basis of Morry's description, one might say that Bradwardine derived God's immensity as if from effect to cause, that is, in an a posteriori manner. Although Bradwardine did not explicitly identify infinite space with God's immensity, others would (see below, Ch. 7, Sec. 2a [Suarez], Sec. 2b [Conimbricenses], Sec. 2d [Amicus], and Sec. 2f [Bona Spes]).

107 They are found on p. 179 of the *De causa Dei*.

108 Of the twenty-four definitions of God in the *Book of the XXIV Philosophers*, this is the second, which Bradwardine immediately followed with the eighteenth ("God is a sphere that has as many circumferences as points") and the tenth ("God is that whose power is not numbered, whose being is not enclosed, [and] whose goodness is not limited"). The translations are drawn from my article "Infinite Void Space," 44, and were made from *De causa Dei*, p. 179(A–B); for the Latin text of the *Book of the XXIV Philosophers*, see Clemens Baeumker, "Das pseudo-hermetische 'Buch der vierundzwanzig Meister' (Liber XXIV philosophorum)" in *Beiträge zur Geschichte der Philosophie des Mittelalters*, band XXV, heft 1/2 (Münster: Aschendorff, 1927), pp. 194–214 (for the three definitions cited here, see pp. 208, 210, and 212).

Among those who cited it before Bradwardine, Dietrich Mahnke mentions Alan of Lille from the twelfth century and Alexander Neckham, Bartholomew the Englishman, Alexander of Hales, St. Bonaventure, and Thomas Aquinas from the thirteenth. See Mahnke's *Unendliche Sphäre und Allmittelpunkt. Beiträge zur Genealogie der mathematischen Mystik* (Halle/Saale: Max Niemeyer Verlag, 1937), pp. 171–176 (to Mahnke's list, we should add Richard of Middleton [see below, n. 134]); for the possible origins of the definition, see pp. 178–195. Mahnke observes (pp. 175–176), however, that prior to Bradwardine, the version cited in the twelfth and thirteenth centuries substituted "intelligible" or "intellectual sphere" for "infinite sphere," a substitution probably derived from Alan of Lille's *Regulae theologicae* (Mahnke, p. 173), usually titled *Regulae de sacra theologia* or *De maximis theologiae* (see Migne, *Patrologiae cursus completus, series Latina* [221 vols.; Paris, 1844–1864], vol. 210, 627a; for Richard of Middleton's use of this version, see below, n. 134). A few years after Bradwardine's *De causa Dei*, Jean de Ripa would employ the expression "Deus spera intelligibilis infinita," thus incorporating both key terms into the famous expression (see Combes, "De Ripa," 237, n. 45, and above, Ch. 6, Sec. 2b and n. 61). Occasionally, the term "infinite" is omitted, as with John Major (see below, n. 112 of this chapter and Ch. 7, Sec. 1 and n. 21).

Despite agreement of Bradwardine's version with that found in the *Book of the XXIV Philosophers*, Mahnke believes it was Nicholas of Cusa's citation of it that made this significant definition widely known. Between the fifteenth and seventeenth centuries, Ficino, Bruno, and Fludd would also cite the very same version (see Yates, *Giordano Bruno and the Hermetic Tradition*, p. 247 and n. 2). Many scholastic authors also found occasion to include it, as for example, the sixteenth-century Coimbra Jesuit commentators in *Commentary on the Physics* (Cologne, 1602), col. 515; Francisco Suarez, *Disputationes Metaphysicae*, vol. 2, p. 110, col. 2, par. 49, who, however substituted "perfect sphere" for "infinite sphere"; and Franciscus Bona Spes, who replaced the latter expression with "intelligible sphere" (see below, Ch. 7, Sec. 2f). For a further description of the history of the metaphor, see Georges Poulet, *The Metamorphoses of the Circle*, trans. from the French by Carley Dawson and Elliott Coleman in collaboration with the author (Baltimore: Johns Hopkins Press, 1966).

Although the author of the *Book of the XXIV Philosophers* may have been the first to describe God as an infinite sphere, finite circles and spheres had long been associated with the divine (for example, see Plato, *Timaeus* 34A–B, 92C, where our spherical world is called a god, and Aristotle, *De caelo* 2.3.286a.10–13).

109 See Karsten Harries, "The Infinite Sphere: Comments on the History of a Metaphor," *Journal of the History of Philosophy 13* (1975), 6.

110 Harries, "The Infinite Sphere," 5, 6, 7. See also Arthur O. Lovejoy, *The Great Chain of Being* (New York: Harper & Brothers, 1960), p. 112.

111 Harries, "The Infinite Sphere," 8.

112 It was similarly used earlier by Richard of Middleton (see below, Ch. 6, Sec. 4 and n. 134) and somewhat later by Jean de Ripa (see above, Ch. 6, Sec. 2b, and n. 108 of this chapter). Neither applied it to an imaginary infinite void; indeed Richard rejected extracosmic void. Gabriel Vasquez, who vigorously opposed the concept of God in an extracosmic imaginary void, also cited the metaphor but insisted that Hermes Trismegistus did not intend to suggest that God would fill an imaginary empty place, but simply sought to show that God is not limited by the universe. ("Ad quod etiam alludens *Trismegist.*, qui communiter refertur, Deum sphaerae comparavit, cuius centrum esset ubique et circumferentia nullibi, non quia vastitatem illam imaginariam impleret, sed quia toto universo non definiretur, ut praedictum est." Vasquez, *Disputationes Metaphysicae*, p. 381.) In this instance, Vasquez sought to combat any spatial connotations that may have been associated with the metaphor. Somewhat earlier in the sixteenth century, at least one scholastic, John Major of the University of Paris, implied a spatial sense to the metaphor when he interpreted Hermes Trismegistus to mean that God's circumference, which is nowhere, is not terminated by the last heaven or sphere. Major invoked the metaphor to support his claim that a place that has the capacity to receive body can lie outside the heaven, just as it can within a hall or room. And because God is assumed capable of acting by His power outside the heaven, why can He not also be there by His essence? (The full passage on which this interpretation is based follows: "Preterea arguitur sic: Deus est in hac bibliotheca et in qualibet parte ipsius. Vel igitur Deus potest destruere aerem huius bibliothece ipso Deo remanente infra latera studioli, vel non. Si primum, pari ratione potest esse extra celum nulla existente creatura. Secundum autem videtur absurdum.

"Ad hoc respondet doctor quidam quod si per impossibile ponatur locum evacuari, tunc remanere intelligitur capacitas loci privati. Capacitas enim aliquid est et in illa Deus est. Privatio autem nihil est et in illa Deus esse non potest. Istud non sufficit. Deus enim potest facere vacuum (ut palam est).

"Item inquiro quid est illa capacitas passiva et quocunque dato talis reperietur extra celum quia ita bene natus est locus esse extra celum sicut in hac aula. Et quia ponitur Deus extra celum per potentiam quare ibi negabitur esse per essentiam. Hoc tenebat Hermes Trismegistus quod 'Deus est sphera cuius centrum est ubique, circumferentia vero nusquam.' Ergo circumferentia eius non terminatur in ultimo celo." John Major, *In Primum Sententiarum ex recognitione Io. Badii* [Paris; apud Badium, 1519], distinction 37, fol. 93r, col. 2). Like Bradwardine, however, John Major also cited numerous biblical authorities immediately after the metaphor, and thus the weight to be assigned to it is left ambiguous. Although Bradwardine and Jean de Ripa associated extracosmic space with God and both cited the famous pseudo-Hermetic metaphor of the infinite sphere, Nicole Oresme, who also believed in extracosmic void, which he identified with God's immensity, made no mention of it (see above, Ch. 6, Sec. 1, and further references in n. 17).

113 "Et cum non sit mundus infinitus, tamen non potest concipi finitus, cum terminis careat, intra quos claudatur." *Nicolai de Cusa Opera Omnia*, ed. by Ernest Hoffmann and Raymond Klibansky, vol. 1: *De docta ignorantia* (Leipzig: Felix Meiner, 1932), bk. 2, ch. 11, p. 100, lines 13–14. For a translation of the *De docta ignorantia*, see Nicolas Cusanus, *Of Learned Ignorance*, trans. Fr. Germain Heron, O. F. M., with an introduction by Dr. D. J. B. Hawkins (London: Routledge & Kegan Paul, 1954). The passage above appears on p. 107.

114 From his remarks that "God is the center and the circumference of all the stellar regions" and that "in every region inhabitants of diverse nobility of nature proceed from Him, in order that such vast regions of the skies and of the stars should not remain void," one may infer that Cusa considered the universe an unlimited plenum filled with matter and spirit. The translation from the *De docta ignorantia* is by Koyré, *From the Closed World*, p. 22. There is nothing here, however, about space as such.

115 Although he had the Stoics in mind, Giordano Bruno thought the concept of a finite cosmos surrounded by an infinite void quite strange because, as he put it, this kind of infinite void "hath no measure nor outer limit, though it hath an inner; and this is harder to imagine than is an infinite or immense universe." From Bruno's *On the Infinite Universe and Worlds*, trans. Dorothea Waley Singer, *Giordano Bruno, His Life and Thought* (New York: Henry Schuman, 1950), p. 254. The passage is also quoted in Koyré, *From the Closed World*, p. 48; see also Greenberg, *The Infinite in Giordano Bruno*, p. 48. On the Stoic conception, see above, Ch. 5, Sec. 1; for more by Bruno on the Stoics, see below, Ch. 8, Sec. 2.

116 Bradwardine, *De causa Dei*, pp. 179 (E) – 180(A). The translation is by Vernon J. Bourke, *Saint Augustine, Confessions* (New York: Cima Publishing, 1953), vol. 5, p. 168. Among those who used Augustine's figure, we mention the sixteenth-century Coimbra Jesuits (Conimbricenses), *Physics* (Cologne, 1602), col. 518 (see below, Ch. 7, Sec. 2b) and three seventeenth-century scholastics: Franciscus Bona Spes in a discussion of "Whether God is Actually in Imaginary Spaces" ("An Deus sit actu in spatiis imaginariis"; see *R. P. Francisci Bonae Spei, Carmelitae Reformati, Commentarii Tres in Universam Aristotelis Philosophiam* [Brussels: apud Franciscum Vivienum, 1652], Disputatio quinta: De spatiis imaginariis, p. 177, col. 2); Bartholomeus Amicus in a question "On Imaginary Space" ("De spatio imaginario"); see Amicus, *Physics*, vol. 2, p. 763, col. 2[C]; and Emanuel Maignan, *Cursus philosophicus* [Lyon, 1673], p. 243. In *De fide orthodoxa*, John Damascene (ca. 675–749) says that God is like a "sea of substance, infinite and indeterminate." See *Saint John Damascene, De fide orthodoxa*. Versions of Burgundio and Cerbanus, ed. by Eligius M. Buytaert, O. F. M., Franciscan Institute Publications, Text Series No. 8 (St. Bonaventure, N.Y.: Franciscan Institute, 1955), bk. 1, ch. 9, p. 49, line 17. See also Leo J. Sweeney, S. J., "John Damascene and Divine Infinity," *The New Scholasticism 35* (1961), 77, where St. Bonaventure's citation of this passage is mentioned.

117 My translation, with a minor alteration, from Grant, *Source Book in Medieval Science*, p. 559, col. 1. See also Grant, "Infinite Void Space," 46, where the passage was first translated, and n. 33 for the Latin text from *De causa Dei*, p. 179 (A); Koyré, "Le vide et l'espace infini," 90, n. 2, also gives the Latin text.

118 *De causa Dei*, p. 180 (A–B). For my translation of this passage, see Grant, *Source Book in Medieval Science*, pp. 559–560.

119 *De causa Dei*, p. 180(B–C). For my translation, see Grant, *Source Book in Medieval Science*, p. 560.

120 "Verum est tamen quod argumenta sua nequaquam de irrefragabili necessitate convincunt, quin Deus posset de omni potentia sua absoluta facere vacuum ubi vellet in mundo vel extra, quin etiam nunc de facto sit situs imaginarius vacuus extra mundum, vacuus, inquam, a corpore et a quolibet alio praeter Deum, sicut praecedentia suaserunt." *De causa Dei*, p. 180(C). I have altered my translation as it appears in Grant, *Source Book in Medieval Science*, p. 560.

121 See above, Ch. 5, Sec. 4. Although Bradwardine did not directly cite, or allude to, the passage from the *Asclepius* quoted above in Ch. 5, Sec. 4, he found occasion to cite the *De aeterno verbo* (the title by which he knew the *Asclepius* [see above, Ch. 5, Sec. 4]) elsewhere in the *De causa Dei* (see, for example, pp. 135[B], 146[D], 175[D], 176[A], 182[A], and 183[A] and [C]).

122 In *De causa Dei*, p. 131 (A–B), Bradwardine argued that if it were conceded that God could create numerical and intensive infinites, it would follow that "He is able to make something that would have all powers at once, really, equivalently, or supereminently; and that He would have all of them infinitely or have an infinite [power] absolutely intensively. Therefore, God can make another God; and, indeed, many other gods, . . ." ("Amplius autem per viam omnipotentiae et supereminentiae summae Dei, si Deus potest facere infinitum numerose et etiam intensive, potest facere aliquid semel habere, realiter, aequivalenter, vel supereminenter omnes virtutes; et quamlibet earum infinite, sive infinitam simpliciter intensive. Potest ergo Deus unum alium Deum; imo et alios deos multos. . . ."). Because he thought the consequences of the creation of an absolute infinite absurd, Bradwardine denied that God could create such infinites. Although he did not include the extensive infinite along with the numerical and intensive infinites (for the kinds of infinites that could be attributed to God, see below, Ch. 6, Secs. 3f and 2b), it is probable that Bradwardine would have denied to God the power to create an infinite void space. For not only did he deny the possibility of a separate, infinite void space distinct from God, but he also chose to avoid any suggestion that God did, or could, create such a space. The most plausible interpretation is that the imaginary infinite void of which Bradwardine speaks is uncreated because it is identified with God's immensity.

123 Although Bradwardine did not expressly declare this, it seems implied by all that he said. Later in the fourteenth century, Nicole Oresme unhesitatingly identified God's immensity with extramundane void, when he declared that "if it is asked what is that vacuum outside the heaven, one should reply that it is nothing but God Himself, Who is His own indivisible immensity and His own eternity as a whole and all at once." Claudia Kren (ed.), "The Questiones super De celo of Nicole Oresme" (Ph.D. dissertation, University of Wisconsin, 1965), bk. 1, question 19, p. 288. The translation is Kren's. Much the same opinion appears in his *Le Livre du ciel et du monde*, where, in 1377, he wrote that extramundane void space "is infinite and indivisible, and is the immensity of God, and is God Himself." My translation from *Nicole Oresme Le Livre du ciel et du monde* (eds. Menut and Denomy), bk. 1, ch. 24, p. 176. See also my article "Infinite Void Space," 48.

124 For the reasons given here, I believe Koyré was mistaken when he judged that Bradwardine formulated the paradoxical conception of the reality of imaginary space by uniting in the same mind "la notion théologique de l'infinité divine avec la notion géométrique de l'infinité spatiale" ("Le vide et l'espace infini," 91). It would appear that Bradwardine allowed theological considerations to suppress completely whatever geometrical instincts and intuitions he may have had about space. Furthermore there is little reason to believe that physical space was geometrized by the application of an alleged infinite Euclidean space to physics and cosmology (see above, Ch. 2, Sec. 2a). The eventual attempt to resolve the tensions between a dimensional space and the demands of theology would prove troublesome indeed, as can be seen below, for example, in our discussion of Francisco Suarez. Bradwardine's denial of a positive

nature to imaginary space would be echoed in the late sixteenth and seventeenth centuries by Bartholomeus Amicus and Pedro Fonseca (see below, Ch. 7, Secs. 2b, d).

125 On this, see Leo J. Sweeney, S. J., "Divine Infinity: 1150–1250." *The Modern Schoolman* 35 (1957–1958), 38–51. It comes as a surprise to learn that God is not called infinite anywhere in the Old and New Testaments and that Plotinus may have been the first to assign infinity as a perfection to the First Principle (see 47–48).

126 My references will be to the text of this question published by Charles J. Ermatinger, "Richard Fishacre: <Commentum in librum I Sententiarum>," *The Modern Schoolman* 35 (1957–1958), 213–235. See also the preceding article by Leo J. Sweeney, S. J. and Charles J. Ermatinger, "Divine Infinity According to Richard Fishacre," *The Modern Schoolman* 35 (1957–1958), 191–211.

127 Ermatinger, "Richard Fishacre," 217–218, lines 41–52. See also Grant, "The Concept of *Ubi* in Medieval and Renaissance Discussions of Place," 73. The same concept was expressed at approximately the same time – that is, between 1236 and 1245 – in the *Summa Theologica* wrongly attributed to Alexander Hales and sometimes titled *Summa Halesiana* (see Rev. Adrian Fuerst, O. S. B., *An Historical Study of the Doctrine of the Omnipresence of God in Selected Writings between 1220–1270* (Ph.D dissertation; *The Catholic University of America, Studies in Sacred Theology* [second series], No. 62; Washington, D.C.: Catholic University of America Press, 1951), p. 56 (for the dates, see pp. 24–25; although Fuerst's study is of importance for the problem of God's omnipresence, it contains little of relevance for the concepts of space and vacuum). Nicole Oresme said much the same thing when he declared that "God in his infinite grandeur without any quantity and absolutely indivisible, which we call immensity, is necessarily all in every extension or space or place which exists or can be imagined" (*Le Livre du ciel et du monde*, bk. II, ch. 2, p. 279 of Menut's edition and translation).

Fishacre's explanation of God's ubiquity by the assumption of His total existence in every part of a magnitude or extension plays a significant part in what follows (for example, see below, Ch. 7, Sec. 2a (Suarez), Ch. 7 Sec. 2e (Maignan), Ch. 8, Sec. 2a (Bruno), Ch. 8, Sec. 4b (Gassendi), and Ch. 8, Sec. 4m (Newton); for Henry More's attack on the doctrine, see Ch. 8, Sec. 4d; also see above, Ch. 6, n. 67). For convenience, the concept may be described as the "whole in every part" doctrine. It was already used by Plotinus in *Enneads* IV.2.1 with respect to the soul, which is both divisible and indivisible, for "its divisibility lies in its presence at every point of the recipient, but it is indivisible as dwelling entire in the total and entire in any part" (Plotinus, *The Enneads*, trans. by Stephen MacKenna [4th ed. revised by B. S. Page; London: Faber and Faber, 1969], p. 257; for a reference to the Greek text, see below, Ch. 8, n. 239). Although Christians employed the explanation for the soul they extended it to God Himself. Two significant sources guaranteed its wide dissemination in the Middle Ages. One was the *De fide orthodoxa* of St. John Damascene (ca. 675–749), which was translated from Greek into Latin at least four times during the Middle Ages. In considering the place of God, John explained: "Sciendum autem quoniam Deus impartibilis est, totus totaliter ubique ens et non pars in parte corporaliter divisus, sed totus in omnibus et totus super omne." *Saint John Damascene, De fide orthodoxa.* Versions of Burgundio and Cerbanus, ed. by Eligius M. Buytaert, O.F.M. (St. Bonaventure, N.Y.: Franciscan Institute, 1955), Ch. 13.3, p. 58 (the translation is that of Burgundio of Pisa). Substantially the same idea had been formulated by Saint Augustine and was quoted by Peter Lombard in the *Sentences*, bk. 1, distinction 37, ch. 9 (Petrus Lombardus, *Sententiae in IV libris distinctae*, 3d ed., vol. 1, part 2, p. 273), the second major source. In the passage cited by Peter, Augustine declared (in a letter to Dardanus) that "Deus, sine labore regens et continens mundum, in caelo totus est, in terra totus, et in utroque totus, et nullo contentus loco, sed in se ipso ubique totus."

128 *Ibid.*, p. 225, lines 271–273.

129 What follows is based on *ibid.*, pp. 230–231, lines 425–450.

130 Although he did not associate God with extracosmic space, Robert Holkot, an English Dominican friar (d. ca. 1349), allowed that a finite vacuum might exist beyond our world. Engaged in a lengthy consideration of whether God knew from eternity that He

would produce our world ("Utrum Deus ab eterno sciverit se producturum mundum"), Holkot sought to determine what would follow if God could make another world than ours, which he believed possible. Should God choose to make another world, it would necessarily exist somewhere with its parts separated from our world. Within this context, Holkot then inquired whether or not something exists beyond our world. If it does, then de facto something exists beyond our world. But if nothing (*nihil*) exists beyond our world, and yet some body could possibly exist there (because it was assumed that God could create another world there), it follows that "beyond [our] world there is a vacuum because where a body can exist but does not there we find a vacuum. Therefore a vacuum is [there] now." ("Preterea, si Deus posset modo facere alium mundum ab isto, posset facere illum esse alicubi, sicut iste est modo ita quod partes illius mundi distarent abinvicem extra mundum istum. Quero ergo quid est ibi modo: an aliquid an nihil? Si aliquid, ergo extra mundum de facto est aliquid. Si nihil, tunc arguitur sic: extra mundum nihil est et extra mundum potest esse corpus; ergo extra mundum est vacuum, quia ubi potest esse corpus et nullum est, ibi vacuum est. Ergo vacuum modo est." Robertus Holkot, *In quatuor libros Sententiarum quaestiones* [Lyon, 1518, reprinted in facsimile Frankfurt: Minerva, 1967], Bk. 2, question 2, sig. hii, recto, col. 2 [unfoliated].) By adopting Aristotle's definition of vacuum (see above, Ch. 2, n. 31, and Ch. 5, Sec. 1) and by merely conceding the possibility that God could create another world beyond ours, Holkot could conclude that in the place where God could create that other world, either the world itself exists there now or a vacuum does. Obviously, Holkot was here concerned only with a finite vacuum.

131   Richard of Middleton, *Super quatuor libros Sententiarum*, vol. 1, bk. 1, distinction 37, question 4 ("Utrum Deus sit extra omnem locum"), p. 325, col. 1.

132   Bradwardine used a similar argument (see above, Ch. 6, Sec. 3b).

133   *Super quatuor libros Sententiarum*, vol. 1, p. 325, cols. 1-2.

134   "Sed septem propositio *De maximis theolog.* dicitur quod Deus est sphaera intelligibilis, cuius centrum ubique, circumferentia nusquam. Et sicut dicit Commentum per centrum intelligitur creatura; per circumferentiam divina immensitas. Ergo divina immensitas est extra mundum." *Ibid.*, p. 325, col. 1. If *De maximis theolog.* is *De maximis theologiae*, as is likely, Richard's source for the metaphor was probably Alan of Lille's treatise of the same title (see above, n. 108), where the seventh proposition is this very definition. See also Etienne Gilson, *History of Christian Philosophy in the Middle Ages* (London: Sheed and Ward, 1955), pp. 173-174 and p. 636, n. 130.

135   *Super quatuor libros Sententiarum*, p. 325, col 2.

136   Thus Richard rejected the following argument: "There is nothing outside the world. But God is not in nothing. Therefore God is not outside the world." (". . . extra mundum nihil est. Sed Deus non est in nihilo; ergo Deus non est extra mundum." *Ibid.*, p. 325, col. 1.) Even before Richard of Middleton wrote, William of Auxerre (Guillelmus Altissiodorensis) (d. 1231) had already adopted a similar position. In bk. 1, ch. 16 of his *Summa aurea*, William declared that it is false to hold that "God is not beyond the world; for He is beyond the world in power because if there were infinite worlds, He would fill them all. He is also actually beyond the world in a certain manner, for He is in Himself, which is greater than the world, since God is properly the place of the world." ("Et ad primo obiectum dicimus quod hec est falsa: Deus non est extra mundum. Est enim extra mundum potentia quia si infiniti tales mundi essent omnes repleret. Est etiam actu quodammodo extra mundum, est enim in seipso qui maior est mundo; Deus enim sibi proprie est locus." *Summa aurea in quattuor libros Sententiarum a doctore Magistro Guillermo Altissiodorensi edita* . . . [Paris, n. d.; reprinted in facsimile Frankfurt: Minerva, 1964], fol. 31v, col. 1.) Because William did not discuss the possibility that God might be in a space beyond the world, it seems that this possibility was considered only later, sometime between the periods of William and Richard of Middleton, with Richard as a distinct candidate for the first significant discussant.

137   Although Richard allowed that God could create worlds, and did create our world, within His infinite, divine immensity, thus suggesting some kind of spatiality or extension to God's immensity, the temptation to attribute to Richard the opinion that God's

immensity is extended must be resisted. Indeed even with Bradwardine, who actually placed God in an imaginary infinite void, the attribution of real extension to that space would also seem unwarranted, as we saw above (see Ch. 6, Sec. 3f).

138 "However powerful an agent be its action can only reach to distant things by using intermediaries. The omnipotence of God, though, is displayed by his acting in everything without intermediary, for nothing is distant from him in the sense of God not being in it." *Summa theologiae*, part I, question 8, article 1 (Ad tertium), trans. by Timothy McDermott, O. P., vol. 2: "Existence and Nature of God" (Blackfriars, 1964), p. 113 (Latin text, p. 112).

139 Phillip Faber, a seventeenth-century Franciscan, would have denied my inference, because he believed that Aquinas denied that any operation or action is possible in a vacuum. Faber argued that because Aquinas assumed that God operates only in the place where He is, and because no action is possible in a vacuum, Aquinas would have denied that God could be in a vacuum (". . . et colligitur etiam ex hac ratione D. Thomae in p. p. q. 8, art. 1, ubi colligit Deum esse in loco ex operatione, quam habet in loco; at in vacuo non potest fieri ulla operatio. Ergo Deus, ex sententia D. Thomae, non est in vacuo." Faber, *Disputationes theologicae*, p. 338, col. 2, sec. 25).

140 "Quaestio Unica: Utrum Deum esse praesentem ubique secundum potentiam inferat ipsum esse ubique secundum essentiam, hoc est, utrum omnipotentia inferat immensitatem?" Scotus, *Opera*, vol. 5, part 2: *Quaestiones in Lib. I Sententiarum*, p. 1281. In an apparently earlier version of his *Sentence Commentary*, bk. 1, distinction 37 bears the title "Utrum Omnipotentia Dei includat eius immensitatem." In this version, the discussion on infinite void is absent. See *Ioannis Duns Scoti Ordinis Fratrum Minorum Opera Omnia*, ed. by Carl Balić, vol. 17 (Rome: Vatican City, 1966), p. 477.

141 *Ibid.*, p. 1282.

142 *Ibid.*, p. 1282.

143 *Ibid.*, p. 1282. Scotus' argument against the existence of a pre-creation void was generally adopted by Scotists, as it was, for example, by Paul Scriptor, who, in discussing Scotus' argument in bk. 1, distinction 37, declared: "Ante mundi creationem non fuit aliquod magnum vacuum cui Deus fuerit presens ut passo et ex illo creaverit mundum. Sed Deus pro tunc nulli fuit presens et tamen creavit mundum. Ergo potentia Dei non requiritur ad hoc quod creet. Ideo videtur quod Deus sit prius presens per potentiam creativam quam per essentiam. Sed quod dicit Philosophus VII *Physicorum*, quod omne agens est presens passo vel immediatum, verum est de agentibus naturalibus quorum principia non possunt in remotum vel non possunt in remotum nisi prius agant in propinquum; sic non est de Deo." *Lectura fratris Pauli Scriptoris, Ordinis Minorum, de observantia quam edidit declarando subtilissimas doctoris subtilis sententias circa Magistrum in primo libro* (no place, 1498), fol. 170r, col. 2.

CHAPTER 7
Extracosmic, infinite void space in sixteenth- and seventeenth-century scholastic thought

1 Absence of citations to a printed scholastic treatise does not allow the inference that it was unread and therefore without influence on scholastic and nonscholastic authors of the early modern period.

2 "Cap. I. Deum esse extra caelum et in vacuo esse posse docuerunt Maior et Caietanus." *Disputationes Metaphysicae*, pp. 357, 358. Disputatio XIX is titled (p. 357): "An Deus extra caelum, vel in vacuo intra caelum esse possit, aut ante mundi creationem alicubi fuerit?"

3 Although Vasquez indicated as his source for Caietanus a commentary on John 12 ("Caietanus in illud Ioannis 12: Ego lux veni in mundum," *Disputationes Metaphysicae*, p. 358), I have been unable to locate such a treatise, or any discussion of extracosmic space. For a bibliography of his works, see P. Isnardus M. Marega, O. P. (ed.), *Thomas de Vio Cardinalis Caietanus (1469–1534), Scripta philosophica* (Rome: apud Institutum 'Angelicum,' 1934), pp. LXVII–LXIX. For the role played by Major, George Lokert, and John Dullaert in fostering interest in medieval physical discussions at the Collège

de Montaigu in the first half of the sixteenth century, see Clagett, *Science of Mechanics in the Middle Ages*, pp. 653–654.

4 For the Latin text with French translation, see *Le Traité 'De l'infini' de Jean Mair* (ed. Elie). The treatise was first published in 1506 and in subsequent editions of Major's logical works. For its history, see pp. xix–xxiii.

5 *In Primum Sententiarum*, fols. 93r, col. 1–93v, col. 1. Major's commentary on the first book of the *Sentences* was first published in 1510, then in 1519, and again in 1530 (see Elie [ed.], *Le Traité 'D l'infini,'* p. xx).

6 *Ibid.*, p. 94. The translation of the passage from Kings is that of the King James Version. Major, and all medieval and early modern scholastic authors, cite it as Kings, bk. 3, ch. 8 (verse 27) in the Vulgate edition. Major also included a prayer to the Virgin, which according to Elie (*Le Traité*, p. 95, n. 3) "est tirée du graduel de la Messe d'un Office commun de la Vierge." Bradwardine had also cited the same two authorities, as did many other later authors. Appeals to authorities of all kinds became commonplace as the controversy raged on through the sixteenth and seventeenth centuries. Bradwardine himself provided an impressive array of philosophical, biblical, and patristic authorities (see Bradwardine, *De causa Dei*, pp. 179–180).

7 "Quero hunc questionis titulum: an Deus sit ubique et in infinito loco imaginario extra celum." John Major, *In primum Sententiarum*, bk. 1, distinction 37, fol. 93r, col. 1.

8 "Prima [conclusio] est: Deus est ubique per essentiam, presentiam, et potentiam." *Ibid.* Major was here repeating the common medieval opinion drawn from Peter Lombard's *Sentences*, bk. 1 distinction 37, ch. 1, where, in considering the ways in which God may be said to be in things, Peter declared, "Quod Deus in omni re est essentia, potentia, praesentia, et in sanctis per gratiam, et in homine Christo per unionem." Petrus Lombardus, *Sententiae in IV libris distinctae*, 3d ed., vol. 1, part 2, p. 263.

9 "Secunda conclusio: Deus potest esse extra celum creatura ibidem existente." *Ibid.*

10 "Tertia conclusio: probabile est quod Deus est extra celum in spatio infinito imaginario." *Ibid.*

11 For the numerous quotations invoked by Bradwardine, see *De causa Dei*, p. 179. Bradwardine's omission of the Augustinian dictum that God is in Himself before the world was created is probably to be explained by the fact that he viewed it as contrary to his contention that prior to creation, God was everywhere in an imaginary infinite void, and therefore not in Himself. The Augustinian dictum was likely invoked subsequently as a counterargument against those who adopted Bradwardine's opinion. The Augustinian dictum then had to be made compatible with the possibility of extracosmic space, which Major attempted, as we shall see below. It was subsequently cited by both sides in the controversies of the sixteenth and seventeenth centuries. For example, see below, n. 110 of this chapter.

12 "Et patet ratione quia Deus est alicubi et non est maior ratio quod sit in uno loco quam in alio, igitur." Major, *In Primum Sententiarum*, fol. 93r, col. 1.

13 In response to an objection to the first conclusion that God is everywhere, some were concerned that God must then also be in vile and evil places. Major conceded that God would be in vile places but would be uncontaminated. ("Contra primam conclusionem arguitur sic: ex ea sequitur quod Deus esset in locis malis et vilibus. Hoc autem videtur inconveniens, igitur. Respondetur concedendo illatum, nec Deus contaminatur per loca vilia, ut explicat Magister in littera de anima in corpore sordido vel radio solis. *Ibid.*, fol. 93v, col 1.)

14 "Secunda conclusio probatur. Possibile est angelum vel corpus esse extra celum et cum illo Deus est. Et Deus potest angelum vel corpus destruere ibidem manendo, igitur. Ibi inquam vel in loco imaginario mirabile esset quod Deus potest ibi existere cum creatura et ea destructa non posset sic permanere." *Ibid.*, fol. 93r, cols. 1–2.

15 "Tertia conclusio probatur: Deus potest esse extra celum nulla creatura existente ultra celum ex premissa et nihil est aliud movens quin ibi nunc existat. Sed perfectionis est sicut enim est infinitus in vigore et per infinitum tempus existit; sic per locum imaginarium infinitum existat alioquin circumscriberetur loco contra dicta." *Ibid.*, fol. 93r, col. 2.

16 "Secundo arguitur ad idem: Deus erat in spatio infinito imaginario extra celum ante creationem mundi et ubicumque vere vel imaginarie erat nunc est. Maior patet quia vel erat precise ubi est nunc iste mundus et non alibi vere vel imaginarie. Et quero causam illius. Si tu dicas quod hic poterat creare mundum sic de facto poterat creare alium mundum eccentricum et concentricum et infinitos. Ergo pari ratione debes dixisse eum fuisse in loco imaginario infinito." *Ibid.*

17 Because of its length, I omit Major's argument and quotation of Augustine's text. It should be noted, however, that the chapter number is incorrectly given as 25 instead of 5. To illustrate a correct interpretation of Augustine's position, I translate Petrus Barbay's argument, which formed part of his attack against the conception of extracosmic space. "[Now] you object that in bk. 11 of *On the City of God*, ch. 5, St. Augustine says that he who denies that God is absent from infinite spaces outside the world speaks in vain.

"I reply that the Holy Doctor argues there with respect to man, for he disputes against certain ancient philosophers who contended that the world was eternal. [St. Augustine's argument was that] it would be absurd that God should exist in vain through infinite stretches of time. He rejects this argument because if they conceive infinite times before the world was created, by an equal reason (*pari jure*) infinite spaces ought to be conceived outside the world in which He could create infinite worlds. [Now] if it is not absurd that God produced nothing in these [spaces], then it ought not to be seen as absurd that He will have produced nothing in those infinite stretches of times. That [Augustine] was, indeed, not of a proper mind to admit imaginary spaces and times is obvious from these words with which he concludes the chapter: 'But they may say that it is idle for man to contemplate infinite regions of space, since there is no place outside the universe. Our reply to them is that by the same token it is idle for man to contemplate bygone eras in which God did nothing, since there is no time before a universe exists.' Augustine also proves this *ex professo* in the next chapter." For Barbay's Latin text, see Barbay, *Commentary on the Metaphysics*, pp. 408–409. The translation of the passage from Augustine is from St. Augustine, *The City of God Against the Pagans* in seven volumes: vol. 3, books 8–11, with an English translation by David S. Wiesen (Loeb Classical Library; London: William Heinemann; Cambridge, Mass.: Harvard University Press, 1968), pp. 445–447. For further references, see above, Ch. 5, n. 53.

18 "Preterea arguitur sic: Deus est in hac bibliotheca et in qualibet parte ipsius. Vel igitur Deus potest destruere aerem huius bibliothece ipso Deo remanente infra latera studioli, vel non. Si primum, pari ratione potest esse extra celum nulla existente creatura. Secundum autem videtur absurdum." Major, *In Primum Sententiarum*, fol. 93r, col. 2.

19 "Ad hoc respondet doctor quidam quod si per impossibile ponatur locum evacuari, tunc remanere intelligitur capacitas loci privati. Capacitas enim aliquid est et in illa Deus est; privatio autem nihil est et in illa Deus esse non potest." *Ibid.* For Bonaventure's version, which Major faithfully reproduced, see *Opera Omnia*, edita studio et cura PP. collegii a S. Bonaventura Ad Claras Aquas (Quaracchi: ex typographia Collegii S. Bonaventurae, 1882–1901), vol. 1: *Commentaria in quatuor libros Sententiarum, in primum librum Sententiarum* (1882), bk. 1, distinction 37, article 1, question 2, p. 655. In contrast to Saint Bonaventure, Phillip Faber insisted that although "a capacity denotes a certain aptitude, or property, which necessarily presupposes a certain subject in which it is based," this does not apply to vacuum because "in a vacuum there is no body or being; nor is vacuum itself any being. Therefore there can be no such capacity in it." (". . . capacitas denotat quamdam aptitudinem sive proprietatem, que necessario presupponit aliquod subiectum in quo fundetur; at in vacuo nullum est corpus, neque ullum ens, neque ipsum vacuum est ens. Ergo in illo nulla esse potest huiusmodi capacitas." *Disputationes Theologicae*, p. 339, col. 2, sec. 28.)

20 "Istud non sufficit. Deus enim potest facere vacuum (ut palam est). Item inquiro quid est illa capacitas passiva et quocunque dato talis reperietur extra celum quia ita bene natus est locus esse extra celum sicut in hac aula. Et quia ponitur Deus extra celum per potentiam quare ibi negabitur esse per essentiam." Major, *In Primum Sententiarum*, fol. 93r, col. 2.

21 "Hoc tenebat Hermes Trismegistus quod 'Deus est sphera cuius centrum est ubique, circumferentia vero nusquam.' Ergo circumferentia eius non terminatur in ultimo celo."

Ibid. On this important Hermetic metaphor, see above, Ch. 6, Sec. 3c and the relevant notes.

22 "Propterea alio modo arguunt quidam sic: nihil debet Christianus attribuere Deo nisi quod deducitur ex sacris literis ultra illa que gentiles philosophi ponerent. Sed quod Deus posset esse extra celum sine creatura non ostenditur ex sacris literis et apud pure naturaliter loquentes hoc non admitteretur, igitur." To this Major replied, "quod doctores illi insufficienter assumunt sed deberent dicere: nihil debet Catholicus ponere nisi quod sequatur ex sacris literis, vel poneretur apud naturaliter loquentes supposito quod naturaliter loquentes Deo attribuant que nos attribuimus. Fateor nunc quod mere naturaliter loquens Deum negaret esse ubique secundum essentiam. Hermes autem non videtur pure naturaliter locutus fuisse." *Ibid.*, fol. 93v, col. 1.

23 "Contra tertiam conclusionem arguitur sic: ex ea sequitur quod Deus posset facere infinitum actu. Hoc est falsum quia esset tante perfectionis quante Deus." *Ibid.* Bradwardine would have agreed with this argument (see above, Ch. 6, n. 122), as would Buridan in his *Questions on the Physics* (above, Ch. 6, Sec. 2b).

24 See Elie (ed.), *Le Traité 'De l'infini,'* pp. xxii–xxiii. On the controversy over the possibility of the creation of an actual infinite, see above, Ch. 6, Sec. 2b.

25 "Ad primum, concedo quod Deus potest facere infinitum, sed nego quod illud esset infinite perfectionis vel infinite resistentie respectu Dei; potest enim a Deo subito vel successive corrumpi." Major, *In Primum Sententiarum*, fol. 93v, col. 1.

26 Francisco Suarez, *Disputationes Metaphysicae* (2 vols.; Paris, 1866, reprinted in facsimile Hildesheim: Georg Olms Verlag, 1965; the work was first printed in 1597), vol. 2, Disputatio XXX: "Quid sit Deus, sectio VII: An Deum esse immensum demonstrari posssit," pp. 95, col. 1–113, col. 1.

27 *Ibid.*, p. 95, col. 1.

28 *Ibid.*, p. 96, col. 1. The biblical authorities, as well as Aristotle, are cited immediately after.

29 Suarez, *Disputationes Metaphysicae*, vol. 2, p. 96, col. 2. As an opponent of those who believed in the existence of an imaginary infinite void space, Phillip Faber (1564–1630), a Franciscan Conventual and Scotist, rightly understood that the major issue underlying the dispute was the question "Whether from God's operation in all things it may rightly be inferred that He is immense" ("Utrum ex operatione Dei in omnibus rebus recte inferatur ipsum esse immensum"). He noted that his opponents said that "God is present by essence in an actual, imaginary infinite space in which God is able to create things and because of this is able to operate to create and conserve things in any part of this space. For if He were not actually present there by essence, He could not operate there. Thus [with regard] to the argument, they deny that God is not in a vacuum, or in this imaginary space, before the creation of the world. This reply demands a long discussion. . . ." *Disputationes Theologicae*, p. 338, col. 1. Faber's *Disputationes* was actually a commentary on the *Sentences* of Peter Lombard, where Faber's lengthy treatment of the question cited above formed the substance of his commentary on bk. 1, distinction 37. See also Ch. 6, Sec. 4, and Ch. 7, Sec. 2a. For bibliography on Faber, who was a professor of metaphysics in the *via Scoti* at the University of Padua from 1603 to 1606 and professor of theology from 1606 on, see Charles H. Lohr, "Renaissance Latin Aristotle Commentaries: D–F," *Renaissance Quarterly* 29, no. 4 (1976), 726.

30 Suarez, *Disputationes Metaphysicae*, vol. 2, p. 97, col. 1.

31 *Ibid.*, p. 97, col. 2.

32 *Ibid.*, pp. 99, col. 2–100, col. 1, par. 15.

33 Because of the significance of this passage, I cite the whole of it. "Respondetur, difficultatem hanc (de qua plura statim dicemus) magis videri sumptam ex modo loquendi nostro et verbis quam ex re ipsa. Nam de re dicendum est, ante actionem Dei praesupponi necessario ex parte Dei talem modum existendi, seu talem dispositionem (ut modo nostro loquamur) substantiae suae, ut ex parte sua ita existat, ut sine sui mutatione possit intime et realiter esse in quacunque re, si illam velit creare, et hunc modum essendi habet Deus ex vi suae immensitatis. Quia vero nos non possumus hanc dispositionem in substantia spirituali concipere, nisi per ordinem ad spatium, eo quod Deus ex

vi praedictae dispositionis de se aptus est ad existendum in quibuscunque corporalibus spatiis, etiamsi in infinitum protendantur. Ideo non possumus illam divinae substantiae dispositionem et immensitatem concipere, nisi per modum cujusdam extensionis, quam necessario explicamus per ordinem ad corpora; et quando vel reipsa, vel mente realia corpora separamus, necessario apprehendimus veluti quoddam spatium aptum repleri corporibus, cui tota divina substantia sit praesens, et tota in toto, et tota in singulis partibus ejus, per quam praesentiam nihil aliud significamus, quam praedictam divinae substantiae dispositionem." Suarez, *Disputationes Metaphysicae*, vol. 2, p. 100, col. 1, par. 16. Although Suarez assumed God's immutability in this passage, he made little of it during the rest of the argument. Indeed it is not until the next section that he considered whether God's immutability can be demonstrated. ("Sectio VIII: An demonstrari possit Deum esse immutabilem et aeternum." *Ibid.*, vol. 2, pp. 113–115.) Bradwardine had used God's immutability as the context of his discussion of extracosmic imaginary space. Some of Suarez's contemporaries and successors would also make use of God's immutability.

In this passage, Suarez employed the fundamental medieval concept that God is wholly in the whole of a space and wholly in every part of that space, a concept that was also applied to other spiritual substances, such as souls and angels. It was the common medieval manner of explaining how a spiritual substance could occupy a body or place and yet remain indivisible despite the divisibility of the body or place. For the source of the idea, see above, Ch. 6, n. 127.

34 Here is the passage in which Suarez declared the problem and the need to make a distinction. "Una superest difficultas circa modum quo declaravimus divinam immensitatem; nam videmur attribuere divinae substantiae dispositionem quamdam seu modum existendi, habentem actu infinitum quamdam quasi extensionem, quam nos per comparationem ad infinitum spatium imaginarium declaramus. Hic autem modus existendi videtur repugnare divinae simplicitati, quia non potest intelligi sine aliqua distinctione et compositione partium componentium illum existendi modum, et consequenter necessariam esse aliquam distinctionem ex natura rei inter illum modum praesentiae et substantiam cujus est modus." *Ibid.*, vol. 2, p. 111, col. 1, par. 49.

35 *Ibid.*

36 *Ibid.*, p. 111, cols. 1–2, par. 49. In this passage, Suarez explained that, although the loss of a hand signifies that the soul no longer has a partial union with that part of the body, it should not imply that a part of the soul has been lost.

37 *Ibid.*, p. 111, col. 2, par. 49. In Peter Lombard's quotation from St. Augustine's letter to Dardanus (see above, Ch. 6, n. 127), Augustine declared that God is wholly in heaven and wholly in the earth.

38 *Ibid.*, p. 107, cols. 1–2, par. 37.

39 This is but one of the numerous possible meanings of "imaginary" described in Ch. 6, Sec. 1. For Martin Pereyra's similar conception, see below, n. 110 of this chapter. In attacking the concept of imaginary space, Phillip Faber also seemed to equate it with vacuum (see above, n. 29 of this chapter).

40 *Ibid.*, vol. 2, p. 107, col. 1, par. 37.

41 *Ibid.*, p. 109, col. 2.

42 *Ibid.*

43 *Ibid.*

44 *Ibid.*, p. 110, col. 1.

45 *Ibid.*, p. 100, col. 2, par. 18. Strictly speaking, Suarez denied that the cause must be present in the effect in order to act, but he insisted that every agent has a "necessary fundament for such a presence," and it is from this that we understand how God's presence precedes His action, especially with respect to creative action.

46 According to Charles B. Schmitt, "these remarkably useful volumes are probably the best example of a fusion of the humanist approach to Aristotle with that of the long established scholastic approach. Each text is presented in the Greek original, accompanied by a parallel Latin translation and followed first by a series of explanatory notes and then by a series of *quaestiones*, somewhat along the lines of the medieval model, though fully up-to-date and cognizant of recent philosophical and scientific developments."

"Towards a Reassessment of Renaissance Aristotelianism," *History of Science* 11 (1973), 169.

47 *Petri Fonsecae Commentariorum in Metaphysicorum Aristotelis Stagiritae Libros,* Tomus I–II (2 vols; Hildesheim: Georg Olms Verlag, 1964; reprint of *Commentariorum Petri Fonsecae Lusitani, Doctoris Theologi Societatis Iesu, in libros Metaphysicorum Aristotelis Stagiritae* (4 vols; Cologne: Lazari Zetzneri Bibliopolae, 1615), bk. 5, ch. 13, question 7: Locum pro spatii non esse veram quantitatis speciem," which is placed under the general heading "Sitne locus vera species quantitatis," col. 700. For a list of the numerous editions of Fonseca's commentary on the *Metaphysics,* see Carlos Sommer-vogel, S. J., *Bibliothèque de la Compagnie de Jésus,* vol. 3 (Brussels: Oscar Schepens; Paris: Alphonse Picard, 1892), cols. 838–840. In Sommervogel's list, the commentary on the fifth book of the *Metaphysics* appears first in a Rome edition of 1589, although publication of the commentary on the first four books appeared in 1577. Because Sommervogel may have missed earlier editions, the dates are only approximate. For a brief sketch of Fonseca's life and works, see Lohr, "Renaissance Latin Aristotle Commentaries: Authors D–F," 739–741.

48 Fonseca, *Metaphysics,* bk. 5, ch. 13, question 7, col. 700. Later (col. 701) Fonseca explained that, in the discussion on quantity in his *Predicamenta,* Aristotle assumed that space was three-dimensional, but in the fourth book of the *Physics* he held that the true place of a thing is the surface of the containing body. See above, Ch. 1, Sec. 1 and the relevant notes.

49 "Quod denique sit quantitas externa ex eo probant: quia non inhaeret corporibus a quibus occupatur, alioqui vel recederet illis recedentibus ac proinde non esset fixum et immobile; vel ex aliis in alia migraret quod fieri non potest si est interna sive inhaerens quantitas." Fonseca, *Metaphysics,* col. 701.

50 *Ibid.,* col. 701.

51 "Ad hoc argumentum Philoponus entitatis spatii defensor respondet duplicem esse quan-titatem trinae dimensionis: unam materialem et eam indigere spatio nec posse simul esse cum alia materiali quantitate; aliam immaterialem et eam esse spatium, quod quia immateriale est adaequatum sibi corpus secum admittit." *Ibid.,* col. 702. On Philoponus' distinction, see above, Ch. 2, Sec. 3; for Crescas' use of it, see Ch. 2, Sec. 4.

52 *Ibid.,* col. 703. Fonseca was mistaken in his attribution to Philoponus of the opinion that space inheres in body ("Itaque videtur credere Philoponus quantitatem spatii rebus locatis inhaerere"). Void space is distinct and separate from body but always occupied by it (see Ch. 2, Sec. 3).

53 *Ibid.,* col. 701. Thus did Fonseca take a significant step and deny that space is either a substance or an accident. In this he would be followed by both scholastics and non-scholastics (for Bartholomeus Amicus, see below, Ch. 7, Sec. 2d; for Francesco Patrizi, who used a similar argument and arrived at the same conclusion at approximately the same time as Fonseca, see below, Ch. 8, Sec. 4a; for Gassendi, Ch. 8, Sec. 4b). Although space was neither substance nor accident, all of the authors just mentioned were agreed that space was an existent entity of some kind. It was quite otherwise with certain scholastic authors, primarily Scotists, who bitterly opposed the existence of any kind of separate space within or beyond the world. For them, the arguments demonstrating that space could be neither substance nor accident were of great utility, because they could infer from them the absolute nonexistence of space. Thus Franciscus Pitigianus, a seventeenth-century Scotist, would argue that "either this [void] space (*spatium*) is something in itself, or it is nothing; if nothing, therefore it is not a place. If it is something, therefore it is a substance or an accident; if a substance it is therefore, corporeal or incorporeal; if incorporeal, it will not be extended or have depth (*profunda*) with the located body because this is a function [or character-istic] of a corporeal thing. If it is corporeal, therefore it has quantity and a penetra-tion of bodies will result. If, [however,] it is an accident, it is either a quantity or a quality; if it is a quantity, there will be a penetration of bodies in all dimensions; if it is a quality, it will be an accident without a subject and will not be extended to permit a body to be extended in it." *R. P. F. Ionnis Duns Scoti . . . in VIII libros Physicorum Aristotelis Quaestiones, cum annotationibus R. P. F. Francisci Pitigiani Arre-*

tini . . . (Lyon, 1639), vol. 2, p. 228, col. 2. This edition of Scotus' works by Luke Wadding has been reprinted in facsimile as Johannes Duns Scotus, *Opera Omnia* (Hildesheim: Georg Olms Verlag, 1968). Although ascribed to Duns Scotus, the *Questions on the Physics*, on which Pitigianus commented, is actually by Marsilius of Inghen and was cited above in Ch. 2, n. 24. The passage also appears in my article, "Place and Space in Medieval Physical Thought," 139. For a similar Scotist approach, see Phillip Faber, *Disputationes Theologicae*, p. 339, col. 2, sec. 29.

54 Fonseca, *Metaphysics*, col. 703.

55 *Ibid.*, cols. 703–704.

56 *Ibid.*, col. 704.

57 The similarity between the position adopted by Vasquez and that formulated some centuries earlier by John of Jandun has been indicated above in Ch. 2, n. 28.

58 I have found no evidence that the Conimbricenses commented upon the *Metaphysics*. No such commentary is listed in the bibliography of their commentaries in Sommervogel, *Bibliothèque de la Compagnie de Jésus*, vol. 2, cols. 1273–1278, and vol. 9, cols. 62–63.

59 "Sit-ne Deus extra coelum, an non?" Conimbricenses, *Physics*, (Cologne, 1602), bk. 8, ch. 10, question 2, col. 514. Of the four articles composing this question and extending over cols. 514–519 (in the Lyon edition of 1602, see cols. 580–586), I have translated and annotated all of the first and fourth and part of the second in Grant, *Source Book in Medieval Science*, pp. 560–562. For an earlier discussion, see my article "Infinite Void Space," 52–53. The point of departure for the question proposed here is the final chapter of Aristotle's *Physics* (bk. 8, ch. 10), where Aristotle discussed the Prime Mover.

60 Capreolus is Johannes Capreolus, O. P. (d. 1444), sometimes called "the Prince of Thomists," who taught theology at Paris from 1408–1411 (see Gilson, *History of Christian Philosophy in the Middle Ages*, p. 800, n. 76). Durandus is probably Durandus de S. Porciano, O. P. (ca. 1275–1334). The reference is cited only as "Quaest. 1" and is perhaps to Durandus' *Sentence Commentary*, bk. 1, distinction 37, where in the Lyon edition of 1556, the first question considers "whether God is in all things" (*Utrum Deus sit in omnibus rebus*). See *In Sententias theologicas Petri Lombardi commentariorum libri quatuor* (Lyon: apud Gasparem a Portonaris, 1556), fols. 85v–86r.

61 "Pro qua haec potissimum sunt argumenta." Conimbricenses, *Physics*, bk. 8, ch. 10, question 2, article 1, col. 514.

62 "Extra coelum nihil est. Sed Deus non potest esse in nihilo. Non est igitur extra coelum." *Ibid.* This argument was already cited in the late thirteenth century by Richard of Middleton who rejected it (see above, Ch. 6, n. 136). Two who adopted and defended the argument were Gabriel Vasquez (*Disputationes Metaphysicae*, p. 368) and Phillip Faber (*Disputationes Theologicae*, p. 345, col. 1, sec. 55). Vasquez insisted that the alleged vast imaginary space supposedly lying beyond the heaven and said to have existed before the creation of the world is absolutely nothing. But God and angels can no more exist in nothing than they can in a chimera. (". . . Deum non esse extra caelum nec in vacuo probatur in hunc modum. Vastitas illa imaginaria, quam extra caelum nunc esse et ante mundi creationem fuisse et inanitas illa quam intra latera vacui corporis esse dicimus, est omnino nihil; ergo vere et re ipsa in ea Deus aut Angelus esse non potest; aut si vere in ea esse dicitur, cur etiam in chimaera non erit? Neque enim inanitas illa est ens minus fictum et nihil quam chimaera.")

63 "Item si Deus esset in spatio quod ultra coeli complexum imaginamur, cum spatium istud revera nihil positivum aut reale sit, pari ratione fas esset dicere Deum esse in caecitate aliisque privationibus, quod absurdum est." Conimbricenses, *Physics*, bk. 8, ch. 10, question 2, article 1, col. 514.

64 "Deinde sicuti eiusmodi spatii sola mente concipimus, ita consequens est ut Deus non re ipsa, sed nostra tantum conceptione inibi existere dicatur. Non ergo Deus revera extra coelum existit." *Ibid.*

65 "Accedit quod substantiae expertes materiae non sunt in loco nisi per operationem. Deus vero extra mundum nihil operatur, sicuti neque ante mundum conditum quicquam operabatur quod D. Augustin. in libro 11 *Confes.*, cap.12, hisce verbis ait: 'Dicenti

quid faciebat Deus antequam faceret coelum et terram, respondeo non illud quod quidam respondisse perhibetur, iocularirter eludens quaestionem violentiam, "alta," inquit, "scrutantibus gehennas parabat." ' Et paulo post: 'Antequam faceret Deus Coelum et terram non faciebat aliquid; si enim faciebat, quid nisi creaturam faciebat?' " *Ibid.* Scotus, of course, insisted that God acts in distant places by the operation of His will and not by His omnipresence. There is an echo of that argument here. See above, Ch. 6, Sec. 4, and Ch. 7, Sec. 2a.

66 "Denique huic sententiae suffragatur idem D. Augustinus lib.3 *Contra Maximum*, cap.21, quo loco sciscitantibus, ubinam Deus esset antequam mundum fabricaret, respondet fuisse in seipso, quod repetit super Psal.122 in illa verba: 'Levavi oculos meos in montes.' Nec ibi ullam imaginarij spatij in quo Deus extiterit mentionem facit." *Ibid.* On this argument, see above, Ch. 5, Sec. 4.

67 These references are distributed over articles 2 and 3 in cols. 515–518 (cols. 581–584 in the Lyon 1602 edition). From the specific citations and passages cited or paraphrased, the following identifications can be made: the *Asclepius* attributed to Hermes Trismegistus; the pseudo-Hermetic *Book of the XXIV Philosophers* or Alan of Lille's *De maximis theologiae* (see above, Ch. 6, Sec. 3c and n. 108); Aristotle, *De caelo* 1.9.279a.19–23, where Aristotle allowed that unchangeable entities that are not in place might be beyond the heaven (see above, Ch. 5, n. 57), and *Metaphysics*, bk. 12, where Aristotle is said to have opposed this position; Clement of Alexandria (ca. 150–ca. 215), *Stromata* (or *Miscellanies*); Eugubinus, *De perenni philosophia*, bk. 4, chs. 1, 2; Gregory Nazianzenus (ca. 330–389), *De theologia*, bk. 2; St. Basil (ca. 331–379), *Homilies*, 16; St. Denis (Pseudo-Dionysius) (fl. 450 A.D.), *De divinis nominibus*, ch. 1; Athanasius (ca. 295–373), "a letter to the Synod of Niceae" (*epist. synodi Niceae*); St. Hilary of Poitiers (ca. 315–366), "on Psalm 118" (*in Psal.* 118); St. Ambrose (340–397), "letter to the Ephesians" (*epist. ad Ephes.*), ch. 3; Marius Victorinus Afer (ca. 280–ca. 363), *Adversus Arium*, bk. 4; St. Augustine (354–430), *De civitate Dei*, bk. 11, ch. 5 (see above, Ch. 5, Sec. 4) and *Confessions*, bk. 7, ch. 5 (the text has "ch. 7" but corresponds to ch. 5 of the modern editions; see above, Ch. 6, Sec. 3c); William of Auxerre, *Summa aurea*, bk. 1, ch. 16 (see above, Ch. 6, n. 136); Thomas Aquinas, *Quodlibet* 11, article 1 (the identification of this quodlibetal question is uncertain; on Aquinas' quodlibetal questions, see James A. Weisheipl, O. P., *Friar Thomas d'Acquino His Life, Thought and Work*, pp. 123–128); Marsilio Ficino, *De immortalitate animae*, bk. 2, ch. 6; and John Major, *In primum Sententiarum*, bk. 1, Distinction 37 (see Ch. 7, Sec. 1). No specific reference is given for Plato, who is cited here through Eugubinus and Ficino.

68 The scriptural references are Kings 3, 8 (Vulgate; 1, 8, 27, in the King James version; see above, Sec. 1, for the passage) and Baruch 3, where the House of the Lord is described: "Magnus est et non habet finem, excelsus et immensus" (in Bradwardine's version the following precedes it: "O Israel, quam magna est domus Domini et ingens locus possessionis eius . . ." [*De causa Dei*, p. 179]). The prayer to the Virgin reads: "Quem coeli capere non possunt, tuo gremio contulisti." Although it differs textually from the prayer mentioned above in n. 6 of this chapter, which was the version used by Bradwardine and John Major, the two prayers are substantially the same. On the tradition of authoritative citations, see above, Ch. 7, Sec 1.

69 "Deus supra verticem summi coeli consistens ubique, omniaque circumspicit. Est enim ultra coelum spatium sine stellis, ab omnibus rebus corpulentis alienum." Conimbricenses, *Physics* (Cologne, 1602), col. 515. Because the text of the passage cited by the Coimbra Jesuits differs somewhat from that in the edition of the *Hermetica* by Walter Scott, vol. 1, p. 324, I have altered the latter's translation to fit the proper sense of the Coimbra text. As the most significant alteration, it should be noted that where Scott's text has "huic est enim ultra caelum locus" (translated by Scott as "his abode is beyond the heaven"), the Coimbra text has "Est enim ultra coelum spatium," which I have rendered as "For there is beyond the sky a space." The difference is important because in the Coimbra version, God is not only said to lie beyond the world, but the existence of a space beyond the world, in which God presumably exists, is expressly stated. In Scott's version, however, God is said only to dwell, or have a place, beyond

the heaven. Nothing is said, or even implied, about a space existing there. On the other significant passage from *Asclepius*, see above, Ch. 5, Sec. 4.

70 Boethius' definition of eternity (*Consolation of Philosophy*, bk. 5, sec. 6) as "the complete, simultaneous and perfect possession of everlasting life" ("Aeternitas igitur est interminabilis vitae tota, simul et perfecta possessio"; for the English translation, see *The Consolation of Philosophy*, trans. with an introduction by V. E. Watts [Harmondsworth,, Middlesex: Penguin Books, 1969], p. 163, which is a better rendering than that of H. F. Stewart in the Loeb Classical Library; the Latin text, ed. by E. K. Rand, is in that same Loeb edition, which bears the title *The Theological Tractates*, with an English translation by H. F. Stewart and E. K. Rand; *The Consolation of Philosophy* with the English translation of "I. T." [1609] revised by H. F. Stewart [Cambridge, Mass.: Harvard University Press, 1953], p. 400, lines 9–11) was frequently cited as the basis for the scholastic concept of eternity. It was cited by the Conimbricenses earlier in their commentary on the *Physics*, bk. 4, ch. 14, question 3, article 1 ("De Aeternitate, aevo, et quibusdam aliis durationibus"; Lyon, 1602), col. 159, where they explained it as a "duration that is absolutely invariable and independent" (". . . hoc est, duratio omnino invariabilis atque independens"). In a much lengthier discussion, Suarez (*Disputationes Metaphysicae*, vol. 2, pp. 922–926 [disputation 50, sec. 3], cited the Boethian definition with approval and characterized the major features of eternity as the absence of succession without past or future. Such a perfection can only belong to God. Thus when God creates and then ceases to create and causes things to move temporally and successively in time, this does not in any way diminish or affect God's eternity "because the succession in these external actions or denominations does not infer variety in any internal actions, or proper perfection, of God Himself." ("Sic enim Deus creare incipit, et a creando desistit, et res in tempore movet actione etiam temporali et successiva, absque variatione vel diminutione suae aeternitatis, quia haec successio in his actionibus vel denominationibus externis non infert varietatem in aliquo actu interno, aut propria perfectione ipsius Dei, . . ." *Ibid.*, p. 924, col. 2.) Thus the analogy between eternity and immensity must be interpreted as follows: Just as God's eternity signifies that He exists perfectly and simultaneously without succession or change through an infinite imaginary time, so also must His immensity, which is an attribute equal in power to that of His eternity, signify that He exists simultaneously and totally in every part of an infinite imaginary space. The analogy between eternity and immensity was frequently drawn, and the two attributes were often mentioned together (for example, see Oresme's statement above, Ch. 6, n. 123, and Maignan's below, Ch. 7, Sec. 2e and n. 148; for a rejection of the analogy, see below, Ch. 7, Sec. 2f).

71 "Secunda ratio: Deus est id quo nihil maius potest concipi; at si extra coelum non esset, posset aliquid eo maius concipi – nimirum ipse Deus consistens in infinitis spatiis in quibus nunc non esset. Impossibile est ergo Deum non esse actu extra mundum." Conimbricenses, *Physics*, col. 517. Anselm's ontological argument, which appeared in his *Proslogium*, has been nicely paraphrased in syllogistic form and explained by Frederick Copleston:

"God is that than which no greater can be thought: But that than which no greater can be thought must exist, not only mentally, in idea, but also extramentally:

"Therefore God exists, not only in idea, mentally, but also extramentally.

"The *Major Premiss* simply gives the idea of God, the idea which a man has of God, even if he denies His existence.

"The *Minor Premiss* is clear, since if that than which no greater can be thought existed only in mind, it would not be that than which no greater can be thought. A greater could be thought, i.e., a being that existed in extramental reality as well as in idea." Frederick Copleston, *A History of Philosophy*, vol. 2: *Mediaeval Philosophy. Augustine to Scotus* (Westminster, Md.: Newman Press, 1957), p. 162.

72 "Corroboraturque argumentum quia substantia immaterialis, quo nobilior [corrected from "mobilior"] est, eo ampliorem existentiae sphaeram habet. Quare si infinite perfecta sit, sphaeram obtinebit infinitam, id est, nullius sphaerae sive spatii limitibus concludetur." Conimbricenses, *Physics*, col. 517.

73 "Tertia ratio: constituat Deus extra coelum duo corpora, quae aliqua ab se intercapedine spatii distent, negari non poterit Deum esse ubi ea corpora existent cum oporteat in qualibet creatura esse Deum. Sed absurdum est affirmare Deum in iis corporibus existere, neque tamen esse in spatio quod inter illa iacet. Ergo Deus actu iam est in toto spatio quod extra mundum tenditur. Probatur minor: quia alioqui sequeretur Deum esse quodammodo a se abiectum et divisum; itemque ab una parte spatii ad aliam commeare et migratorium motum subire, si videlicet eiusmodi corpora in quibus esset ab ipso Deo, vel ab Angelo, hinc illuc transferrentur. Quare cum nulla mutatio in Deum cadat, fatendum erit existere Deum in infinito spatio extra mundum." *Ibid.*, col. 517. For Bradwardine's argument that God's motion would imply mutability, see above, Ch. 6, Sec. 3b.

Similar arguments in defense of God's omnipresence in extracosmic space were reported by Phillip Faber. One such imagines a vacuum in the space between the sphere of water and the lunar orb. Should God now move a stone within that vacuum, and if it is further assumed that God can only be in positive, created things and not in void space, it would follow that God is confined to the moving stone and would consequently be in local motion (*Disputationes Theologicae*, p. 338, col. 2, sec. 24). Another argument that Faber described (*ibid.*) reveals the impact of article 49 of the Condemnation of 1277 (see above, Ch. 5, Sec. 2) and shows how that article could be applied to the controversy on extracosmic space: "If God were only in the world and not outside it, [then], since the whole world could, by the power of God, be transported through that space to different imaginary spaces, it follows that God could be moved" (Gabriel Vasquez reported the same argument, *Disputationes Metaphysicae*, p. 362). Faber and Vasquez, who denied the existence of extracosmic space and God's omnipresence therein, rejected these arguments. Faber denied that God can be changed by motion of the positive things in which He is present because God does not have a real relation to creatures (*ibid.*, p. 345, col. 2, section 56, reply to the fourth argument). Vasquez allowed that it is not impossible that God could move the world rectilinearly (*ibid.*, p. 383); and because He is present in the world, one might conclude that God is moved because He will have acquired new external relations. But such relations are only accidental and secondary, because God does not suffer any alteration of His internal relations (*ibid.*, pp. 384–385).

74 "Articulus IV. Quidnam sit imaginarium spatium, quo pacto Deus in eo existat." *Ibid.*, col. 518.

75 What has been cited here forms part of the first argument: "Primum sit: hoc spatium non esse veram quantitatem trina dimensione praeditam; alioqui non possent recipi in eo corpora, cum plures eiusmodi dimensiones in eodem situ naturae viribus simul esse nequeant." *Ibid.* Here, once again, is the famous Aristotelian argument denying dimensionality to place or space on the basis of the impenetrability of separate dimensions (see above, Ch. 1; Ch. 2, Sec. 2b; Ch. 6, Sec. 2a; Ch. 7, Sec. 2b). We may rightly assume that the Coimbra Jesuits would have allowed that by supernatural means, two dimensions could indeed occupy the same place simultaneously (see above, Ch. 2, n. 56).

76 "Item nec esse ullum aliud reale ac positivum ens, cum nihil tale praeter Deum ab aeterno fuerit, hoc vero spatium semper extiterit, semperque esse debeat." *Ibid.*, cols. 518–519. On this important idea, see Bradwardine's discussion above, Ch. 5, Sec. 3, and Ch. 6, n. 98; also see below, Ch. 7, Sec. 2d for Amicus' defense of this position.

77 What follows is subsumed under the heading of a second major opinion about imaginary space that the Coimbra Jesuits proclaimed: "Secundum est: spatium hoc non esse ens rationis, cum ab eo re ipsa absque opera intellectus intra mundum corpora recipiantur et extra mundum recipi queant, si illic a Deo creentur. Quare eius dimensiones non iccirco imaginariae dici consueverunt quod fictitiae sint, aut a sola mentis notione pendeant, nec extra intellectum dentur, sed quia imaginamur illas in spatio proportione quadam respondentes realibus ac positivis corporum dimensionibus." *Ibid.*, col. 519.

78 Suarez did not quite make this explicit, but surely implied it in his explanation of the manner in which God's immensity, which is equated with infinte void space, relates to bodies (see above, Ch. 7, Sec. 2a).

79 "In hoc igitur imaginario spatio afferimus actu esse Deum, non ut in aliquo ente realis, sed per suam immensitatem quam, quia tota mundi universitas capere non potest, necesse est etiam extra coelum in infinitis spatiis existere." *Ibid.*, col. 519.

80 "Ad primum inficiamur non posse Deum esse in nihilo, id est, in spatio, quod ens reale et positivum non est; alioqui nec lapis divina virtute extra coelum esse posset." *Ibid.*, col. 519.

81 "Ad secundum negandum, eandem esse rationem in caecitate aliisque privationibus qua in spatio. Sicut enim spatio convenit aptitudo ad recipienda corpora, quae non convenit aliis negationibus nec privationibus, ita eidem competit, ut in eo Deus esse dicatur, etsi in aliis negationibus vel privationibus non dicatur esse." *Ibid.*

82 "Ad tertium ita concipi a nobis illud spatium extra mundum ut illic vere sit; non tamen ut ens reale et positivum." *Ibid.*

83 "Ad quartum: Deum non proprie dici in loco esse cum nullis spatiis concludatur; sed neque substantias creatas materiae expertes esse in loco per operationem, sed per suam substantiam certo modo spectatam, quod enucleatius explicare non est huius loci." *Ibid.* Here we have an attack on the Scotist conception of the operation of God's omnipotence (see above, Ch. 6, Sec. 4; also Ch. 7, Sec. 2a).

84 "Ad postremum: licet D. August. nullam in eo libro imaginarii spatii fecerit mentionem, fecisse tamen et quidem apertis verbis ac luculentis lib. 11 *De civitate Dei*, loco a nobis citato." *Ibid.* See above, n. 66 of this chapter for the text of the rejected argument. As we have already seen, the passage in Augustine's *City of God* does not substantiate the claim made by the Conimbricenses. Augustine did not approve the existence of imaginary spaces beyond the world (see above, Ch. 5, Sec. 4 and n. 53, and Ch. 7, n. 17).

85 See above, Ch. 6, Sec. 2b. Major offered no suggestion as to the possible manner of creation for infinite space but hinted at its possible dimensionality (see above, Ch. 7, Sec. 1).

86 See above, Ch. 7, Sec. 1, for Major's description of Bonaventure's interpretation of privation as nothing; see also the first two arguments presented by the Conimbricenses in Ch. 7, Sec. 2b and the discussions by Vasquez (with relevant notes) in Ch. 2, Sec. 1 and John of St. Thomas in Ch. 6, n. 14. In Ch. 5, n. 35, the anonymous author of the *Summa Philosophiae* described extracosmic void as "nonbeing," in contrast to the plenum of our world. It is, however, a description derived from the position of Averroes rather than one held by the author. For John of Jandun's medieval conception of the conditions under which a vacuum could and could not be conceived as a privation, see above, Ch. 2, Sec. 1.

87 *In Aristotelis libros De physico auditu dilucida textus explicatio et disputationes in quibus illustrium scholarum Averrois, D. Thomae, Scoti, et Nominalium sententiae expenduntur earumque tuendarum probabiliores modi afferuntur* (2 vols.; Naples, 1626–1629), vol. 2, Tractatio XXI: De vacuo in 4. Phys., pp. 732–766. The two volumes on the *Physics* represented the third and fourth volumes of a seven-volume work by Amicus titled: *In universam Aristotelis philosophiam notae et disputationes, quibus illustrium scholarum Averrois, D. Thomae, Scoti et Nominalium sententiae expenduntur; earumque tuendarum probabiliores modi afferuntur.* The first two volumes on logic were published in 1623; the fifth, on Aristotle's *De caelo*, appeared in 1626; and the sixth and seventh, on *De generatione et corruptione*, in 1648. For this, and other information on the works of Amicus, see Sommervogel, *Bibliothèque de la Compagnie de Jésus*, vol. 1, cols. 279–280. The five questions on vacuum, each subdivided into articles and doubts (dubitationes), covered the following topics: (1) "What is understood by the term 'vacuum' and whether it pertains to physics"; (2) "On the existence of vacuum"; (3) "On the motion of bodies"; (4) "On motion in a vacuum"; and (5) "On imaginary space." Within the articles of each question, Amicus distinguished "the opinions of the doctors" (*opiniones doctorum*) from his own opinion (*opinio auctoris*).

88 Amicus, *Physics*, vol. 2, pp. 759–766.

89 "Dubitatio I: *An sit et quid sit spatium imaginarium.* Omissa opinione antiquorum dicentium spatium imaginarium esse corpus, vel quantitatem, in quibus recipiuntur alia corpora, contra quos plura Arist. 4. phys. ostendens tale spatium esse impenetrabile cum

alio corpore atque adeo non posse habere rationem loci, inter recentiores adhuc manet controversia quid illud sit." *Ibid.*, p. 759, col. 1(A).

90 The three major opinions are as follows: "The first opinion is of those who say that imaginary space has no being as a thing, whether this be taken according to essence or existence, but is absolutely nothing, just as fictitious beings [are absolutely nothing] but have the nature of a being only from our [mental] conception, as do other beings of the mind." ("Prima opinio est dicentium spatium imaginarium non habere aliquam entitatem a parte rei, sive secundum essentiam sive existentiam, sed esse omnino nihil sicut entia fictitia et habere solum rationem entis ex nostra conceptione, sicut caetera entia rationis." *Ibid.*, p. 759, col. 1[B].) As the major proponent of this opinion, Amicus cited Gabriel Vasquez, who in turn mentioned, among other supporters, William of Auxerre, St. Bonaventure, Ricardus (presumably Richard of Middleton), Duns Scotus, Durandus de Saint Porciano, Capreolus, Thomas of Strasbourg (Thomas de Argentina), Aegidius Romanus, Albertus (whether Albertus Magnus or Albert of Saxony is unclear), St. Thomas Aquinas, and others. For Vasquez's lengthy attack on the notion of God's extracosmic existence and extracosmic space, see his *Disputationes Metaphysicae*, disputatio XIX (Amicus cited it incorrectly as 29), pp. 357–388, which considers the question "whether God could exist outside the heaven [or world], or in a vacuum inside the heaven [or world]; or [whether God] was somewhere before the creation of the world?" ("An Deus extra caelum vel in vacuo intra caelum esse possit, aut ante mundi creationem alicubi fuerit?" *Ibid.*, p. 357.) Vasquez insisted that God was only in Himself before the world and not in any imaginary space and offered many arguments to show that such a space is impossible and absolutely nothing.

"The second major opinion is of those who say that imaginary space is indeed negative with respect to existence, but positive as to essence because they believe that it [i.e., imaginary space] is nothing other than positive extension." ("2. Opin. est dicentium spatium imaginarium esse quidem negativum quoad existentiam, sed positivum quoad essentiam quia putant non esse aliud quam positivam extensionem." Amicus, *Physics*, p. 759, col. 2[A].) Citing Philip Mongeus and Hurtado (probably the sixteenth-century polymath Diego Hurtado de Mendoza) as major supporters of this opinion, Amicus explained that those who hold this opinion deny existence to imaginary space but consider it necessary to assume a possible positive extension, as if it were a true and real being by virtue of the existence of intervals between bodies. That is, we can make affirmative assertions about the locations of bodies in relation to an imaginary space as if that space were real. For example, if there is an interval of two palms between two bodily locations, there is an aptitude to receive a body of two palms in that interval. Although the interval may only be imaginary, it behaves as if it were something real by virtue of its capacity to receive a body of two palms. Moreover, it cannot receive two bodies in that same place. Thus although this space is only imaginary, it acts as if it were real. This opinion, and the arguments used to indicate that imaginary space is somehow real, bears a striking resemblance to Jean de Ripa's approach described above, Ch. 6, Sec. 2a.

"The third opinion is of those who say that imaginary space is indeed a certain negation, but not one that is impossible, but [rather] one that is possible, just as is blindness in the eye, or the negation of a stone in a man." ("3. Opinio est dicentium spatium imaginarium esse quidem negationem quandam, sed non impossibilem, sed possibilem, sicut est caecitas in oculo, vel negatio lapidis in homine." *Ibid.*, p. 760, col. 1[A–B].) Here we find the names of Fonseca, Suarez, and the Coimbra Jesuits. Amicus called this "the common [opinion] of the moderns" ("est communis recentiorum"). Because it relates intimately to the position that Amicus would adopt, we shall say more about it later.

91 *Ibid.*, p. 760, col. 2(A). For the Latin text of the first conclusion (*prima conclusio*) see above, Ch. 5, n. 40.

92 *Ibid.*, p. 760, col. 2(A–B). For the Latin text and translation, see above, Ch. 5, n. 40. For Bradwardine's similar argument, see above, Ch. 5, Sec. 3 and Ch. 6, n. 98.

93 "Secundo, vel esset substantia vel accidens. Non primum quia vel substantia spiritualis

et repugnat esse diffusum et extensam; vel corporalis et haec non reperitur sine quantitate atque adeo incapax alterius corporis quanti. Non secundum quia accidens non potest esse sine subiecto." *Ibid.*, p. 760, col. 2(B). For much the same kind of analysis by Franciscus Pitigianus and Pedro Fonseca, see above, n. 53 and Sec. 2b of this chapter, respectively. Because the remaining arguments in the first conclusion are somewhat lengthy, the Latin texts will be omitted and only an interpretive summary given below.

94 The position described here resembles Suarez's interpretation of the relations between God's immensity, identified with imaginary space, and the bodies scattered through that space. See above, Ch. 7, Sec. 2a.

95 For Philoponus, see above, Ch. 2, Sec. 3a. Although Philoponus denied that a three-dimensional immaterial dimension or void space is a substance, his description of void space as an immaterial dimension accords well with the opinion rejected by Amicus.

96 "2. Conclusio: spatium imaginarium non est formaliter quantitas positiva non existens realiter, sed per intellectum." *Ibid.*, p. 761, col. 1(A).

97 "Primo quia spatium consistit in capacitate corporis sed quantitas sic concepta non est capacitas, sed potius ob rationem impenetrabilitatis incapax corporis." *Ibid.* In connection with this first argument, Amicus declared that in order to exist, a quantity must be in a space. But that space cannot be nonexistent, because no thing can be in something that is nonexistent. Moreover, as something that has existed from eternity, space is distinct from every intellectual creature. Now, if through all eternity a single stone were created, it would necessarily be in that eternal space. As a contingent quantity, however, the existence of the stone may be negated. The space in which that stone exists, however, cannot be negated with respect to existence because it is eternal. Hence space cannot be a quantity. My interpretation of these difficult arguments, the Latin texts of which are omitted, is uncertain.

98 Secundo probatur ex incorruptibilitate spatii, nam spatium non destruitur per existentiam quantitatis; sed quod includit negationem existentiae corrumpitur per positionem existentiae quia idem repugnat simul existere et non existere." *Ibid.*, p. 761, col. 1(B). Although the Latin texts will be omitted, Amicus offered two additional brief supporting arguments for the corruptibility of a quantitative space. If space were a quantitative extension of body, it could not be immobile because every quantity is absolutely mobile by its nature. Mobility, however, implies corruptibility. Furthermore, in a quantitative space, bodies could not succeed each other in the same identical space, because, on the implied assumption that two quantities cannot occupy the same place simultaneously, each body would corrupt the space it occupied and a newly generated space would then be occupied and destroyed by the succeeding body. But if space is really incorruptible, the same identical space must serve as the location of every body that occupies it.

99 "3. Conclusio: spatium imaginarium formaliter est quid negativum et concomitanter positivum." *Ibid.*, p. 761, col. 1(D).

100 "Prima pars probatur quia quid existit ab aeterno ante illam creationem et nunc est extra orbem universum, non potest esse nisi negativum: nam positiva omnia extra Deum sunt ex efficientia Dei; sed tale est spatium imaginarium, ut infra cum de eius existentia dicetur, ergo. Confirmatur ex dictis in prima conclusione, nam cum formaliter non sit positivum, necessario debet esse negativum." *Ibid.*, p. 761, col. 1(E)–col. 2(A).

101 "Secunda pars probatur quia a nobis apprehenditur ut quid positivum per modum quantitatis positivae diffusae in infinitum. Hinc aliqui dixerunt spatium esse ens rationis quatenus scilicet apprehenditur sub ratione positiva. . . ." *Ibid.*, p. 761, col. 2(A). Although Amicus affirmed that it is natural for us to conceive all negations as if they were positive things, he insisted that we must use reason to correct our imagination and natural tendencies. ("Ad secundum respondetur concedendo a nobis sic concipi quia omnes negationes concipimus ad modum rerum positivarum; sed imaginatio corrigenda est per rationem quae probat illud quod subest positivo esse vere negativum. *Ibid.*, p. 764, col. 1[B–C].)

102 In replying to an argument against his position, Amicus said the following: "Respon-

detur secundo negando illud spatium imaginarium, licet concipiatur per modum extensionis positivae, esse ens reale; nam est ens rationis positivum, quia concipitur ut ens positivum, nam licet quantitas secundum se sit ens reale, tamen negatio non potest esse quantitas realiter, sed tantum conceptibiliter.' *Ibid.*, p. 761, col. 2(B).

103 This is made clear in the first response to the same objection to which Amicus replied in the preceding note. "Respondetur primo essentiam realem posse habere tum obiectivam quando scilicet non existit realiter, sed cognoscitur; nam obiectum illius cognitionis habet esse fundamentaliter et ex vi cognitionis reflexae habet existentiam obiectivam. Neque idcirco est ens rationis quia hoc dicitur quod solam habet existentiam obiectivam et est incapax existentiae realis." *Ibid.*, p. 761, col. 2(A–B).

104 By "real space" Amicus probably understood the space that is associated only with bodies, just as Suarez described it (see above, Ch. 7, Sec. 2a). His interpretation of "place" is probably the standard Aristotelian conception wherein the place of a contained body is the innermost immobile surface of a containing body in contact with the outermost surface of the contained body.

105 "4. Conclusio: haec negatio, quae constituit spatium imaginarium, non est negatio corporis, vel quantitatis, vel spatii seu loci realis; sed negatio repugnantiae ad capiendum extensum." *Ibid.*, p. 761, col. 2(B–C).

106 "Prima pars probatur quia negatio corporis non stat cum ipso corpore, sed spatium existit simul cum corpore replente illud, nam capacitas est simul cum re quae in re illa continetur." *Ibid.*, p. 761, col. 2(C).

107 "Secunda pars probatur quia cum sit negatio, qua respicit, debet habere terminum; sed non potest habere alium terminum quam repugnantiam ad capiendum, ergo. Minor probatur quia non negat corpus repletivum, ut dictum est, neque assignatur alius; ergo negatur repugnantia." *Ibid.*, pp. 761, col. 2(E)–762, col. 1(A). Omitted here is a lengthy discussion as to whether negation of resistance ought to be considered a positive something (*quid positivum*) on the assumption that only something positive can be opposed to a negation such as resistance. Amicus denied this, arguing that a negative can be opposed to a negative, and concluded that space and resistance are negatives, with space being the negation of resistance. Space is thus the negation of a negation.

108 "Secundo patet recte a Fons. [i.e., Fonseca], Molin. [i.e., Molina] et aliis, dici spatium non esse aliquid positivum reale, sed capedinem, quae est non repugnantia ad capienda corpora, eamque esse extensam et maiorem et maiorem iuxta quantitatem corporis capibilis, nam commensuratur corporibus ibi locandis, ita ut spatium sextarii sit non repugnantia recipiendi sextarium aquae et est repugnantia naturaliter recipiendi cadum." *Ibid.*, p. 763, col. 1(A). The assumption that imaginary space has a capacity for receiving bodies because it has the property of nonresistance was attacked by Phillip Faber. If space had a property of nonresistance, Faber asked (*Disputationes Theologicae*, p. 345, col. 1, sec. 55) whether that nonresistance would be "positive, privative, or negative? If the first [i.e., positive], then something is assigned which is outside of God and which was not made by God, which is erroneous. If it is the second [i.e., privative], every privation has a foundation (*fundamentum*) and a positive and real subject in which it is based (*fundatur*). Therefore this capacity and nonresistance (*non repugnantia*) were founded in some positive being and this is false because no such thing was conceded, as we said; therefore [etc.]. If it is the third [i.e., negative], I concede this. But a simple negation is nothing; however, to say that God was in nothing is to say that He was nowhere but in Himself." For Faber, then, nonresistance is at best a negation, and because that is equivalent to nothing, to say that God is in nothing is to say that God is in Himself and not in any kind of space.

109 In the second of three arguments to demonstrate that space is distinguished from vacuum (". . . patet hoc spatium distingui a vacuo"), Amicus declared: "Secundo ratione significati formalis quia licet utrumque dicat negationem, tamen vacuum dicit negationem corporis replentis quia aufertur adveniente corpore replente; negatio autem aufertur per positionem termini qui negatur. At spatium non potest dicere negationem corporis replentis quia non aufertur per positionem corporis replentis, ut patet, nam concipitur ut receptaculum et capedo corporis; at recipiens non destruitur per receptum, sed potius perficitur." *Ibid.*, p. 763, col. 1(B–C). That a void is destroyed when filled with body was

denied by Sextus Empiricus (vol. 3: *Against the Physicists* II.20–24, pp. 221–223). For "if it perishes, it becomes in a state of change and motion [and if it perishes it is generable]; but that which becomes in a state of change and motion is a body [both generable and perishable]; so that the void does not perish." With arguments formulated by the Sceptics, Sextus sought to demonstrate that place could be neither body nor void. During the seventeenth century, however, it is likely that some scholastics and nonscholastics who accepted the existence of an absolute, immutable, eternal, infinite space also assumed, at least implicitly, Amicus' distinction between vacuum and space.

110 To these we may add the seventeenth-century Coimbra theologian Martin Pereyra, who explained that theologians understand by imaginary space void and empty spaces that lie beyond the heavens and in which God could create other bodies. Indeed when we say that before the creation of the world God was in Himself, we can understand that He was in this imaginary space. ("Ut autem intelligatur quomodo Deus in seipso fuerit antequam crearetur mundus, licet fuerit in spatio imaginario. Notandum est quod per spatium imaginarium intelligunt theologi spatia quaedam vacua et inania, quod concipimus extra caelos capacia ut repleantur corporibus quae Deus ibi potest creare, sicut etiam ante molem hujus mundi concipimus vastitatem quandam inanem et vacuum in spatio quod nunc occupat hic mundus aequalis extensionis cum illo." *Commentariorum in primum librum Magistri Sententiarum tomus secundus* . . . authore Reverendissimo ac Sapientissimo Patre Magistro D. Fr. Martino Pereyra [Coimbra: ex typographia in Regali Collegio Artium Societatis Jesu, 1715], pp. 89–90, par. 64.) Pereyra was apparently a Jesuit and either a contemporary of Amicus, whom he mentioned (p. 90, par. 65) and who died in 1649, or who lived during the second half of the seventeenth century. Amicus, it should be noted, shared Pereyra's opinion that the phrase "God is in Himself" signifies God's omnipresence in an imaginary space (see below, Ch. 7, Sec. 2d and n. 125); so, it seems, did Maignan (below, n. 143). Two scholastics who vehemently opposed this interpretation were Gabriel Vasquez and Phillip Faber. Vasquez insisted that "God, or other spirits, cannot be outside the heaven in another [thing] other than in Himself, since beyond the heaven absolutely nothing, neither space nor interval, exists." (". . . Deus, aut alias spiritus, esse non potest extra caelum in alio quam in se ipso, cum extra caelum nihil prorsus sit, nec spatium aut intervallum, ut dictum est." *Disputationes Metaphysicae*, p. 374.) Faber adopted much the same opinion when he declared that "before the creation of the world, God was only in Himself. For God is only in Himself and in creatures. Thus since there were no creatures before the creation of the world, God was not able to be in them; therefore He was only in Himself. And equally, since a vacuum is nothing, and is, consequently, not a creature . . . God could not be in it at all." *Disputationes Theologicae*, p. 388, col. 2, sec. 25. Like Vasquez, Faber allowed that "God is outside the heaven, but not in a vacuum or imaginary space." *Ibid.*, p. 345, col. 1, sec. 53. Indeed God is only in Himself.

111 This definition appears as the first of three ways in which Amicus distinguished vacuum from space: "Primo ratione connotati, nam vacuum connotat latera concava corporis inter quae dicitur vacuum." Amicus, *Physics*, p. 763, col. 1(B). By contrast, imaginary space is not associated with concavities. As an eternal entity existing before the creation of the world and stretching to infinity, space cannot be enclosed by body. ("At spatium imaginarium nihil tale connotat quale dicitur fuisse ante mundi creationem et nunc est extra universum." *Ibid.*)

112 ". . . quia si Deus destruat omnia sublunaria, conservatis lateribus orbis Lunae ubi nunc sunt et collocet in medio illo vacuo lapidem, tunc spatium lapide occupatum distabit a concavo Lunae per spatium carens corpore, aptum tamen repleri, ergo spatium et concavum continentis non sunt idem." *Ibid.*, p. 763, col. 1 (C–D).

113 ". . . totus mundus dicitur occupare tantum spatium quanta est diametri longitudo, at non datur concavum quod tale spatium in se contineat totum mundum." *Ibid.*, p. 763, col. 1(C).

114 "Quinta conclusio: haec negatio constituens spatium non debet concipi per modum inhaerentis in subiecto, sed per modum per se subsistentis. Patet quia ibi nullum est subiectum in quo possit esse vel intelligi inesse." *Ibid.*, p. 763, col. 1(D).

115 "Ex quo patet hanc negationem non posse habere rationem privationis sed purae nega-

tionis quia privatio est negatio formae in subiecto apto ad formam, quare aufertur adveniente forma negata. At haec negatio non tollitur sed permanet adveniente corpore; sed [sine] posita repugnantia, quae non potest poni in spatio quia essentialiter dicit negationem repugnantiae." *Ibid.*, p. 763, col. 1(E)–col. 2(A). The insertion of "sine" seems essential if the sense of the argument is to be preserved.

116 The equation of imaginary space and the divine immensity and the consequence of spatial infinity were declared by Amicus in response to an opposing argument that imaginary space can exist only in our minds, for if it had real existence it would of necessity be infinite. "Ad quartam respondetur spatium illud esse infinitum quia adaequari debet immensitati Divinae." *Ibid.*, p. 766, col. 1(C). For the argument to which this is a response, see *ibid.*, p. 764, col. 2(D) (the fourth confirmatory argument). To those who rejected the actual existence of an infinite space for fear that an actually infinite quantity or body would have to be assigned to fill it, Amicus responded that just because God is of infinite omnipotence, it does not follow that He could create an actual infinite quantity, although He could make a syncategorematic infinite quantity that would always be capable of further increase. ("Dices hinc sequitur posse dari infinitam quantitatem, nam si datur infinitum spatium capax quantitatis, ergo debet dari terminus illius, scilicet quantitas capibilis. Respondetur negando consequentiam, nam sicut ex eo quod omnipotentia Dei sit infinita non sequitur posse dari creaturam actu infinitam, sed satis esse infinitam syncategorematice, ut supra dictum est; ita ex spatio actu infinito non sequitur dari posse infinitam quantitatem actu." *Ibid.*, p. 766, col. 1[C–D].) Because Amicus believed that God did not create infinite imaginary space and also seemed to deny that God could create an actual infinite quantity, it would appear that Amicus was not an actual infinitist (see above, Ch. 6, Sec. 2b).

In an earlier section of his lengthy discussion of vacuum, Amicus made an even more emphatic statement that infinite space is filled by the infinite immensity of God, in support of which he cited with approval John Major's adoption of the same position in the latter's *Sentence Commentary*, which has been described in Ch. 7, Sec. 1.

117 The immobility of space is declared unequivocally in Amicus, *Physics*, p. 764, col. 1(C–D).

118 This was the third argument against Amicus' position in support of the claim that space has no real existence but is only a mental concept: "Confirmatur tertio quia daretur aliquod existens independens a Deo quia illud spatium existeret ante ullam Dei productionem; imo esset coaeternum et coexistens Deo." *Ibid.*, p. 764, col. 2(D).

119 The text of this argument reads as follows: "Ad tertiam respondetur quidem [corrected from "quidam"] concedendo illud spatium esse aeternum et coexistens Deo et independens per se ab actione Dei quia hoc est proprium negationum, ut tunc existant quando circa terminos negatos non est actio, nam si esset actio non esset negatio sed potius positio existentiae termini. Ego respondeo esse aeternum et suo modo abusivo coexistens Deo sicut dicunt de omnibus negationibus creaturarum quae ab aeterno abusive extiterunt et posita creatione deserunt suo modo existere. Nec mirum quia non habent rationem entis, sed non entis. Unde duplex est dependentia a Deo: una per se et haec non potest convenire nisi rebus positivis quae indigent quoad sui conservationem influxu divino; sed potius negatione influxus huiusmodi sunt negationes. Sed quia tam influxus in terminum quam negatio illius est in libera potestate Dei, ideo huiusmodi negationes dicuntur pendere a Deo quia Deo nolente influere in terminos, existunt; et Deo volente influere, non existunt, sed existunt termini. Hinc negationes rerum existentium dici possent contineri per accidens sub obiecto omnipotentiae Dei . . ." *Ibid.*, p. 766, col. 1(A–C).

120 These are the ten attributes identified by Aristotle in the *Categories*, or *Predicaments*, as it was known in the Middle Ages.

121 "Neque est absurdum aliquid existere in spatio negativo existente et hinc etiam posse dici corpus esse illi praesens; non quidem per praesentiam quae sit relatio predicamentalis, quia haec requirit terminum positivum realiter existentem; sed quae sit relatio transcendentalis, quae potest esse ad terminum negativum." *Ibid.*, p. 765, col. 2(C).

122 In a brief statement on negations, Honoré Fabri (c. 1607–1688) a Jesuit who emphasized mathematics and physics as well as theology, explained that God knows negations in themselves, for which reason apparently "even non-being, which is a negation, has objective existence." ("14. Quaeres quomodo Deus videat negationes? Respondeo in se

ipsis; nam etiam non-ens, quod est negatio, habet esse obiectivum; sic malum cognoscit, id est peccatum, propter eandem rationem. Nec enim opus est ut ea Deus in se contineat quae cognoscit nisi tantum obiective, cum ipsa essentia divina sit forma essentialiter expressiva omnis entis obiectivi, id est omnis scibilis." *Summula theologica in qua quaestiones omnes alicuius momenti, quae a scholasticis agitari solent, breviter discutiuntur ac definiuntur. Auctore Honorato Fabri Societatis Iesu, cum indicibus tractatuum et rerum notabilium. Nunc primum in lucem prodit* (Lyon: sumptibus Laurentii Anisson, 1679), p. 27, col. 1.

123 Amicus explained that negations are not properly beings and have no proper essence or existence. Hence they possess no perfection, but only imperfection. But they do have an improper existence that depends on God. ("Ad quintam: cum negationes non sint entia proprie, non possunt habere nec essentiam nec existentiam proprie atque adeo nullam perfectionem habent, sed potius imperfectionem quia perfectio vel convertitur cum ente vel est passio supponens entitatem, ex quo fit ut negationes nequeant habere existentiam per essentiam quia habere hoc modo existentiam supponit rem habere essentiam et existentiam qualem solum habet Deus.

"At inquies saltem habebit existentiam abusivam per suam essentiam abusivam quia illam non habet ab alio. Respondetur neque id dici posse, nam ut diximus negationes quoad ipsarum existentiam abusivam habere aliquam dependentiam ab alio, tum a libera negatione influxus divini, tum a termino et fine; et hinc patet." Amicus, *Physics*, p. 766, col. 2 [B–C].)

124 "Respondetur [ad] secundam: sicut non omne receptum in alio est positivum reale, ut patet ex caecitate quae recipitur in oculo, ita neque omne recipiens debet esse positivum." *Ibid.*, p. 764, col. 1(A).

125 Amicus' rebuttal here was in reply to those who described imaginary space as nothing but a figment of our imaginations and who cited in confirmation "the authority of the saints who say that God was in Himself before the creation of the world and in no way in another thing because there was nothing before the creation of the world." ("Confirmatur ex auctoritate Sanctorum dicentium Deum ante mundi creationem fuisse in seipso et nullo modo in alio ex eo quia ante creationem nihil erat." *Ibid.*, p. 759, col. 1[C]. In support of this opinion, Amicus mentioned Tertullian, Augustine, and Bernard of Clairvaux.) It is to this specific passage that Amicus replied in our summary above, the Latin text of which follows: "Ad primam confirmationem respondetur: cum Sancti Patres dicunt Deum ante mundi creationem fuisse in seipso et negant fuisse in alio, intelligunt de alio tamquam in effectu vel tanquam in loco qui dicit continentiam activam et passivam; non tamen negant quin fuerit in spatio imaginario in quo fuit mundus productus et in alio spatio infinito in quo possunt alii mundi produci, non quidem localiter, sed repletive et praesentialiter praesentialitate negativa . . ." *Ibid.*, p. 763, col. 2(A–B). That God is said to be in an imaginary space *repletive* is probably a reference to *ubi repletive*, which was one of the four kinds of *ubi* traditionally distinguished in scholastic discussions of the ways things or substances could be located. In a scholium to bk. 1, distinction 37, part 1, article 1, question 2 of St. Bonaventure's *Sentence Commentary*, the editors explain that for St. Bonaventure and Alexander of Hales *ubi repletive* signified the special mode by which God inheres in all things, fills and contains all things, and immensely superexceeds all things and yet remains distinct from, and unmixed with, them. See St. Bonaventure, *Opera Omnia*, vol. 1, p. 641, and above, Ch. 6, Sec. 2b, for *ubi circumscriptivum* and *ubi definitivum*; on the Augustinian dictum that God was in Himself before the world, see above, Ch. 5, Sec. 4; for Pereyra's agreement with Amicus that God's being in Himself before the creation is compatible with His also being in an imaginary space beyond the world, see above, n. 110 of this chapter.

Gabriel Vasquez sharply disagreed with those who, like Pereyra, interpreted the dictum that God is in Himself as signifying God's omnipresence in some kind of extracosmic space. Vasquez argued that even if a body were created beyond the world, it would not be in a space distinct from itself, but would only be in itself. In the same manner, we ought to understand that if God or an angel existed beyond our finite world, neither God nor angel should be conceived to be in another but only in themselves. From this, however, it would be improper to infer that God is in Himself as a body is in itself. Although a body beyond the world would not be in a space distinct from itself, it would yet contain a space within itself (i.e., an internal space, concerning which, see above, Ch. 2, Sec. 2a), have a position, and be distant from other things. Moreover, adverbs of place could also be properly applied to such a body. Such terms and concepts cannot, however, be applied to God, who is in Himself in a manner wholly different than a body. Indeed God does not exist beyond the heaven. ("Quare si corpus extra caelum non dicitur esse in aliquo spatio, nec in alio quam in se ipso, nulla ratione dici potest Deus, aut Angelus, extra caelum in alio quam in se ipso esse. Quod si *Doctores* oppositae sententiae contendant Deum in se ipso existere extra caelum sicut existeret corpus, ita ut dicatur in se habere spatium et situm, distantiam, et adverbia loci ei convenire, concedemus quidem Deum esse in seipso; at caetera omnia, nempe situm, distantiam, adverbia loci, et extra caelum esse, ei tribui nullo modo, ut subinde patebit." *Disputationes Metaphysicae*, p. 372). Phillip Faber, who was in general agreement with Vasquez, argued in a similar vein. The occasion for his discussion is a truncated version of the old Stoic argument about the extension of a hand beyond the world (see above, Ch. 5, Sec. 1). In this argument, the fifth in a series defending the opinion "that God is outside the world in a vacuum" (*Disputationes Theologicae*, p. 338, col. 1), the proponents of this opinion argued that "a void space is given outside the heaven. Thus if a man were at the extremity of the heaven, he could extend his hand outside the heaven. But in that space there are incorporeal angelic spirits; therefore nothing prevents God from also being there" (*ibid.*, p. 338, col. 2, sec. 24). Faber's response followed (*ibid.*, p. 345, col. 2, sec. 57): "To the fifth [argument], it must be noted, as we said above with respect to the authority of Aristotle, that there is nothing truly and really outside the heaven, nor is a vacuum present [there]. Thus if something is posited outside the heaven, either it is a body or a spirit. But [whether a body or a spirit] it cannot be there as in a place, since the nature of place requires that two things be present, namely a container and the contained, which are really distinct [things]. Therefore, since there is no real body outside the heaven there cannot be a container [there]. And so, if [a body] were there, it could not, in truth, be said to be in a place; nor would differences of place – as up, down, distance, and such things – be appropriate to it because all these things are appropriate to a contained thing by reason of the surfaces of a container, which are not there. Hence if [a body] were really there, it would be in itself only and an arm could be extended [to it] by considering the arm and hand in relation to the body and the position (*situs*) which it had before in its body and not in relation to a surrounding space, which is nothing except by our imagination. Moreover, there is no capacity for real local motion, [since] this body would be in itself. The same thing must be said about God, angels, and intelligences that are outside the heaven, with this difference [however]: because if we imagine that God is in an imaginary space, we cannot imagine that He is moved from place to place because we have stated that He is immense. But if we imagine an angel, we can imagine that it could be moved because it is limited (*deffinitus*) [in extent] and not immense so that we could determine imaginary points between which it is moved. But since these are fictions they are to be rejected and, in truth, it must be said that they are in themselves only and we ought to imagine them in this way." Because of its length, the Latin text is omitted. Like Suarez, Faber would conceive any body lying beyond the world as being in its own internal space. Unlike Suarez, however, Faber denied that in the absence of body there would only be imaginary space (see above, Ch. 7, Sec. 2a).

126 Amicus invoked Job 11, verses 8–9 and Psalm 8, verse 1. Of those who explicated one or the other of these texts, Amicus mentioned Thomas Aquinas, Jerome, Bede, Gregory, and Augustine. See Amicus, *Physics*, p. 763, col. 2(B–C). For similar references by the Coimbra Jesuits, see above, n. 67 of this chapter.

127 "Dico secundo hoc spatium imaginarium, quod est pura negatio, est diffusum per totum orbem et extra et praecessit omnem creaturam per aeternitatem." *Ibid.*, vol. 2, p. 748, col. 1(A–B).

128 Under "opinion of the author" (*opinio auctoris*), Amicus defined vacuum in the first conclusion of his lengthy discussion on vacuum: "Prima conclusio: vacuum est locus privatus corpore." *Ibid.*, p. 739, col. 2(E).

129 It is not the negation of a spiritual substance because the latter can coexist with a vacuum.

Only a body can destroy a vacuum. Thus if the water is evacuated from a container or vessel, and we do not perceive another body entering, we can assume the existence there of a vacuum. If a spiritual substance remained there, however, the vacuum would not be destroyed. Thus it is that in defense of the definition of vacuum as "place deprived of body," Amicus declared: "Patet ex communi conceptione, cum enim ex phiala aquam eductimus et concipimus non subintrare aliud corpus, intelligimus phialam esse vacuum. Sed quia duo possunt intelligi intra superficiem illius phialae quibus repleatur: vel corpore vel substantia aliqua spirituali. Vacuum tollitur non ex repletione substantiae spiritualis, sed corporis, sive sit substantia quanta sive sola quantitas quae subeat vicem materiae, ut in Eucharistia patet." *Ibid.*, pp. 739, col. 2(E)–740, col. 1(A).

130 This is given as the second conclusion: "Secunda conclusio: vacuum non datur quod sit spatium reale separatum per quod moveantur corpora . . ." *Ibid.*, p. 741, col. 1(C). A real space, for Amicus, is a dimensional quantity, whether material or immaterial. By assuming the hypothetical existence of a three-dimensional vacuum following the destruction of the body that formerly occupied it, Amicus unwittingly violated the Aristotelian principle of impenetrability, which he and all other scholastics accepted almost axiomatically. If a dimensional vacuum is left behind by the departure of the body that occupied it, we must assume that when the body was in that vacuum, two three-dimensional quantities were in the same place simultaneously.

131 *Perspectiva horaria, sive de horographia gnomonica tum theoretica tum practica* (Rome, 1647). In my description of the life and works of Maignan, I have followed P. J. S. Whitmore, *The Order of Minims in Seventeenth-Century France* (The Hague: Martinus Nijhoff, 1967), part III, ch. 6 ("Emanuel Maignan"), pp. 163–186.

132 Maignan's *Cursus philosophicus* was published in two editions, the first in Toulouse, 1653, the second in Lyon, 1673. Only the latter is used here. For references to Maignan's discussions of ballistics and vacuum, see Whitmore, *Order of Minims*, pp. 171–175. On p. 241 of the *Cursus philosophicus* Maignan declared his belief in the possibility of micro- and macrovacua as follows: "His ita depulsis adversariorum rationibus, non solum dico possibile esse parvum vacuum, de quo est prima pars propositionis, sed etiam intrepide assero magnum, de quo est secunda pars, non ita esse impossibile. . . ."

133 Whitmore, *Order of Minims*, p. 174.

134 In emphasizing that Aristotelians of the Renaissance "do not form a single compact school, in any but the vaguest of senses," Charles B. Schmitt takes as an example Emanuel Maignan, who "prepared a *cursus philosophicus* in which the structure, vocabulary, and subject headings are clearly in the traditional Aristotelian mould. But the *content* is often very un-Aristotelian and, in places, even anti-Aristotelian, as when Maignan discusses various experiments and rational arguments demonstrating that there must be a void in nature!" "Towards a Reassessment of Renaissance Aristotelianism," p. 160.

135 Whitmore, *Order of Minims*, p. 174.

136 For all three, see the *Dictionary of Scientific Biography*.

137 "Dum cogitamus rem aliquam esse in loco tali vel tali et multum sive parum distare ab hac aut illa existente in hoc aut illo loco, concipimus (non sane asserimus esse, sed simpliciter cogitamus quasi esset) spatium aliquod permanens ad modum expansi aeris vel aquae, aut similis alterius spatii realis, diffusum quaquaversum, etiam absque fine; item invariabile, quasi nec secundum totum, nec secundum partes moveri, contrahi, vel distrahi possit in quo tamen fiant motus omnes ab una eius parte ad aliam et in quo sint mobilia omnia, in una scilicet eius parte vel in alia: et hoc vocamus spatium imaginarium." Maignan, *Cursus philosophicus*, ch. 8, prop. 3, p. 230. See also n. 139 below.

138 *Ibid.*, ch. 8, prop. 3, p. 230.

139 "Propositio XI: Deus Opt. Max. vi suae immensitatis substantialiter est non modo in parte hac spatii imaginarii in qua mundum creavit, sed etiam in spatiis extramundanis quaquaversus infinitis." *Ibid.*, p. 242.

140 After declaring that God exists beyond real space in what he called "imaginary space," Maignan asserted that this is true "non tantum philosophica ratione, sed etiam theologica fide certum; fides enim Deum docet esse immensum. At quomodo erit immensus si sub mensuram cadere potest? Evidens autem est cadere posse si mundi huius extensionem sua

magnitudine non excedit, eodem penitus modo quo anima rationalis intra corpus, cuius (ut est informans), non excedit quantitatem." *Ibid.*, p. 244.

141 "Imo a priori supponit mundus creatus eiusque creatio Deum in eo spatio fuisse, nisi enim fuisset ibi Deus per prius non ibi potuisset mundum efficere." *Ibid.*, p. 243. Contrary to the position adopted by Duns Scotus and his followers (see above, Ch. 6, Sec. 4), Maignan believed that God must be present where He operates.

142 "Nunc demum probo primi superius facti argumenti secundam consequentiam in qua consistit prima pars propositionis. Sic argumentor: Deus potest existere in spatio extramundano, ergo et iam actu in eo existit. Antecedens patet ex dictis, tum quia spatium extramundanum et istud intramundanum sunt eiusdem rationis, tum quia potest Deus in eo alium mundum condere, sicut hunc in isto; et potuisset hunc ipsum in illo condere, ut diserte asserit D. Aug. libro 11, *De civit. Dei*, cap.5; imo et posset hunc in illud spatium transferre sive localiter movendo, sive destruendo prius hic et postmodum reproducendo ibi. Atque hoc totum est evidens; patet ergo antecedens." *Ibid.*, pp. 243–244. For Bradwardine's similar arguments, see above, Ch. 6, Sec. 3b.

143 This is the import of the following passage: "Deus iam de praesenti est secundum se, ac non solum ratione alterius in ea parte imaginarii spatii in quo mundus hic est; ergo ante mundum conditum, Deus in eadem parte existebat; ergo et nunc est pariter secundum se in spatio extra mundum." *Ibid.*, p. 242. On St. Augustine, see above, Ch. 5, Sec. 4. Where this was clearly implied by Maignan, Martin Pereyra explicitly declared it when he asserted that everything that exists must be somewhere and that God is not only in Himself but also negatively in an imaginary space. ("Respondetur 2: omne quod est debere esse alicubi et sic Deus est in seipso et est in spatio imaginario, ut explicavimus negative." *Commentariorum in primum librum Magistri Sententiarum tomus secundus*, p. 90, col. 1, par. 67; see also above, Ch. 7, Sec. 2d and notes 110 and 125 of this chapter.

144 Here is the full text of all that has been described in this passage: "Explicatur id quod usu communi vocamus 'extendi,' non aliud esse puto (quantum quidem fert vis ipsa vocabuli) quam extra se tendi, ut etiam alibi memini me annotasse, presertim cap. 2, prop. 3. Inde autem est ut extendi conveniat proprie solis quantis habentibus nimirum partes impenetratas, seu, quod idem est, extra se invicem positas. Et totum quantum non dicitur extendi nisi ratione partium quarum una est extra aliam. Quia itaque res simplex nullas hoc ipso partes habet, id enim est esse simplex, sequitur rei simplici non proprie convenire extensionem, nam idem tendi extra se ipsum dici nequit, sicut nec esse extra se ipsum; ac proinde res simplex non proprie imo nullatenus est in loco per dictam illam sibi ratione suarum partium intrinsecam extensionem, quae formalis extensio dicitur. Sed nihilominus in loco est alio quodam modo qui aequivalet eidem dictae extensioni ex quo et vocatur extensio virtualis, videlicet per suam substantiam ut indivisibiliter se tota respondentem toti extensioni loci et singulis eius partibus quod quidem prop. 8 dixi esse id quod communiter vocatur esse definitive in loco: ita rationalis anima est in toto corpore quod informat; et angelus est in toto spatio quod est aequale sphaerae praesenti ipsius; ita et Deus est in toto hoc universo." Maignan, *Cursus philosophicus*, p. 242.

145 The following passage is a continuation of the Latin text cited above in n. 144: "Et cum hoc sit in confesso apud omnes addo Deum substantialiter (nempe dicto modo definitive) esse vi suae immensitatis non modo etc, ut habetur in propositione, id est, (1) non coarctari Deum angustiis huius universi, sed sicut intra eius fines, ita et extra existere; (2) eundem non solum in aliqua determinata ac definita spatii extramundani latitudine, sed etiam absque termino in omnibus atque omni ex parte infinite patentibus spatiis existere." *Ibid.* Although he did not use any version of the term *definitivum*, Richard Fishacre, in the thirteenth century, described the same mode of existence for God (see above, Ch. 6, Sec. 3f and n. 127).

146 ". . . cum, ut ostendi, immensitas, sive quod idem est, virtualis extensio sine termino, sit attributum non solum quoad entitatem sed etiam quoad exercitium summam habens necessitatem sicut habet aeternitas, intellectus, etc. Igitur Deus spatia omnia extramundana actualiter implet sua substantia." *Ibid.*, p. 245.

147 "Quia vero spatia illa omnia quaquaversum adaequate sumpta ut quisque facile potest intelligere non tantum sunt indefinita, quasi diceres actu finita, infinita autem potentia;

sed etiam sunt infinita actu, id est nullum actu habent terminum suae extensionis, ac proinde habent extensionem (imaginariam illam) actu interminatam. Et revera si aliqui sunt in spatio imaginario termini ultra quos actu non sit aliud spatium imaginarium, ego quaererem an sint reales, an imaginarii, et sive hi sive illi dicuntur, ego ultra illos possum concipere spatium aequale ei quod citra esse intelligitur. Et si aequale, possum duplum; possum et triplum, etc. Et iterum si ibi figantur termini reales vel imaginarii, eodem modo possum spatium ultra eos imaginario extendere absque fine, non solum per alios atque alios imaginationis actus quorum posteriores aliquid addant extensionis supra priores, sed etiam per unicum actum quo simul concipiatur extensio nullum habens actu finem, illa vero erit consequenter ut sic actu infinita quia inquam spatia illa quaquaversum adequate sumpta infinita sunt et Deus in est in omnibus, cum sit ubique; sequitur, ut fuit propositum, Deum substantialiter vi suae immensitatis in infinitis extramundanis spatiis existere." *Ibid.*, pp. 245–246.

148 "Certe ex parte Dei tam est fundamentum immensitas ad existendum in spatio imaginario extra omne spatium reale, quam ex parte eiusdem Dei fundamentum est aeternitas ad existendum in tempore imaginario ante omne tempus reale. At spatium imaginarium non est minus aptum ex parte sui ad denominandum Deum existentem extra spatium reale. Ergo etc." *Ibid.*, p. 244. In his commentary on bk. 1, distinction 37 of the *Sentences*, Martin Pereyra made the same distinction, noting that the argument about the eternity of God with respect to imaginary time is the same as the immensity of God with respect to imaginary space. ("Eadem enim ratio est de aeternitate Dei respectu temporis imaginarii ac de immensitate illius respectu spatii imaginarii quia tam infinita est immensitas in ratione extensionis ad spatium, quantum aeternitas in ratione durationis ad tempus. Quare sicut Deus, quia habet durationem infinitam, existit ante omne tempus ab aeterno occupans omne tempus imaginarium, non cum habitudine reale ad illud, cum nihil sit reale, sed tanquam comprehendens sua infinita duratio ne omne illud spatium temporis excogitabile ante tempus reale, ita similiter explicato modo quia est immensus occupat illum spatium loci imaginarium. Et sicut ante et extra tempus reale erat durans in se ipsa realiter, ita etiam ante et extra omne spatium reale hujus mundi existit realiter in se ipso." *Commentariorum in primum librum Magistri Sententiarum tomus secundus*, p. 90, par. 66.) See also above, Ch. 7, Sec. 2b and n. 70. For a rejection of the space-time analogy, see below, Ch. 7, Sec. 2f and n. 164.

149 ". . . quod enim tu vocas existere vere extra spatium reale, hoc ipsum ego intelligo quando dico existere vere in spatio imaginario extramundano et solum utor spatio imaginario ut rem captui meo melius accommodem, sicut cum utor tempore imaginario antecedente omne tempus reale." Maignan, *Cursus philosophicus*, p. 244.

150 Following immediately upon the passage cited in n. 149, Maignan declared: "Neque velim putes a me cum dico Deum existere vere in spatio imaginario, hanc particulam *vere* referri ad spatium imaginarium, quasi existimem illud esse vere et realiter aliquid velut continens Deum; illam siquidem refero ad Deum qui vere existat extra spatium reale, extra quod ego concipio aliud spatium quod voco imaginarium." *Ibid.*

151 See above, Ch. 7, Secs. 2b, c, and d. Like most scholastics, Maignan characterized vacuum as the negation or privation of a body in a place. ("At vero vacuum in ratione vacui est privatio vel negatio locati in loco, sicut coecitas est privatio vel negatio facultatis visivae in subiecto, tenebrae privatio vel negatio lucis, etc." *Ibid.*, ch. 20, p. 490.) But unlike Amicus, who adopted the same definition of vacuum (see above, Ch. 7, Sec. 2d), Maignan did not interpret imaginary space as a pure negation.

152 Cf. Bradwardine, Ch. 6, Sec. 3c above; for a *possibly* more forceful interpretation, see the discussion of Jean de Ripa, Ch. 6, Sec. 2b; see also the descriptions of imaginary space in this chapter.

153 Bona Spes, *Commentary on Aristotle*, tractatus III, Physics, part 1, pp. 177–178 ("De spatiis imaginariis"). Franciscus Bona Spes, whose secular name was Crespin, was born in Belgium. In 1634, at the age of seventeen, he entered the Carmelite Order, taking solemn vows in 1635. For many years he taught philosophy and theology at Louvain and was the author of numerous treatises on those subjects. A brief biography with accompanying bibliography appears in Cosmas de Villiers, *Bibliotheca Carmelitana* (Aurelianis: M. Couret de Villeneuve, 1752), vol. 1, cols. 482–485. I am indebted to Dr.

Charles H. Lohr (Raimundus-Lullus Institut der Universität Freiburg), who not only brought de Villiers's biography to my attention but also supplied a photocopy of the article on Bona Spes.

154 "Constat inprimis spatia illa, quia imaginaria, non esse realia positiva aut negationes ubicationis realis non tollibiles, . . . quia, per impossibile, omni cessante et impossibili facta imaginatione, negationes ubicationis realis non tollibiles suo modo essent; et tamen eo casu spatia imaginaria cessarent et impossibilia essent ac proinde spatia imaginaria esse non possunt." Bona Spes, *Commentary on Aristotle*, p. 177, col. 1.

155 "Denique . . . spatia imaginaria debere esse actu, quod intelligi non potest si sint negationes; quia negationes actu esse, idem est ac *non esse actu*, esse, *esse actu*, quia negationes sunt non esse actu, ut consideranti patebit, dicta supra Disp. quinta 'de privatione'; ista autem sunt contradictoria." *Ibid*. The italics are provided by Bona Spes. Here, and in the preceding note, the arguments are drawn from Comptonus (probably Thomas Compton-Carleton, S. J. [ca. 1591–1666]) but accepted by Bona Spes.

156 Under the heading of "Resolutio," Bona Spes declared: "Dico spatia imaginaria non nisi imaginarie. Imaginaria dicuntur et attentis proprietatibus, quae illis a theologis et philosophis communiter tribuuntur, ut, quod sint spatium infinite extra coelos diffusum, aeternum, capax (puta per non repugnantiam) corporis, etc., omnino cum spatiis realibus possibilibus confundi debent." *Ibid*.

157 "Dixi non nisi imaginarie quia revera non nisi quia haec spatia possibilia imaginamur nobis ad instar nebulae obscurae aut aeris in infinitum diffusi imaginaria dicuntur. Unde S. Augustinus *De civit*., lib. primo, cap. quinto, insinuat spatia imaginaria 'inanes esse hominum cogitationes quibus infinita imaginantur loca,' . . ." *Ibid*. On the passage from St. Augustine, see above, Ch. 5, Sec. 4 and n. 53; also see Ch. 7, n. 17.

158 In treating the second question, or *dubium*, Bona Spes proceeded directly to the "Resolutio," where he declared: "Dico Deus non est actu in spatiis imaginariis (ita Bonaventura, Scotus, Vasquez, Capreolus, Oviedus, D. Fromondus [*De anima*, lib. 3, c. 10] et plures alii, apud Comptonum hic contra Suarez, Majorem, Sotum, Hurtadum, Arriagam, Fonsecam, Comptonum et plures alios, quos citat." *Ibid*. The list of authors was apparently drawn from a work by Comptonus. The brackets are mine, although the reference is provided by Bona Spes.

159 "Objicies primo: Deus est extra mundum realiter immensus et juxta Trimegistum *sphaera intelligibilis, cujus centrum ubique, circumferentia nusquam*." *Ibid*., p. 177, col. 2. The italics are supplied by Bona Spes. On the popular pseudo-Hermetic dictum that Bona Spes quoted, see above, Ch. 6, Sec. 3c.

160 Following immediately upon the quotation in n. 159, we read: "alioqui producto novo mundo ultra coelos non esset illi praesens; vel si esset illi praesens, esset etiam in medio, vel non? si primum, ergo esset in spatiis imaginariis, uti ista sunt; si secundum, ergo esset a seipso divisus uti illi mundi, quod est absurdum, ergo, etc." *Ibid*.

161 Although Bona Spes did not specifically address both arguments, his response was applicable to each. Indeed he introduced his rebuttal by taking up the claim that God would be divided if He were not in the intervening distances between worlds: "Respondeo negando ultimam consequentiam et dico quod Deus, si produceretur novus mundus, foret illi praesens et in medio inter utrumque; non quidem tamquam in spatio imaginario, sed tamquam spatium immensum reale a se et sua immensitate Divina realiter indistinctum." *Ibid*.

162 "Hinc optime August. 7, *Confess*., cap. quinto . . . : mundus in Deo est velut spongia parva in oceano immenso, qui ultra spongiam late fuseque secundum propriam substantiam distenditur." *Ibid*. For the passage, see above, Ch. 6, Sec. 3c. Two popular scriptural passages were also cited to show God's immense extent, namely Job 11 (probably verses 8 and 9 of the King James version) and Kings 3, 8 (in the King James version 1, 8). For the text of the latter, see above, Ch. 7, Sec. 1, and n. 6, 67, and 68 of this chapter for further typical references to God's immensity and extent.

163 "Objicies secundo: ante mundum conditum Deus erat actu in temporibus imaginariis, ergo et in spatiis imaginariis et consequenter etiam jam est in spatiis imaginariis ultra mundum." *Ibid*.

164 "Respondeo negando antecedens quia ista actu tempora imaginaria haud minus sunt

chymaerica quam spatia imaginaria. Unde sicut Deus ultra coelos jam est spatium reale suae immensitatis, ita ante mundum conditum fuit tempus reale suae aeternitatis et in ea sola existens." *Ibid.* For typical acceptances of this analogy, see above, Ch. 7, Sec. 2e, and n. 148 of this chapter.

165 Although those who believed that God was omnipresent in an infinite space prior to the creation of the world and is in such a space now were accused by Gabriel Vasquez of treating God "as if He were a quantity and extended," this charge must be dismissed as merely polemical and without proper substance. Despite occasional use of quantitative language to describe the relations between God and infinite space, we have seen how careful Vasquez's opponents were to avoid the attribution of quantity or extension to God. (Here is the text of Vasquez's serious accusation: "Reijcienda ergo est *illorum Doctorum* sententia, qui Deum ante mundi creationem et nunc extra caelum ita cogitant, ut sua praesentia repleat spatium aliquod infinitum, ac si ipse quantus et extensus esset." *Disputationes Metaphysicae*, p. 377.)

CHAPTER 8
Infinite space in nonscholastic thought during the sixteenth and seventeenth centuries

1 By Otto von Guericke, about whom more will be said below. Occasionally an indirect compliment might even be paid, as when John Locke was said to have admitted that in his consideration of "simple ideas," he had "not treated his Subject in an Order perfectly Scholastick, having not had much Familiarity with those sorts of Books during the Writing of his, and not remembering at all the Method in which they were written; and therefore his Readers ought not to expect Definitions regularly placed at the Beginning of each new Subject." This statement appears in a footnote in the fifth edition (1706) of Locke's *Essay Concerning Human Understanding* and was apparently placed there by the editors (see p. 201, n. in Peter Nidditch's edition of Locke's *Essay*, which is cited in full below, n. 318.

2 When Henry More assigned space as an attribute of God, he supported his claim by appeal to the Cabbalists who numbered "place" (*locus*) among the numerous attributes of God (*Enchiridion metaphysicum* [London, 1671], part 1, ch. 8, p. 74; as his source for the claim, More cited Cornelius Agrippa's *De occulta philosophia*, which, according to Flora Isabel MacKinnon, *The Philosophical Writings of Henry More* [New York: AMS Press, 1969; first published 1925], p. 293, is a reference to bk, 3, ch. 2). On the rabbinic tradition identifying God with place, see above, Ch. 5, Sec. 4, and Ch. 6, n. 20; and for the practice of citing early authors, see above, Ch. 5, Sec. 4.

3 Ironically, many early modern scholastics also tended to ignore the medieval scholastic tradition and emphasize not only the Church Fathers but also many of the same Greek authorities, quite a number of which became available only during the fifteenth and sixteenth centuries. And like their nonscholastic contemporaries, they tended to cite their own early modern scholastic predecessors and contemporaries, only occasionally mentioning arguments from the nonscholastic domain.

4 In his otherwise splendid book *From the Closed World*, Alexandre Koyré operated on this implicit assumption and simply ignored the entire scholastic tradition wherein the concept of a finite world surrounded by an infinite space was discussed. That this was a deliberate and conscious policy may be determined from the fact that Koyré was himself the author of a momentous study on medieval concepts of void space in which Thomas Bradwardine's views were central (see above, Ch. 6, Sec. 3 and n. 90 for the title of Koyré's article). In his book, however, Koyré chose to omit Bradwardine and begin with Nicholas of Cusa, who appears to have had no explicit position on the doctrine of space (see above, Ch. 6, Sec. 3c).

5 For a variety of Greek Stoic and Neoplatonic interpretations of void space, including those by Philoponus, Simplicius, Iamblichus, Proclus, Damascius, and Cleomedes, see Duhem, *Le Système du monde*, vol. 1, part 1, ch. 5, pp. 297–350 and Sambursky, "Place and Space in Late Neoplatonism," 173–181. Virtually all of this material became available in the sixteenth century.

6 See E. N. Tigerstedt, *The Decline and Fall of the Neoplatonic Interpretation of Plato*.

An Outline and Some Observations. Commentationes Humanarum Litterarum, vol. 52 (Helsinki: Societas Scientiarum Fennica, 1974), p. 19.

7 Although Epicurus is frequently mentioned, none of his works are extant. During the period discussed here, his atomistic and spatial concepts were known largely through the *De rerum natura* of Lucretius, certain philosophical works of Cicero, the *Lives and Opinions of the Philosophers* (bk. 10) of Diogenes Laertius, and *Against the Physicists* by Sextus Empiricus (for the full titles of the works by Diogenes and Sextus, see below, n. 21). Articles by David Furley on Lucretius (*Dictionary of Scientific Biography*, vol. 8 [New York: Scribner, 1973], pp. 536–539) and Epicurus (*loc. cit.*, vol. 4 [New York: Scribner, 1971], pp. 381–382) provide useful brief summaries and bibliographies.

8 The Stoic conception of the universe was conveyed to the sixteenth century not only by Simplicius (see above, Ch. 5, Sec. 2) but also by Cleomedes (above, Ch. 5, Sec. 1), Philoponus (*Commentary on Aristotle's Physics*; see p. 244, col. 1 of Rasario's Latin translation cited above, Ch. 2, n. 61), and Plutarch (*Moralia V: De defectu oraculorum*, 425D–E; *Moralia XI: De placitis philosophorum*, 874D; and *Moralia XIII: De Stoicorum repugnantis*, 1054C–D).

9 See above, Ch. 2, Sec. 3a. The idea that body can coexist with space could also have been derived from a brief passage in Sextus Empiricus' *Against the Physicists*, where, in reporting on Hesiod's Chaos, Sextus explained that it was really the container of all things, without which the universe could not have been constructed (for the passage, see above, ch. 2, n. 67). In his *Commentary on the Physics*, Simplicius reported a similar view by Proclus (see Duhem, *Le Système du monde*, vol. 1, pp. 339–340).

10 In *De l'infinito universo et mondi*, Bruno said that the infinite universe is as a mere point with respect to God's infinity (Giordano Bruno, *Opere Italiane*, ed. Giovanni Gentile, 2d ed. [2 vols.; Bari, Gius. Laterza & Figli, 1925], vol. 1: *Dialoghi Metafisici*, Dialogo Primo, p. 294: for the English translation, see p. 257 of Singer's translation, cited below in n. 13) and thus seems to avoid an equation of God and universe. But elsewhere Bruno also tended to erase any distinction between them, as when he spoke of the infinite universe as an image of an infinite God (see Bruno, *De l'infinito universo et mondi, Opere Italiane*, vol. 1, pp. 294–295, and Singer (trans.), p. 257; also Michel, *The Cosmology of Giordano Bruno*, p. 88, and Paul Oskar Kristeller, *Eight Philosophers of the Italian Renaissance* [Stanford, Calif.: Stanford University Press, 1964], pp. 134, 136).

11 For the discussion of these ideas and references to Bruno's works, especially the *De la causa, principio et uno*, see Paul Henri Michel, *The Cosmology of Giordano Bruno*, trans. by Dr. R. E. W. Maddison (Paris: Hermann; London: Methuen; Ithaca, N.Y.: Cornell University Press, 1973; original French text published 1962), pp. 87–88, 91.

12 Michel, *Cosmology of Giordano Bruno*, p. 159, and Koyré, *From the Closed World*, pp. 42, 52.

13 From the First Dialogue of Bruno's *De l'infinito universo et mondi* of 1584 in the translation by Dorothea Waley Singer, *Giordano Bruno, His Life and Thought*, with annotated translation of His Work On the Infinite Universe and Worlds (New York: Henry Schuman, 1950), p. 260; Bruno, *Opere*, vol. 1, p. 297. The full passage is quoted and discussed by Koyré, *From the Closed World*, pp. 52–54.

14 Singer (trans.), *On the Infinite Universe and Worlds*, p. 261; *Opere*, vol. 1, p. 298.

15 Michel, *Cosmology of Giordano Bruno*, p. 48.

16 Singer (trans.), *On the Infinite Universe and Worlds*, p. 251; *Opere*, vol. 1, p. 289.

17 Singer (trans.), *On the Infinite Universe and Worlds*, p. 252; *Opere*, vol. 1, p. 290.

18 *Ibid.*; *ibid.* Burchio, the Aristotelian in the dialogue, was prepared to argue that "if a person would stretch out his hand beyond the convex sphere of heaven, the hand would occupy no position in space, nor any place, and in consequence would not exist" (Singer [trans.], *On the Infinite Universe and Worlds*, p. 253; *Opere*, vol. 1, p. 290). Here we have Bruno's conception of the Aristotelian response to the Stoic argument described above (Ch. 5, Sec. 1; Bruno also described the Stoic argument on p. 231 [Singer trans.; *Opere*, vol. 1, p. 271] and in his *De immenso et innumerabilibus* [*Jordani Bruni Nolani Opera Latine Conscripta*, ed. F. Fiorentino (Naples, apud Dom.

Morano, 1879, 1884), bk. 1, ch. 7 in vol. 1, part 1, pp. 227–228]; although he attributed the argument in both places to Epicurus he probably derived it from Lucretius [see above, Ch. 5, n. 10]), a conception that does not fit those scholastic Aristotelians who accepted the existence of infinite extracosmic space. Indeed Buridan who did not accept such a space, judged that a hand extended beyond the world would be in its own space (see above, Ch. 2, Sec. 2a, and Ch. 6, Sec. 2).

19 Singer (trans.), *On the Infinite Universe and Worlds*, p. 252; *Opere*, vol. 1, p. 290. Bruno's specific rejection of Aristotle's definition of place as a containing surface appears in Singer (trans.), *On the Infinite Universe and Worlds*, p. 253 and in *Opere* vol. 1, pp. 290–291.

20 Singer (trans.), *On the Infinite Universe and Worlds*, p. 252; *Opere*, vol. 1, p. 290. As we saw earlier, scholastics who accepted the existence of extracosmic space found no difficulty accepting the views Bruno rejected. For a further discussion of Bruno's conception of God's relation to space, see below, Ch. 8, Sec. 2a.

21 Singer (trans.), *On the Infinite Universe and Worlds*, p. 272; *Opere*, vol. 1, p. 309. Bruno's attribution to the Stoics of the distinction between world and universe may have been derived from either Sextus Empiricus or Diogenes Laertius, whose works were available in the sixteenth century. From Sextus he could have learned that "the philosophers of the Stoic school suppose that 'the Whole' differs from 'the All'; for they say that the Whole is the Cosmos, whereas the All is the external void together with the Cosmos, and on this account the Whole is limited (for the Cosmos is limited) but the All unlimited (for the void outside the Cosmos is so)" and from Diogenes that "By the totality of things, the All, is meant, according to Appollodorus, (1) the world, and in another sense (2) the system composed of the world and the void outside it. The world is finite, the void infinite." See *Sextus Empiricus* with an English translation by the Rev. R. G. Bury (4 vols.; Loeb Classical Library; London: William Heinemann; Cambridge, Mass.: Harvard University Press, 1933–1949), vol. 3: *Against the Physicists*, bk. 1.332, p. 161, and Diogenes Laertius, *Lives of Eminent Philosophers* with an English translation by R. D. Hicks (2 vols.; Loeb Classical Library; Cambridge, Mass.: Harvard University Press; London: William Heinemann, 1925; revised 1938, 1942, 1950), vol. 2, bk. 7.143 (Zeno), p. 247. In his *Physiologiae Stoicorum*, Justus Lipsius (1547–1606), regarded as the founder of Neo-Stoicism, cited the distinction from Diogenes (see Jason Lewis Saunders, *Justus Lipsius, the Philosophy of Renaissance Stoicism* [New York: Liberal Arts Press, 1955], p. 185). The highly abbreviated medieval version of Diogenes' *Lives* by Walter Burley in the fourteenth century lacks the passage quoted above from the section on Zeno the Stoic (see Hermann Knust [ed.], *Gualteri Burlaei Liber De Vita et moribus philosophorum . . . Bibliothek des Literarischen Vereins in Stuttgart*, vol. 177 [Tübingen, 1886], p. 304.

22 Singer (trans.), *On the Infinite Universe and Worlds*, p. 254; *Opere*, vol. 1, p. 291. See also Ch. 6, n. 115 above. We shall see below how Patrizi justified this conception.

23 Singer (trans.), *On the Infinite Universe and Worlds*, p. 257; *Opere*, vol. 1, pp. 294–295. See also Koyré, *From the Closed World*, pp. 42, 44, 52.

24 See Singer (trans.), *On the Infinite Universe and Worlds*, p. 245 (*Opere*, vol. 1, p. 282), where he praised Democritus and Epicurus. Although it is likely that Bruno was inspired by Lucretius' *De rerum natura*, which he quoted directly, he may also have read Diogenes Laertius' life of Epicurus in the former's *Lives of Eminent Philosophers* (see Singer [trans.], *On the Infinite Universe and Worlds*, p. 273, n. 8, and Diogenes Laertius, *Lives of Eminent Philosophers*, vol. 2, bk. 10. In Walter Burley's medieval version of the *Lives*, the section on Epicurus includes nothing on the atomic theory or infinite space [see pp. 272–276 of Knust's edition cited in n. 21 of this chapter]).

25 Singer (trans.), *On the Infinite Universe and Worlds*, pp. 273, 370–371; *Opere*, vol. 1, pp. 310, 410–411. For a recent discussion of Bruno's conception of a plurality of worlds, see Steven J. Dick, "Plurality of Worlds and Natural Philosophy: An Historical Study of the Origins of Belief in Other Worlds and Extraterrestrial Life" (Ph.D. diss., Indiana University, 1977), pp. 122–130, and his article "The Origins of the Extraterrestrial Life Debate and Its Relation to the Scientific Revolution," *Journal of the History of Ideas 41* (1980), 4–6, 26.

26 Despite their "diversity," Bruno's worlds are all composed of the four Aristotelian elements (Dick, "Plurality of Worlds," p. 126).

27 On the fundamental role of Bruno's concept of unity, see Dick, "Plurality of Worlds," pp. 124–130, and Singer (trans.), *On the Infinite Universe and Worlds*, p. 229 (*Opere*, vol. I, p. 270), where Bruno exclaimed that "It is Unity that doth enchant me."

28 As, for example, in the discussions of Bruno by Koyré, *From the Closed World*, pp. 39–54; Michel, *Cosmology of Giordano Bruno*; and Jammer, *Concepts of Space*, pp. 87–90.

29 All references to the *De immenso* are to Fiorentino's edition in the Latin *Opera*, vol. I, part I, cited above in n. 18 of this chapter. See also Michel, *Cosmology of Giordano Bruno*, pp. 49–50.

30 *De immenso*, *Opera*, vol. I, part I, pp. 231–233.

31 "Est ergo spacium quantitas quaedam continua physica triplici dimensione constans in qua corporum magnitudo capiatur" (*ibid.*, p. 231). After citing this sentence up to the word "constans" (and omitting "continua"), Michel (*Cosmology of Giordano Bruno*, p. 176, n. 48), in his only reference to the section under discussion here, remarked that "the continuation of the passage stresses the abstract and 'indifferent' character of space which is nothing more than Euclidean space, though Bruno regards it as a physical reality." But Bruno made no mention of Euclid or geometry and did not have a geometrical space in mind. He was in no sense making a mathematical space physical (see above, Ch. 2, Sec. 2a), for, as Michel himself explained (p. 133), Bruno was more concerned "with the realities of nature than with the imaginings of geometers." On Philoponus' concept of space, see above, Ch. 2, Sec. 2a.

32 With but a few exceptions, the Latin texts will not be cited.

33 In Bruno's natural philosophy there is no hierarchy of being. Matter and form are equal and emanate directly from God without intermediaries. See Michel, *Cosmology of Giordano Bruno*, pp. 86–87.

34 "Quod septimo immiscibile, clarum est, quia corporum invicem cedentium secundum partes mixtio est: spacii vero non est cedere, sed sibi invicem cedentia suscipere." *De immenso*, *Opera*, vol. I, part I, p. 232.

35 "Eadem octavo ratione impenetrabile: id enim penetratur cuius partes a partibus distantiores fiunt, vel discontinuae sunt aut discontinuabiles, tales, per harum conditionem secundam, spacii nequeunt esse parteis." *Ibid.*

36 Bruno denied continuity and infinite divisibility to matter and was thus compelled to assume its discontinuity. Prime matter, which in Bruno's scheme represented absolute possibility or potentiality (see Michel, *Cosmology of Giordano Bruno*, p. 128), was composed of discrete atoms (*ibid.*, p. 134). Perhaps Bruno thought that bodies move through prime matter, which in some sense "occupies" space. But how prime matter might be conceived to occupy an impenetrable space, he did not explain.

37 Bruno thus agreed with Aristotle that a place does not require a place. According to Aristotle (*Physics* 4.1.209a.24–26), it was Zeno who raised the problem of a potential infinite regression of places. For Aristotle's reply, see *Physics* 4.3.210b.23–26.

38 "Quintodecimo, neque substantia neque accidens, quia non est ex quo res sunt, neque quod est in rebus, sed potius in quo res localiter sunt, natura, (quidquid sit de duratione), ante locata, cum locatis, post locata." *De immenso*, *Opera*, vol. I, part I, pp. 232–233. In his earlier *Camoeracensis Acrotismus seu Rationes Articulorum Physicorum adversus Peripateticos Parisiis Propositorum, etc.* (Wittenberg, 1588) (*Opera*, vol. I, p. 141), Bruno had also declared that vacuum was neither substance nor accident.

39 In the introductory epistle to the *De l'infinito universo et mondi*, where Bruno described the argument of the First Dialogue, he declared, as part of his seventh point, that "Beyond our world then, one space is as another; therefore the quality (*attitudine*) of one is also that of the other." Singer (trans.), *On the Infinite Universe and Worlds*, p. 232; *Opere*, vol. I, p. 272; a similar statement appears on p. 255 of the Singer translation and on p. 293 of *Opere*, vol. I. See also Michel, *Cosmology of Giordano Bruno*, p. 240. In defense of homogeneity, Bruno attacked Marcellus Palingenius (Pietro Manzoli, 1502–1543), who in his *Zodiacus vitae* assumed an infinite universe but subdivided it into three different parts: subcelestial, the region of change and

decay; celestial, the region of the unchangeable ether wherein the planets move; and, finally, the region beyond the *primum mobile*, which is infinite and filled with the purest light, apparently identified with God. On Palingenius, see Michel, *Cosmology of Giordano Bruno*, pp. 240–241 and Francis R. Johnson, *Astronomical Thought in Renaissance England* (New York: Octagon Books, 1968; first published by Johns Hopkins Press 1937), pp. 145–149. The *Zodiacus vitae* was first published around 1531 in Venice and was reprinted many times thereafter. An English translation by Barnabe Googe, which first appears complete in 1565 (the first three books were apparently printed in 1560), has been reprinted from the 1576 edition under the title *The Zodiake of Life by Marcellus Palingenius*, trans. by Barnabe Googe with an introduction by Rosemond Tuve (New York: Scholars' Facsimiles & Reprints, 1947). For the threefold division of the universe, see bk. 12 (Pisces), p. 229.

40 As we shall see below, Telesio seemed to foreshadow the idea, with Patrizi making it explicit no later than 1587, some four years before Bruno.

41 See above, Ch. 3, Sec. 4. It is ironic that Bruno's spatial theory may be interpreted as the existential equivalent of Aristotle's hypothetical argument that if a separate three-dimensional space existed, it would be like a body and therefore impenetrable with respect to material bodies (see above, Ch. 1). This was sufficiently absurd to prompt Aristotle to reject the existence of such a space. With Bruno, however, Aristotle's absurdity became reality. Despite his probable knowledge of John Philoponus' anti-Aristotelian argument (Bruno mentioned Philoponus; see above, Ch. 8, Sec. 2), Bruno did not employ Philoponus' conception of a penetrable incorporeal spatial magnitude that was capable of receiving bodies (see above Ch. 2, Sec. 2a).

42 Joseph Raphson, *De spatio reali seu ente infinito* (London, 1702), p. 76, where Raphson said, "It is patent that space is not penetrated by anything: being infinite and undivided it penetrates everything by its innermost essence, and therefore cannot itself be penetrated by anything, nor even can it be conceived as penetrated" (trans. by Koyré, *From the Closed World*, p. 195). Raphson's *De spatio reali* is appended to the second edition of his *Analysis aequationum universalis* (for full title, see below, n. 277; Raphson is discussed in Ch. 8, Sec. 4f).

43 "Spatium est omni-continens et omni-penetrans" (*De spatio reali*, p. 75), which Koyré (*From the Closed World*, p. 195) has translated as "*Space* is all-containing and all-penetrating."

44 Singer (trans.), *On the Infinite Universe and Worlds*, p. 273; *Opere*, vol. 1, pp. 309–310. To reinforce this point, Bruno explained, some lines below, that "the ancients like ourselves regarded the Void as that in which a body may have its being, that which hath containing power and doth contain atoms and bodies" (Singer trans., pp. 273–274; *Opere*, vol. 1, p. 310).

45 Singer (trans.), *On the Infinite Universe and Worlds*, p. 255; *Opere*, vol. 1, pp. 292–293. Greenberg observes that "For Bruno total absence of matter is a contradiction in terms" (*The Infinite in Giordano Bruno*, p. 49), whereas Koyré explains that "Bruno's space is a void; but this void is nowhere really void; it is everywhere full of being. A vacuum with nothing filling it would mean a limitation of God's creative action and, moreover, a sin against the principle of sufficient reason which forbids God to treat any part of space in a manner different from any other" (*From the Closed World*, p. 283, n. 25).

46 Singer (trans.), *On the Infinite Universe and Worlds*, p. 372 (*Opere*, vol. 1, p. 412), where Bruno said that "the ether is of his own nature without determined quality, but it receiveth all the qualities offered by neighbouring bodies. . . ."

47 Singer (trans.), *On the Infinite Universe and Worlds*, p. 363; *Opere*, vol. 1, p. 404.

48 Singer (trans.), *On the Infinite Universe and Worlds*, p. 236; *Opere*, vol. 1, p. 275. In *De immenso*, bk. 1, ch. 10, Bruno said that all of space is filled ("Dicendum nihilum spacii non esse repletum"; *Opera*, vol. 1, part 1, p. 236).

49 Singer (trans.), *On the Infinite Universe and Worlds*, p. 373; *Opere*, vol. 1, p. 413. On Bruno's attitude toward void and ether and which is to be understood as space, see Michel, *Cosmology of Giordano Bruno*, pp. 139, 239–240.

50 See above, Ch. 2, Sec. 3. In his *Camoeracensis Acrotismus*, Bruno discussed the vacuum as space but made no mention of ether (perhaps because he was specifically rebutting

Aristotle's arguments against the vacuum). In this treatise, Bruno insisted that the dimensions of a material cube are distinct from the dimensions of the space or vacuum in which the cube may be placed (*Opera*, vol. 1, part 1, p. 134; Bruno was repudiating Aristotle's argument as described above in Ch. 1). Thus void space is separate from bodies and functions as "the necessary receptacle of all things" ("et necessarium omnium receptaculum," *ibid.*, p. 137). It follows that "the vacuum is not absolutely nothing" ("vacuum enim non est absolute nihil," *ibid.*, p. 136). Indeed it is also an infinite continuum (*ibid.*, pp. 140-141).

51 In the *Camoeracensis Acrotismus* (p. 142), Bruno explained that by means of the intellect we understand that vacuum is physically real and distinct from all other things. Despite its independence and separate nature, however, void space does not exist apart from bodies, which always occupy it. From the earlier *De l'infinito universo et mondi*, we learn that it is the material ether that fills it.

52 See Michel, *Cosmology of Giordano Bruno*, p. 139, for a statement drawn from Bruno's *De minimo* (*Opera*, vol. 1, part 3, p. 140).

53 For references to Bruno's works, see Michel, *Cosmology of Giordano Bruno*, p. 239, and for the notes, p. 248. Bruno would by no means be the sole "vacuist" to rely on an imponderable ether (or *anima mundi* or *spiritus mundi*). For a discussion, see C. de Waard, *L'Expérience barométrique*, pp. 28-33, where Telesio, Patrizi, and Bruno are considered.

54 See Michel, *Cosmology of Giordano Bruno*, pp. 87-88, and Kristeller, *Eight Philosophers*, p. 134.

55 See *De la causa, principio e uno* in Gentile, *Bruno Opere Italiane*, vol. 1, p. 176, and the translation by Sidney Greenberg, *The Infinite in Giordano Bruno with a Translation of His Dialogue "Concerning the Cause Principle, and One"* (New York: King's Crown Press, 1950), p. 109. For a summary, see Kristeller, *Eight Philosophers*, p. 132.

56 *De l'infinito universo et mondi*, *Opere*, vol. 1, p. 298 as translated in Michel, *Cosmology of Giordano Bruno*, p. 162, which captures the sense of this important passage better than Singer does on p. 261 of her translation.

57 See above, Ch. 6, n. 127. By employing the medieval explanation, Bruno also avoided total pantheism because God's infinity as described here is greater than the infinity of the universe (see Michel, *Cosmology of Giordano Bruno*, p. 162, and Singer (trans.), *On the Infinite Universe and Worlds*, p. 261).

58 Steven Dick has shown that few in the seventeenth century adopted the Brunonian "plurality of worlds" cosmology. Newton refused to commit himself to a belief in other world systems (Dick, *Plurality of Worlds*, pp. 254-256). Not until Richard Bentley declared a positive belief in a plurality of worlds in 1693 did the implications of the Newtonian system lead cosmologists, astronomers, and proponents of natural religion to postulate a Bruno-like universe, which became popular in the eighteenth century with authors like William Whiston, William Derham, and John Keill (see *ibid.*, pp. 2 8-272). On Bruno's direct influence, see Singer (trans.), *On the Infinite Universe and Worlds*, ch. 8, pp. 181-201, and Michel, *Cosmology of Giordano Bruno*, pp. 287-302.

59 For the passage, see above, Ch. 2, n. 63. On pp. 1186-1194 of the *Examen vanitatis*, Pico described the opinions of Philoponus from the fourth book of the latter's *Physics* commentary. Indeed Pico also devoted a special chapter to the anti-Aristotelian arguments of Hasdai Crescas ("Quid pro vacuo adversus Aristotelem attulerit Rabi Hasdai," pp. 1194-1195), where he also mentioned, but did not endorse, Crescas' conception of an infinite vacuum beyond the world. Wolfson (*Crescas' Critique of Aristotle*, p. 34) suggests that Pico, who lacked sufficient knowledge of Hebrew to have read Crescas' *Or Adonai* directly, may have learned about Crescas' doctrines from a Jewish intermediary. Because of a number of striking similarities between Bruno and Crescas, Wolfson also conjectures (pp. 35-36) that the former may have been influenced by the latter in his anti-Aristotelian arguments on vacuum and a plurality of worlds. As with Pico, the plausibility of such an influence depends on the postulation of a Jewish intermediary. That Spinoza, however, read Crescas' views on the infinite directly from the Hebrew is indubitable (pp. 36-37).

60 I have used the edition of Vincenzo Spampanato, *Bernardini Telesii De rerum natura*

(3 vols.; Modena/Genoa/Rome: A. F. Formiggini, 1910–1923). Despite the title, which suggests that Telesio was an atomist following Lucretius' work of the same title, such an inference would be mistaken, because Telesio was neither an atomist nor a corpuscularian (see Neal W. Gilbert, "Telesio, Bernardino," *Dictionary of Scientific Biography*, vol. 13 [New York: Scribner, 1976], pp. 277–280, especially p. 278).

61 See *De rerum natura*, bk. 1, ch. 28, p. 101, lines 2–15. A brief but useful description of Telesio's concept of space appears in Jammer, *Concepts of Space*, pp. 85–86.

62 *De rerum natura*, p. 100, lines 10–12. We observe that although Telesio assumed that space penetrates body, he did not attribute to space the property of impenetrability, as did Raphson and Bruno (see above, Ch. 8, Sec. 2).

63 Jammer (*Concepts of Space*, pp. 85–86) neglects to mention this important feature of Telesio's theory of space. In his important article "Experimental Evidence for and Against a Void: The Sixteenth-Century Arguments," *Isis* 58 (1967), 352–366, Charles Schmitt describes Telesio's arguments in favor of the existence of artificial vacua produced by various means – for example, by use of a bellows or clepsydra, or by freezing water in a closed vessel (see above, Ch. 4, Sec. 5). But Telesio did not believe in the natural occurrence of vacua.

64 For the text in support of this paragraph, see Telesio, *De rerum natura*, bk. 1, ch. 28, p. 101, lines 14–26.

65 *Ibid.*, bk. 1, ch. 25, pp. 90, lines 11–12 and 97, lines 1–3. Whether space was created before the world or simultaneously with it is left open (*ibid.*, bk. 1, ch. 28, p. 101, lines 6–8).

66 *Ibid.*, p. 90, lines 6–7.

67 *Ibid.*, p. 90, lines 4–6.

68 As early as 1588, when Campanella read the first two books of Telesio's *De rerum natura*, he became, and remained, an admirer of Telesio. See Bernardino M. Bonansea, *Tommaso Campanella, Renaissance Pioneer of Modern Thought* (Washington, D.C.: Catholic University of America Press, 1969), p. 25, and p. 68 of Schmitt's article cited in the next note.

69 Koyré (*From the Closed World*) omits him entirely, whereas Jammer allots him a brief paragraph (*Concepts of Space*, p. 90) but is twice mistaken when he includes Campanella among those who assumed the existence of an infinite space (*ibid.*, pp. 93, 110) and as one who "develops Patritius' theory of space still further" (*ibid.*, p. 90). There is little evidence of any direct influence by Patrizi. In the parts of Campanella's *Universalis philosophiae seu metaphysicarum rerum iuxta propria dogmata, partes tres* (Paris, 1638), which have been edited with an Italian translation by Giovanni di Napoli under the title *Metafisica* (3 vols.; Bologna: Zanichelli, 1967), the name of Telesio appears fairly frequently and Patrizi's not at all (see the index of names in vol. 3). Although Campanella may have known of Patrizi and perhaps mentioned him elsewhere in the metaphysics or in other treatises, his influence on Campanella's spatial doctrines was nonexistent or minimal. Indeed we shall see that they disagreed on fundamental points. Much has been written on Campanella, but a good thumbnail sketch with useful bibliography appears in Charles B. Schmitt, "Campanella, Tommaso," *Dictionary of Scientific Biography*, vol. 15, supplement 1 (New York: Scribner, 1978), pp. 68–70.

70 The *Universalis philosophiae*, or *Metafisica*, went through at least five known drafts in both Latin and Italian (see Bonansea, *Tommaso Campanella, Renaissance Pioneer*, p. 30). The 1638 edition will be cited when it contains relevant sections that have been omitted from di Napoli's partial three-volume edition cited in the preceding note.

71 Unless otherwise indicated, what follows on the five worlds appears in the *Universalis philosophiae* (Paris, 1638), bk. 10, ch. 1, article 11, pp. 248–249 (because the eighteen books of the *Universalis philosophiae* are numbered in sequence, it is superfluous to cite also the part number). Campanella arranged the five worlds in a diagram with nine concentric circles. Proceeding from the outermost to the innermost circle, we find (1) the *mundus archetypus*; (2) *mundus mentalis sive angelicus*; (3) *mundus mathematicus seu spatium*; (4) *mundus materialis sive corporea moles*; and (5) *mundus situalis*. After the fifth world, Campanella repeated the worlds in the order 4, 3, 2, 1 and thus reached the innermost, or ninth, circle with the repetition of the archetypal world. Central to all of Campanella's speculations on metaphysics, physics, and logic are the three

"primalities" (*primalitates*): power (*potentia*), wisdom (*sapientia*), and love (*amor*). Although the three primalities are constitutive of being, they will be ignored here because they are not germane to Campanella's concept of space (for a description of the primalities, see Bonansea, *Tommaso Campanella, Renaissance Pioneer*, pp. 144–163).

72 See Di Napoli (ed.), *Metafisica*, vol. 2, p. 368 (bk. 10, ch. 1, article 3). In this same passage, Campanella allowed that God could make innumerable worlds but did not infer that God has indeed done so. But even if God had made other worlds, they would not be constituted of atoms, as Epicurus believed. Elsewhere Campanella made this explicit by conceding that God could create an infinity of worlds "but that there actually are such, we cannot know unless it is revealed to us by God." The passage appears in Campanella's "On the Sense and Feeling in All Things and on Magic," as translated from the Italian edition in *Renaissance Philosophy*, vol. 1: *The Italian Philosophers, Selected Readings from Petrarch to Bruno*, ed., trans. and introduced by Arturo B. Fallico and Herman Shapiro (New York: Modern Library, 1967), p. 369. The Italian edition, *Del senso delle cose e della magia*, was composed by Campanella in 1604 and edited by Antonio Bruers, *Del senso delle cose e della magia* (Bari: Gius. Laterza & Figli, 1925). For the passage cited above, see p. 32. In 1609, Campanella himself translated the Italian version into Latin, which was subsequently published by Tobias Adami under the title *De sensu rerum et magia* (Frankfort: apud Egenolphum Emmelium, 1620). In his Italian edition, Bruers also cites the Latin variants from the 1620 edition, as well as from the edition of Paris, 1637. For the history of the Italian and Latin versions, see Bruers, *Del senso delle cose*, pp. vii–xi.

73 *Universalis philosophiae*, p. 249, col. 1 (bk. 10, ch. 1).

74 Campanella described the *mundus situalis* more fully in bk. 10, article 7, p. 246 of the Paris edition. The influence of Telesio on Campanella is apparent in the role assigned by the latter to the qualities hot and cold (a glance at the table of contents of Telesio's *De rerum natura* as listed in the first volume of Spampanato's edition, pp. 325–332, will immediately convey the pervasive role played by hot and cold in Telesio's natural philosophy).

75 Campanella described the interrelationships between the five worlds in bk. 10, ch. 1, article 8 ("Qua ratione mundi quinque colligati sunt et priores sunt in posterioribus et posteriores in prioribus secundum esse") of his *Universalis philosophiae* (see Di Napoli, *Metafisica*, vol. 2, pp. 374–376).

76 On these interrelationships, see Léon Blanchet, *Campanella* (New York: Burt Franklin, n. d.; first published Paris, 1920), p. 313 and Bonansea, *Tommaso Campanella, Renaissance Pioneer*, p. 220.

77 Fallico and Shapiro, *Renaissance Philosophy*, vol. 1, p. 366; Campanella, *Del senso delle cose* (Bruers ed.), bk. 1, ch. 12, p. 29.

78 "Est autem locus omnium divinitas substentans. . . . In ipsa enim vivimus, movemur, et sumus." *Universalis philosophiae*, bk. 2, ch. 13, p. 288, col. 1. The phrase "in ipsa enim vivimus, movemur, et sumus" is from St. Paul, Acts 17, 28 and would be frequently cited in discussions on space (for example Newton included it in his General Scholium to the second edition of his *Principia Mathematica*; for the quotation of that passage and the appropriate references to Newton, see below, Ch. 8, Sec. 4m and n. 375; Guericke also used it [see below, Ch. 8, Sec. 4c and n. 216]).

79 Di Napoli (ed. and trans.), *Metafisica*, vol. 1, p. 292 (bk. 2, ch. 5, article 2).

80 *Del senso delle cose* (Bruers ed.), bk. 1, ch. 12, p. 29; Fallico and Shapiro, "On the Sense and Feeling in All Things and on Magic," *Renaissance Philosophy*, vol. 1, p. 367. The Arabs Campanella had in mind are left unmentioned.

81 In view of its significance, the relevant text is cited in full: "Finitum quidem spatium est, non modo ex hoc, quod multis caret entitatibus: neque enim est angelus, neque mens, neque corpus; sed etiam ex hoc quod non porrigitur in infinitum, uti putant qui ipsum esse Deum putant." Di Napoli (ed.), *Metafisica*, vol. 2, p. 372 (bk. 10, ch. 1, article 5). Thus Campanella offered two reasons for spatial finitude: (1) space lacks many entities such as angels, mind, and body and so cannot be infinite in what it contains or embraces; and (2) it cannot be extended into infinity "as those believe who think that it [i.e., space] is God Himself." Among those who mistakenly include Campanella as an advocate of

infinite space, we may mention Max Jammer, *Concepts of Space*, pp. 93, 110, and W. von Leyden, *Seventeenth-Century Metaphysics, An Examination of Some Main Concepts and Theories* (London: Gerald Duckworth, 1968), p. 258 (on the same page, von Leyden compounds his error by the inclusion of Telesio among spatial infinitists). Perhaps these authors had in mind Campanella's statement in the *Epilogo magno*, where he said that God made a space that was "almost infinite" in extent (". . . il primo Architetto . . . steso uno spatio presso che infinito. . . ."; see *Epilogo magno* [*Fisiologia Italiana*], testo Italiano inedito con le varianti dei codici e delle edizioni Latine, ed. by Carmelo Ottaviano [Rome: Reale Accademia d'Italia, 1939], pp. 187 and 60; also cited by Bonansea, *Tommaso Campanella, Renaissance Pioneer*, p. 364, n. 166). "Almost infinite" is nonetheless finite.

82 In the *Del senso delle cose*, Campanella explained that "Space is very great – not in terms of material quantity, but rather in incorporeal quantity; . . . it is believed to be infinite in extension beyond the world." Fallico and Shapiro, *Renaissance Philosophy*, vol. 1, p. 366, and Bruers' Italian text, p. 29 (bk. 1, ch. 12). Indeed in the same chapter, Campanella even spoke of a space that encompasses the sky (Fallico and Shapiro, *Renaissance Philosophy*, vol. 1, p. 368; Bruers [ed.], *Del senso delle cose*, p. 31). Moreover, in the very next chapter, after first admitting that God could create an infinite number of worlds beyond ours and then insisting that without divine revelation we cannot know whether He did in fact create such worlds, Campanella mentioned "the famous argument centering around the question of what would happen if one were located on the last sphere and fired an arrow" and concluded that this "proves only that if the arrow does not proceed, there must be a resistant body, while if it does proceed, there must be a space and a yielding body. And if the sphere itself is limited by nothing, then this nothing would be circular like the sphere. Hence it is being; and cannot be understood as being nothing – as obtaining, that is, neither in the mind nor outside of it, neither in God nor outside of Him – for if there is nothing, both being and God would be finite." Fallico and Shapiro, *Renaissance Philosophy*, vol. 1, p. 369; Bruers (ed.), *Del senso delle cose*, p. 32 (bk. 1 ch. 13), and n. 2 of that page, where Bruers indicates that in the Latin edition of 1620, Campanella specifically cited the Stoics as the authors of "the famous argument." From these passages, we may, I believe, rightly conclude that although Campanella did not explicitly proclaim an infinite extracosmic space beyond the world, he did allow for the possibility of its existence. Whatever the merit of this interpretation, there can be no doubt that Campanella would ultimately reject infinite space in the *Universalis philosophia seu metaphysicarum rerum iuxta propria dogmata* of 1638, where his most mature and near final, if not final, opinions on space appear.

83 Di Napoli (ed.), *Metafisica*, vol. 2, p. 372 (bk. 10, ch. 1, article 5).

84 Campanella denied that space is the surface of the surrounding body, as Aristotle defined it, because such a surface is nothing more than the extremity of the surrounding body itself (see *Universalis philosophiae*, p. 288, col. 2 [bk. 2, ch. 13]).

85 At one point, Campanella said that "space is immobile and the whole is similar to itself" ("Spatium autem immobile est totumque sibi simile . . ."; see *Universalis philosophiae*, p. 124 [bk. 13, ch. 2, article 6]), by which homogeneity is probably intended.

86 See Di Napoli (ed.), *Metafisica*, vol. 1, p. 294 (bk. 2, ch. 5, article 2), and *Universalis philosophiae*, p. 288, col. 1, (bk. 2, ch. 13).

87 With space clearly in mind, Campanella declared "Mediumque est inter divina entia et corporalia" (Di Napoli [ed.], *Metafisica*, vol. 1, p. 294 [bk. 2, ch. 5, article 2]), a description similar to one proclaimed earlier by Francesco Patrizi, who conceived space as intermediate between incorporeal substance and body (see Ch. 8, Sec. 4a).

88 "Habere spatium dimensiones scimus ex corporum dimensionibus, quae in eis extenduntur, et tamen ratione modo habitas." Di Napoli (ed.), *Metafisica*, vol. 2, p. 370 (bk. 10, ch. 1, article 5).

89 Translated from Di Napoli (ed.), *Metafisica*, vol. 1, p. 294 (bk. 2, ch. 5, article 2).

90 "Mens autem dividit spatium, et delineat, et superficiat, et profundat, quoniam superioris est ordinis in mundo metaphysico. . . ." De Napoli (ed.), *Metafisica*, vol. 2, pp. 370–372.

91 *Ibid.*, vol. 2, p. 370.

92 *Ibid.*, vol. 2, p. 372.

93 "Spatium enim nequit extendi, neque ampliari, neque in punctum acuminari, neque

ingrossari per linearum et superficierum compositionem; sed intellectu ita in eo consideramus." *Ibid.*, vol. 2, p. 370.

94 See below, Ch. 8, Sec. 4a. On the basis of this significant difference and another whereby Patrizi assumed space infinite and Campanella finite, there is little basis for the claim that "Campanella develops Patritius' theory of space still further" (Jammer, *Concepts of Space*, p. 90).

95 See Di Napoli (ed.), *Metafisica*, vol. 2, p. 368 (bk. 10, ch. 1, article 3).

96 Despite his unequivocal description of space as indivisible and unalterable, where only the mind is capable of dividing space and imposing geometric figures on it, the mind requires an external space, because "all mathematical propositions are verified in space, otherwise they would be chimerical and nothing would correspond to them in reality" (*Universalis philosophiae*, p. 124, col. 1 [bk. 13, ch. 2, article 6]). When astronomers draw axes in the heavens and distances are measured between cities, such lines are in space, not in matter. Somehow the ideas of geometrical figures in our minds can be represented in space, but do not actually divide space itself because they are not integral to it and possess their real existence only in the mind. Thus Campanella said that "The intellect makes them [i.e., geometric figures] in space, since it knows them in the ideal divine light: stimulated in a certain hidden manner by the likeness of sensible things. . . . For the physical and mathematical world is founded on a prior mental one. Ideas are therefore as a sign in us. . . . In this way do the things which are defined exist in mathematics: they can be assumed, because the idea is in us and space is in nature, in which they are represented." Trans. in John Herman Randall, Jr., *The Career of Philosophy from the Middle Ages to the Enlightenment* (New York: Columbia University Press, 1962), p. 219, from *Universalis philosophiae*, p. 125, col. 1 (bk. 13, ch. 2, article 6; I have added the qualification in brackets). Although his solution to the problem was radically different than Campanella's, Patrizi also resolved the same dilemma by the assumption of an interplay between mind and space (see below, Ch. 8, Sec. 4a).

97 Campanella, "On the Sense and Feeling in all Things and on Magic," trans. in Fallico and Shapiro, *Renaissance Philosophy*, vol. 1, p. 366; Bruers (ed.), *Del senso delle cose*, p. 29 (bk. 1, ch. 12). The same opinion is expressed in the later *Universalis philosophiae* (see the edition of Di Napoli, *Metafisica*, vol. 2, p. 98 [Bk. 6, ch. 7, article 1]).

98 The world has a spherical figure because of its motion ("Ma al mondo, per il moto basta la figura tonda; . . ." Bruers [ed.], *Del senso delle cose*, p. 34 [bk. 1, ch. 13]). That the motion intended is rotatory follows from Campanella's earlier statement that although one cannot determine by experiential means whether the earth or the stellar sphere rotates, he was convinced that it is the latter. He believed it impossible, however, to determine whether the world also has a rectilinear motion (see *ibid.*, pp. 32–33, and Fallico and Shapiro, *Renaissance Philosophy*, vol. 1, pp. 369–370).

99 "It is an error to think that the world does not feel just because it does not have legs, eyes, and hands"; and later, in the same paragraph, Campanella declared that "It is just as foolish to deny sense to things because they have no eyes, no mouth, and no ears as it is to deny motion to the wind because it has no legs, or eating to fire because it has no teeth, . . ." Fallico and Shapiro, *Renaissance Philosophy*, p. 371; Bruers (ed.), *Del senso delle cose*, pp. 34–35 (bk. 1, ch. 13).

100 Fallico and Shapiro, *Renaissance Philosophy*, vol. 1, p. 362; Bruers (ed.), *Del senso delle cose*, p. 26 (bk. 1, ch. 9).

101 Campanella proclaimed this in *Del senso delle cose*, bk. 1, ch. 10, the title of which is "That a vacuum comes about by violence and not by nature; sensible proof of this, contra Aristotle" (Fallico and Shapiro, *Renaissance Philosophy*, vol. 1, p. 363; Bruers [ed.], *Del senso delle cose*, pp. 27–28). Campanella thus followed Telesio and Patrizi in the conviction that artificial vacua could be produced (see above, Ch. 4, Sec. 5). He explained that "just as our body, by nature, abhors any division and yet can be made to suffer such division by violence, so it is with the body of the world" (Fallico and Shapiro, *Renaissance Philosophy*, vol. 1, p. 364, and Bruers [ed.], *Del senso delle cose*, p. 28; see also Di Napoli [ed.], *Metafisica*, vol. 2, p. 98 [bk. 6, ch. 7, article 1 of the *Universalis philosophiae*]).

102 See Bruers (ed.), *Del senso delle cose*, p. 31 (bk. 1, ch. 12); Fallico and Shapiro, *Renais-*

sance Philosophy, vol. 1, pp. 367–368. Fire, for example, would voluntarily convert to earth and thus destroy itself than permit a vacuum. In the *Universalis philosophiae*, Campanella explained that in order to fill a vacuum, bodies would move contrary to their natural motions (see Di Napoli [ed.], *Metafisica*, vol. 2, p. 98; this was, of course, a common medieval explanation [see above, Ch. 4, Secs. 1 and 2]).

103 Fallico and Shapiro, *Renaissance Philosophy*, vol. 1, p. 368; Bruers (ed.), *Del senso delle cose*, p. 31 (bk. 1, ch. 12).

104 Fallico and Shapiro, *Renaissance Philosophy*, vol. 1, p. 368; Bruers (ed.), *Del senso delle cose*, p. 31 (bk. 1, ch. 12). See also Bonansea, *Tommaso Campanella, Renaissance Pioneer*, p. 158.

105 Fallico and Shapiro, *Renaissance Philosophy*, vol. 1, p. 367; Bruers (ed.), *Del senso delle cose*, p. 30 (bk. 1, ch. 12).

106 Fallico and Shapiro, *Renaissance Philosophy*, vol. 1, p. 368; Bruers (ed.), *Del senso delle cose*, p. 31 (bk. 1, ch. 12). Despite emphasis on matter's desire to spread out and expand itself, Campanella did not altogether abandon his earlier belief that matter is also drawn together, though he tended to confine it to parts of the world that are possessed of a similar nature (Fallico and Shapiro, *Renaissance Philosophy*, vol. 1, p. 368, and Bruers [ed.], *Del senso delle cose*, p. 31 [bk. 1, ch. 12]).

107 Here again we see that Campanella still retained the idea that bodies seek to preserve material continuity (also see above, n. 106 of this chapter).

108 "Habet ergo locus sensum suae perfectionis; sed sive fiat abstractio ab ipso spatio, sive accurant sponte corpora, ne dividantur, sive ut dilatent regnum suum, ubi a nullo patiuntur resistentiam, sicut ibidem dixi, cuncta enim sese amplificare student: necesse est asserere hoc ex sensu rerum locatarum aut loci provenire; et mundum esse animal maxime sensitivum." *Universalis philosophiae*, bk. 6, ch. 7, article 1, in the edition of Di Napoli, *Metafisica*, vol. 2, p. 100.

109 For references, see Bonansea, *Tommaso Campanella, Renaissance Pioneer*, p. 189, and the relevant notes on p. 364.

110 *Universalis philosophiae*, bk. 10, ch. 1, article 5 in Di Napoli (ed.), *Metafisica*, vol. 2, p. 372.

111 The space of the *mundus mathematicus* adopted by Campanella must not be construed as an illustration of Koyré's claim that space was geometrized in the seventeenth century by the acceptance of the "essentially infinite and homogeneous extension" of Euclidean geometry (see above, Ch. 2, Sec. 2a). Not only was Campanella's space finite, rather than infinite, but it also lacked objective dimensionality, which was imposed on it by mind (mens).

112 *Universalis philosophiae*, p. 288, col. 1 (bk. 2, ch. 13, article 1); see also Bonansea, *Tommaso Campanella, Renaissance Pioneer*, p. 189.

113 ". . . siquidem et spatium ipsum intimius est corpori quam corpus sibi." Di Napoli (ed.), *Metafisica*, vol. 2, p. 368 (*Universalis philosophiae*, bk. 10, ch. 1, article 3).

114 Fallico and Shapiro, *Renaissance Philosophy*, vol. 1, p. 368; Bruers (ed.), *Del senso delle cose*, p. 31 (bk. 1, ch. 12).

115 See Bonansea, *Tommaso Campanella, Renaissance Pioneer*, p. 189, and the notes on p. 364 for references.

116 *Universalis philosophiae*, p. 288, col. 1 (bk. 2, ch. 13, article 1).

117 In Bernardino Bonansea's chapter on "Transcendentals, Predicaments, and Causes" (*Tommaso Campanella, Renaissance Pioneer*, ch. 8, pp. 164–195), space is not mentioned, nor does it appear in a table of "Transcendental Predicaments" and "Generic Predicaments" (p. 360, n. 74).

118 See *Francisci Patricii Nova De Universis Philosophia. In qua Aristotelica methodo non per motum, sed per lucem, et lumina, ad primam causam ascenditur* . . . (Ferrara: apud Benedictum Mammarellum, 1591), Pancosmia (Part Four), bk. 2 (*De spacio mathematico*), fol. 68v, col. 2. In Benjamin Brickman's translation, "On Physical Space, Francesco Patrizi," *Journal of the History of Ideas* 4 (1943), see 245.

119 *Franc. Patricii Philosophiae, De rerum natura libri II priores. Alter de spacio physico; alter de spacio mathematico* (Ferrara: Victor Baldinus, 1587).

120 All references to Patrizi's doctrine of space will be to the 1591 version of the *Nova philosophia* (another edition was issued in Venice in 1593, from which Brickman made his translation; because the two versions seem almost identical, Brickman's translation will also be cited here). The first three parts of Patrizi's cosmology were called, in order: *Panaugia, Panarchia,* and *Pamsychia.* For a brief description of all four parts, see Kristeller, *Eight Philosophers,* ch. 7 ("Patrizi"), pp. 118-125.

121 Brickman (trans.), "On Physical Space," p. 225; *Nova philosophia,* fol. 61r, col. 1. It is ironic that the anti-Aristotelian Patrizi should support the priority of space by observing that "Aristotle himself maintained that that in the absence of which nothing else exists, and which can exist without anything else, is necessarily prior to all other things" (Brickman [trans.], "On Physical Space," p. 225; *Nova philosophia,* fol. 61r, col. 1; the passage to which Patrizi referred appeared in Aristotle's discussion on place in *Physics* 4.1.208b.34-209a.1). In Patrizi's judgment, the only entity that meets these specifications is three-dimensional space, a claim that Aristotle would have emphatically denied. Whether separate or occupied, no such space could exist. Perhaps Aristotle would have identified two-dimensional surfaces, or places, as prior to the things that occupied them. In Aristotle's system, however, place (*locus*) depends on the inner surfaces of material things, and because the latter are eternal, without beginning or end, it is not likely that Aristotle could have attributed anything more than a logical priority to place. Patrizi was, however, aware of his differences with Aristotle, even as he attempted a feeble reconciliation. "For what is his [Aristotle's] '*locus*'" Patrizi explained, "other than Space, with length and breadth, even if in *locus* he himself foolishly overlooked depth (*profundum*), which is more properly *locus*?" Brickman (trans.), "On Physical Space," p. 226; Patrizi, *Nova philosophia,* fol. 61r, col. 1. As a Platonist and Neoplatonist, Patrizi's emphasis on the priority of space was probably derived from Plato's receptacle in *Timaeus* 52. For Patrizi, however, in contrast to Plato, space is God's creation and not an uncreated, coeternal entity.

122 Following arguments to indicate that space cannot be bounded by corporeal or incorporeal entities and that it must therefore be infinite, Patrizi asked whether it is a potential or actual infinite: "If it were said to be potential, it would necessarily follow that it is now finite, and that later it would become infinite, but still only potentially infinite. But if that is an absurdity, we conclude that it is actually infinite." Brickman (trans.), "On Physical Space," p. 237; Patrizi, *Nova philosophia,* fol. 64r, col. 2.

123 See above, Ch. 6, Sec. 2b, and Ch. 7, n. 116. Although Jean de Ripa and John Major may have believed in actual created infinite spaces (see above, Ch. 6, Sec. 2b, and Ch. 7, Sec. 1), their discussions are too vague to stand as proper precursors to Patrizi. Scholastics were almost unanimous in denying a creation to space (for example, see Fonseca's statement in Ch. 7, Sec. 2b above).

124 Patrizi, *Nova philosophia,* fols. 61r, col. 2-61v, col. 1; Brickman (trans.), "On Physical Space," p. 226.

125 Patrizi specifically mentioned the Stoics as proponents of an infinite extracosmic void space but added that their reasons for so believing were unclear (see *Nova philosophia,* fol. 64r, col. 1, and Brickman [trans.], "On Physical Space," p. 236). Ultimately, as will be seen below, Patrizi's world and surrounding space would differ from the Stoic version by assuming the existence of interstitial vacua within the world and an infinite space filled with light beyond it. For clear evidence that Patrizi's world was geocentric, see Patrizi, *Nova philosophia,* fol. 65v, col. 2, and Brickman (trans.), "On Physical Space," p. 243. In what follows, I have made use of relevant material from my article "Place and Space in Medieval Physical Thought," 160-161.

126 Patrizi *Nova philosophia,* fol. 64v, col. 1; Brickman (trans.), "On Physical Space," p. 238.

127 Brickman (trans.) "On Physical Space," p. 240; Patrizi, *Nova philosophia,* fol. 65r, col. 1. Here, and in the lines following, Patrizi probably had in mind the Stoic doctrine of recurrent fiery destructions and subsequent re-creations of the world. Although he spoke of God hypothetically destroying the world and re-creating it, much as medieval scholastics did, his inspiration must be linked with pagan Greek rather than medieval Chris-

tian thought. From the tone of the discussion, it seems likely that Patrizi believed in the doctrine of successive divine re-creations of the world. Perhaps the creation of our world, as well as divine re-creations, were made from a pre-creation matter, which Patrizi described as being either atoms, chaos, or unformed matter (Patrizi, *Nova philosophia*, fol. 65r, col. 1; Brickman [trans.], "On Physical Space," p. 240). We may also ask how the existence of pre-creation matter is to be reconciled with Patrizi's statement that if God destroyed our world, the place where the world now is, that is, its *locus*, would remain entirely empty (Patrizi, *Nova philosophia*, fol. 65r, col. 1; Brickman [trans.], "On Physical Space," p. 240). Indeed Patrizi appears to have left unexplained the relationship of pre-creation matter and successive worlds with the traditional Christian account of a unique creation *ex nihilo*.

128 Patrizi, *Nova philosophia*, fols. 64r, col. 2–64v, col. 1; Brickman (trans.), "On Physical Space," p. 238.

129 Brickman (trans.), "On Physical Space," pp. 239–240; Patrizi, *Nova philosophia*, fols. 64v, col. 2–65r, col. 1. I have placed brackets around "essential" in Brickman's translation because no Latin equivalent of that term appears in Patrizi's text. Earlier in the treatise, Patrizi explained that "*vacuum, spacium, plenum,* and *locus*" are really the same thing, for "When it is filled with a body, it is *locus*; without a body it is a vacuum. And on this account this vacuum, like *locus*, must have the three common dimensions – length, width, and depth. And the vacuum itself is nothing else than three-dimensional Space." Brickman (trans.), "On Physical Space," p. 231; Patrizi, *Nova philosophia*, fol. 62v, col. 2. Some years after Patrizi, Bartholomeus Amicus would sharply distinguish between vacuum and space (see above, Ch. 7, Sec. 2d).

130 Patrizi, *Nova philosophia*, fols. 64r, col. 1, and 64v, col. 1; Brickman (trans.), "On Physical Space," pp. 236, 238.

131 Patrizi, *Nova philosophia*; Brickman (trans.), "On Physical Space." Patrizi's "proofs" for the infinity of space appear on fols. 64r, cols. 1–2 (Brickman, pp. 236–237). They are similar to those attributed to Archytas of Tarentum (above, Ch. 5, Sec. 1), drawing their effect from the demonstrated impossibility that either body or void could function as limits to spatial extension.

132 Brickman (trans.), "On Physical Space," p. 238; Patrizi, *Nova philosophia*, fol. 64v, col. 1.

133 Also see my article "The Principle of the Impenetrability of Bodies in the History of Concepts of Separate Space," *Isis* 69 (1978), 569. Patrizi stressed the absolute immobility of space (see Patrizi, *Nova philosophia*, fols. 62v, col. 1 and 65r, col. 2; Brickman, "On Physical Space," pp. 230, 241).

134 See above, Ch. 3, Sec. 4. The position described by Burley, and the earlier views of Duns Scotus, assume, as was customary in the Middle Ages (see Sec. 4), that three-dimensional void space is a body. On that assumption, a body can occupy space only by "displacing" it; otherwise it could not penetrate it at all.

135 Patrizi explained that "When, however, it is said that *locus* is different from the *locatum*, this is to be taken to mean that every *locatum* is a body, while *locus* is not a body, otherwise two bodies will interpenetrate. Hence, *locus*, not being a body, will of necessity be a Space (*spacium*) provided with three dimensions – length, breadth, and depth – with which it receives into itself and holds the length, breadth, and depth of the enclosed body." Brickman (trans.), "On Physical Space," p. 231; Patrizi, *Nova philosophia*, fol. 62v, col. 1. Even though occupied, the void space, or *locus*, that receives the body continues to exist. The continued existence of void is what Bartholomeus Amicus would later deny (see above, Ch. 7, Sec. 2d and n. 129 of this chapter); for him void and space differ.

136 Brickman (trans.), "On Physical Space," p. 229; Patrizi, *Nova philosophia*, fol. 62r, col. 1.

137 Brickman (trans.), "On Physical Space," p. 225; Patrizi, *Nova philosophia*, fol. 61r, col. 1.

138 Brickman (trans.), "On Physical Space," p. 227; Patrizi, *Nova philosophia*, fol. 61v, col. 1.

139 Later in the *Pancosmia*, in the chapter "On Air" (*De aere*), Patrizi declared that "Among bodies, space is the most incorporeal of all because it is the rarest" ("Inter corpora enim incorporeum maxime omnium est spacium quia est rarissimum"; Patrizi, *Nova philos-*

ophia, fol. 122r, col. 1). At this point, where he called space a body, Patrizi seems to have forgotten that he had earlier sharply distinguished space from body.

140 See *ibid.,* fols. 63r, col. 1–63v, col. 1; Brickman (trans.), "On Physical Space," pp. 232–234. For more on Patrizi's description of interstitial vacua, see John Henry, "Francesco Patrizi da Cherso's Concept of Space and Its Later Influence," *Annals of Science 36* (1979), 563–564.

141 The section on light, which is the fourth book of the *Pancosmia* (the books on physical and mathematical space constitute the first two books), bears the title *De primaevo lumine* and is found on fols. 73v–75r. What I have described here appears on fol. 74v, col. 1.

142 Later in the *Pancosmia,* in the section *De aere,* Patrizi explained that light (*lumen*) is the rarest of bodies after space. Unfortunately, in that same section, he also described space as a body (see above, n. 139 of this chapter). Only God, it seems, is completely incorporeal, whereas among bodies, the greater the rarity, the closer to the deity is that body. From this standpoint, space is closest to God, followed by light, heat (*calor*), and fluidity (*fluor*), in that order.

For more on Patrizi's concept of light, see John Henry, "Francesco Patrizi da Cherso's Concept of Space," p. 556. Henry observes that Patrizi's ideas on space and light were strikingly similar to those of Proclus (410–485), who seemed to identify space and light. Proclus' ideas were probably conveyed to Patrizi via Simplicius' commentary on Aristotle's *Physics* (see H. Diels [ed.], *In Aristotelis physicorum libros quattuor priores commentaria* [Berlin, 1882], p. 612; for a Latin translation, see the Venice 1566 edition [apud Hieronymum Scottum]: *Simplicii philosophi perspicacissimi, clarissima commentaria in octo libros Arist. de Physico Auditu,* pp. 221, col. 2–222, col. 1. In *Le Système du monde,* vol. 1, p. 339, Duhem describes Proclus' theory of space).

In the later chapters of the *Pancosmia,* Patrizi distinguished three major parts of the universe: (1) an empyrean, which is the extracosmic infinite void filled with light; (2) the aetheric world, which embraces everything between stars and moon; and (3) an elementary region that contains everything between moon and earth. See Kristeller, *Eight Philosophers,* pp. 119, 124.

143 C. de Waard observes (*L'Expérience barométrique,* p. 29) that many "vacuists" of the seventeenth century invoked an imponderable *spiritus mundi* that filled the void spaces between particles of matter and between celestial bodies. Consequently, when they spoke of "void," they meant void of ponderable matter but not of weightless *spiritus mundi.*

144 Patrizi, *Nova philosophia,* fol. 64v, col. 1; Brickman (trans.), "On Physical Space," p. 238.

145 Patrizi, *Nova philosophia,* fol. 64v, col. 2; Brickman (trans.), "On Physical Space," p. 239. For Telesio, see above, Ch. 8, Sec. 3a and n. 67.

146 Brickman (trans.), "On Physical Space," p. 240; Patrizi, *Nova philosophia,* fol. 65r, col. 1.

147 Patrizi, *Nova philosophia,* fol. 65r, cols. 1–2; Brickman (trans.), "On Physical Space," p. 240.

148 Brickman (trans.), "On Physical Space," pp. 240–241; Patrizi, *Nova philosophia,* fol. 65r, cols. 1–2. In the passage cited, I have altered Brickman's translation as follows: for the various forms of *mundanus,* I have substituted "worldly" for Brickman's "earthly"; and for *philosophandum,* I have replaced his "treated" with "philosophized about."

149 It is ironic that the category of place in Aristotle's *Categories* was actually described as three-dimensional. It was not, however, interpreted as Aristotle's true opinion (see above, Ch. 1).

150 For Amicus' denial that space is substance or accident, see above, Ch. 7, Sec. 2d; for Franciscus Pitigianus, Ch. 7, n. 53. In speaking of Gassendi, who would adopt the same opinion, as we shall see below, Bernard Rochot explains that for Gassendi space and time "are neither substance nor accident. They exist when their content disappears and when nothing is happening" ("Gassendi [Gassend], Pierre," *Dictionary of Scientific Biography,* vol. 5 [New York: Scribner, 1972], p. 287). It is obvious, however, that Rochot is mistaken when he characterizes this as "contrary to the scholastic view" and when he identifies Gassendi as "one of the first to state this universal, categorial law of space and time." Scholastics who accepted the existence of extracosmic space were well aware that it did not fit the traditional categories.

151 See Ch. 6, n. 46, and Grant, "The Condemnation of 1277," 233–234 and n. 80, 83.

152 Brickman (trans.), "On Physical Space," p. 241; Patrizi, *Nova philosophia*, fol. 65r, col. 2. Patrizi went on to emphasize that space "is not quantity. And if it be a quantity, it is not that of the categories, but prior to it, and its source and origin. Nor can it be called an accident, for it is not the attribute of any substance."

153 Brickman (trans.), "On Physical Space," p. 241; Patrizi, *Nova philosophia*, fol. 65r, col. 2. In the Greek text of his *De motu circulari corporum caelestium*, Cleomedes also described infinite extracosmic space as a "substance" (ὑπόστασις). See H. Ziegler's edition (Leipzig: Teubner, 1891), pp. 4, lines 14–15 and 8, lines 10–11 and also Duhem, *Le Système du monde*, vol. 1, pp. 311–312. Cleomedes did not consider how his "void substance" might relate, if at all, to Aristotle's categories. The Greek text and a Latin translation of Cleomedes' treatise were both available to Patrizi.

154 After explaining that space is an independent substance existing in itself (*in se existens*), Patrizi said that "it always remains fixed *per se* and in itself, nor is it ever or anywhere moved, nor does it change its essence or *locus* in any of its parts or in its entirety. . . . It is therefore entirely unmoved and immovable." Brickman (trans.), "On Physical Space," p. 241; Patrizi, *Nova philosophia*, fol. 65r, col. 2–65v, col. 1.

155 Brickman (trans.), "On Physical Space," p. 241; Patrizi, *Nova philosophia*, fol. 65r, col. 2

156 Brickman (trans.), "On Physical Space," p. 244; Patrizi, *Nova philosophia*, fol. 68r, col. 2.

157 Although Patrizi assumed only a finite world, and therefore a finite body within infinite space, we have already seen that he assumed that light was a body that God created to fill infinite empty space (see above, Ch. 8, Sec. 4a and n. 142).

158 Patrizi, *Nova philosophia*, fol. 68r, col. 1; Brickman (trans.), "On Physical Space," p. 243. Patrizi went on to say that "The mind does not separate these spaces from bodies by abstraction, as some contend, since these spaces are not primarily and *per se* in worldly (*mundanis*) bodies, but are prior to bodies, and are actualized in primary space." Brickman (trans.), "On Physical Space," pp. 243–244; Patrizi, *Nova philosophia*, fol. 68r, col. 1. Here, as in note 148 above, I have altered Brickman's translation of *mundanus* from "earthly" to "worldly."

159 Because this important passage is omitted by Brickman, I also give the Latin: "Continuum autem asserimus esse id quod extenditur; discretum vero quod de continuo desecatur. Et continuum semper actu est; a mente vero nec actu, nec [corrected from "ne"] potentia dividi, sed divisum imaginari tantum." Patrizi, *Nova philosophia*, fol. 68r, col. 1. Cf. Newton's similar approach below, Ch. 8, Sec. 4l.

160 Brickman (trans.), "On Physical Space," p. 244; Patrizi, *Nova philosophia*, fol. 68r, col. 2.

161 For a brief description of Patrizi's significant influence in the seventeenth century, see John Henry, "Francesco Patrizi da Cherso's Concept of Space," pp. 566–571.

162 Among other important matters concerning Newton's concept of space, J. E. McGuire also considers the influence of Gassendi and More. See his "Existence, Actuality and Necessity: Newton on Space and Time," *Annals of Science 35* (1978), 463–508. Gassendi's influence on Newton was apparently through Walter Charleton's *Physiologia Epicuro-Gassendo-Charletoniana* (London, 1654), which Newton seems to have read early in his career (see McGuire, "Newton on Space and Time," p. 469; for the general influence of Gassendi on the young Newton, see Richard S. Westfall, "The Foundation of Newton's Philosophy of Nature," *British Journal for the History of Science 1* [1962], 171–182).

163 Olivier René Bloch (*La Philosophie de Gassendi, Nominalisme, Matérialisme et Métaphysique* [The Hague: Martinus Nijhoff, 1971], pp. 173–174) rightly argues that Gassendi presented a clear and vigorous conception of both space and time that would prove a model for their classic formulation by Newton. He observes (p. 173) that in his influential treatise *From the Closed World to the Infinite Universe*, Alexandre Koyré accorded only a minimal place (*place infime*) to Gassendi. Not only did he ignore Gassendi, but we should also observe that Patrizi is mentioned only once and then merely to record him as in agreement with Bruno that planets are animated beings. Koyré thus eliminated

from his otherwise important study two of the most significant contributors to the concept of infinite space in the seventeenth century.

164 Bloch, *La Philosophie de Gassendi*, pp. 197–200. On Charleton, see notes 162, 167 of this chapter.

165 Among the *recentiores* who had sought to forge a new physics, Gassendi, in his posthumously published *Syntagma philosophicum* (Lyon, 1658), mentioned Telesio, Patrizi, Campanella, and Kenelm Digby. Of these authors, it was only Patrizi's spatial doctrine that was described by Gassendi. The *Syntagma philosophicum*, which will be the major source for our discussion of Gassendi's spatial theory, appears in vol. 1 of Gassendi's posthumously published *Opera Omnia* (6 vols.; Lyon, 1658; cited here from the facsimile reprint published in Stuttgart-Bad Canstatt: Friedrich Frommann Verlag, 1964). For the discussion of the *recentiores*, see *Syntagma*, Part 2 (*Physica*), sec. 1, bk. 3 (*De materiali principio, sive materia prima rerum*), pp. 245, col. 2–247, col. 2. What influence Campanella may have had is unclear, although Gassendi personally welcomed him to France after the former's flight from Rome in 1634 (see Schmitt, "Campanella, Tommaso," *Dictionary of Scientific Biography*, vol. 15, supplement 1, p. 69). Bloch allows (*La Philosophie de Gassendi*, pp. 194–196) that Gassendi might have been influenced by Bradwardine on imaginary space (more on this below) and by Patrizi on the exclusion of space from the traditional categories (also to be discussed below). Campanella's influence appears far less significant. Bloch also observes (p. 196, n. 109) that Gassendi's section on the *recentiores* does not appear in the earlier *Animadversiones* (on the relationship between the *Animadversiones* and the *Syntagma*, see n. 168 below).

166 *Syntagma philosophicum*, Physics, sec. 1, bk. 3, ch. 8 in *Opera*, vol. 1, p. 280, col. 1, as translated by Craig B. Brush, *The Selected Works of Pierre Gassendi* (New York: Johnson Reprint Corporation, 1972), p. 399; for the reference to Guillaume de Conches, see p. 398.

167 Charleton, *Physiologia*. Charleton's fundamental source for Gassendi's concept of space was probably the latter's *Animadversiones in decimum librum Diogenis Laertii* (for full title and further discussion, see below, n. 168). Charleton's treatise is significant because "it was widely read in the early 1650's as a convenient substitute for Gassendi's scarce works" (Robert Kargon, "Charleton, Walter," *Dictionary of Scientific Biography*, Vol. 3 [New York: Scribner, 1971], p. 209; also see above, n. 162 of this chapter). It would also be read by Robert Boyle and Newton.

168 Translated by Brush, *Selected Works of Pierre Gassendi*, p. 386, from Gassendi, *Opera*, vol. 1; *Syntagma philosophicum*, Physics, part 2, sec. 1, bk. 2 ("De loco et tempore, seu spatio, et duratione rerum"), ch. 1, p. 182, col. 2. Unless otherwise indicated, my references to Gassendi's discussion of space are to this book of the *Syntagma philosophicum* (cited hereafter as "*Opera* 1, S. P."), which extends over pp. 179–228 (most of what is relevant to our study, however, appears in bk. 2, ch. 2, pp. 185–191, which is titled: "That a void space is assigned, and first what they call separate and extramundane" [Dari inane spatium, ac primum quod vocant separatum extraque mundum]). For Charleton's version of Gassendi's imaginary annihilation, see the former's *Physiologia*, bk. 1, ch. 6 ("Of Place"), article 4, p. 63. Later in this same book, Gassendi not only explained that the Stoics especially admitted the void beyond the world (p. 186, col. 2) but described in some detail the views of Cleomedes the Stoic (pp. 187, col. 2–188, col. 1; 189, col. 1; on Cleomedes see above, Ch. 5, Sec. 1). Indeed Gassendi also saw fit to reproduce (from the *Physics* commentary of Simplicius) the famous argument of Archytas of Tarentum (p. 188, col. 1), which was usually invoked to justify belief in an infinite extracosmic space (see above, Ch. 5, Sec. 1 of this work). In the *Animadversiones*, Gassendi discussed the nature of void space in at least two places: (1) in a chapter titled "Dari praeter corpora etiam inane in rerum natura," pp. 169–177; and (2) in an appendix (Appendix altera) titled "Philosophae Epicuri Syntagma" (this must not be confused with the later *Syntagma philosophicum*), pp. xcvii–cclxii, where, in ch. 2, he considered that the universe is infinite, immobile and immutable ("Esse universum infinitum, immobile, immutabile"). In addition to these sections, on pp. 424–444, in a chapter titled "De nupero experimento circa inane coacervatum," Gassendi described and analyzed experiments devised to establish the

existence of void space. There is virtually nothing here, however, on the nature of void space.

Although my description and assessment of Gassendi's spatial doctrine have relied exclusively on his posthumously published *Syntagma philosophicum*, Gassendi also discussed infinite space in his *Animadversiones in decimum librum Diogenis Laertii, qui est de vita, moribus, placitisque Epicuri. Continent autem placita, quas ille treis statuit philosophia parteis*: I. Canonicam nempe, habitam Dialecticae loco: II. Physicam, *ac imprimis nobilem illius partem* Meteorologiam: III. Ethicam, cuius gratia ille excoluit caeteras (3 tomes in 2 vols.; Lyon: apud Guillelmum Barbier, 1649). Gassendi's *Animadversiones* on the tenth book of Diogenes' life of Epicurus is cued to the Greek text, which, with a Latin translation, occupies the first 100 pages of this continuously paginated two-volume work. It was presumably from these sources that Charleton derived his knowledge of Gassendi's concept of infinite void space, a concept that is in close agreement with the basic ideas in the later *Syntagma philosophicum*. There can be no doubt, however, that the later *Syntagma philosophicum* not only contains much more on the nature of space than the earlier *Animadversiones* but is much better organized. To date, no systematic comparison has been made between the physical ideas and concepts in the *Animadversiones* and the *Syntagma philosophicum*. Bernard Rochot (*Les Travaux de Gassendi sur Epicure et sur l'atomisme 1619–1658* [Paris: Librairie Philosophique J. Vrin, 1944], ch. 8, pp. 167–202), however, observes that there are significant similarities and differences between them. Not only does he cite specific differences between the sections on physics in the two treatises, but he estimates (p. 168) that approximately half of the *Syntagma* is new and half repeats what is in the *Animadversiones*. Indeed the structures of the two treatises are sufficiently different that wholesale transference of lengthy sections from the *Animadversiones* to the *Syntagma* would have been simply unfeasible. Rochot explains (*Les Travaux de Gassendi*, p. 172) that a commentary on Greek atomism, to which a large part of the *Animadversiones* is devoted, was not a convenient vehicle for the introduction of modern ideas on the subject. Thus contemporary debates about imaginary spaces were better and more appropriately discussed in the *Syntagma*, as was much else.

169   Gassendi was keenly aware of the importance of his use of imaginary conditions, especially those situations in which all or part of the world is annihilated. As he explained, "it is frequently necessary to proceed in this fashion in philosophy, as when they tell us to imagine matter without any form in order to permit us to understand its nature. . . . Therefore there is nothing that prevents us from supposing that the entire region contained under the moon or between the heavens is a vacuum, and once this supposition is made, I do not believe that there is anyone who will not easily see things my way" (Brush, *Selected Works of Pierre Gassendi*, p. 386; *Opera* 1, S. P., p. 182, col. 2). Thomas Hobbes was no less impressed by the methodological importance of such a procedure when he declared in *De corpore* (1655) that "In the teaching of natural philosophy, I cannot begin better (as I have already shewn) than from *privation*; that is, from feigning the world to be annihilated" (*Elements of Philosophy, the first section: "Concerning Body," written in Latin by Thomas Hobbes of Malmesbury and translated into English* in *The English Works of Thomas Hobbes of Malmesbury*, ed. by William Molesworth [16 vols.; London, 1839–1845], vol. 1, p. 91). Because Hobbes spoke highly of Gassendi as one of those (along with Kepler and Mersenne [*ibid.*, p. ix of Hobbes's "Epistel Dedicatory"]) who had advanced astronomy and natural philosophy "extraordinarily," it is by no means farfetched to suppose that Hobbes may have derived his methodology of "*privation*, that is, from feigning the world to be annihilated," from Gassendi, or perhaps through Charleton's *Physiologia* of 1654, where, in the course of summarizing Gassendi's example on annihilation (pp. 63–64, article 5), Charleton explained that "nothing is more usual, nor laudable amongst the noblest order of *Philosophers*" than the assumption of "natural impossibilities." Perhaps Charleton influenced Newton, who also used this device (see below, Ch. 8, Sec. 4l; for Berkeley's use of it, Sec. 4m; also above, Ch. 5, n. 30). The methodological tactic described here was, as we have seen, commonplace in the Middle Ages, when it was generally described as proceeding *secundum imaginationem*. In addition to the in-

stances of imaginary annihilations of parts of the world mentioned earlier in this volume, see also Grant, "The Condemnation of 1277," especially 239–244.

170 Although a priest, Gassendi was undoubtedly a Copernican. The procedures against Galileo and the heliocentric system, however, had compelled him, on grounds of faith, to speak for geocentrism (see Bloch, *La Philosophie de Gassendi*, pp. 326–334). In the hypothetical example cited above, Gassendi described an Aristotelian geocentric world, with water, air, and fire ranged around the earth at center.

171 Brush, *Selected Works of Pierre Gassendi*, p. 387; *Opera* 1, S. P., p. 183, col. 1. This passage, along with other parts of Gassendi's discussion on space in the *Syntagma*, has been translated by Milič Čapek and Walter Emge in Čapek (ed.), *Concepts of Space and Time*, p. 92. In these two excellent translations, Brush has included more that is directly relevant to my purposes and has usually been cited here. For Charleton's summary of the passage cited above, see *Physiologia*, p. 64, article 6.

172 Brush, *Selected Works of Pierre Gassendi*, p. 387; *Opera* 1, S. P., p. 183, col. 1.

173 "Et quia si mundus maior maiorque in infinitum praeexstitisset, Deo deinceps pariter totum redigente in nihilum, intelligimus dimensiones spatialeis ampliores semper amplioresque in infinitum superfore; ideo concipimus hoc spatium fore quoquoversum cum suis dimensionibus prolatatum in infinitum." *Opera* 1, S. P., p. 183, col. 1.

174 As the first of three conclusions, Gassendi declared that " spatia immensa fuisse antequam Deus conderet mundum. . . ." *Opera* 1, S. P., p. 183, col. 1. That Gassendi intended an infinite space is likely from the use of the expression *in infinitum* in the passage cited in n. 173. In his interpretation of this passage, Bloch has unhesitatingly rendered *spatia immensa* as infinite spaces ("les espaces infinis"; *La Philosophie de Gassendi*, p. 177). And yet the use of *immensa* is somewhat unsettling and troublesome. If Gassendi had wished to delare space unambiguously infinite, he had only to write *spatia infinita* or *spatium infinitum*. But he did not. It is almost as if he wished to emulate Descartes, who called the universe and therefore space "indefinite," that is, potentially infinite, rather than actually infinite (see Koyré, *From the Closed World*, pp. 106–109). Just as most of his contemporaries assumed that Descartes really intended an actually infinite space but spoke of an indefinite one to avoid a confrontation with the theologians (Koyré, p. 109), so perhaps did Gassendi also really believe in an infinite space but, for theological reasons, spoke of an immense space.

175 The supernatural movement of the world through space was discussed in the Middle Ages in connection with article 49 of the Condemnation of 1277 (see above, Ch. 5, Sec. 2).

176 Gassendi, *Opera* 1, S. P., p. 183, col. 2; Brush, *Selected Works of Pierre Gassendi*, p. 388; Čapek, *Concepts of Space and Time*, p. 93.

177 Brush, *Selected Works of Pierre Gassendi*, p. 389; *Opera* 1, S. P., p. 183, col. 2. I have changed Brush's translation from "has the negative quality (*repugnantia*) of allowing" to "lacks resistance (*repugnantia*), which allows . . ."

178 "Anticipatio, seu notio, quam de corpore habemus est ut sit quid dimensiones habens et capax resistentiae." *Opera* 1, S. P., Pars Prima, Quae est Logica, bk. 1, ch. 7, canon 7, p. 55, col. 1; see also Bloch, *La Philosophie de Gassendi*, p. 177, n. 29, where this line is quoted. In the very next paragraph (canon 8), Gassendi explained that "void (*inane*) . . . is conceived as the lack of capacity to resist (*resistentiae incapax*); nor does it suffer anything or have the power to act. It is merely an existing space (*existens spatium*) in which a body can be received or through which a body could traverse." This distinction between void space and body had already been made by Patrizi (see above, Ch. 8, Sec. 4a).

179 In a summary of the major ideas of Patrizi's *Nova philosophia*, Gassendi (*Opera* 1, S. P., part 2 [*Physics*], sec. 1, bk. 3 [*De materiali principio sive materia prima rerum*], p. 246, cols. 1–2), after mentioning Patrizi's assumption that God had produced space outside Himself, explained that "with respect to this space, with which the three dimensions, length, breadth, and depth coincide, he [Patrizi] presents nothing other than what we ourselves have reasoned about it above." Gassendi's discussion formed part of his account of the ideas of the *recentiores* (see above, n. 165 of this chapter and also John Henry, "Francesco Patrizi da Cherso's Concept of Space," pp. 568–569.

180 Brush, *Selected Works of Pierre Gassendi*, p. 384; *Opera* 1, S. P., p. 182, col. 1. Franciscus Pitigianus used such an approach to demonstrate the impossibility of void space (see above, Ch. 7, n. 53).

181 *Ibid.; ibid.*

182 *Ibid.; ibid.* The italics are mine. During the seventeenth century, time and space were usually conceived as the same kind of being. Indeed Walter Charleton would call time the "twin-brother" of space (*Physiologia*, p. 66, article 10). Scholastics also linked them (see above, Ch. 7, Secs. 2e, f and n. 148, 163, and 164). Aristotle had, of course, devoted the fourth book of his *Physics* exclusively to place, void, and time.

In denying that space is substance or accident, Gassendi was not the immediate non-scholastic successor to Patrizi. There was at least one, David Gorlaeus, or Van Goorle, and perhaps others, who adopted that position before Gassendi. Gorlaeus, who died at an early age sometime in the second decade of the seventeenth century, considered the problem of space under the rubric "De loco" in his *Exercitationes philosophicae*, published posthumously in 1620. (Gorlaeus' anti-Aristotelianism is apparent from the subtitle of the treatise, which declared that "many dogmas, especially of the Peripatetics, are overthrown.") According to Gorlaeus (what follows appears in *Exercitatio decima*, pp. 214–216), place (*locus*), or space, is not something real, but a "pure, unmixed nothing, for it is neither substance nor accident." ("Locus non est aliquid reale, sed purum putum nihil. Neque enim est substantia neque accidens.") It is not a substance, because space is neither a spirit (*spiritus*), nor an "intelligent thing" (*res intelligens*). Nor is space a body, because a body requires a place, and we would have the absurd situation in which a place needs a place or a space a space. Space cannot be body because it is separable from body and remains behind when a body is removed. Space is also not an accident because it can have no subject.

Prior to the creation of the world, space existed as an "unmixed nothing" (*putum nihil*). It was not, however, another being separate from and coexistent with God; God exists alone. And yet this "unmixed" and "pure nothing" is an infinite, three-dimensional space within which God would create the world. "Therefore that which was previously a vacuum, wholly void of any body, has now been partially filled" and constitutes the place of the world, because a place is nothing other than "a space filled with body." With the creation of the world, a void space exists beyond it, although Gorlaeus confessed ignorance as to whether anything else might exist within it.

Although Gorlaeus agreed with Patrizi and Gassendi that space is neither substance nor accident, he diverged from them on a number of points. He stood alone in his conception of infinite space as a three-dimensional, uncreated, pure, unmixed nothing, which coexists with God but is not a real being and thus no threat to God. Unfortunately, Gorlaeus' discussion is brief and avoids the difficult theological problems. His cosmological ideas are discussed in a few pages by William H. Donahue, "The Dissolution of the Celestial Spheres 1595–1650" (Ph.D. diss., University of Cambridge, 1972), pp. 171–175. In the few paragraphs devoted to Gorlaeus' spatial ideas, Donahue concludes, mistakenly it seems, that Gorlaeus "very nearly identifies God and space" (p. 174). In fact Gorlaeus seemed deliberately to avoid such an identification. And small wonder! Although it seems a fair inference that space must somehow be associated with God – for Gorlaeus denied reality to space *and* also assumed its coeternality with God – it is difficult to imagine how something that is neither a spirit nor an "intelligent thing" could be linked with God. Without additional evidence, Gorlaeus' thoughts on these monumental puzzles must remain forever unknown.

183 Corrected from "of" in Brush's translation.

184 Brush, *Selected Works of Pierre Gassendi*, p. 384; *Opera* 1, S. P., p. 182, col. 1.

185 *Ibid.*, pp. 384–385; *ibid.* For Charleton's summary of the relations between substance, accident, space, and time, see *Physiologia*, pp. 65–66, articles 9, 10. In a letter to Le Pailleur in February 1648, Blaise Pascal allowed that space and time need not be body, spirit, substance, or accident (see Blaise Pascal, *Oeuvres Complètes*, ed. by Jean Mesnard [2 vols.; Paris: Desclée de Brouwer, 1970], vol. 2, pp. 564–565; the passage is quoted by Bloch, *La Philosophie de Gassendi*, p. 197, who suggests that Pascal may

have derived the idea from Gassendi, either from a manuscript version of the *Animadversiones*, which was not printed until 1649, from a letter, or perhaps from direct correspondence with Gassendi or with friends of the latter).

186 Bloch rightly emphasizes the importance of the indifference of space (*La Philosophie de Gassendi*, p. 177).

187 Whatever Gassendi's attitude on theological matters, Koyré's explanation for the omission of Gassendi from his study is surely unacceptable, namely that "Gassendi is not an original thinker and does not play any role in the discussion I am studying. He is a rather timorous mind and accepts, obviously for theological reasons, the finitude of the world immersed in void space; . . ." (*From the Closed World*, p. 290, n. 5). Because Koyré's major concern in his study was the development of the concept of an infinite void space, his deliberate elimination of Gassendi is untenable, as should be evident from what has been said here. Other authors who also ignored Gassendi in their studies on infinite space in the seventeenth century are Edwin Arthur Burtt, *The Metaphysical Foundations of Modern Physical Science* (rev. ed.; London: Routledge & Kegan Paul, 1932) and Markus Fierz, "Ueber den Ursprung und die Bedeutung der Lehre Isaac Newtons vom absoluten Raum," *Gesnerus 11* (1954), Fascicule 3/4, 62–120. Among those who have recognized Gassendi's significance we may mention Max Jammer, *Concepts of Space*, pp. 92–94, and Milič Čapek, "Was Gassendi a Predecessor of Newton?" *Actes du Dixième Congrès International d'Histoire des Sciences* (Paris: Hermann, 1964), vol. 2, pp. 705–709 (for others, see above, n. 163 of this chapter). If it was a "timorous mind" and "theological reasons" that led Gassendi to propose a finite world "immersed in a void space," what, we may well ask, prompted Barrow, Newton, Locke, Clarke, Otto von Guericke, and other seventeenth-century luminaries to adopt that very same cosmology? Indeed, as we have repeatedly emphasized in this volume, Gassendi and most, if not all, in the seventeenth century who accepted the actual existence of an infinite void space did so within the framework of the by-then widely accepted Stoic cosmology that located a finite world within that infinite space. As for Gassendi's timidity, it will become apparent below that his spatial theory was daring to the point of heresy.

188 Čapek and Emge, in Čapek, *Concepts of Space and Time*, p. 94; *Opera* 1, S. P., p. 191, col. 1. Gassendi's description of God's indivisible omnipresence is, of course, the medieval whole in every part doctrine, which is discussed above, Ch. 6, n. 127.

189 Čapek, *Concepts of Space and Time*, p. 94; *Opera* 1, S. P., p. 191, col. 1.

190 Gassendi thus aligned himself with a large body of scholastic opinion; see above, Ch. 5, Sec. 4 and Ch. 7, n. 110, 125, and 143.

191 Gassendi, *Opera* 1, S. P., pp. 183, col. 2–184, col. 1; Brush, *Selected Works of Pierre Gassendi*, pp. 389–390. That the concept of an uncreated space independent of God was theologically troublesome was noted by Charleton, who insisted, however, that "though we concede them [i.e., spatial dimensions] to be *improduct* by, and *independent* upon God, yet cannot our Adversaries therefore impeach us of impiety, or distort it to the disparagement of our theory" (*Physiologia*, p. 68, article 16; the brackets are mine, but the italics are Charleton's; how Charleton and Gassendi thought they had avoided charges of impiety will be seen below). Charleton even used the English cognates of the very terminology employed by Gassendi. Indeed Isaac Barrow would also use the same terminology – *improductum* and *independens* – when he repudiated as impious the idea that space could be uncreated and independent of God (see Barrow's *Lectiones mathematicae*, Lectio X in *The Mathematical Works of Isaac Barrow, D.D.*, ed. by W. Whewell [Cambridge: Cambridge University Press, 1860], vol. 1, p. 149; the *Lectiones mathematicae* were given at Cambridge University during the years 1664, 1665, and 1666, as we learn from the title page of the London edition of 1683, which was reprinted by Whewell; see also Burtt, *Metaphysical Foundations of Modern Physical Science*, p. 149). For Bradwardine's typically hostile medieval reaction to a space independent of God, see above, Ch. 6, n. 98.

192 By "sacred doctors" Gassendi probably meant the Church Fathers, but I know of none who held such a view. See below, n. 200.

193 Brush, *Selected Works of Pierre Gassendi*, p. 389; *Opera* 1, S. P., p. 183, col. 2. For

the concluding word *mundum*, I have substituted "world" for Brush's "universe."

194 See above, Ch. 7, Secs. 2b, d and n. 77. Bloch (*La Philosophie de Gassendi*, p. 472, n. 148) remarks that Gassendi was favorably disposed toward the Jesuits.

195 *Opera* 1, S. P., p. 183, col. 2; Brush, *Selected Works of Pierre Gassendi*, p. 389. See also Charleton, *Physiologia*, p. 68, article 16.

196 *Opera* 1, S. P., pp. 183, col. 2–184, col. 1; Brush, *Selected Works of Pierre Gassendi*, pp. 389–390.

197 *Physiologia*, p. 68, article 16.

198 For all this, see *Opera* 1, S. P., p. 183, col. 2; Brush, *Selected Works of Pierre Gassendi*, p. 389. It should be noted that in her study of Gassendi, Lillian Unger Pancheri ("The Atomism of Pierre Gassendi: Cosmology for the New Physics" [Ph.D. diss., Tulane University, 1972], p. 105) makes a serious error in her translation of the same passage on which my analysis is based. Instead of the dichotomy that Gassendi plainly made between incorporeal substances such as God, intelligences, and human minds, on the one hand, and an incorporeal substance such as space, on the other, Pancheri translates the passage so that space has the same properties as God and "is a genuine substance, a genuine nature, to which there correspond real faculties and activities." Gassendi's true conception of space as completely inactive and passive is thus destroyed, with detrimental consequences for the interpretation of his spatial theory.

199 For example, by Henry More and Spinoza. Félix Thomas mistakenly assigns to Gassendi the view that immensity and eternity are attributes of God (*La Philosophie de Gassendi* [Paris: Ancienne Librairie Germer Baillière, 1889], p. 56). Immensity is associated with space, which in turn is distinct from God, whom we can only imagine *as if* He were extended (see above, Ch. 8, Sec. 4b).

200 As was customary with most nonscholastics in the seventeenth century, Gassendi mentioned no scholastic contemporaries or recent predecessors in his spatial discussion. Instead he spoke of the "sacred doctors" (see above, Ch. 8, Sec. 4b), by whom he appears to have meant the Church Fathers. In this he was probably following scholastic tradition wherein it was common to cite St. Augustine and others as proponents or opponents of this or that opinion about imaginary space (for example, see Amicus' discussion above, Ch. 7, Sec. 2d). In this sense, the sacred doctors were important and were regularly invoked even though they actually contributed little to spatial theory proper. It is in this sense that Gassendi invoked them (see above, Ch. 5, Sec. 4). But he could have learned about the role attributed to the sacred doctors only by reading to some extent in sixteenth- and seventeenth-century scholastic discussions on imaginary space. As we shall see below, the Coimbra Jesuits and Bartholomeus Amicus held opinions that were quite similar to what Gassendi presented as the scholastic concept of imaginary space.

201 For references to these and other scholastics, see Otto von Guericke, *Experimenta nova (ut vocantur) Magedeburgica de vacuo spatio* (Amsterdam, 1672, reprinted in facsimile Aalen: Otto Zeller, 1962), pp. 51–52, 61–65, and my translation of much of this material in Grant, *Source Book in Medieval Science*, pp. 563–568. It was the Jesuits in particular who seemed to have influenced Guericke.

202 For a general account with brief bibliography, see Fritz Krafft, "Guericke (Gericke), Otto von," *Dictionary of Scientific Biography*, vol. 5 (New York: Scribner, 1972), pp. 574–576.

203 In his monograph, *Otto von Guericke Philosophisches über den leeren Raum* (Berlin: Akademie-Verlag, 1968), Alfons Kauffeldt not only considers the experiments but also summarizes and analyzes Guericke's second book (*De vacuo spatio*), which will be almost the sole focus of our attention, in ch. 2, pp. 25–70 (in a lengthy appendix, he translates most of the second book of the *Experimenta nova* and parts of the first and fourth). Unfortunately, Kauffeldt's interpretation and analysis are seriously flawed by the pervasive intrusion of Marxist ideology. The powerful scholastic influences on Guericke are perceived as mere reactionary holdovers from an earlier mode of thought with which Guericke was compelled to cope even though he did not really believe what he wrote on the nature of the relationship between God and space (pp. 63–64). These scholastic remnants in Guericke's work merely reflect the fact, as dialectic teaches, that

when we advance from a lower stage of knowledge to a higher, the older and lower stage does not simply vanish without a trace (p. 65). Pioneer thinkers include such materials but no longer really believe them. The real Guericke was a materialist (p. 63) whose ideas about void space Kauffeldt links to the ideas about nothing and the "negation of negations" in the thought of Friedrich Engels (for example, see pp. 34, 56). Despite these reservations about Kauffeldt's interpretations, and ideology aside, he has sought to comprehend the text objectively and his translations appear sound.

Guericke's earliest experiments on air pressure and vacuum were published by the Jesuit Gaspar Schott (1608–1666) as an appendix to the latter's *Mechanica Hydraulico-pneumatica* (Würzburg [Herbipolis]: Henricus Pigrin, 1657), a treatise originally intended as a guide to the hydraulic and pneumatic instruments in Athanasius Kircher's museum. Although some of Guericke's experiments were described and discussed in Schott's edition, none of his ideas on extracosmic space were included. These appear only in Guericke's own edition of his experiments in 1672. The famous Kircher was a Jesuit with whom Schott had studied (see A. G. Keller, "Schott, Gaspar," *Dictionary of Scientific Biography*, vol. 12 [New York: Scribner, 1975], pp. 210–211).

204 Guericke, *Experimenta Nova*, bk. 2, ch. 7, p. 62, col. 1.
205 *Ibid.*
206 Guericke here upheld the scholastic view that space as a negation is not nothing but possesses some kind of existence even though not conceived as a positive entity. See above, Ch. 7, Secs. c, d.
207 Guericke, *Experimenta nova*, bk. 2, ch. 7, p. 63, col. 1.
208 *Ibid.* For Aristotle's argument, see above, Ch. 5, Sec. 1. Roger Bacon also accorded this nothing a certain existential status (above, Ch. 2, Sec. 2, and Ch. 5, Sec. 1).
209 Guericke, *Experimenta nova*, p. 63, col. 2. The translation, somewhat altered, is based upon my earlier version in Grant, *Source Book in Medieval Science*, p. 566. In the *Source Book*, all or part of bk. 1, ch. 35 and bk. 2, chs. 6, 7, and 8 of Guericke's *Experimenta nova* has been translated. The whole work has been translated into German in Hans Schimank (ed. and trans.), *Neue (sogennante) Magdeburger Versuche über den leeren Raum. Nebst Briefen, Urkunden und anderen Zeugnissen seiner Lebens- und Schaffensgeschichte.* With the collaboration of Hans Gossen, Gregor Maurach and Fritz Krafft (Düsseldorf: VDI-Verlag, 1968). Most of the passage above has also been translated into German by Kauffeldt, *Otto von Guericke Philosophisches über den leeren Raum*, p. 56.
210 The full passage proclaims that "the *nothing* (*nihil*) beyond the world and *space* (*spatium*) are one and the same; and so-called *imaginary* space is true space, for imaginary space (in the common opinion of philosophers) is nothing and nothing is space, and the space which they call imaginary is true space (*spatium verum*)." Grant, *Source Book in Medieval Science*, p. 565; Guericke, *Experimenta nova*, bk. 2, ch. 6, p. 62, col. 1. It is noteworthy that this statement followed upon Guericke's claim that a spear or stone projected beyond the world would continue on ad infinitum because there is nothing to stop it. Such a motion could occur only in an infinite space. The indefinite motion of a spear is but a special case of the ancient anti-Aristotelian Stoic argument that "if someone should reach the last confines of the world, he either could, or could not, extend his arm beyond the last surface of the last heaven or hurl a spear and a stone into this nothingness" (Grant, *Source Book in Medieval Science*, p. 565; Guericke, *Experimenta nova*, p. 61, col. 2; see above, Ch. 5, Sec. 1 and n. 20 for medieval scholastics who cited it).

Although Guericke's discussion of extracosmic space is in the Stoic framework, that is, a finite world surrounded by an infinite space, he apparently believed in the possibility of a plurality of worlds, though he definitely rejected the possibility of an infinity of worlds in the sense of Giordano Bruno. Just as God did not create the whole sea for a single fish, or the entire air for a single fly or bird, so also would He not create a single world in an infinite universe, but would rather produce a variety of worlds, one different from the other (see *Experimenta nova*, bk. 7, ch. 5, p. 243, and Krafft, "Guericke [Gericke], Otto von," *Dictionary of Scientific Biography*, vol. 5, p. 575).
211 On this distinction, see above, Ch. 7, Sec. 2a and n. 39 for further references. Although

imaginary space was usually identified with vacuum, as it was for Guericke, Bartholomeus Amicus sharply distinguished between imaginary space and vacuum (Ch. 7, Sec. 2d).

212 Guericke, *Experimenta nova*, bk. 2, ch. 6, p. 62, col. 2; Grant, *Source Book in Medieval Science*, p. 565.

213 Grant, *Source Book in Medieval Science*, p. 565; Guericke, *Experimenta nova*, p. 62, col. 2.

214 *Ibid.*; *ibid.*

215 Guericke, *Experimenta nova*, p. 57, col. 1.

216 *Ibid.*; see also Kauffeldt, *Otto von Guericke Philosophisches über den leeren Raum*, pp. 35–36. The phrase "in which all things exist, live, and are moved" clearly reflects the line from St. Paul, Acts 17, 28 (see above, n. 78 of this chapter). Guericke applied it to space, which he would equate with God's immensity.

217 Unless otherwise indicated, the ideas in this paragraph are drawn from Guericke, *Experimenta nova*, p. 57, cols. 1–2.

218 Perhaps Guericke's position is identical with that of Joseph Raphson and for the same reasons (see below, Ch. 8, Sec. 4f). If so, space yields only in the sense that it penetrates all things, that is, is "everywhere in all things, [and] through all things whether corporeal or incorporeal." Despite yielding, then, space is not itself penetrated, as Patrizi seemed to hold (see above, Ch. 8, Sec. 4a). Although Raphson assigned the property of impenetrability to space and Patrizi assigned it the property of yielding, which seem at first glance radically different, it is conceivable that Patrizi (and Gassendi, for that matter) really agreed with Raphson. To reconcile their views, it is only necessary to assume that for Patrizi, whereas space simultaneously penetrates a body, it is only the latter, not space, that suffers the penetration.

219 Guericke, *Experimenta nova*, p. 58, cols. 1–2, where Descartes is mentioned by name.

220 Grant, *Source Book in Medieval Science*, p. 567; for the full title of Du Bois' treatise, see Grant, *Source Book*, p. 566, n. 76.

221 Grant, *Source Book in Medieval Science*, p. 567; Guericke, *Experimenta nova*, p. 64, cols. 1–2. The passage from Kircher, though substantially correct, deviates somewhat from the text in the second edition. See *R. P. Athanasii Kircheri e Societate Jesu Iter Extaticum Coeleste, . . . interlocutoribus Cosmiele et Theodidacto . . . ipso auctore annuente, P. Gaspare Schotto, Regiscuriano e Societate Jesu . . .* (2d ed.; Würzburg [Herbipolis], 1660), dialogue 2, ch. 9, sec. 4, p. 437. The whole of ch. 9 (pp. 433–440) is devoted to imaginary space and is even titled *De spacio imaginario*.

222 Grant, *Source Book in Medieval Science*, p. 567; Guericke, *Experimenta nova*, p. 64, col. 2. The Lessius referred to here is probably Leonard Lessius (1554–1623), a Jesuit professor of theology at the University of Louvain who wrote *De perfectione divino*, from which Guericke quoted the passage referred to here in bk. 1, ch. 35 (see Grant, *Source Book*, p. 564, for the translation of it). Kircher was a bitter foe of those who identified nothing with a vacuum beyond the world. God's existence beyond the world obliterates any void or emptiness, or nothing. Earlier in the dialogue (p. 434 of the *Iter Extaticum Coeleste* cited in n. 221), Kircher, through his spokesman Cosmiel, defined nothing as follows: "Nothing is indeed nothing. Nothing is not something; [it is] not this or that or another being, but it is no being. Nothing is nowhere, neither in the mind, nor in the nature of things, nor in the intelligible or sensible world; [it is] not in God, nor beyond God in any creatures. Any whatever being exists; any whatever something exists; all full things have being. Nothing [however] is superfluous; a vacuum is nothing; nothing is empty; nothing is banished from the universe." For Kircher imaginary space was God Himself, not a vacuum or nothing. His position resembles that of Franciscus Bona Spes (see above, Ch. 7, Sec. 2f).

223 Guericke, *Experimenta nova*, pp. 64, col. 2–65, col. 2; Grant, *Source Book in Medieval Science*, pp. 567–568. Not only did Guericke identify space with God's immensity, but he apparently also identified God with nature (see Alfons Kauffeldt, "Otto von Guericke on the Problem of Space," *Actes du XIe Congrès International d'Histoire des Sciences*, Aug. 24–31, 1965 [Wroclaw/Warsaw/Cracow: Ossolineum, Maison d'Edition de l'Académie Polonaise des Sciences, 1968], vol. 3, p. 365).

224 Grant, *Source Book in Medieval Science*, p. 567; Guericke, *Experimenta nova*, p. 64, col. 2. I have altered my translation here.

225 As Fritz Krafft wrongly maintains in his article "Guericke (Gericke), Otto von," *Dictionary of Scientific Biography*, vol. 5, p. 575, col. 2. In the final book of the *Experimenta nova*, p. 241, Guericke declared in reaction to further ideas of Kircher that "God alone is infinite; two infinites cannot exist. Thus our author [Kircher] responds correctly in this place [when he says] that God did not create an actual infinite. . . ." Because two infinites do not exist, and space and God's immensity are both described as infinite, they must be one and the same. As if to reinforce this view, Guericke insisted that *immensum, infinitum,* and *aeternum* are identical and belong to the Uncreated (see bk. 2, ch. 9, p. 65, col. 1). Space, however, is independent of and prior to all created things, for which it serves as the universal container.

226 The passage in question appears at the end of bk. 1, ch. 35, where, after reporting numerous interpretations of imaginary space, Guericke concluded: "Finally, if imaginary space is conceded, it follows that it is infinite and immense or infinitely expanded in all its parts, that is, in length, width, and depth; it also follows that it is incorruptible, sempiternal, immobile, fixed, and permanent, so that by no force or reason could it be destroyed, so that it could be the receptacle [or container] of any body whatever, whether great or small. On this matter, see several places in Book II, following." Grant, *Source Book in Medieval Science*, p. 564. Guericke, *Experimenta nova*, p. 52, col. 2. It is, of course, in bk. 2, where Guericke gave his own opinions and where he said that imaginary space, or the Uncreated, is not to be interpreted as an extended magnitude, as is done when we consider space in the vulgar or common way.

227 Guericke, *Experimenta nova*, bk. 2, ch. 4, p. 58, col. 2.

228 See Kauffeldt, "Otto von Guericke on the Problem of Space," vol. 3, p. 366. In this Guericke differed from many seventeenth-century "vacuists" (see above, Ch. 8, Sec. 4a and n. 143), but not from Gassendi (see above, Ch. 8, Sec. 4b).

229 Guericke, *Experimenta nova*, p. 85, col. 1; see also p. 86, col. 2, where he said that "beyond or above the region of air, there begins a pure space void of every body." Void will occur whenever the quantity of matter available is insufficient to fill space.

230 Most of what Guericke utilized from the Coimbra Jesuits on the nature of imaginary space and its relation to God appears in a single paragraph that he quoted in bk. 1, ch. 35, titled *De spatio imaginario extra mundum* (*On imaginary space beyond the world*). The lengthy passage appears on p. 51 and, as Guericke said, is drawn from the Coimbra *Commentary on Aristotle's Physics*, bk. 8, ch. 10, question 2, article 4. For a translation of Guericke's discussion, see Grant, *Source Book in Medieval Science*, p. 563, col. 1. The passage quoted from the Coimbra Jesuits is omitted in my Guericke translation because it was translated on p. 562, col. 1, as part of the relevant selection on imaginary space from the Coimbra Jesuits. The Coimbra ideas on imaginary space cited by Guericke from the fourth article are described above, Ch. 7, Sec. 2b.

231 "Spatium itaque dupliciter intelligendum: vel secundum Vulgi conceptum, vel secundum *Universale rerum omnium Continens.*" Guericke, *Experimenta nova*, bk. 2, ch. 4, p. 57, col. 1. Although I have altered the punctuation, the italics are Guericke's.

232 In light of all that has been said here on Guericke, Alfons Kauffeldt's effort to denigrate and even neutralize scholastic influences on Guericke (see above, n. 203 of this chapter) must be rejected as contrary to the evidence. If Guericke really believed that scholastic ideas on space were silly, untrue, or reactionary, he had only to omit them from his book. He was under no compulsion to discuss such abstract and difficult metaphysical and theological conceptions and under less constraint to accept them. That he included them and accepted much on space and God from the scholastic tradition, whose authors he was not ashamed to cite by name, ought to convince us not only that he took his discussion of imaginary space seriously and believed much, if not all, of it, but also that he considered it of central importance to the overall study of void space.

233 There were, of course, those who denied actual infinity to space and conceived it instead as either indefinite or as a potential infinite in the Aristotelian sense. Such a space was unbounded and without limits and practically speaking, though not conceptually,

was tantamount to an actually infinite space. For Descartes's argument for the indefiniteness of space, see Koyré, *From the Closed World*, pp. 116–124 (Descartes's reasons were theological: Only God is infinite; extended space, which Descartes identified with matter, is separate and independent of God and cannot therefore be infinite; however, because it is unlimited, it functions as an infinite space). Ralph Cudworth would also insist that whether space beyond the world were conceived as body or as mere extension, there is no ground for the assumption of its infinity, for however great the world may be, "it is not Impossible, but that it might be still Greater and Greater, without end. Which *Indefinite Encreasableness* of Body and Space, seems to be mistaken for a *Positive Infinity* thereof." *The True Intellectual System of the Universe* (London, 1678), bk. 1, ch. 4, p. 766; cited also by Flora MacKinnon, *Philosophical Writings of Henry More*, p. 295.

234　See above, Ch. 8, Sec. 4c. Koyré makes the curious claim that Spinoza, Malebranche, and Henry More discovered, at approximately the same time, that "Absolute space is infinite, immovable, homogeneous, indivisible, and unique" (*From the Closed World*, p. 149). These properties had been discovered much earlier by Patrizi, Gassendi, and scholastics such as Bartholomeus Amicus.

235　Nicolas Malebranche, *Dialogues on Metaphysics and on Religion*, trans. Morris Ginsberg (London: Allen & Unwin, 1923), eighth dialogue, pp. 212–213; for the French text, see *Oeuvres complètes de Malebranche*, ed. by André Robinet, vol. 12: *Entretiens sur la Métaphysique et sur la Religion* (Paris: Librairie Philosophique J. Vrin, 1965), pp. 184–185. In the same eighth dialogue, Malebranche explained that "When I tell you that God is in the world and infinitely beyond it, you do not grasp my meaning if you believe that the world and the imaginary space (*les espaces imaginaires*) beyond are, so to speak the space (*le lieu*) which the infinite substance of the Divinity occupies. God is in the world only because the world is in God, for God is only in Himself, only in His immensity." Ginsberg (trans.), *Dialogues on Metaphysics and on Religion*, p. 210; *Oeuvres complètes de Malebranche*, vol. 12, p. 182. Although I have not altered Ginsberg's translation, the French text has been inserted in two places to identify certain imprecisions, which do not, however, affect the meaning of the passage. From this passage, it is obvious that Malebranche held the same opinion as those scholastics who refused to admit extracosmic imaginary spaces because they insisted that God is not in any spaces but is only in Himself (for example, see Phillip Faber's opinion described above in Ch. 7, n. 110; Otto von Guericke was also of this opinion [see above, Ch. 8, Sec. 4c]). For good brief descriptions of Malebranche's ideas on God's immensity, see Beatrice K. Rome, *The Philosophy of Malebranche* (Chicago: Henry Regnery, 1963), pp. 201–203, and Koyré, *From the Closed World*, pp. 156–159. In light of Malebranche's ideas on God and space presented here, it is difficult to determine the basis of Burtt's assertion that for Malebranche "space became practically God Himself" (*Metaphysical Foundations of Modern Physical Science*, p. 140). This is surely false if by space Burtt meant three-dimensional incorporeal extension, but reasonably accurate if by space he intended "intelligible extension." For as Koyré has observed (*From the Closed World*, p. 156), Malebranche did not wish "to spatialize God in the manner in which Henry More or Spinoza did it." To avoid this, he distinguished "the *idea* of space, or 'intelligible extension,' which he places in God, from the gross material extension of the world created by God." Whatever Malebranche had in mind for intelligible extension, it was not three-dimensional incorporeal extension, for that would have linked him immediately with More. Koyré's assertion that for Malebranche intelligible extension is "the space of geometers" is therefore untenable. For how would such a space differ from Henry More's? Malebranche's divine, or intelligible, extension is more akin to what Bradwardine had in mind when he said of God that "He is infinitely extended without extension and dimension" (see above, Ch. 6, Sec. 3c). If so, Malebranche's intelligible extension defies the understanding and is anything but intelligible. Koyré's analysis was based on Malebranche's *Méditations chrétiennes*, méd. IX, sec. 9 (for references, see Koyré, *From the Closed World*, p. 293) in the Paris edition of 1926 (a more recent edition appeared in 1967 by H. Gouhier and A. Robinet and constitutes vol. 10 of the *Oeuvres complètes de Malebranche*).

More generally, Malebranche, as a Cartesian, followed his master in rejecting imaginary spaces. Descartes rightly believed that imaginary spaces were the invention of "the philosophers," by whom he meant scholastic Aristotelians (see Descartes, *Le Monde* in *Oeuvres de Descartes*, eds. Charles Adam and Paul Tannery, vol. 11 [Paris: Léopold Cerf, 1909], ch. 6, pp. 31–32, and also the edition of *Le Monde* in Ferdinand Alquié, *Oeuvres Philosophiques, Tome I (1618–1637)* [Paris: Editions Garnier Frères, 1963], pp. 343–344 and the useful notes that explain Descartes's attitude toward real and imaginary spaces or places). In a letter to Henry More in May 1649, Descartes insisted that it was contradictory to suppose a finite or limited world "because I cannot but conceive a space outside the boundaries of the world wherever I presuppose them. But, for me, this space is a true body. I do not care if it is called by others imaginary. . . ." Trans. by Koyré, *From the Closed World*, p. 123, from Descartes's second letter to More (for the Latin text, see *Oeuvres de Descartes*, vol. 5 [Paris: Léopold Cerf, 1903], p. 345, or *Henrici Mori Epistolae Quatuor ad Renatum Des-Cartes* in *A Collection of Several Philosophical Writings of Dr. Henry More* [2d ed.; London: printed by James Flesher for William Morden, 1662], pp. 83–84). In the *Principia Philosophiae*, part 2, sec. 21 (*Item mundum esse indefinite extensum*), Descartes equated the spaces that we can always imagine beyond the world with real space (see *Oeuvres de Descartes*, vol. 8, pt. 1, p. 52, and Koyré's translation of it in *From the Closed World*, p. 104). Descartes thus attacked the scholastic philosophers because they distinguished between real and imaginary spaces (for at least one scholastic who rejected that distinction, albeit for radically different reasons, see the discussion on Franciscus Bona Spes, above, Ch. 7, Sec. 2f). For Descartes, the spaces we imagine beyond the world are real and, of course, extended.

236 Early in his career, in his first poem (*Psychodia Platonica*), More believed that infinite extension implied a contradiction. Between 1642 and 1646, however, when he published his *Democritus Platonissans*, which was subtitled *An Essay Upon the Infinity of Worlds out of Platonick Principles* (Cambridge, 1646), More underwent a transformation and became a believer not only in infinite extension but in an infinity of worlds as well (though he eventually abandoned this belief; see n. 251 below). See the facsimile reprint by the Augustan Reprint Society, Publication Number 130, with introduction by P. G. Stanwood (Los Angeles: William Andrews Clark Memorial Library, U.C.L.A., 1968); also see Ernest Tuveson, "Space, Deity, and the 'Natural Sublime'," *Modern Language Quarterly* 12 (1951), 23–24, and, on More's concept of a plurality of worlds, Dick, *Plurality of Worlds*, pp. 202–206.

237 Except for the omission of More's attack on the whole in every part doctrine (Holenmerism), both Koyré (*From the Closed World*, chs. 5 and 6, pp. 110–154) and Jammer (*Concepts of Space*, pp. 41–49) have excellent sections on More's ideas about space. For the details of More's controversy with Descartes, see Koyré's fifth chapter. The omission by both Koyré and Jammer of More's discussion of Holenmerism is a direct consequence of their neglect of scholastic ideas, the neglect being virtually total with Koyré.

238 More called Descartes "the Prince of the Nullibists." See the *Enchiridion metaphysicum*, ch. 27, sec. 2 (title in margin), p. 351. The translation is that of Joseph Glanvil, who translated the final two chapters, 27 and 28, of More's *Enchiridion metaphysicum* and included them in his *Saducismus Triumphatus: Or, Full and Plain Evidence Concerning Witches and Apparitions* (London, 1681) under the title "The Easie, True, and Genuine Notion, and Consistent Explication of the Nature of a Spirit." Both chapters from Glanvil's 1681 translation are reprinted in Flora Isabel MacKinnon (ed.), *Philosophical Writings of Henry More*. Glanvil's entire *Saducismus Triumphatus*, including the two chapters by More, has been reprinted in facsimile from the 1689 edition by Coleman O. Parsons (Gainesville, Florida: Scholars' Facsimiles & Reprints, 1966). For convenience, references to both translations, which seem identical, will be given. For the few words cited in this note, see MacKinnon, p. 184, and Parsons, p. 134. When he declared that "the ESSENCE of a Spirit is where its OPERATION is" (MacKinnon, p. 189; Parsons, p. 140; *Enchiridion metaphysicum*, p. 356), More effectively linked the Nullibist position and its denial of location to spirits with the Scotist conception

(though Scotus is not mentioned) that God operates by His will rather than direct presence.

239 MacKinnon, pp. 183–184; Parsons, p. 134; *Enchiridion metaphysicum*, ch. 27, p. 351. In the next line, More explained that the Greeks called it οὐσίαν ϱλενμεϱῆ," *an Essence that is all of it in each part. . . ."* Although the Holenmerian position is the same as that expressed by Plotinus in *Enneads* IV, 2, 1 the Greek word, or words, from which More derived his term "Holenmerian" does not appear in this Plotinian passage (see *Plotini Opera*, Tome 2: *Enneades* IV–V, eds. Paul Henry and Hans-Rudolf Schwyzer [Paris: Desclée de Brouwer; Brussels: L'Edition Universelle, 1959], p. 6, lines 61–66; for the English translation, see above, Ch. 6, n. 127). Apparently More coined the term (see MacKinnon, p. 326, n. 61). Although he may not have been the first to attack the whole in every part doctrine, his criticisms are the first that have come to my attention.

240 The figure is More's and appears in *Enchiridion metaphysicum*, p. 367; also MacKinnon, p. 198, and Parsons, p. 151.

241 MacKinnon, p. 198; Parsons, pp. 151–152; *Enchiridion metaphysicum*, pp. 367–368.

242 MacKinnon, p. 201; Parsons, p. 155; *Enchiridion metaphysicum*, ch. 27, sec. 13, p. 371.

243 MacKinnon, p. 200; Parsons, p. 153; *Enchiridion metaphysicum*, p. 369.

244 MacKinnon, p. 200; Parsons, p. 154; *Enchiridion metaphysicum*, ch. 27, sec. 12, p. 370.

245 MacKinnon, p. 201; Parsons, p. 154; *Enchiridion metaphysicum*, pp. 370–371.

246 Henry More, *The Immortality of the Soul* (London, 1662), p. 3 (Preface). With separate title page and pagination, *The Immortality of the Soul* was published in *A Collection of Several Philosophical Writings of Dr Henry More* (London, 1662). The passage above is also cited by Burtt, *Metaphysical Foundations of Modern Physical Science*, p. 129.

247 Koyré, *From the Closed World*, p. 128.

248 The collection of properties listed here was derived from More's *Antidote Against Atheism* (London, 1662), p. 15, which was printed with separate title page and pagination in *Collection of Several Philosophical Writings of Dr Henry More* (London, 1662), and the *Enchiridion metaphysicum*, ch. 28, pp. 377–380 (in the translation by Glanvil, see MacKinnon, pp. 206–209, and Parsons, pp. 161–164); on p. 128 of *From the Closed World*, Koyré cites the properties of spirit from *Antidote Against Atheism*. Although nothing will be said of it here, More also assigned a fourth dimension to spirit that he called *Essential Spissitude* (*Enchiridion metaphysicum*, ch. 28, sec. 7, pp. 384–385; MacKinnon, pp. 213–214; Parsons, pp. 169–170). Such a concept was needed, in More's judgment, to explain how spirits could contract themselves without any impairment of their essences or extensions and to explain further how two equal extensions, spirit and body or two spirits, could occupy the same place.

249 For More's description see *Immortality of the Soul*, bk. 3, chs. 12, 13, pp. 193–204; also Koyré, *From the Closed World*, pp. 132–133 and Burtt, *Metaphysical Foundations of Modern Physical Science*, pp. 134–135.

250 *Immortality of the Soul*, p. 203.

251 *Enchiridion metaphysicum*, ch. 6, sec. 4, pp. 44–45 and Koyré, *From the Closed World*, pp. 140–141, where part of sec. 4 is translated, including More's reference to Plutarch's report that the Stoics denied void within the world but assumed an infinite one outside. More abandoned his earlier acceptance of an infinity of worlds (see above, n. 236 of this chapter).

252 For More's attack on the Cartesian conception of relative motion, see *Enchiridion metaphysicum*, ch. 7, pp. 52–64; for a description and analysis, see Koyré, *From the Closed World*, pp. 142–145, and Burtt, *Metaphysical Foundations of Modern Physical Science*, p. 139.

253 See above Ch. 2, Sec. 2a; for the anti-Cartesian position represented by Philoponus, see above, Ch. 2, Sec. 3. Virtually all of the seventeenth-century authors considered in this study rejected the Cartesian thesis.

254 See Koyré, *From the Closed World*, p. 145. "For Descartes," as Burtt explains (*Metaphysical Foundations of Modern Physical Science*, p. 138), "space and matter were the same thing, a material body being nothing but a limited portion of extension."

255 In the *De corpore* of 1655, Hobbes explained all this. See his *Elements of Philosophy, the first section: "Concerning Body,"* in *The English Works of Thomas Hobbes,* vol. 1, pp. 92–93. For the distinction between real and imaginary space, see p. 105, sec. 4. In sec. 5, p. 105, Hobbes explained that "space, by which word I here understand imaginary space, which is coincident with the magnitude of any body, is called the place of that body; . . ." For the methodological significance that Hobbes attached to the imagined annihilation of the world, see above, n. 169 of this chapter.

The basic content of Hobbes's ideas on space in the *De corpore* of 1655 had already been formulated in late 1642 and early 1643 in a critique of Thomas White's *De mundo dialogi tres* published in 1642. The Latin text of this work, hitherto known only in a single manuscript copy in the Bibliothèque Nationale, has been published by Jean Jacquot and Harold Whitmore Jones, *Critique du "De Mundo" de Thomas White,* introduction, texte, critique et notes (Paris: Librairie Philosophique J. Vrin, 1973; for the text on space, see ch. 3, secs. 1–3, pp. 116–118). An English translation by Harold Whitmore Jones has also recently appeared with the title *Thomas White's "De Mundo" Examined* (London: Bradford University Press in association with Crosby Lockwood Staples, 1976). For the relevant sections, see pp. 39–43. Here we find definitions of real and imaginary space (p. 41), the former conceived as an internal space in the manner of an accident in a subject, so that real space can be defined as " 'corporeity itself, or the essence of body taken *simpliciter,* inasfar as it is body.' " By contrast, imaginary space involves "the absence of body" as "understood through the mind-picture we have of bodies."

256 With one significant difference, Hobbes's description of imaginary space is much the same as that of the Coimbra Jesuits (see above, Ch. 7, Sec. 2b), Bartholomeus Amicus (see above, Ch. 7, Sec. 2d), and Gassendi (above, Ch. 8, Sec. 4b). For Hobbes, imaginary space has no reality outside the mind; for the others it has some sense of external reality ranging from the negation of Amicus to the three-dimensional existence of Gassendi. What links them is a common derivation of space from the presence and then imagined absence of bodies.

257 Hobbes, *De corpore* in *The English Works of Thomas Hobbes,* vol. 1, p. 106. See also Burtt, *Metaphysical Foundations of Modern Physical Science,* p. 138, and n. 344 below.

258 *An Antidote Against Atheism,* Appendix, ch. 7, p. 163, in *Collection of Several Philosophical Writings of Dr Henry More.* See Burtt, *Metaphysical Foundations of Modern Physical Science,* p. 140, for the quotation, including the term "disimagine."

259 The arguments appear in *Antidote Against Atheism,* Appendix ch. 7, pp. 163–165; see also Burtt, *Metaphysical Foundations of Modern Physical Science,* pp. 140–141.

260 The second reply (p. 164, paragraphs 3–5) bears a striking resemblance to Thomas Hobbes's concept of imaginary space (see above, Ch. 8, Sec. 4d). Here More considered space as "not the imagination of any real thing, but only of the large and immense capacity of the potentiality of the *Matter,* which we cannot free our *Mindes from.* . . ." We can imagine this possible matter to be measurable infinitely in every direction "whether this *corporeal Matter* were actually there or no," but it implies no real essence or being. But what if we should imagine such a space devoid of matter except for three brass balls? Does this not necessarily imply a triangular distance between these balls? And if there is a real distance between the balls, ought we not to infer the existence of a real, rather than imaginary, space? More replied that distance is not a real thing but merely "the privation of tactual union." Distance is not the property of anything because more or less of it may accrue to a thing that has not itself been altered. For example, if ball *A* remained fixed and the other balls were moved farther from it, the distance between *A* and the other balls would have altered so that *A* is now more distant from each of the other two. But *A* itself has in no way been altered. Hence distance cannot be a property of body. In the *Enchiridion metaphysicum,* ch. 8, sec. 10, p. 71, More would reject this opinion by his insistence that space is independent of our imaginations and of everything else.

261 Trans. by Koyré (*From the Closed World,* pp. 145–146) from *Enchiridion metaphysicum,* ch. 8, sec. 6, p. 68.

262 Trans. by Koyré, *From the Closed World,* p. 148, from More, *Enchiridion metaphysicum,*

ch. 8, sec. 8, p. 69. In a letter to Descartes (Dec. 11, 1648), More had declared that "God does indeed seem to be an extended thing . . ." ("Res enim extensa Deus videtur esse . . .;" *Oeuvres de Descartes*, vol. 5, Correspondance, DXXXI, p. 238, line 21).

263 Trans. by Koyré, *From the Closed World*, p. 148. Because of the importance of these terms, I also quote the Latin text: "Ut *Unum, Simplex, Immobile, Aeternum, Completum, Independens, A se existens, Per se subsistens, Incorruptibile, Necessarium, Immensum, Increatum, Incircumscriptum, Incomprehensibile, Omnipraesens, Incorporeum, Omnia permeans et complectens, Ens per Essentiam, Ens actu, Purus Actus.*" (*Enchiridion metaphysicum*, ch. 8, sec. 8, p. 69.)

264 Koyré's translation (*From the Closed World*, p. 151) from *Enchiridion metaphysicum*, ch. 8, sec. 12, p. 72. The twenty attributes are separately discussed on pp. 70–73. As reinforcement for his twenty divine attributes, More mentioned that he did not include among them the very name that the Cabbalists use to designate God: "MAKOM, that is, Place (*locus*)" (Koyré, *From the Closed World*, p. 148; *Enchiridion metaphysicum*, pp. 69–70). Near the end of ch. 8 (sec. 15, p. 74), More mentioned that the Cabbalist Cornelius Agrippa had assigned *locus* as one of God's attributes, by which he surely meant a space really distinct from matter. In his *De spatio reali* (pp. 14–19, 23; for the full title, see below, n. 277), Joseph Raphson would summarize and quote at some length from Cabbalistic sources.

265 Trans. by Burtt, *Metaphysical Foundations of Modern Physical Science*, p. 141, from *Enchiridion metaphysicum*, ch. 8, sec. 15 (incorrectly cited as 14 by Burtt), p. 74.

266 Whether Philoponus himself thought of void space as divisible or not is unclear.

267 For a detailed account of Spinoza's ideas on extension, see the superb study by Harry A. Wolfson, *Philosophy of Spinoza*, vol. 1, chs. 7 ("Extension and Thought"), pp. 214–261, and 8 ("Infinity of Extension"), pp. 262–295.

268 Wolfson, *Philosophy of Spinoza*, vol. 1, p. 224.

269 I have here followed the summary of T. V. Smith and Marjorie Grene (eds.), *From Descartes to Kant* (Chicago: University of Chicago Press, 1940), pp. 272–273. Whether or not Spinoza was a pantheist is of no import here, but Wolfson seems to deny it when he declares that Spinoza's God "is not identical with the physical universe" (*Philosophy of Spinoza*, p. 243; see also p. 325).

270 In *Ethics*, part I, prop. 13, Spinoza declared that "*Substance absolutely infinite is indivisible.*" See *The Chief Works of Benedict de Spinoza*, trans. from the Latin with an introduction by R. H. M. Elwes (New York: Dover Publications, 1951; reprinted from the edition of 1883), p. 54.

271 Not surprisingly, More himself opposed Spinoza's teachings. See MacKinnon (ed.) *Philosophical Writings of Henry More*, pp. 294–295.

272 Cudworth, *True Intellectual System of the Universe*, bk. 1, ch. 5, pp. 769–770.

273 See also above, n. 233 of this chapter, where Cudworth denied the existence of an actually infinite world or universe. Because God alone is actually infinite, it seems that he did not object to the assumption that infinite extension, or space, might be an attribute of God.

274 From *A Discourse of the Divine Omnipresence and its Consequences, Delivered in a Sermon before the Honourable Society of Lincolnes-Inn* (London, 1683) as quoted by Tuveson, "Space, Deity, and the 'Natural Sublime'," p. 29.

275 Turner refers here to the old issue of whether God operates by His will at a distance or by His presence (see above, Ch. 6, Sec. 4).

276 In his informative article "Space, Deity, and the 'Natural Sublime'," Tuveson suggests (pp. 29–30) that a consequence of the new ideas about God's extension and physical omnipresence was that "every man could cultivate the contemplation of the divine; for, it seemed, He is visible and accessible by the most ordinary faculties. No longer were such qualities as infinity, eternity, and omnipresence accessible to the comprehension of only a small group of the spiritually gifted." Thus an obscure country preacher and husbandman in Lancaster "printed, in 1649, a sermon the theme of which is the infinity of space as an attribute of God" (Tuveson, p. 25; the title of the work is *A Week-daies Lecture, or, Continued Sermon to wit, The Preaching of the Heavens*, wherein our theme is discussed on p. 2). Tuveson's quotations from it (p. 26) re-

veal an extraordinary conception of an infinitely immense God filling an infinite space.

277 Joseph Raphson, *Analysis aequationum universalis, seu ad aequationes algebraicas resolvendas methodus generalis et expedita. Ex nova infinitarum serierum methodo deducta ac demonstrata.* Editio secunda . . . cui etiam annexum est *De spatio reali seu ente infinito conamen mathematico-metaphysicum* (London, 1702). The *Analysis aequationum* was first published in 1696 without the *De spatio reali*, which was added to the second edition only in 1702. Koyré presents an excellent description of Raphson's ideas in *From the Closed World*, ch. 8 ("The Divinization of Space: Joseph Raphson"), pp. 190–205.

278 Raphson, *De spatio reali*, p. 26. The *Metaphysics* is presumably More's *Enchiridion metaphysicum*.

279 Raphson (*De spatio reali*, p. 69) quoted the famous metaphor that "God is an infinite sphere whose center is everywhere and circumference nowhere," although his version omits "infinite." The metaphor is embedded in a lengthy quotation from Gassendi's *Syntagma philosophicum*, bk. 2: De loco, et tempore, seu spatio, et duratione rerum, ch. 2, p. 190, col. 2 (for the full title, see above n. 165; for more on the metaphor, see above, Ch. 6, Sec. 3c and n. 108).

280 On p. 90 of *De spatio reali*, one of Raphson's two quotes from Guericke contains the description of space as "the universal container of all things" (*Universale illud omnium rerum Continens*) from the *Experimenta nova*, bk. 7, ch. 5, p. 242. The same description appears in bk. 2 (see above, Ch. 8, Sec. 4c, and n. 231 of this chapter).

281 For Lessius, see *De spatio reali*, p. 85 (Guericke also cited him; see above, n. 222 of this chapter); for references to scholastics, see pp. 37, 40, 78, 83, and 84.

282 *De spatio reali*, pp. 70 and 91–92; Koyré, *From the Closed World*, pp. 192 and 202–203.

283 *De spatio reali*, p. 71; for Koyré's translation of the passage, see *From the Closed World*, pp. 192–193.

284 Koyré translates all of the thirteen propositions and other material (pp. 194–196); for Raphson's text, see *De spatio reali*, ch. 5, pp. 74–80. Because of their importance, however, I shall repeat most of them. Bradwardine's *De causa Dei* and Spinoza's *Ethics* were models of geometrical procedure in essentially theological treatises. The latter was known to Raphson, and the former had been available in England since its publication in 1618.

285 Raphson deduced spatial infinity from indivisibility and immobility (*De spatio reali*, p. 74) and then argued that absolute immobility must follow from infinity because there is no place into which an infinite thing can move (*ibid.*, p. 75).

286 *De spatio reali*, p. 76 for the corollary. Koyré (*From the Closed World*, p. 195) interprets Raphson to mean that indivisible space is impenetrable to spirits as well as bodies, from which he infers Raphson's opposition to More's conflation of space and spirit. But this must surely be qualified, because Raphson subsequently identified space as God's immensity (proposition 13) and thereby also conflated space and spirit. Perhaps Raphson implicitly distinguished between uncreated spirit, which is God and infinite space, and created spirit, the latter no more able to penetrate space than bodies.

287 Perhaps this is also a fair representation of the opinions held by Patrizi and Gassendi (see above, Ch. 8, Sec. 4a for Patrizi and sec. 4b and n. 178 for Gassendi), although their opinions also seem reconcilable with Raphson's (see above, n. 218 of this chapter). Bruno (above, Ch. 8, Sec. 2) and perhaps also Guericke agreed with Raphson. For someone who clearly did not agree with Raphson, we need only mention John Keill, who, in his second lecture given at Oxford in 1700, a few years before the publication of Raphson's *De spatio reali*, described space as "altogether penetrable, receiving all bodies into itself, and refusing Ingress to nothing whatsoever" (John Keill, *An Introduction to Natural Philosophy: or, Philosophical Lectures Read in the University of Oxford 'Anno Dom.' 1700*, 2d ed. [London: Printed for J. Senex . . ., 1726], lect. 2, p. 15; see also Jammer, *Concepts of Space*, p. 128).

288 Trans. by Koyré, *From the Closed World*, p. 196.

289 Trans. by Koyré, *From the Closed World*, pp. 197–198, from Raphson, *De spatio reali*, ch. 6, p. 82.

290 Trans. by Koyré, *From the Closed World*, p. 200; *De spatio reali*, p. 85.

291 *De spatio reali*, p. 84; Koyré, *From the Closed World*, p. 199. Raphson's problem was similar to Spinoza's, who repudiated the idea that a material world could derive from an immaterial God (see above, Ch. 8, Sec. 4e). And just as Raphson (and More) would make God extended, so Spinoza would make Him material.

292 Koyré explains (*From the Closed World*, p. viii) that he first presented this idea in his earlier *Galileo Studies* (*Etudes Galiléennes*, 3 vols. [Paris: Hermann, 1939]).

293 *From the Closed World*, p. viii; see also Koyré's *Newtonian Studies* (Cambridge, Mass.: Harvard University Press, 1965), pp. 12–13.

294 For Campanella's special interpretation, see above, Ch. 8, Sec. 3b.

295 "Extensionem (*scil.*) infinitam, qua talem, id est, abstracte sumptam, veluti genuinum (licet hactenus non ita animadversum) Matheseos objectum, e Geometricis consideravimus in capite tertio." *De spatio reali*, ch. 6, p. 81. Earlier, in ch. 3 (p. 51), Raphson declared: "Spatium, abstracte sumptum (quo melius res intelligatur) concipiatur absolute infinitum, cuius Amplitudinis, Diffusionis, sive Extensionis, nullus omnino Finis, sive Terminus, concipi potest. . . . Similiter longitudo, latitudo, cum utrinque sint in infinitum diffusae seu protensae, absolute (in suis respective generibus) dicantur infinitae; ex parte vero infinitae, cum aliquo modo vel aliquibus modis, ex parte hac, vel illa terminatae; ex altera vere in infinitum extensae concipantur."

296 *Ibid.*

297 For Barrow's remarks in his *Lectiones Mathematicae XXIII*, see p. 162 of Whewell's edition of Barrow's *Works* cited below, n. 302, and pp. 181–182 of Kirkby's translation cited in the same note. Barrow's spatial theory is discussed below, Ch. 8, Sec. 4.

298 Hasdai Crescas argued, against Aristotle, that an infinite could be conceived as divisible in a certain sense and therefore that infinite extracosmic void space was divisible in this same sense. After summarizing Aristotle's arguments in proposition I, part 1 of his *Or Adonai* (see Wolfson, *Crescas' Critique of Aristotle*, p. 137), Crescas insisted that the infinite he assumed is an infinite incorporeal magnitude and therefore an incorporeal quantity. He then added (*ibid.*, p. 179) that "it does not follow that the definition of the infinite would have to be applicable to all its parts, just as such reasoning does not follow in the case of a mathematical line. Nor would there have to be any composition in it except of its own parts." In his explanation of this cryptic passage, Wolfson explains (*ibid.*, pp. 393–394) that for Crescas "the infinite is said to be divisible in the same sense as the mathematical line is said to be divisible, namely into 'parts of itself' . . ., i.e. infinites in the case of the former, and lines in that of the latter. Finally, just as the mathematical line is not composed of the parts into which it is divisible, that is to say, its parts have no actual co-existence with the whole, so the infinite parts of the infinite have no actual co-existence with the whole infinite." The analogy becomes intelligible when one realizes, as Wolfson puts it (*ibid.*, p. 62), that a mathematical line "is infinitely divisible into parts which are linear. Still it is not composed of those parts into which it is divisible, for the linear parts into which it is divisible, by definition, are bounded by points, and consequently if it were composed of these linear parts it would also be composed of points, but a line is not composed of points." As Wolfson explains elsewhere (*Philosophy of Spinoza*, p. 281), "to be divisible does not always mean to be composed." For more on Crescas, see above, Ch. 2, Sec. 4. Aristotle's relevant discussion of infinite magnitude and whether it is divisible or indivisible appears in *Physics* 3.5.204a.20–32.

299 There is, however, a certain ambiguity to the notion of distinguishable parts. In *Introduction to Natural Philosophy*, John Keill, in contrasting space to body, said that space is immovable and capable of no action and is a thing "whose Parts it is impossible to separate from each other, by any Force however great" (p. 15). Acknowledging the existence of distinguishable parts here, Keill would later declare (lecture 6, p. 73) that "Space itself is a similar and uniform Being, whose Parts cannot be seen, or distinguished from one another. . . ." What Keill really meant, however, is that space is actually distinguishable into different sections or parts, even if only in the sense that one can point in different directions and indicate positionally distinct parts, but that those parts are not only inseparable from each other but are identical in all aspects except position or situation.

300 See Newton's *De gravitatione*, pp. 103 (Latin), 136 (English) in the edition of A. Rupert Hall and Marie Boas Hall cited below, n. 338. There is an analog here between the interpretation of an indivisible space with distinguishable, though inseparable, parts and with the nature of an indivisible atom in Greek atomism, where the extended atom has, in principle, distinguishable parts that are, however, inseparable from the indivisible atom itself.

301 Burtt (*Metaphysical Foundations of Modern Physical Science*, p. 149) suggests that, because More and Barrow were in Cambridge at the time each was grappling with the problem of space, "it is likely that the thinking of each was directly influenced by the other." If so, one can only conclude, as will be seen, that their mutual influence was largely negative.

302 *Lectiones Mathematicae XXIII*, the Latin text of which appears in *Mathematical Works of Isaac Barrow* (Whewell ed.). For *Lectio X*, see pp. 149–165. An English translation of the *Lectiones Mathematicae* was made by the Rev. John Kirkby with the title *The Usefulness of Mathematical Learning explained and demonstrated: Being Mathematical Lectures Read in the Publick Schools at the University of Cambridge by Isaac Barrow, D. D. Professor of the Mathematics, and Master of Trinity-College, etc* (London: printed for Stephen Austen, 1734). Whewell ([ed.], *Mathematical Works of Isaac Barrow*, p. viii) judged Kirkby's translation "so badly executed that it cannot be of use to any one." The little that will be presented in translation from Barrow will be my own, although references to Kirkby's translation will be provided along with references to the Latin text.

303 Whewell (ed.), *Mathematical Works of Isaac Barrow, Lectio X*, p. 158; Kirkby trans., p. 175. In Whewell's edition, each work is separately paginated.

304 "Dicerem secundo, spatium non esse quid actu existens, actuque diversum a rebus quantis, nedum ut habeat dimensiones aliquas sibi proprias, a magnitudinis dimensionibus actu separatas." Whewell (ed.), *Mathematical Works of Isaac Barrow, Lectio X*, p. 158; Kirkby trans., pp. 175–176. With this, Barrow adopted a sense of space that is simply antithetical to Henry More's, where space is a separate, three-dimensional, incorporeal extension.

305 ". . . spatium nihil est aliud quam pura puta potentia, mera capacitas, ponibilitas, aut (vocabulis istis veniam) interponibilitas magnitudinis alicujus." Whewell (ed.), *Mathematical Works of Isaac Barrow, Lectio X*, p. 158; see also Kirkby trans., pp. 175–176, and Burtt (*Metaphysical Foundations of Modern Physical Science*, p. 149). Elsewhere Barrow declared that "space . . . is a general and indefinite capacity" (Whewell [ed.], *Mathematical Works of Isaac Barrow, Lectio X*, p. 163; Kirkby trans., pp. 181–182). After concluding that "there is in reality a Space distinct from all Body; which is as a universal Receptacle, wherein all bodies are contained and moved" (cf. Otto von Guericke, above, Ch. 8, Sec. 4c, who spoke of space as the "universal container of all things"), John Keill refused to comment further on the nature of space (he left that to the metaphysicians), "whether," for example, "its Extension arises from the Relation of Bodies existing in it, so that it may be a mere Capacity, Ponibility, or Interponibility, as some love to express themselves, . . ." (*Introduction to Natural Philosophy*, p. 19). Here Keill probably had Barrow in mind. As will be seen below (Ch. 8, Sec. 4k), Locke's earliest description of space is strikingly similar to Barrow's. A comparison of Barrow with Pedro Fonseca (see above, Ch. 7, Sec. 2b), for example, also reveals a scholastic quality to Barrow's interpretation of space. Its nondimensionality makes that relationship even more intimate (see also n. 308 below).

306 The four examples cited below appear in Whewell (ed.), *Mathematical Works of Isaac Barrow, Lectio X*, pp. 158–159; Kirkby trans., p. 176.

307 *Lectiones Geometricae* (London, 1670), p. 161, in the edition of Whewell (ed.), *Mathematical Works of Isaac Barrow*. For more on the relationship of God and space, see below, Ch. 8, Sec. 4i. In this same passage, the only one in the *Lectiones Geometricae* where space is discussed, Barrow made the startling statement that "something might have existed long before the World was made; and there may now be something in the Extramundane Space, capable of such a continuance: Some *Sun* might have given Light long before; and at present this, or some other like it, may diffuse Light thro'

Imaginary Spaces." The translation is by Edmund Stone, *Geometrical Lectures: Explaining the Generation, Nature and Properties of Curve Lines. Read in the University of Cambridge, by Isaac Barrow, D. D.* (London: printed for Stephen Austen, 1735), p. 6.

308 Barrow's sense of space as a "potency" or "capacity" should be compared to Bonaventure's interpretation of the "capacity" of a similarly divinely evacuated room as described by John Major above, Ch. 7, Sec. 1.

309 This suggests that Barrow would deny that space and body coexist. Where body exists, space does not, and vice versa. As the mere capacity for the reception of body, space disappears when it is filled. We are reminded here of Bartholomeus Amicus, who, as we saw (above, Ch. 7, Sec. 2d), denied that vacuum could be space because the former disappeared when occupied by body. Unlike Barrow, however, Amicus then opted for a space that was more than a mere capacity.

310 *Lectiones Mathematicae* in Whewell (ed.), *Mathematical Works of Isaac Barrow*, p. 160; Kirkby trans., p. 177.

311 Whewell (ed.), *Mathematical Works of Isaac Barrow*, p. 160; Kirkby trans., p. 178.

312 "Nec alia praeter substantiam et accidens entia nova refert in censum . . . sed utriusque modum duntaxat aliquem, et possibilitatem connotat." *Ibid.*; *ibid.*

313 "Nec ubiquitati divinae derogat omnino, quae nil aliud significat, quam omni spatio Deum adesse, vel ubicunque res aliqua potest existere." *Ibid.*; *ibid.* For the reference to Barrow's statement on the coexistence of God and space in the *Lectiones Geometricae*, see above, n. 307.

314 R. I. Aaron and Jocelyn Gibb (eds.), *An Early Draft of Locke's Essay Together with Excerpts from His Journals* (Oxford: Clarendon Press, 1936), p. 94.

315 *Ibid.*, p. 96.

316 *Ibid.*, p. 105.

317 For more on the second opinion, see James Gibson, *Locke's Theory of Knowledge and Its Historical Relations* (Cambridge: Cambridge University Press, 1917), p. 249. Gibson gives a brief summary of Locke's theory of space on pp. 248–254.

318 *Essay Concerning Human Understanding*, bk. 2, ch. 15, par. 4 (p. 198), par. 6 (p. 199), and par. 7 (p. 200) in the edition by Peter H. Nidditch (Oxford: Clarendon Press, 1975).

319 *Essay Concerning Human Understanding* (Nidditch ed.), bk. 2, ch. 15, par. 4, p. 198.

320 *Ibid.*, par. 3, p. 197.

321 *Ibid.*, bk. 2, ch. 17, par. 20, p. 221.

322 *Ibid.*, bk. 2, ch. 13, par. 14, p. 173.

323 *Ibid.*, bk. 2,, ch. 13, pars. 12 (p. 172) and 13 (p. 173).

324 *Ibid.*, bk. 2, ch. 15, par. 10, p. 203; see also bk. 2, ch. 13, par. 13, p. 172.

325 *Ibid.*, bk. 2, ch. 18, par. 20, p. 222.

326 *Ibid.*, par. 21, pp. 222–223. Following upon his description of the famous Stoic argument for the existence of an infinite extracosmic void Locke linked space with duration and argued that if the latter is infinite, so is the former, "For I would fain meet with that thinking Man, that can, in his Thoughts, set any bounds to Space, more than he can to Duration; or by thinking, hope to arrive at the end of either: And therefore if his *Idea* of Eternity be infinite, so is his *Idea* of Immensity; they are both finite or infinite alike" (*Essay Concerning Human Understanding*, bk. 2, ch. 13, par. 21, p. 176). There is little doubt that Locke understood Space and Duration to be infinite.

327 *Ibid.*, par. 20, p. 222.

328 *Ibid.*, bk. 2, ch. 15, par. 3, p. 197.

329 *Ibid.*, par. 8, p. 200. That space is God's immensity seems a reasonable inference from the assertion that expansion *belongs* "only to the Deity." More than any other, perhaps, this passage seems to confirm Locke's belief in an absolute, infinite, rather than relative, space. Earlier in the *Essay Concerning Human Understanding*, Locke raised doubts about the absolute nature of space, when in bk. 2, ch. 13, par. 26 (p. 179), after acceptance of the clear distinction between space and solidity, he declared that "whether any one will take Space to be only a relation resulting from the Existence of other Beings at a distance; or whether they will think the Words of the most knowing

King *Solomon, The Heaven, and the Heaven of Heavens, cannot contain Thee;** or those more emphatical ones of the inspired Philosopher St. Paul, *In Him we live, move, and have our Being,*** are to be understood in a literal sence, I leave every one to consider; . . ." Nidditch gives as the first reference I Kings 8:27; II Chronicles 2:6 and 6:18; for the second Acts 17:28 (both citations were frequently invoked in favor of an infinite space that was God's immensity). In this passage, Locke resurrected the relational theory of space, which he described in his journal entry of 1678 (see above, Ch. 8, Sec. 4k). At this point in the *Essay Concerning Human Understanding*, it is offered as an equally plausible alternative to an absolute theory of space. But in the passage above, and in the constant linkage of space with duration, Locke gave reasonable evidence of his belief in an absolutely infinite and immobile space rather than a relational one dependent on distances between real things.

330 *Ibid.*, par. 17, p. 174.

331 *Ibid.*, par. 18, p. 174.

332 *Ibid.*, par. 19, p. 175.

333 *Ibid.*

334 Although Locke admitted that it was difficult to conceive any real being without expansion, he did not affirm Henry More's attribution of extension to spirits and remained content with the opinion that "what Spirits have to do with Space, or how they communicate in it, we know not" (*Essay Concerning Human Understanding*, bk. 2, ch. 15, par. 11, p. 203). By the very nature of his philosophical quest and by natural temperament as well, Locke probably demanded firmer evidence than did Henry More before rendering a judgment on matters that were either philosophical or scientific.

335 With the ready availability of Koyré's excellent chapters on Newton's ideas about space (*From the Closed World*, chs. 7, 9, 10, and 11), we shall concentrate here on those aspects of Newton's thought that are most germane to the major themes of this study.

336 See Alexandre Koyré and I. Bernard Cohen, "Newton & the Leibniz-Clarke Correspondence with Notes on Newton, Conti, & Des Maizeaux," *Archives Internationales d'Histoire des Sciences 15* (1962), 79. In this splendid article, Koyré and Cohen provide much additional evidence for Newton's role in the Clarke-Leibniz controversy; see especially p. 69 for the general background of Newton's involvement. See also Koyré, *From the Closed World*, pp. 300–301, n. 3.

337 Although Berkeley's criticism of Newton's theory of absolute space in the *Principles of Human Knowledge* (1710) may have contributed to Newton's decision to formulate the General Scholium (see Koyré, *From the Closed World*, pp. 222–223), Leibniz was undoubtedly the major catalyst (see the letter from Roger Cotes to Newton in 1713 as published in J. Edelston, *Correspondence of Sir Isaac Newton and Professor Cotes* [London: J. W. Parker, 1850], p. 153 ff.; an excerpt from that letter appears in Koyré, *From the Closed World*, p. 229, n. 4; see also n. 375, below).

338 The Latin text with an excellent English translation appears in *Unpublished Scientific Papers of Isaac Newton. A Selection from the Portsmouth Collection in the University Library, Cambridge*, chosen, edited and translated by A. Rupert Hall and Marie Boas Hall (Cambridge: Cambridge University Press, 1962), pp. 89–121 (Latin); 121–156 (English). Because Newton's treatise bears no title, the one above was assigned by the Halls, who believe it was composed during Newton's student days (*ibid.*, p. 75), perhaps between 1664 and 1668 (*ibid.*, p. 90; Newton wase a student between 1661 and 1668, receiving his B.A. degree in 1665 and his M.A. in 1668; see I. Bernard Cohen, "Newton, Isaac," *Dictionary of Scientific Biography*, vol. 10 [New York: Scribner, 1974], pp. 43, 44). These dates acquire further plausibility from the fact that an earlier unpublished treatise by Newton, *Questiones quaedam philosophiae*, has been dated approximately in early 1664 (see Richard S. Westfall, *Force in Newton's Physics. The Science of Dynamics in the Seventeenth Century* [London: MacDonald; New York: American Elsevier, 1971], p. 327). J. E. McGuire suggests, without comment, the inclusive dates 1666 and 1670 ("Newton on Place, Time and God: An Unpublished Source," *The British Journal for the History of Science 11* [1978], 124).

339 Richard S. Westfall argues that Newton had already become familiar with Charleton's

work, and perhaps even with the works of Gassendi, by the time he composed the *Questiones quaedam philosophiae* (see his *Force in Newton's Physics*, pp. 326–327; 400, n. 3).

340 Hall and Hall, *Unpublished Scientific Papers (De gravitatione)*, pp. 94, 104 (Latin); 132, 138 (English).

341 *Ibid.*, pp. 132; 99 (Latin).

342 *Ibid.*

343 *Ibid.* For the use of "annihilation" as a methodological device, see above, n. 169 of this chapter. Gassendi had, of course, argued that even if all substances (and therefore bodies) and accidents ceased to exist, space would continue on (see above, Ch. 8, Sec. 4b). Many years before, Bartholomeus Amicus, who distinguished between vacuum and space, had insisted that when a body fills a void, the latter disappears but not the space that the body occupies (see above, Ch. 7, Sec. 2d). If a body were annihilated, both void and space would remain.

344 *Ibid.*, pp. 104 (Latin); 137 (English). Thus Newton agreed with Hobbes and More (above, Ch. 8, Sec. 4d) that space cannot be imagined as nonexistent. For an attack on that position, see below, n. 425.

345 *Ibid.*, pp. 132; 99 (Latin).

346 *Ibid.*, pp. 100–105 (Latin); 132–138 (English). Later, in contrasting extension to body, Newton reiterated much of this when he said, "extension is eternal, infinite, uncreated, uniform throughout, not in the least mobile, nor capable of inducing change of motion in bodies or change of thought in the mind" (*ibid.*, pp. 145; 111 [Latin]). For the property of penetrability, see *ibid.*, pp. 108 (Latin); 141 (English).

347 *Ibid.*, pp. 133; 100 (Latin). For Patrizi, see above, Ch. 8, Sec. 4a; also Sec. 4g. Every variety of figure is also represented by those of its kind that are infinite in extent, even as each of them is crisscrossed by the lines, surfaces, and volumes of every other kind of figure (see *ibid.*, pp. 101 [Latin]; 133–134 [English]).

348 *Ibid.*, pp. 133; 101 (Latin).

349 *Ibid.*, p. 134; 101 (Latin). Newton's assertion bears a resemblance to Anselm's conception of God's infinity (see above, Ch. 7, Sec. 2b and n. 71). For a comparison of Newton's position with that of Descartes, see J. E. McGuire, "Existence, Actuality and Necessity: Newton on Space and Time," *Annals of Science* 35 (1978), 469, n. 23. McGuire's fine article should also be consulted for further insights into Newton's *De gravitatione*.

350 Hall and Hall, *Unpublished Scientific Papers (De gravitatione)*, pp. 138; 104 (Latin).

351 *Ibid.*, pp. 136; 103 (Latin).

352 *Ibid.*

353 *Ibid.*, pp. 104 (Latin); 137.

354 *Ibid.*, pp. 135; 102 (Latin).

355 ". . . nam infinitas non est perfectio nisi quatenus perfectionibus tribuitur." *Ibid.*, pp. 135–136 (English); 102 (Latin). Where the Halls have "an attribute of perfect things" I have substituted "attributed to perfections."

356 ". . . et infinitas extensionis talis est perfectio qualis est extendi." *Ibid.*, pp. 103 (Latin); 136 (English). Here again I have slightly altered the translation of the Halls in order to link the two passages and to convey the fact that Newton here made infinite space an attribute of God.

357 *Sir Isaac Newton's Mathematical Principles of Natural Philosophy and his System of the World*, trans. into English by Andrew Motte in 1729. The translations revised, and supplied with an historical and explanatory appendix by Florian Cajori (Berkeley: University of California Press, 1947), p. 13 (hereafter cited as "Motte-Cajori translation"). See also Jammer, *Concepts of Space*, pp. 101–102, and generally his discussion of Newton's physical space and the theological influences that shaped it (ch. 4: "The Concept of Absolute Space," pp. 95–116).

358 Motte-Cajori trans., p. 6.

359 *Ibid.*; see also p. 9.

360 *Ibid.*, p. 8.

361 *Ibid.*, p. 9.

362 *Ibid.*, p. 7.
363 *Ibid.*, p. 6.
364 *Ibid.*, pp. 419 (hypothesis I) and 420 (corollary to proposition XII); see also Jammer, *Concepts of Space*, p. 103. For Patrizi, see above, Ch. 8, Sec. 4a.
365 See the Motte-Cajori translation, pp. 10–11. For a good description and analysis of the experiment, see Jammer, *Concepts of Space*, pp. 106–109.
366 According to Frank E. Manuel, *The Religion of Isaac Newton* (Oxford: Clarendon Press, 1974), p. 31.
367 For the relations between these two "Books" in the seventeenth century, see *ibid.*, ch. 2 ("God's Word and God's Works"), pp. 27–49.
368 These thoughts appear in the Portsmouth Collection of Newton's manuscripts at University Library, Cambridge, and have been edited and translated by J. E. McGuire, "Newton on Place, Time, and God," 118 (Latin), 119 (English). Other relevant remarks may lie as yet undetected in these unpublished manuscripts.
369 W. G. Hiscock (ed.), *David Gregory, Isaac Newton and Their Circle: Extracts from David Gregory's Memoranda, 1677–1708* (Oxford: Printed for the editor, 1937), pp. 29–30. The passage is quoted by Koyré, *From the Closed World*, p. 297, n. 2, as it is by Westfall, *Force in Newton's Physics*, p. 421 n. 184.
370 *Optice: sive de reflexionibus, refractionibus, inflexionibus et coloribus lucis libri tres. Authore Isaaco Newton. Latine reddidit, Samuel Clarke* . . . (London: impensis Sam. Smith et Benj. Walford, 1706). In the second English edition of 1717 (and the third [1721] and fourth [1730] as well), query 20 of the Latin edition becomes query 28 (and query 23 becomes query 31). My quotations are from queries 28 and 31 as they appear in Isaac Newton, *Opticks*, based on the fourth edition, London, 1730 (New York: Dover Publications, 1952). For a brief summary of the relationships between Newton's queries in the different editions of the *Opticks*, see *Isaac Newton's Papers & Letters on Natural Philosophy and Related Documents*, ed., with a general introduction, by I. Bernard Cohen assisted by Robert E. Schofield (Cambridge, Mass.: Harvard University Press, 1958), pp. 14–15.
371 *Opticks* (4th ed.), query 28, p. 370; Latin ed. (1706), p. 315. The italics are mine and serve to identify the words that render the Latin term *tanquam*, apparently missing from the initial copies of the 1706 Latin edition. In subsequent copies, Newton added the word *tanquam*, so that God would see things "in infinite Space, as it were in his Sensory," thus assuring the reader that space is not to be taken literally as God's sensorium, but only analogically. For the detailed arguments and evidence, see Alexandre Koyré and I. Bernard Cohen, "The Case of the Missing *Tanquam*: Leibniz, Newton & Clarke," *Isis* 52 (1961), 555–566. The authors conjecture (p. 566) that Leibniz may have read a copy without the *tanquam*, which would explain his charge that Newton believed that space was literally God's organ (for Leibniz's words, see below, Ch. 8, Sec. 4m). They speculate further that perhaps Newton's true conviction is to be found in the discarded version and that he really did believe space is God's sensorium. Frank Manuel insists (*Religion of Isaac Newton*, pp. 77–78) that whether or not Newton troubled to add the missing *tanquam*, Leibniz should have been fully aware that Newton did not conceive space literally as God's organ, but intended that description analogically. Manuel is probably right, although in query 23 of the 1706 Latin edition, Newton spoke of God moving all bodies by His will "in His infinite Sensorium" ("in infinito suo Sensorio"; p. 346 of the 1706 edition), and not until the second English edition of 1717, after Leibniz's attack in the Leibniz-Clarke correspondence of 1715–1716, did Newton add (in query 31) an explicit denial that God's "uniform Sensorium" is His organ (thus Leibniz, who based his attack partly on the 1706 edition, had no such denial before him). But even if Newton did not conceive space as God's organ literally, the conjecture by Koyré and Cohen remains plausible and compatible with all else that is known of Newton's conception of the deity. Although not an organ, Newton may yet have conceived space as God's infinite "sensorium," the means by which He is omnipresent and in touch with everything. Because it will be shown below that Newton probably believed in the actual tridimensionality of God, with space as its manifestation, his conception of space as God's sensorium is consistent with his overall

conception of God's nature and operation. For an early version of Newton's statement in query 28, see McGuire, "Newton on Place, Time, and God," 127.

372 *Opticks* (4th ed.), query 31, p. 403; Latin ed. (1706) query 23, p. 346. Although query 31 of the English edition is the counterpart of query 23 of the Latin edition, Newton made changes in his English version of 1717, adding, for example, the part cited here about God's sensorium not being His organ (see n. 371).

373 *Ibid.* This sentence was not in the 1706 Latin edition. Immediately following these words, Newton made the startling declaration that space is infinitely divisible, a judgment that contradicts its declared indivisibility elsewhere. The remark occurred when Newton allowed that God can create particles of matter "of several Sizes and Figures, and in several Proportions to Space." At this point, Newton declared that "space is divisible *in infinitum*," thus enabling God to make infinitely small, as well as larger, particles of matter. Has Newton here abandoned the indivisibility of absolute infinite space, or is this intended to apply only to relative space? Whatever the merits of the latter alternative, Newton did not abandon the indivisibility of absolute space, which was one of its cardinal features. Because we shall see immediately below that God is, among other things, our space, the divisibility of space would imply the divisibility of God, an untenable consequence. Newton's momentary lapse here was probably occasioned by the need to proportion any possible particle of matter, however small, to an equal quantity of space. This apparent contradiction in Newton's conception of space was also noted by Christopher Browne Garnett, Jr., *The Kantian Philosophy of Space* (New York: Columbia University Press, 1939), pp. 18–23, whose explanation of it, however, seems mistaken. Garnett believes that in the *Principia* Newton considered space "divisible into parts or particles which are indivisible" (p. 19), unlike Clarke, who would hold that "space is only figuratively divisible." In the *Opticks*, however, Newton "continued to hold that space (the whole of space) is divisible; but instead of claiming that it is divisible into indivisible parts or particles he held that there is no limit to the divisibility." This obviously had disastrous consequences for Newton's conception of the relation of space to God. Newton, who held that space is a property of God (p. 17), was thus committed to the untenable position that space was divisible and God was not. "The omnipresence of space cannot be arbitrarily considered to be God's omnipresence, yet its divisibility, not to be God's. Along such lines as these, Leibniz attacked the theory that space is divine omnipresence" (p. 20). According to Garnett, it was Clarke who was more consistent than Newton and made space and God indivisible. "Believing that if space is an attribute of God and also divisible, then God would be divisible, and possibly realizing that, in spite of these considerations, Newton had made space divisible, Clarke went on to hold that space is indivisible . . . Space is indivisible, but it may be figuratively treated as divisible. It has no parts, but it may be treated as if it had them" (pp. 20–21). The view that Garnett attributes to Clarke must also be attributed to Newton, from whom Clarke probably derived it. Garnett, of course, was unaware of Newton's role in the Leibniz-Clarke correspondence. Despite Newton's attribution of infinite divisibility to space in the 1706 edition of the *Opticks*, his considered opinion was that absolute space is indivisible.

374 For Newton's use of similar language, see below, n. 415. James P. Ferguson (*The Philosophy of Dr. Samuel Clarke and Its Critics* [New York/Washington/Hollywood: Vantage Press, 1974], p. 98) observes that Samuel Clarke did much the same thing. That is, when Clarke wished to avoid the characterization of space as an attribute, he would speak of it as "a consequence or effect of God's existence" or declare "that the self-existent Being is the substratum of space and duration." In these passages, Newton also described space as a consequence or effect of God's existence. For similar illustrations for Clarke, and perhaps for Newton as well, see the Clarke-Leibniz correspondence, Clarke's fourth reply, pars. 10 (where he said that "space and duration are not *hors de Dieu*, but are caused by, and are immediate and necessary consequences of his existence") and 15 ("even as necessarily as space and duration, which depend not on the will, but on the existence of God"). See Alexander (ed)., *Leibniz-Clarke Correspondence*, pp. 47, 49 (for the full title, see below, n. 383; because Newton was apparently

directly involved with Clarke in drawing up the latter's replies to Leibniz [see below, Ch. 8, Sec. 4m], these illustrations may be assigned to Newton as well).

375 All the passages cited here from the 1713 edition appear in Motte-Cajori, p. 545. The italics are Newton's. For the Latin text of these passages, see vol. 2, pp. 761–762, of the Koyré and Cohen edition of the *Principia Mathematica*. In the paragraph quoted above, Newton added a note in which he mentioned standard biblical and pagan sources for the belief that all things are moved and contained in God (see above, n. 78 of this chapter). Koyré and Cohen ("Newton & the Leibniz-Clarke Correspondence," 68) explain that "a portion of the concluding *Scholium Generale* to the second edition of the *Principia* was a direct response to Leibniz's letter to Hartsoeker and was the first statement in print by Newton himself in defense of his own philosophy against Leibniz." Leibniz's letter to Hartsoeker was printed in 1711.

376 In what follows, I rely on Westfall, *Force in Newton's Physics*, pp. 395–400.

377 Cited by Westfall, *Force in Newton's Physics*, p. 399, from Newton's Add. MS. 3965.6, f.266v (University Library, Cambridge). See also, p. 422, n. 186, for another passage dated around the time of the first edition of the *Principia Mathematica*.

378 Westfall, *Force in Newton's Physics*, p. 396.

379 *Ibid.*, p. 397. On p. 420, n. 183, Westfall cites a draft from about 1715 intended for the queries wherein Newton inquired, "is there not something diffused through all space in & through wch bodies move without resistance & by means of wch they act upon one another at a distance in harmonical proportions of their distances"?

380 George Berkeley, *A Treatise Concerning the Principles of Human Knowledge*, part I, par. 110, in *The Works of George Berkeley D. D.; Formerly Bishop of Cloyne Including his Posthumous Works*, with prefaces, annotation, and an account of his life by Alexander Campbell Fraser (4 vols.; Oxford: Clarendon Press, 1901), vol. 1, pp. 318–319. That Newton's *Principia Mathematica* was intended is obvious from Berkeley's introductory remarks to par. 110, where he refers to a "treatise of *Mechanics*, demonstrated and applied to Nature, by a philosopher of a neighbouring nation, whom all the world admire." Because Berkeley's *Principles of Human Knowledge* was first published in Ireland, Newton could be cited as the "philosopher of a neighbouring nation." In 1709, in *An Essay Towards a New Theory of Vision*, par. 126, Berkeley rejected the idea that "pure space, vacuum, or trine dimension" are "equally the object of sight and touch." See *Works of George Berkeley*, vol. 1, p. 189.

381 Berkeley, *Principles of Human Knowledge*, par. 116 (*Works of George Berkeley*, vol. 1, p. 323). In a note to this passage, Fraser, the editor, explains, "in short, empty Space *is* the sensuous idea of unresisted motion" (p. 323, n. 1). For the widely used concept of the imaginary annihilation of bodies and worlds, see above, n. 169 of this chapter.

382 *Ibid.*, par. 117 (*Works of George Berkeley*, vol. 1, p. 323). Berkeley here identified the real dilemma for those who, from the Middle Ages to the early eighteenth century, believed in the existence of an external space. If it is not somehow associated with God, then it must be independent of Him as a rival uncreated entity. Although most who conceived such a separate space associated it with God, Berkeley would have neither alternative. John Locke is probably one of the "philosophers of great note" whom Berkeley had in mind. As for those who "have of late set themselves particularly to shew that the incommunicable attributes of God agree to it," Berkeley may have intended Henry More and Joseph Raphson.

Even after Newton's death, Berkeley continued to attack those who identified or associated God with space. In a letter of March 24, 1730, he denounced Raphson and his *De spatio reali* for virtually deifying space and pretending "to find out fifteen of the incommunicable attributes of God in space" (see A. A. Luce [ed.], *Philosophical Commentaries, Generally Called the Commonplace Book* [London: Thomas Nelson, 1944], p. 378; also cited by Tuveson, "Space, Deity and the 'Natural Sublime'," 29). Indeed in entry 298 of his Commonplace Book, Berkeley declared that "Locke, More, Raphson etc seem to make God extended" (Luce, *Commonplace Book*, p. 97), and in entry 825 (*ibid.*, p. 293), he asserts that "Hobbs & Spinosa make God extended. Locke also seems to do the same." Finally, in paragraph 270 of his *Siris: A Chain of Philosophical Reflex-*

ions and Inquiries Concerning the Virtues of Tar-Water, first published in 1744, Berkeley used a somewhat different tactic and showed that the alleged attributes that were believed common to God and space could just as appropriately be assigned to nothing. "The doctrine of real, absolute, external space," Berkeley explained, "induced some modern philosophers to conclude it was a part or attribute of God, or that God himself was space; inasmuch as incommunicable attributes of the Deity appeared to agree thereto, such as infinity, immutability, indivisibility, incorporeity, being uncreated, impassive, without beginning or ending; not considering that all these negative properties may belong to nothing. For, nothing hath no limits, cannot be moved, or changed, or divided, is neither created nor destroyed." *Works of George Berkeley*, vol. 3, p. 253.

383 See *The Leibniz-Clarke Correspondence, Together with Extracts from Newton's "Principia" and "Opticks,"* ed. with introduction and notes by H. G. Alexander (Manchester: University of Manchester Press, 1956), p. 11. On pp. lv–lvi, Alexander lists all the previous editions of the correspondence, the first of which appeared in London, 1717. In this edition, Clarke translated Leibniz's French letters into English, which is the translation Alexander reproduces with modernized spelling and punctuation. For an edition in which the letters appear in their original form, see André Robinet, *Correspondance Leibniz-Clarke présentée d'après les manuscrits originaux des bibliothèques de Hanovre et de Londres* (Paris: Presses Universitaires de France, 1957). Fundamental to our understanding of this significant correspondence and its aftermath is Koyré and Cohen, "Newton & the Leibniz-Clarke Correspondence," 63–126. Useful interpretations and summaries also appear in two works by Koyré: *From the Closed World*, ch. 11, and *Newtonian Studies* (Cambridge, Mass.: Harvard University Press, 1965), pp. 166–167. A good brief summary of some of the highlights of the correspondence appears in Ferguson, *Philosophy of Dr. Samuel Clarke*, pp. 45–51. For the background to Leibniz's attack on Newton, see Alexander (ed.), *Leibniz-Clarke Correspondence*, pp. xv–xvi, where Alexander also cites Addison's explication of Newton's position (*Spectator*, No. 565, July 1714). Here Addison declared Newton's conception of infinite space as God's sensorium superior to that which regards it merely as a receptacle. For "as God Almighty cannot but perceive and know everything in which he resides, infinite Space gives Room to infinite Knowledge, and is, as it were, an Organ to Omniscience"; also see above, n. 370–372, for queries 20 and 23 and the versions that Leibniz may have read.

384 On Clarke, see Koyré and Cohen, "Newton & The Leibniz-Clarke Correspondence," 64–66.

385 *Ibid.*, 78–80. In his correspondence with Clarke, Leibniz was convinced that Clarke was but a stand-in for Newton, who was his real opponent. See Leibniz's letter to Rémond (August 15, 1716), which is reprinted by Robinet, *Correspondance Leibniz-Clarke*, p. 120.

386 Alexander(ed.), *Leibniz-Clarke Correspondence*, Clarke's fourth reply, par. 7, p. 46. In his fourth letter (par. 7), Leibniz had described all space as imaginary, by which he meant nonexistent (*ibid.*, p. 37). In his fifth letter, Leibniz declared that "since space in itself is an ideal thing, like time; space out of this world must needs be imaginary, as the schoolmen themselves have acknowledged" (par. 33, p. 64). Indeed he went on to argue that the vacuum within the world is also imaginary, despite claims for its existence by Otto von Guericke. Of course, Leibniz was wrong to suppose that by imaginary scholastics meant only nonexistent or a figment of the mind. In the sixth and seventh chapters of this study, it is obvious that despite its lack of extension, many thought imaginary space was something quite real because of its association with God. And at least one scholastic, Franciscus Bona Spes, who denied the existence of imaginary space, insisted that God is His own *real* immense space (see above, Ch. 7, Sec. 2f), much as did Samuel Clarke, though the latter's space is dimensional and the former's is not.

387 *Ibid.*, Clarke's third reply, par. 2, p. 31.

388 *Ibid.*, Clarke's third reply, par. 3, p. 31; see also Clarke's second reply, par. 4, p. 22.

389 *Ibid.*, Leibniz's fourth letter, par. 8, p. 37.

390 *Ibid.*, Leibniz's fourth letter, par. 9, p. 37. Leibniz would draw a variety of absurdities

from Clarke's notion that space is a property of God. See Leibniz's third letter, par. 3, p. 25; fourth letter, pars. 8–10, p. 37; fifth letter, pars. 36–46, pp. 66–69.

391 *Ibid.*, Clarke's fourth reply, par. 9, p. 47. Perhaps with this claim in mind, Leibniz, in his fifth letter, par. 45 (*ibid.*, p. 68), asserted that if "God's immensity makes him actually present in all spaces," then "if God is in space, how can it be said that space is in God, or that it is a property of God? We have often heard that a property is in its subject; but we never heard, that a subject is in its property."

392 *Ibid.*, Clarke's third reply, par. 3, p. 31.

393 *Ibid.*, Clarke's fourth reply, par. 10, p. 47.

394 A caution must be raised with respect to the term "attribute," which was invoked by Clarke in direct response to Leibniz, who had used the French "attribut" in his fourth letter (see Robinet [ed.], *Correspondance Leibniz-Clarke*, par. 8, p. 84). Ordinarily, Clarke and Newton studiously avoided its use in the correspondence with Leibniz, probably because Spinoza had used the term in assigning material extension to God (Koyré and Cohen, "Newton & the Leibniz-Clarke Correspondence," 93 and n. 70). Because Spinoza's philosophy was construed as pantheistic and was therefore condemned, Clarke and Newton substituted the terms "property" or "quality," although it is quite clear that these were synonymous with "attribute" (*ibid.*). It is likely that Newton converted Clarke to this approach, because Clarke had already used the term in 1705, when he wrote that Eternity and Immensity, from which latter notion the idea of infinite space was abstracted, "seem both to be but Attributes of an Essence Incomprehensible to Us." Samuel Clarke, *A Demonstration of the Being and Attributes of God* (London, 1705; reprinted in facsimile Stuttgart-Bad Cannstatt: Friedrich Frommann Verlag, 1964), pp. 78–79.

395 Alexander (ed.), *Leibniz-Clarke Correspondence*, Leibniz's third letter, par. 4, pp. 25–26; for a lengthier presentation of his views on space, see Leibniz's fifth letter, par. 47, pp. 69–72. For descriptions of Leibniz's theory of space, see Jammer, *Concepts of Space*, pp. 116–123, and Nicholas Rescher, *Leibniz, An Introduction to His Philosophy* (Totowa, N.J.: Rowman and Littlefield, 1979), ch. 10 ("Space and Time: Motion and Infinity"), pp. 84–109; Leibniz's early views on space are described and analyzed by Hector-Neri Castañeda, "Leibniz's Meditation on April 15, 1676 About Existence, Dreams, and Space," in *Leibniz à Paris (1672–1676)*, Symposion de la G. W. Leibniz-Gesellschaft (Hannover) et du Centre National de la Recherche Scientifique (Paris) à Chantilly (France) du 14 au 18 Novembre 1976, vol. 2: *La Philosophie de Leibniz* (Wiesbaden: Franz Steiner Verlag, 1978), pp. 91–129.

396 Alexander (ed.), *Leibniz-Clarke Correspondence*, par. 5, p. 26. For a clear statement of the "principle of sufficient reason," see Leibniz's second letter, par. 1, p. 16. Not only did Leibniz reject the attribution of absolute space to God as a property, but he followed the traditional scholastic argument that if a separate space existed as the place of God, it would "be a thing co-eternal with God, and independent upon him; nay, he himself would depend upon it, if he has need of place" (fifth letter, par. 79, p. 81).

397 In *Le Livre du ciel et du monde*, bk. II, ch. 8 (pp. 368, 369 for the French and English texts in Menut ed.), Nicole Oresme assumed the existence of an imaginary infinite space beyond the world in which God could, if He wished, move the entire world with a rectilinear motion, a possibility that had to be conceded by virtue of article 49 of the Condemnation of 1277. In Oresme's judgment, such a motion would be absolute because "no other body exists with which the world could vary with respect to place." Indeed in this absolute infinite space, a body would "bear different relationships with respect to the imagined immobile space, for it is with regard to this space that the speed of motion and of its parts are measured." For references to my discussion of Oresme and others in the medieval and early modern periods, see above, Ch. 6, n. 76. The translation is mine from Grant, "The Condemnation of 1277," 230.

398 Alexander (ed.), *Leibniz-Clarke Correspondence*, Clarke's third reply, par. 4, p. 32.

399 *Ibid.*, par. 13, p. 38.

400 *Ibid.*, Clarke's fourth reply, par. 13, p. 48. Clarke had already said virtually the same thing in his third reply, par. 2, p. 31 (for the text, see below, Ch. 8, Sec. 4m).

401 Clarke next appealed to definition 8 of Newton's *Principia Mathematica*, by which he seemed to mean the scholium following that definition. Jammer (*Concepts of Space*, p. 119) observes that Leibniz responded well to the kinematic implications of Clarke's hypothetical assumption of the motion of the world, but did not reply to the dynamic aspects concerned with the sudden cessation of the world's motion or that of the ship, perhaps because he had no ready reply. With respect to Clarke's argument, James P. Ferguson observes that "if the universe stopped suddenly, everything in it would stop suddenly, and no effects would be perceptible. It is only when bodies come to a stop with unequal rates of speed that there are discernible effects, but in such a case the effects arise because some bodies continue to move while others have stopped, and are thus entirely relative" (*Philosophy of Dr. Samuel Clarke*, p. 46).

402 Alexander (ed.), *Leibniz-Clarke Correspondence*, Leibniz's fifth letter, par. 29, pp. 63–64.

403 *Ibid.*, Clarke's fifth reply, pars. 26–32, pp. 100–101.

404 *Ibid.*, par. 3, p. 25.

405 The italics are mine.

406 Koyré, *From the Closed World*, p. 244.

407 Alexander (ed.), *Leibniz-Clarke Correspondence*, Leibniz's fourth letter, par. 18, p. 39.

408 *Ibid.*, Clarke's fourth reply, par. 13, p. 48. For Newton, see Hall and Hall (eds. and trans.), *Unpublished Scientific Papers of Isaac Newton, De Gravitatione*, pp. 103, sec. 3 (Latin), 136, sec. 3 (English) and *Principia Mathematica*, Motte-Cajori trans., p. 9, where Newton explained that only immovable places "retain the same given position one to another."

409 Alexander (ed.), *Leibniz-Clarke Correspondence*, Clarke's third reply, par. 2, p. 31.

410 *Ibid.*, Clarke's third reply, par. 3, p. 31.

411 *Ibid.*, Clarke's second reply, par. 4, p. 22.

412 "To say that infinite space has no parts," Leibniz declared in his fourth letter, par. 11 (ibid., p. 38), "is to say that it does not consist of finite spaces." Here again, Leibniz was speaking of distinguishable, not movable parts. In his fourth reply, pars. 11 and 12 (*ibid.*, p. 48), Clarke referred once again to his discussion of movable parts and replied to Leibniz with the assertion that "infinites are composed of finites, in no other sense, than as finites are composed of infinitesimals." Corporeal, movable, and divisible parts are contrasted to parts as they are conceived in infinite space. The latter are parts "(improperly so called) being essentially indiscernible and immoveable from each other, and not partable without an express contradiction in terms" (here once again, Clarke cited his basic argument in his third reply, par. 3). By "essentially indiscernible," Clarke meant, of course, merely distinguishable, for otherwise he could not have said that "two places, though exactly alike, are not the same place" or that "different spaces are really different or distinct one from another, though they be perfectly alike" (see above, Ch. 8, Sec. 4m; Leibniz was aware of this in his fourth letter, par. 12, p. 38, because Clarke had made the same distinction in earlier works that Leibniz mentioned).

As his final declaration, Leibniz objected to the notion that space, which has (distinguishable) parts, could be a property of God and therefore "belong to the essence of God." Under these circumstances, there would be "parts in the essence of God" (fifth letter, par. 42, p. 68). To this Clarke responded by contrasting the parts of God's immensity to corporeal parts, which are, of course, separable, divisible, and movable. Parts in God, however, "do no more hinder immensity from being essentially one" — an obvious reference to distinguishable but inseparable and immobile parts — "than the parts of duration hinder from being essentially one" (fifth reply, pars. 36–48, p. 103). And once again, Clarke sent the reader to his earlier argument on movable and separable parts in the third reply, par. 3. Thus Clarke never responded to Leibniz's major question: Would God not be a being constituted of parts if infinite space, with its distinguishable and inseparable parts, is God's property or immensity? The closest to a response Clarke came occurred in his fourth reply, pars. 11 and 12 (see the quotation above), where he glossed over the problem, dismissing the notion of parts in God's immensity by describing them as "essentially indiscernible." They are essentially indiscernible, however, only in the sense that infinite space is homogeneous and all parts are alike in every way except position. *But they are distinguishable,* as Clarke himself admitted when he said that "two places, though exactly alike, are not the same place."

413 Samuel Clarke was also reluctant to make such an explicit assertion. But he came much closer than Newton to an open admission of it in a consideration of whether immaterial substances can be extended. In a "Defence of an Argument Made Use of in a Letter to Mr. Dodwell to Prove the Immateriality and Natural Immateriality of the Soul," a letter that was known to Leibniz, who referred to it in the Leibniz-Clarke correspondence (fourth letter, par. 12, p. 38), Clarke insisted that immaterial beings are necessarily indivisible ("indiscerpible") because they possess the property of consciousness. But how far "such *Indiscerpibility* can be reconciled and be consistent with some kind of Expansion . . . is another Question of considerable Difficulty, but of no Necessity to be resolved in the present Argument." See *The Works of Samuel Clarke, D. D. Late Rector of St. James's Westminster,* vol. 3 (4 vols.; London: printed for John and Paul Knapton, 1738), p. 763. Clarke was more forthcoming in "A Second Defence of an Argument Made Use of in a Letter to Mr. Dodwell, To Prove the Immateriality and Natural Immateriality of the Soul" (*ibid.,* pp. 794–795). Clarke now argued that merely because there are serious difficulties in the assumption that immaterial beings have extension, this does not of itself weaken the claim that they are indeed extended. "For there are many demonstrations even in abstract Mathematicks themselves, which no Man who understands them can in the least doubt of the Certainty of, which yet are attended with difficult Consequences that cannot perfectly be cleared, The *infinite Divisibility of Quantity,* is an instance of this kind. Also the *Eternity of God,* than which nothing is more self-evident; . . ." Although beset with great difficulties, Clarke did not abandon the notion that an immaterial, extended substance is indivisible. To concede its division would destroy its essential unity and oneness.

414 For the texts of the French preface and Newton's five drafts, see Koyré and Cohen, "Newton & the Leibniz-Clarke Correspondence," 83, 96–103. That Newton should have drawn up this qualification to the use of certain terms in the Leibniz-Clarke correspondence lends considerable support to the claim that he was intimately involved with Samuel Clarke in the original exchange of letters (on this, see *ibid.,* 80–81, who consider the *Avertissement au Lecteur* as one of four pieces of evidence for Newton's direct participation in the Leibniz-Clarke correspondence). For why should Newton have drawn up this brief clarification for Clarke, to whom it was publicly ascribed, if he were not directly involved in the original correspondence? Whatever his reasons, Newton wished to avoid public involvement.

415 *Ibid.,* p. 101 (Draft E). Newton went on to say that *Ubi* and *Quando,* in the Aristotelian categories, are closer to what he meant by property and quality when these are applied to the ubiquity and eternity of God. Although to Ubi, "where?", Newton may have answered "everywhere" (see Koyré and Cohen, "Newton & the Leibniz-Clarke Correspondence," 99), the next few lines of Draft E also indicate Newton's belief that infinite space as a property or quality of God had the sense of being a consequence "of the Existence of a Being wch is really necessarily & substantially Omnipresent. . . ." Here again, Newton seemed to declare that God is literally omnipresent in an infinite extended space. Indeed infinite space is the consequence of His actual omnipresence (for similar language employed by both Newton and Clarke, see above, n. 374 of this chapter).

416 See McGuire, "Newton on Space and Time," 504–505.

417 See David Gregory's statement above, Ch. 8, Sec. 4l.

418 From the General Scholium in the 1713 edition of the *Principia Mathematica* (see p. 545 of the Motte-Cajori translation). In the *De gravitatione,* Newton declared that "no being exists or can exist which is not related to space in some way. God is everywhere, created minds are somewhere, and body is in the space that it occupies; and whatever is neither everywhere nor anywhere does not exist" (Hall and Hall, *Unpublished Scientific Papers of Isaac Newton,* pp. 136; 103 [Latin]). Newton never abandoned this conviction.

419 For More's rejection of it, see above, Ch. 8, Sec. 4d. Leibniz not only rejected the actual

extension of the soul in the body but also denied that the whole of the soul could be "in every part of the body" (third letter, par. 12, pp. 28–29 of Alexander [ed.], *Leibniz-Clarke Correspondence*). For scholastic usage of this principle, see above, Ch. 6, Sec. 3f and n. 127; for a nonscholastic who used it, see above, Ch. 8, Sec. 4b (Gassendi).

420 J. E. McGuire however, believes otherwise ("Newton on Space and Time," 506) and conjectures that Newton actually adopted it as the best explanation of God's omnipresence in infinite space. It enabled Newton to hold at one and the same time that the "Divine presence is everywhere with respect to the expanses of infinite space" and "God is everywhere the *same* unique individual." Because McGuire has no evidence for his claim, he believes "we can plausibly reconstruct the following argument. To say that God is everywhere with respect to space, is to say that one and the same individual exists at *each* place in extended space. To make such a claim is a contradiction only with respect to the manner in which extended things exist spatially. For they cannot as complete beings exist at once in each 'part' of the place they occupy. But there is no contraditcion in holding that an essentially non-extended being is capable of so existing." But there surely is a contradiction here. How can one and the same nonextended being be wholly in one point and wholly in all of infinite space at one and the same time? The concept is unintelligible. McGuire cites Cudworth's *True Intellectual System* as employing the same approach. But we have seen that the whole in every part doctrine is an old scholastic tradition with roots in Plotinus and St. Augustine.

421 Koyré and Cohen, "Newton & the Leibniz-Clarke Correspondence," 101 (Draft E); see also p. 103 and Koyré, *From the Closed World*, p. 248. For the *Avertissement au Lecteur* of 1720, see above, Ch. 8, Sec. 4m.

422 See above, Ch. 6, Sec. 4; Ch. 7, Sec. 2a. Leibniz would declare that "God is not present to things by situation, but by essence: his presence is manifested by his immediate operation" (Alexander [ed.], *Leibniz-Clarke Correspondence*, Leibniz's third letter, par. 12, p. 28). Clarke's reply beautifully reflected Newton's sentiments: "God being omnipresent, is really present to every thing, essentially and substantially. His presence manifests itself indeed by its operation, but it could not operate if it was not there" (*Ibid.*, Clarke's third reply, par. 12, pp. 33–34).

423 See W. von Leyden, *Seventeenth-Century Metaphysics*, pp. 275–278.

424 Koyré, *From the Closed World*, p. 274.

425 An excellent account of these disputes is furnished by Ferguson, *Philosophy of Dr. Samuel Clarke*, ch. 3 ("The Argument for the existence of God"), pp. 22–121. In arguments for the existence of God, the nature of space and its relation to God played an important role. Focusing on Clarke, who was in fact the storm center of disputes about the existence of God and space in the first half of the eighteenth century, Ferguson describes the positions of Clarke's defenders and opponents. Among those who attacked the doctrine of absolute space of Clarke and Newton were Berkeley, Leibniz, Joseph Butler, Edmund Law, Joseph Clarke (Fellow of Magdalene College, Cambridge), Samuel Colliber, Isaac Watts and David Hume; its defenders were John Clarke, brother of Samuel, and the Rev. John Jackson.

During the first half of the eighteenth century, but extending somewhat beyond with the inclusion of David Hume, the debate about the nature of space and its relation to God reached its most intensive stage and achieved its most extensive expression. Most of the central arguments from the Leibniz-Clarke correspondence were reexamined and subjected to penetrating and exhaustive analysis. By 1750, it would appear that the defenders of Newton and Clarke had been vanquished.

Although the post-Newtonian period lies beyond the scope of this study, it will be useful to describe briefly some of the major eighteenth-century thrusts against the Newton-Clarke conception of a God-filled, absolute, infinite, void space. Because Ferguson supplies all references, only the page numbers of his summaries will be cited here.

Edmund Law and other Englishmen would come to agree with Leibniz that "real absolute space" had become "an idol of some modern Englishmen" (Leibniz's third letter, par. 2 in Alexander [ed.], *Leibniz-Clarke Correspondence*, p. 25). Law considered space a negative concept derived from the imagined absence of matter. From such a negative notion, one could not properly argue to an external space. Space is really nothing and therefore cannot be hypostasized into a real entity by the attribution to it of the property, or capacity, to receive bodies. Such a move would be tantamount to "saying that darkness must be a real existence because it has the property of receiving light, or that absence is a real something because it has the property of receiving presence" (Ferguson's summary, p. 60). Clarke's assumption of spatial infinity and the denial to space of parts also came in for criticism. Law denied true infinity to space and argued that our notion of its alleged infinity is derived from the very parts that Clarke denied. For our concept of spatial infinity is an extrapolation from the indefinite addition of one part of space to another; consequently space must have parts.

Samuel Clarke's idea that our alleged inability to eliminate space from our minds (More would say to "disimagine" space [see above. Ch. 8, Sec. 4d]) counts as evidence for its real existence, an idea that Henry More propounded (and Newton as well; see above, Ch. 8, Sec. 4l and n. 344), was attacked by Joseph Clarke (Ferguson, pp. 65–66), who argued that space is the idea of nothing and corresponds to nothing outside. To say that we cannot imagine space away is of no more consequence than to say that we cannot imagine nothing away. It was blasphemous, in his judgment, to call God an infinite vacuum, which is nothing more than a negation or diffused nothing. For him, the word "nothing" has two senses (Ferguson, p. 73): Either it means "not any existent" or "not any thing at all." Only the first sense is properly applicable to space, because space is an idea with no external archetype to which the idea of it corresponds.

Edmund Law seized upon yet another weakness in the concept of absolute space. Samuel Clarke's brother, John, had argued that if space were simply the negation or absence of matter, as Law insisted, no difference could obtain between touching and nontouching. For "if there is nothing between two walls, they must touch each other. If, however, they do not touch there must be something between them, otherwise the contradiction would result that touching and not touching were identical" (Ferguson's summary, pp. 62–63; for medieval consideration of this problem, see above, Ch. 6, Sec. 2). Law replied that, although it is true that when two things touch there is nothing between them, "we cannot conclude from this that when there is nothing between things they must touch. If they are distant from each other, they will not touch, but distance itself is an abstract idea and denotes nothing real." In joining this attack, Joseph Clarke would argue against John and Samuel Clarke that their view of the absolute and immutable nature of space committed them to the belief that "when two bodies touch, there is space between and hence there is no difference between touching and not touching" (Ferguson, p. 65).

As expected, Samuel Clarke's interpretation of space as God's property evoked intense debate. Law was convinced (Ferguson, p. 70) that if extended space were really a property of God, the divine substance would be divided into parts, because one could then say that a part of God is in one place and another part in another place. Joseph Clarke and Isaac Watts would argue that in the identification of space with God's extension, Newton had deified space. In their judgment, Newton had actually identified space with God (Clarke had denied such an identification; see above, Ch. 8, Sec. 4m).

And finally, we should observe that David Hume would deny the existence of an external absolute space as an unwarranted inference from sense data (Ferguson, p. 113). For Hume, our idea of space derives from the order of our sense impressions (*ibid.*, p. 112).

The literature on the nature of space generated in the eighteenth century as a result of the reaction to the ideas of Newton and Clarke was philosophically and theologically significant. Not only did much of it emphasize the relational and psychological aspects of space, but it also revealed that God's assumed omnipresence in an absolute infinite space was no longer a tenable concept. Four centuries of debate on that monumental issue were drawing to a close.

426 See Koyré, *From the Closed World*, p. 276.
427 Even with the elimination of God, the concept of absolute space remained controversial. Jammer (*Concepts of Space*, p. 139) insists that its acceptance in the eighteenth and

nineteenth centuries was largely as a working hypothesis without deep concern for its theoretical implications. In this connection, he observes "how little the actual progress of the science of mechanics was affected by general considerations concerning the nature of absolute space." Although Clerk Maxwell (Jammer, *Concepts of Space*, p. 140) and many others would accept it in the nineteenth century, Ernst Mach (1838–1916), in the first edition of his *Science of Mechanics* in 1883, would describe absolute space and motion as "pure things of thought, pure mental constructs, that cannot be produced in experience" (*The Science of Mechanics: A Critical and Historical Account of Its Development*, trans. by Thomas J. McCormack; new introduction by Karl Menger; 6th ed. with revisions through the 9th German edition [LaSalle, Ill.: Open Court Publishing, 1960], p. 280). In Mach's judgment, absolute space was not required for the law of inertia (*ibid.*, p. 288). The view that absolute motion, and therefore absolute space, is "a conception which is devoid of content and cannot be used in science struck almost everybody as strange thirty years ago," Mach explained, "but at the present time it is supported by many and worthy investigators" (*ibid.*, p. 293). Mach's rejection of absolute space and adoption of a "relativist" position were destined to influence Einstein's new theory of space in this century.

Bibliography

Primary sources

Adelard of Bath. "Die *Quaestiones Naturales* des Adelardus von Bath." Ed. by Martin Müller. *Beiträge zur Geschichte der Philosophie und Theologie des Mittelalters*, vol. 31, heft 2. Münster: Aschendorff, 1934.

Dodi Venechdi (Uncle and Nephew), the Work of Berachya Hanakdan, now edited from MSS. at Munich and Oxford, with an English Translation, Introduction, etc., to which is added the first English Translation from the Latin of Adelard of Bath's 'Quaestiones Naturales' trans. by Hermann Gollancz. London: Oxford University Press, 1920.

Aegidius Romanus. *Egidii Romani in libros De physico auditu Aristotelis commentaria accuratissime emendata et in marginibus ornata quotationibus textuum et commentorum ac aliis quamplurimis annotationibus. Cum tabula questionum in fine. Eiusdem questio De gradibus formarum.* Venice, 1502.

Commentationes Physicae et Metaphysicae: in Physicorum libros octo; De coelo libros quatuor; De generatione et corruptione libros duos; Meteororum lib.; De anima libros tres; Parva naturalia; Metaphysicorum libros duodecim tradita a . . . Fratre Aegidio Romani. . . . Ursellis, 1604. Falsely ascribed to Aegidius.

Albert of Saxony. *Questiones et decisiones physicales insignium virorum: Alberti de Saxonia in octo libros Physicorum; tres libros De celo et mundo; duos lib. De generatione et corruptione; Thimonis in quatuor libros Meteororum; tres lib. De anima; Buridani in lib. De sensu et sensato; librum De memoria et reminiscentia; librum De somno et vigilia; lib. De longitudine et brevitate vite; lib. De iuventute et senectute Aristotelis. Recognitae rursus et emendatae summa accuratione et iudicio Magistri Georgii Lokert Scotia quo sunt Tractatus proportionum additi.* Paris, 1518.

Albertus Magnus. *Omnia Opera.* Ed. by Peter Jammy. 21 vols. Lyon, 1651. Vol. 2: *Physicorum lib. VIII; De coelo et mundo lib. IV; De generatione et corruptione lib. II; De meteoris lib. IV; De mineralibus lib. V.*

Alexander of Aphrodisias (Pseudo). *Problemata Alexandri Aphrodisei e graeco in latinum traducta per Georgium Valla. Et Aristotelis Problemata Theodoro [Gaza] interprete. Et Plutarchi problemata, etc.* Venice, 1488.

"Problemata Alexandri Aphrodisiensis." In Julius Ludwig Ideler (ed.), *Physici et Medici Graeci Minores.* 2 vols. Amsterdam: Adolf M. Hakkert, 1963. Reprint of 1841–1842 edition. Vol. 1, pp. 3–80.

Alexander, H. G. See Leibniz, Gottfried.

Alhazen. *Opticae Thesaurus Alhazeni Arabis libri septem nunc primum editi. Eiusdem liber De crepusculis et nubium ascensionibus. Item Vitellonis Thuringopolonis libri X.* Ed. by Frederick Risner. Basel, 1572.

Ambrose, St. *Hexameron, Paradise and Cain and Abel.* Trans. by John J. Savage. New York: Fathers of the Church, 1961.

Amicus, Bartholomeus. *In Aristotelis libros De physico auditu dilucida textus explicatio et disputationes in quibus illustrium scholarum Averrois, D. Thomae, Scoti, et Nominalium sententiae expenduntur earumque tuendarum probabiliores modi afferuntur.* 2 vols. Naples, 1626–1629. It appears that Amicus's *Physics* commentary formed part of a seven-volume set of commentaries with the following title: *In universam Aristotelis philosophiam notae et disputationes, quibus illustrium scholarum Averrois, D. Thomae,*

Scoti et nominalium sententiae expenduntur earumque tuendarum probabiles modi afferuntur. Naples, 1623–1648.

Aquinas, St. Thomas. *Summa Theologiae.* Latin text and English translation, introductions, notes, appendices, and glossaries. New York and London: Blackfriars in conjunction with McGraw-Hill and Eyre & Spottiswoode.

Vol. 2: Existence and Nature of God (1a.2–11). Trans. by Timothy McDermott, O. P. Blackfriars, 1964.

Vol. 8. Creation, Variety and Evil (1a.44–49). Trans. by Thomas Gilby, O. P. Blackfriars, 1967.

S. Thomae Aquinatis in octo libros De physico auditu sive Physicorum Aristotelis commentaria. Editio novissima. Ed. by P. Fr. Angeli-M. Pirotta, O. P. Naples: M. D'Auria Pontificus Editor, 1953.

St. Thomas Aquinas, Siger of Brabant, St. Bonaventure, On the Eternity of the World (De Aeternitate Mundi). Trans. from the Latin with an introduction by Cyril Vollert, Lottie H. Kendzierski, and Paul M. Byrne. Milwaukee, Wis.: Marquette University Press, 1964.

Aristotle. *The Works of Aristotle translated into English* under the editorship of J. A. Smith and W. D. Ross. 12 vols. Oxford: Clarendon Press, 1908–1952.

Aristotle On Coming-To-Be and Passing-Away (De Generatione et Corruptione), revised text with introduction and commentary by Harold H. Joachim. Oxford: Clarendon Press, 1922.

On the Heavens. With an English translation by W. K. C. Guthrie. Loeb Classical Library. London: William Heinemann; Cambridge, Mass.: Harvard University Press, 1939.

Arnobius of Sicca. *The Case Against the Pagans.* Newly translated and annotated by George McCracken. 2 vols. *Ancient Christian Writers: The Works of the Fathers in Translation,* nos. 7, 8. Westminster, Md.: Newman Press, 1949.

Augustine, St. *Corpus Scriptorum Ecclesiasticorum Latinorum.* Vienna Academy.

Vol. 25, sec. 6, part 2: *Contra felicem de natura boni; Epistula Secundini contra Secundinum accedunt; Evodii de fide contra Manichaeos; et Commonitorium Augustini quod fertur.* Ed. by Joseph Zycha. Prague/Vienna/Leipzig: F. Tempsky/G. Freytag, 1892.

Vol. 33, sec. 3, part 1: *De Genesi ad litteram libri duodecim; eiusdem libri capitula; De Genesi ad litteram imperfectus; Locutionum in Heptateuchum libri septem.* Ed. by Joseph Zycha. Prague/Vienna/Leipzig: F. Tempsky/G. Freytag, 1894.

Vol. 40, parts 1, 2: *De civitate Dei.* Prague/Vienna/Leipzig: F. Tempsky/G. Freytag, 1899, 1900.

Corpus Christianorum, series Latina: Aurelii Augustini Opera. Vol. 40, part 10, 3: *Enarrationes in Psalmos CI-CL.* Turnholt: Brepols, 1956.

Confessions. Trans. by Vernon J. Bourke. *The Fathers of the Church, A New Translation,* vol. 5. New York: Cima Publishing, 1953.

The City of God. Trans. and ed. by Marcus Dods. 2 vols. New York: Hafner Publishing, 1948. First published Edinburgh: T & T Clark, 1872.

The City of God Against the Pagans. 7 vols. Vol. 3, books 8–11, with an English translation by David S. Wiesen. Loeb Classical Library. London: William Heinemann; Cambridge, Mass.: Harvard University Press, 1968.

Averroes. *Aristotelis opera cum Averrois commentariis.* 9 vols. in 11 parts plus 3 supplementary vols. Venice, Junctas ed., 1562–1574. Reprinted in facsimile Frankfurt: Minerva, 1962.

Vol. 4: *Aristotelis De physico auditu libri octo cum Averrois Cordubensis variis in eosdem commentariis.*

Vol. 5: *Aristotelis De coelo; De generatione et corruptione; Meteorologicorum; De plantis cum Averrois Cordubensis variis in eosdem commentariis.*

Vol. 8: *Aristotelis Metaphysicorum libri XIII cum Averrois Cordubensis in eosdem commentariis, et Epitome.*

Averroes' Tahafut al-Tahafut (The Incoherence of the Incoherence). Trans. from the Arabic with introduction and notes by Simon van den Bergh. 2 vols. Printed Oxford: Oxford University Press; published London: Luzac, 1954.

Averroes' "Destructio Destructionum Philosophiae Algazelis" in the Latin Version of Calo Calonymos. Ed. with an introduction by Beatrice H. Zedler. Milwaukee, Wis.: Marquette University Press, 1961.

Avicenna. *Avicenne perhypatetici philosophi ac medicorum facile primi opera in lucem redacta ac nuper quantum ars niti potuit per canonicos emendata: Logyca; Sufficienta; De celo et mundo; De anima; De animalibus; De intelligentiis; Alpharabius de intelligentiis; Philosophia prima.* Venice: Octavianus Scotus, 1508. Reprinted in facsimile Frankfurt: Minerva, 1961.

Bacon, Roger. *Opera hactenus inedita Rogeri Baconi.*

  Fasc. 3: *Liber primus Communium naturalium Fratris Rogeri,* partes tertia et quarta. Ed. by Robert Steele. Oxford: Clarendon Press, 1911.

  Fasc. 8: *Questiones supra libros quatuor Physicorum Aristotelis.* Ed. by Ferdinand M. Delorme, O. F. M., with the collaboration of Robert Steele. Oxford: Clarendon Press, 1928.

  Fasc. 13: *Questiones supra libros octo Physicorum Aristotelis.* Ed. by Ferdinand M. Delorme, O. F. M., with the collaboration of Robert Steele. Oxford: Clarendon Press, 1935.

The Opus Majus of Roger Bacon. Trans. by Robert B. Burke. 2 vols. Philadelphia: University of Pennsylvania Press, 1928. Reprinted New York: Russell and Russell, 1962.

Baeumker, Clemens. "Das pseudo-hermetische Buch der vierundzwanzig Meister (Liber XXIV philosophorum)." *Studien und Charakteristiken zur Geschichte der Philosophie inbesondere des Mittelalters. Gesammelte Vorträge und Aufsätze von Clemens Baeumker. Beiträge zur Geschichte der Philosophie des Mittelalters,* band XXV, heft 1/2. Münster: Aschendorff, 1927, pp. 194–214.

Barbay, Petrus. *Commentarius in Aristotelis Metaphysicam authore Petro Barbay celeberrimo quondam in Academia Parisiensi Philosophiae Professore. Editio quarta emendatior multo, schematibus philosophicis adornata et figuris ad sphaeram spectantibus aucta.* Paris, 1684.

Bardenhewer, Otto (ed.). *Die pseudo-aristotelische Schrift ueber das reine Gute bekannt unter dem Namen "Liber de causis."* Freiburg: Herderhandlung, 1882.

Barrow, Isaac. *The Mathematical Works of Isaac Barrow, D. D.* Ed. by W. Whewell. Cambridge: Cambridge University Press, 1860.
  1. *Lectiones Mathematicae XXIII.*
  2. *Mathematici Professoris Lectiones.*
  3. *Lectiones Opticae XVIII.*
  4. *Lectiones Geometricae XIII.*

The Usefulness of Mathematical Learning explained and demonstrated: Being Mathematical Lectures Read in the Publick Schools at the University of Cambridge by Isaac Barrow, D. D. Professor of the Mathematics, and Master of Trinity-College, &c. To which is prefixed, The Oratorical Preface of Our Learned Author, spoke before the University on his being elected Lucasian Professor of the Mathematics. Trans. by the Rev. Mr. John Kirkby. London: printed for Stephen Austen, 1734.

Geometrical Lectures: Explaining the Generation, Nature and Properties of Curve Lines. Read in the University of Cambridge, by Isaac Barrow, D. D. Professor of the Mathematicks and Master of Trinity-College, &c. Trans. from the Latin edition, revised, corrected, and amended by the late Sir Isaac Newton by Edmund Stone. London: printed for Stephen Austen, 1735.

Bartholomeus Mastrius and Bonaventura Bellutus. *In Arist. Stag. libros Physicorum quibus ab adversantibus tum veterum tum recentiorum iaculis Scoti philosophia vindicatur a P.P. Magistris Bartholomeo Mastrio de Meldula, . . . et Bonaventura Belluto de Catana . . .* 2d ed. Venice: typis Marci Ginammi, 1644.

Berkeley, George. *The Works of George Berkeley D. D.; Formerly Bishop of Cloyne Including his Posthumous Works.* With prefaces, annotation, and an account of his life by Alexander Campbell Fraser. 4 vols. Oxford: Clarendon Press, 1901.
  Vol. 1: *An Essay Towards a New Theory of Vision; A Treatise Concerning the Principles of Human Knowledge.*

Vol. 3: *Siris: A Chain of Philosophical Reflexions and Inquiries Concerning the Virtues of Tar-Water, And Divers Other Subjects Connected Together and Arising One From Another.*

Philosophical Commentaries, Generally Called the Commonplace Book. Ed. by A. A. Luce. London: Thomas Nelson, 1944.

Blasius of Parma. *Questio Blasij de Parma De tactu corporum durorum* in *Questio de modalibus Bassani Politi; Tractatus proportionum introductorius ad calculationes suisset; Tractatus proportionum Thoma Bradwardini; Tractatus proportionum Nicholai Oren; Tractatus De latitudinibus formarum Blasij de Parma; Auctor sex inconvenientum.* Venice, 1505. Although omitted from the title page, Blasius' *De tactu corporum durorum*, along with another treatise by Johannis de Casali (*De velocitate motus alterationis*), is listed on the 74th and final folio of the volume.

Boethius. *The Theological Tractates*, with an English translation by H. F. Stewart and E. K. Rand; *The Consolation of Philosophy*, with an English translation by "I. T." (1609), revised by H. F. Stewart. Cambridge, Mass.: Harvard University Press, 1953.

The Consolation of Philosophy. Trans. with an introduction by V. E. Watts. Harmondsworth, Middlesex: Penguin Books, 1969.

Bona Spes, Franciscus. *R. P. Francisci Bonae Spei, Carmelitae Reformati, Commentarii Tres in Universam Aristotelis Philosophiam.* Brussels: apud Franciscum Vivienum, 1652.

Bonaventure, St. *Opera Omnia.* Edita studio et cura PP. Collegii a S. Bonaventura. Ad Claras Aquas (Quaracchi): ex typographia Collegii S. Bonaventurae, 1882–1901.

Vol. 1: *Commentaria in quatuor libros Sententiarum, in primum librum Sententiarum* (1882).

Bonetus, Nicholas. *Habes Nicholai Bonetti viri perspicacissimi quattuor volumina: Metaphysicam, videlicet naturalem phylosophiam, predicamenta, necnon theologiam naturalem in quibus facili calle et per brevi labore omnia fere scibilia comprehenduntur. . . .* Venice, 1505.

Bradwardine, Thomas. *De causa Dei contra Pelagium et de virtute causarum ad suos Mertonenses libri tres . . . opera et studio Dr. Henrici Savili, Collegii Mertonensis in Academia Oxoniensis custodis, ex scriptis codicibus nunc, primum editi.* London: apud Ioannem Billium, 1618. Reprinted in facsimile Frankfurt: Minerva, 1964.

Thomas of Bradwardine, His "Tractatus de proportionibus." Ed. and trans. by H. Lamar Crosby, Jr. Madison, Wis.: University of Wisconsin Press, 1955.

Bruno Giordano. *De immenso et innumerabilibus. Jordani Bruni Nolani Opera Latine Conscripta*, vols. 1 (bks. 1–3) and 2 (bks. 4–8). Ed. F. Fiorentino. Naples: apud Dom. Morano, 1879, 1884.

Opere Italiane, 2d. ed. 2 vols. Vol. 1: *Dialoghi Metafisici.* Ed. Giovanni Gentile. Bari: Gius. Laterza & Figli, 1925.

The Infinite in Giordano Bruno With a Translation of His Dialogue "Concerning the Cause Principle, and One" by Sidney Greenberg. New York: King's Crown Press, 1950.

Giordano Bruno, His Life and Thought. With annotated translation of his work *On the Infinite Universe and Worlds* by Dorothea Waley Singer. New York: Henry Schuman, 1950.

Buridan, John. *Acutissimi philosophi reverendi Magistri Johannis Buridani subtilissime questiones super octo Phisicorum libros Aristotelis diligenter recognite et revise a Magistro Johanne Dullaert de Gandavo antea nusquam impresse.* Paris, 1509. Reprinted in facsimile with the title *Johannes Buridanus, Kommentar zur Aristotelischen Physik*, Frankfurt: Minerva, 1964.

Iohannis Buridani Quaestiones super libris quattuor De caelo et mundo. Ed. by Ernest Addison Moody. Cambridge, Mass.: Mediaeval Academy of America, 1942.

Burley, Walter. *Questiones circa libros Physicorum.* MS Basel Universitätsbibliothek F. V. 12, fols. 108r–171v.

Burleus super octo libros Phisicorum. Venice, 1501. Reprinted in facsimile as Walter Burley, *In Physicam Aristotelis expositio et quaestiones*, Hildesheim/New York: Georg Olms Verlag, 1972.

Gualteri Burlaei Liber De vita et moribus philosophorum mit einer altspanischen Ueber-

setzung der Eskurialbibliothek. Ed. by Hermann Knust. *Bibliothek des Literarischen Vereins in Stuttgart,* vol. 177. Tübingen, 1886.

Cajetan. See Thomas de Vio.

Campanella, Tommaso. *Universalis philosophiae seu metaphysicarum rerum iuxta propria dogmata, partes tres, libri 18.* Paris, 1638.

Metafisica. Ed. by Giovanni di Napoli with Italian translation. 3 vols. Bologna: Zanichelli, 1967.

Del senso delle cose e della magia. Ed. by Antonio Bruers. *Classici della Filosofia Moderna* collana di testi e di traduzioni a cura di B. Croce e G. Gentile, vol. 24. Bari: Gius. Laterza & Figli, 1925.

Epilogo magno (Fisiologia Italiana). Testo Italiano inedito con le varianti dei codici e delle edizioni Latine. Ed. by Carmelo Ottaviano. Rome: Reale Accademia d'Italia, 1939.

Čapek, Milič (ed.). *The Concepts of Space and Time, Their Structure and Their Development.* Dordrecht/Boston: D. Reidel Publishing, 1976.

See Cornford, F. M.

Charleton, Walter. *Physiologia Epicuro-Gassendo-Charltoniana.* London, 1654.

Clarke, Samuel. *The Works of Samuel Clarke, D. D. Late Rector of St. James's Westminster.* 4 vols. London: printed for John and Paul Knapton, 1738.

A Demonstration of the Being and Attributes of God 1705; A Discourse concerning the Unchangeable Obligations of Natural Religion 1706. Facsimile reprint of the London editions. Stuttgart-Bad Cannstatt: Friedrich Frommann Verlag, 1964.

See Leibniz, Gottfried.

Cleomedes. *De motu circulari corporum caelestium.* Greek text with Latin translation by H. Ziegler. Leipzig: Teubner, 1891.

Kleomedes Die Kreisbewegung der Gestirne. Trans. by Arthur Czwalina. *Ostwald's Klassiker der exakten Wissenschaften.* Leipzig: Engelmann, 1927.

Cohen, Morris R., and Israel E. Drabkin. *A Source Book in Greek Science,* 2d. ed. Cambridge, Mass.: Harvard University Press, 1958. 1st ed. 1948.

Coimbra Jesuits. See Conimbricenses.

Combes, André. See Jean de Ripa.

Complutenses. *Collegii Complutensis Discalceatorum Fratrum Ordinis B. Mariae de Monte Carmeli Disputationes in octo libros Physicorum Aristotelis iuxtam miram Angelici Doctoris D. Thomae et scholae eius doctrinam, eidem communi magistro et Florentissimae scholiae dicatae; nunc primum in Galliis excusae.* Paris, 1628.

Conimbricenses. *Commentarii Collegii Conimbricensis Societatis Iesu in octo libros Physicorum Aristotelis Stagiritae.* Lyon: sumptibus Horatii Cardon, 1602. The Cologne and Lyon editions are nearly identical.

Commentariorum Collegii Conimbricensis Societatis Iesu in octo libros Physicorum Aristotelis Stagiritae. Cologne: sumptibus Lazari Zetzneri, 1602. First published at Coimbra, 1591.

Coronel, Luis. *Physice perscrutationes Magistri Ludovici Coronel Hispani Segoviensis.* Paris, after 1511.

Crescas, Hasdai. *Crescas' Critique of Aristotle, Problems of Aristotle's 'Physics' in Jewish and Arabic Philosophy* by Harry Austryn Wolfson. Cambridge, Mass.: Harvard University Press, 1929.

Cudworth, Ralph. *The True Intellectual System of the Universe: The First Part: Wherein, All the Reason and Philosophy of Atheism is Confuted; and Its Impossibility Demonstrated.* London: printed for Richard Royston, Bookseller to His Most Sacred Majesty, 1678.

Cyprian. *S. Thasci Caeceli Cypriani Opera Omnia.* Ed. by William Hartel. *Corpus Scriptorum Ecclesiasticorum Latinorum,* vol. 3, part 1. Vienna: apud C. Geroldi Filium Bibliopolam Academiae, 1868.

Denifle, Heinrich, and Emil Chatelain. *Chartularium Universitatis Parisiensis.* 4 vols. Paris: Fratrum Delalain, 1889–97.

Descartes, René. *Oeuvres de Descartes.* Ed. by Charles Adam and Paul Tannery. Vol. 5: *Correspondance* (Paris: Léopold Cerf, 1903); vol. 8, part 1: *Principia Philosophiae* (Paris:

Librairie Philosophique J. Vrin, 1964); vol. 11: *Le Monde* (Paris: Léopold Cerf, 1909).
Oeuvres Philosophiques, Tome I (1618–1637). Textes établis, présentés et annotés par
Ferdinand Alquié. Paris: Editions Garnier Frères, 1963.
The Philosophical Works of Descartes. Rendered into English by Elizabeth S. Haldane and
G. R. T. Ross. 2 vols. New York: Dover Publications, 1955. 1st ed. 1911; reprinted
with corrections 1931 by Cambridge University Press.
Diogenes Laertius. *Lives of Eminent Philosophers* with an English translation by R. D.
Hicks. 2 vols. Loeb Classical Library. Cambridge, Mass.: Harvard University Press;
London: William Heinemann, 1925; revised 1938, 1942, 1950.
Dullaert, Johannes. *Questiones super octo libros Physicorum Aristotelis necnon super libros
De coelo et mundo*. Paris: Oliverius Senant, 1506.
Durandus de S. Porciano. *In Sententias theologicas Petri Lombardi commentariorum libri
quatuor*. Lyon: apud Gasparem a Portonaris, 1556.
Ermatinger, Charles J. "Richard Fishacre: <Commentum in librum I Sententiarum>."
The Modern Schoolman 35 (1957–1958), 213–235.
Euclid. *The Thirteen Books of Euclid's Elements*. Trans. from the text of Heiberg with intro-
duction and commentary by Sir Thomas L. Heath. 2d ed. revised with additions. 3 vols.
New York: Dover Publications, 1956. Reprinted from the Cambridge edition of 1926.
Faber, Phillip. *F. Philippi Fabri Faventini, Ordinis Min. Conventualium . . . Disputationes
Theologicae Librum Primum Sententiarum Complectentes . . . secundum seriem distinc-
tionum Magistri Sententiarum et questionum Scoti paucis exceptis ordinate. . . .*
Venice, 1613.
Fabri, Honoré. *Summula theologica in qua quaestiones omnes alicuius momenti, quae a
scholasticis agitari solent, breviter discutiuntur ac definiuntur. Auctore Honorato Fabri
Societatis Iesu, cum indicibus tractatuum et rerum notabilium. Nunc primum in lucem
prodit*. Lyon: sumptibus Laurentii Anisson, 1679.
Fallico, Arturo B., and Herman Shapiro. *Renaissance Philosophy*, vol. 1: *The Italian Philoso-
phers, Selected Readings from Petrarch to Bruno*. Ed., trans., and introduced by Arturo B.
Fallico and Herman Shapiro. New York: Modern Library, 1967.
Federicus Pendasius *Federici Pendasii Mantuani, philosophi acutissimi, in antiquissimo
Bononiensium Gymansio e supremo loco unice profitentis Physicae auditionis texturae libri
octo. Nunc primum in lucem editi in quibus Aristotelis ars observatur, vetus Graecorum
doctrina restauratur, Arabum etiam, atque inter Latinos primatum sensa expenduntur.
Et difficilia quaeque ac ardua explanatur. . . .* Venice, 1604.
Fonseca, Pedro. *Commentariorum Petri Fonsecae Lusitani, Doctoris Theologi Societatis Iesu,
in libros Metaphysicorum Aristotelis Stagiritae*. 4 vols. Cologne: Lazari Zetzneri Bibliopolae, 1615. Reprinted in facsimile with the title *Petri Fonsecae Commentariorum in
Metaphysicorum Aristotelis Stagiritae Libros*. 2 vols. Hildesheim: Georg Olms Verlag,
1964.
Fortin, Ernest L. and O'Neill, Peter D. (trans.). "Condemnation of 219 Propositions" in
Ralph Lerner and Muhsin Mahdi (eds.), *Medieval Political Philosophy: A Source Book*.
New York: Free Press of Glencoe, 1963, pp. 337–354. Reprinted in Arthur Hyman
and James J. Walsh (eds.), *Philosophy in the Middle Ages: The Christian, Islamic, and
Jewish Traditions*. Indianapolis: Hackett Publishing, 1973, pp. 540–549.
Francisus de Mayronis. *Tria principia Antonii Andree Ordinis Minorum; Expositio Fran.
Mayronis Ordinis Minorum super octo libros Phisicorum; Formalitates eiusdem Francisci
Mayronis; Tractatus eiusdem De principio complexo; Tractatus eiusdem De terminis
theologicis; Tractatus De ente et essencia secundum Thomam*. Ferrara, 1490.
Fridugis the Deacon. *Letter on Nothing and Darkness (Epistola de nihilo et tenebris)*. Trans.
by Hermigild Dressler in *Medieval Philosophy from St. Augustine to Nicholas of Cusa*.
Ed. by John F. Wippel and Allan B. Wolter. New York: Free Press; London: Collier
Macmillan Publishers, 1969, pp. 103–108.
Galileo Galilei. *Le Opere di Galileo Galilei. Edizione Nazionale*. Ed. by Antonio Favaro. 23
vols. Florence, 1891–1909.
Dialogues Concerning Two New Sciences by Galileo Galilei. Trans. from the Italian
and Latin into English by Henry Crew and Alfonso de Salvio, with an introduction
by Antonio Favaro. New York: Macmillan, 1914; reprinted by Dover Publications, n.d.

Galileo Galilei Two New Sciences, Including Centers of Gravity & Force of Percussion. Trans. with introduction and notes by Stillman Drake. Madison, Wis.: University of Wisconsin Press, 1974.

Galileo Galilei On Motion and On Mechanics comprising *De motu* (c. 1590), trans. with introduction and notes by Israel E. Drabkin, and *Le Meccaniche* (c. 1600), trans. with introduction and notes by Stillman Drake. Madison, Wis.: University of Wisconsin Press, 1960.

Gassendi, Pierre. *Animadversiones in decimum librum Diogenis Laertii, qui est de vita, moribus, placitisque Epicuri. Continent autem placita, quas ille treis statuit philosophia parteis:* I. *Canonicam nempe, habitam Dialecticam loco:* II. *Physicam, ac imprimis nobilem illius partem Meteorologiam:* III. *Ethicam cuius gratia ille excoluit caeteras.* 3 tomes in 2 vols. Lyon: apud Guillelmum Barbier, 1649.

Syntagma philosophicum. Opera Omnia, vol. 1. Lyon, 1658. Reprinted in facsimile Stuttgart-Bad Cannstatt: Friedrich Frommann Verlag, 1964.

Syntagma philosophicum. Opera Omnia, vol. 1. Florence, 1727.

The Selected Works of Pierre Gassendi. Ed. and Trans. by Craig B. Brush. New York: Johnson Reprint Corporation, 1972.

Giles of Rome. See Aegidius Romanus.

Glanvil, Joseph. *Saducismus Triumphatus: Or, Full and Plain Evidence Concerning Witches and Apparitions (1689).* A facsimile reproduction with an introduction by Coleman O. Parsons. Gainesville, Fla.: Scholars' Facsimiles & Reprints, 1966.

Godfrey of Fontaines. *Les quatre premiers Quodlibets de Godefroid de Fontaines.* Ed. by M. de Wulf and A. Pelzer. *Les Philosophes Belges.* Textes et Etudes, vol. 2. Louvain: Institut Supérieur de Philosophie de l'Université, 1904.

Gorlaeus, David. *Exercitationes philosophicae quibus universa fere discutitur philosophia theoretica. Et plurima ac praecipua peripateticorum dogmata evertuntur.* Post mortem auctoris editae cum gemino judice. Westerhuysen: impensis Johannis Ganne & Harmanni, 1620.

Grant, Edward (ed.). *A Source Book in Medieval Science.* Cambridge, Mass.: Harvard University Press, 1974.

Grosseteste, Robert. *Die philosophischen Werke des Robert Grosseteste. Beiträge der Geschichte der Philosophie des Mittelalters,* vol. 9. Ed. by Ludwig Baur. Münster: Aschendorff, 1912.

Roberti Grosseteste Episcopi Lincolniensis Commentarius in VIII libros Physicorum Aristotelis. Ed. by Richard C. Dales. Boulder, Colo.: University of Colorado Press, 1963.

Guericke. Otto von. *Experimenta nova [ut vocantur] Magdeburgica de vacuo spatio.* Amsterdam, 1672. Reprinted in facsimile Aalen: Otto Zeller, 1962.

Neue (sogenannte) Magdeburger Versuche über den leeren Raum. Nebst Briefen, Urkunden und anderen Zeugnissen seiner Lebens-und Schaffensgeschichte. Ed. and trans. Hans Schimank. With the collaboration of Hans Gossen, Gregor Maurach, and Fritz Krafft. Düsseldorf: VDI-Verlag, 1968.

Guillelmus Altissiodorensis. See William of Auxerre.

Henry of Ghent. *M Henrici Goethals a Gandavo, Doctoris Solemnis, Socii Sorbonici, Ordinis Servorum B. M. V. et Archid. Tornacensis, Aurea Quodlibeta . . . Hac postrema editione commentariis doctissimis illustrata M. Vitalis Zuccolii Patavini Ordinis Camaldulensis Theologi Clarissimi.* 2 vols. Venice, 1613.

Hero of Alexandria. *The Pneumatics of Hero of Alexandria from the Original Greek.* Trans. for and ed. by Bennet Woodcroft. The translation was made by J. G. Greenwood. London: Taylor, Walton and Maberly, 1851. Reprinted in facsimile with introduction by Marie Boas Hall. London: Macdonald; New York: American Elsevier, 1971.

See Philo of Byzantium.

Hiscock, W. G. (ed.). *David Gregory, Isaac Newton and Their Circle: Extracts from David Gregory's Memoranda 1677–1708.* Oxford: Printed for the editor, 1937.

Hobbes, Thomas. *De corpore ("Concerning Body"). The English Works of Thomas Hobbes of Malmesbury.* Ed. by William Molesworth. 16 vols. London, 1839–1845. Vol. 1: *Elements of Philosophy, the first section: "Concerning Body," written in Latin by Thomas Hobbes of Malmesbury and translated into English.*

Critique du "De Mundo" de Thomas White. Introduction, texte critique et notes. Ed. by

Jean Jacquot and Harold Whitmore Jones. Paris: Librairie Philosophique J. Vrin, 1973. *Thomas White's "De Mundo" Examined.* The Latin translated by Harold Whitmore Jones. London: Bradford University Press in association with Crosby Lockwood Staples, 1976.

Holkot, Robert. *Robertus Holkot, in quatuor libros Sententiarum quaestiones.* Lyon, 1518. Reprinted in facsimile Frankfurt: Minerva, 1967.

Javelli, Chrysostomus, O. P. *Totius rationalis, naturalis, divinae ac moralis philosophiae compendium, innumeris fere locis castigatum et in duos tomos digestum.* 2 vols. Lyon, 1568.

Jean de Ripa. "Jean de Ripa I Sent. Dist. XXXVII: De modo inexistendi divine essentie in omnibus creaturis." Edition critique par André Combes et Francis Ruello. Présentation de Paul Vignaux. *Traditio 23* (1967), 191–267.

Johannes Canonicus. *Ioannis Canonici questiones super VIII. lib. Physicorum Aristotelis perutiles; nuperrime correcte et emendate additis textibus. Commentorum in margine una cum utili reportorio cunctorum auctoris notabilium indice.* Venice, 1520.

Johannes Quidort Parisiensis. *Jean de Paris (Quidort) O. P., Commentaire sur les Sentences, Reportation, Livre I.* Ed. by Jean-Pierre Muller, O. S. B. *Studia Anselmiana, Philosophica, Theologica,* edita a professoribus Instituti Pontificii S. Anselmi de Urbe, fasc. 47. Rome: Orbis Catholicus, Herder, 1961.

John Damascene. *Saint John Damascene, De fide orthodoxa.* Versions of Burgundio and Cerbanus. Ed. by Eligius M. Buytaert, O. F. M., Franciscan Institute Publications, Text Series No. 8. St. Bonaventure, N.Y.: Franciscan Institute, 1955.

John Major. *In Primum Sententiarum ex recognitione Io. Badii.* Paris: apud Badium, 1519. *Le Traité 'De l'infini' de Jean Mair.* Nouvelle édition avec traductions et annotations par Hubert Elie. Paris: Librairie Philosophique J. Vrin, 1938.

John of Jandun. *Questiones in libros Physicorum Aristotelis. Acced. Heliae Cretensis annotationes.* Venice: Joannes Lucilius Santritter et Hieronymus de Sanctis, impensis Petri Benzon et Petri (de Piasiis), 1488.

Questiones Ioannis de Janduno De physico auditu noviter emendate. Helie Hebrei Cretensis questiones: De primo motore; De efficientia mundi; De esse essentia et uno; annotationes in plurima dicta commentatoris. Venice, 1519.

John of St. Thomas (John Poinsot). *Cursus Philosophicus Thomisticus.* New ed. by P. Beatus Reiser, O. S. B. 3 vols. Turin: Marietti, 1933. It appears that the various parts were first published between 1633 and 1635.

Keill, John. *An Introduction to Natural Philosophy: or, Philosophical Lectures Read in the University of Oxford, 'Anno Dom.' 1700. To Which are Added the Demonstrations of Monsieur Huygens's 'Theorems', concerning the Centrifugal Force and Circular Motion,* 2d ed. Trans. from the last edition of the Latin. London: printed for J. Senex, W. and J. Innys, J. Osborn, and T. Longman, 1726.

Kircher, Athanasius. *R. P. Athanasii Kircheri e Societate Jesu Iter Extaticum Coeleste, quo mundi opificium, id est, coelesti expansi, siderumque . . . interlocutoribus Cosmiele et Theodidacto . . . ipso auctore annuente, P. Gaspare Schotto, Regiscuriano e Societate Jesu . . . ,* 2d ed. Würzburg [Herbipolis], 1660.

Leibniz, Gottfried Wilhelm. *Nouveaux Essais sur l'Entendement Humain.* Chronologie et introduction par Jacques Brunschwig. Paris: Garnier-Flammarion, 1966.

Correspondance Leibniz-Clarke présentée d'après les manuscrits originaux des bibliothèques de Hanovre et de Londres par André Robinet. Paris: Presses Universitaires de France, 1957.

The Leibniz-Clarke Correspondence, Together with Extracts from Newton's "Principia" and "Opticks." Ed. with introduction and notes by H. G. Alexander. Manchester: University of Manchester Press, 1956.

Liber sex principiorum. See Minio-Paluello, L.

Locke, John. *An Essay Concerning Human Understanding.* Ed. with an introduction, critical apparatus, and glossary by Peter H. Nidditch. Oxford: Clarendon Press, 1975.

An Early Draft of Locke's Essay Together with Excerpts from His Journals. Ed. by R. I. Aaron and Jocelyn Gibb. Oxford: Clarendon Press, 1936.

Lucretius. *De rerum natura libri sex.* Ed. by Cyril Bailey. Editio altera. Oxford: Clarendon Press, 1922.

On the Nature of the Universe. Trans. by R. E. Latham. Harmondsworth, Middlesex: Penguin Books, 1951.

Mach, Ernst. *The Science of Mechanics: A Critical and Historical Account of Its Development.* Trans. by Thomas J. McCormack; new introduction by Karl Menger. 6th ed. with revisions through the 9th German ed. LaSalle, Ill.: Open Court Publishing, 1960.

Maignan, Emanuel. *Cursus philosophicus.* Lyon, 1673.

Maimonides, Moses. *The Guide of the Perplexed.* Trans. with an introduction and notes by Shlomo Pines, with an introductory essay by Leo Strauss. Chicago: University of Chicago Press, 1963.

Malebranche, Nicholas. *Oeuvres complètes de Malebranche.* Ed. by André Robinet. 21 vols. Paris: Librairie Philosophique J. Vrin, 1958–1970. Vol. 12: *Entretiens sur la Métaphysique et sur la Religion,* 1965.

Dialogues on Metaphysics and on Religion. Trans. by Morris Ginsberg, with a preface by Prof. G. Dawes Hicks. London: Allen & Unwin, 1923.

Marsilius of Inghen. *Questiones subtilissime Johannis Marcilii Inguen super octo libros Physicorum secundum nominalium viam. Cum tabula in fine libri posita suum in lucem primum sortiuntur effectum.* Lyon, 1518. Reprinted in facsimile. Frankfurt: Minerva, 1964.

Minio-Paluello, Laurence (ed.). *Aristoteles Latinus I 6–7, Categoriarum Supplementa: Porphyrii Isagoge translatio Boethii et Anonymi Fragmentum vulgo vocatum 'Liber sex principiorum.'* Ed. by Laurence Minio-Paluello with the assistance of Bernard G. Dod. Bruges and Paris: Desclée de Brouwer, 1966.

More, Henry. *Democritus Platonissans (1646).* Introduction by P. G. Stanwood. The Augustan Reprint Society, Publication Number 130. Los Angeles: William Andrews Clark Memorial Library, U.C.L.A., 1968. First published, 1646.

A Collection of Several Philosophical Writings of Dr Henry More, Fellow of Christ's Colledge in Cambridge. As Namely, His "Antidote against Atheism."
"Appendix to the said Antidote."
"Enthusiasmus Triumphatus."
"Letters to Des-Cartes, &c."
"Immortality of the Soul."
"Conjectura Cabbalistica."
The 2d ed. more correct and much enlarged. London: printed by James Flesher for William Morden, 1662.

Enchiridion metaphysicum: sive De rebus incorporeis succincta et luculenta dissertatio. Pars prima: De existentia et natura rerum incorporearum in genere. . . . London, 1671.

Philosophical Writings of Henry More. Ed. with introduction and notes by Flora Isabel MacKinnon. New York: AMS Press, 1969; first published 1925.

Neckham, Alexander. *Alexandri Neckam De naturis rerum libri duo; with the poem of the same author, De laudibus divinae sapientiae.* Ed. by Thomas Wright. London: Her Majesty's Stationery Office, 1863; reprinted by Kraus, 1967.

Nemesius of Emesa. *Nemesii Episcopi Premnon Physicon . . . a N. Alfano Archiepiscopo Salerni in latinum translatus.* Ed. by Karl Burkhard. Leipzig: Teubner, 1917.

Némésius d'Emèse, De natura hominis. Traduction de Burgundio de Pise. Edition critique avec une introduction sur l'anthropologie de Némésius. Ed. by G. Verbeke and J. R. Moncho. Leiden: E. J. Brill, 1975.

Newton, Isaac. *Sir Isaac Newton's Mathematical Principles of Natural Philosophy and His System of the World.* Trans. into English by Andrew Motte in 1729. The translations revised and supplied with an historical and explanatory appendix by Florian Cajori. Berkeley: University of California Press, 1947.

Isaac Newton's Philosophiae Naturalis Principia Mathematica. The 3d ed. (1726) with variant readings assembled by Alexandre Koyré and I. Bernard Cohen with the assistance of Anne Whitman. 2 vols. Cambridge: Cambridge University Press, 1972.

Optice: sive de reflexionibus, refractionibus, inflexionibus et coloribus lucis libri tres. Authore Isaaco Newton. Latine reddidit, Samuel Clarke . . . Accedunt Tractatus duo ejusdem Authoris de speciebus et magnitudine figurarum curvilinearum, Latine scripti. London: impensis Sam. Smith et Benj. Walford, 1706.

Opticks. Based on the 4th ed., London, 1730. With a foreword by Albert Einstein; an introduction by Sir Edmund Whitaker; a preface by I. Bernard Cohen; and an analytical table of contents prepared by Duane H. D. Roller. New York: Dover Publications, 1952.

Correspondence of Sir Isaac Newton and Professor Cotes, including letters of other eminent men. . . . Now first published from the originals in the library of Trinity College, Cambridge . . . by J. Edelston. London: J. W. Parker, 1850. Reprinted in facsimile, London: F. Cass, 1969.

Isaac Newton's Papers & Letters on Natural Philosophy and Related Documents. Ed., with a general introduction, by I. Bernard Cohen assisted by Robert E. Schofield. Cambridge, Mass.: Harvard University Press, 1958.

Unpublished Scientific Papers of Isaac Newton. A Selection from the Portsmouth Collection in the University Library, Cambridge. Chosen, edited, and translated by A. Rupert Hall and Marie Boas Hall. Cambridge: Cambridge University Press, 1962.

Nicholas of Autrecourt. *Exigit ordo executionis.* Ed. by J. Reginald O'Donnell, C. S. B., in "Nicholas of Autrecourt." *Mediaeval Studies 1* (1939), 179–280.

The Universal Treatise of Nicholas of Autrecourt. Trans. by Leonard A. Kennedy, Richard E. Arnold, and Arthur E. Millward, with an introduction by Leonard A. Kennedy. Milwaukee, Wis.: Marquette University Press, 1971.

Nicholas of Cusa. *Nicolai de Cusa Opera Omnia iussu et auctoritate Academiae Litterarum Heidelbergensis ad codicum fidem edita.* Vol. 1: *De docta ignorantia.* Ed. by Ernest Hoffmann and Raymond Klibansky. Leipzig: Felix Meiner, 1932.

Of Learned Ignorance. Trans. by Fr. Germain Heron, O. F. M., with an introduction by Dr. D. J. B. Hawkins. London: Routledge & Kegan Paul, 1954.

Nock, A. D., and A.-J. Festugière. *Corpus Hermeticum.* 4 vols. Paris: Société d'Edition "Les Belles Lettres," 1945 and 1954.

Ockham, William. *Guillelmus de Occam, Quotlibeta septem; Tractatus de sacramento altaris.* Strasbourg, 1491. Reprinted in facsimile. Louvain: Editions de la Bibliothèque S. J., 1962.

The "Tractatus de successivis" Attributed to William Ockham. Ed. with a study on the life and works of Ockham by Philotheus Boehner, O. F. M. St. Bonaventure, N.Y.: Franciscan Institute, 1944.

Oresme, Nicole. *Nicole Oresme and the Kinematics of Circular Motion: Tractatus de commensurabilitate vel incommensurabilitate motuum celi.* Ed. with an introduction, English translation, and commentary by Edward Grant. Madison, Wis.: University of Wisconsin Press, 1971.

Nicole Oresme: De proportionibus proportionum and Ad pauca respicientes. Ed. with introductions, English translations, and critical notes by Edward Grant. Madison, Wis.: University of Wisconsin Press, 1966.

Nicole Oresme: Le Livre du ciel et du monde. Ed. by Albert D. Menut and Alexander J. Denomy, C. S. B.; trans. with an introduction by Albert d. Menut. Madison, Wis.: University of Wisconsin Press, 1968.

"The Questiones super De celo of Nicole Oresme." Ed. by Claudia Kren. Ph.D. dissertation, University of Wisconsin, 1965.

Palingenius, Marcellus. *The Zodiake of Life by Marcellus Palingenius.* Trans. by Barnabe Googe with an introduction by Rosemond Tuve. New York: Scholars' Facsimiles & Reprints, 1947.

Pascal, Blaise. *Oeuvres Complètes.* 2 vols. Ed. by Jean Mesnard. Paris: Desclée de Brouwer, 1970.

Physical Treatises of Pascal: The Equilibrium of Liquids and the Weight of the Mass of Air. Trans. by I. H. B. Spiers and A. G. H. Spiers, with introduction and notes by Frederick Barry. New York: Columbia Univesity Press, 1937.

Patrizi, Francesco. *Franc. Patricii Philosophiae, De rerum natura libri II priores. Alter de spacio physico; alter de spacio mathematico.* Ferrara: Victor Baldinus, 1587.

Francisci Patricii Nova De Universis Philosophia. In qua Aristotelica methodo non per motum, sed per lucem, et lumina, ad primam causam ascenditur; deinde propria Patricii methodo, tota in contemplationem venit Divinitas. Ferrara: apud Benedictum Mammarellum, 1591.

"On Physical Space, Francesco Patrizi." Trans. from *De spacio Physico* and *De spacio mathematico* by Benjamin Brickman. *Journal of the History of Ideas* 4 (1943), 224–245.

Pattin, Adriaan. *Le "Liber de causis."* Louvain: Tijdschrift voor Filosofie, 1966.

Paul of Venice. *Expositio Pauli Veneti super octo libros Phisicorum Aristotelis super commento Averrois cum dubiis eiusdem.* Venice, 1499.

Paulus Aegineta. *The Seven Books of Paulus Aegineta.* Trans. from the Greek with a commentary embracing a complete view of the knowledge possessed by Greeks, Romans, and Arabians on all subjects connected with medicine and surgery by Francis Adams. 3 vols. London: printed for the Sydenham Society, 1846.

Paulus Scriptor. *Lectura fratris Pauli Scriptoris, Ordinis Minorum, de observantia quam edidit declarando subtilissimas doctoris subtilis sententias circa Magistrum in primo libro.* [No place], 1498.

Pereyra, Martin. *Commentariorum in primum librum Magistri Sententiarum tomus secundus* . . . authore Reverendissimo ac Sapientissimo Patre Magistro D. Fr. Martino Pereyra. Coimbra: ex typographia in Regali Collegio Artium Societatis Jesu, 1715.

Peter of Abano. *Conciliator differentiarum philosophorum et precipue medicorum.* Mantua, 1472.

Petrus Lombardus. *Magistri Petri Lombardi, Parisiensis Episcopi, Sententiae in IV libris distinctae.* Editio tertia ad fidem codicum antiquiorum restituta. Tom I, pars II, liber I et II. Grottaferrata (Rome): Editiones Collegii S. Bonaventurae ad Claras Aquas, 1971.

Philo Judaeus. *Philo with an English Translation.* Ed. and trans. by F. H. Colson, G. H. Whitaker, and R. Marcus. 12 vols. Loeb Classical Library. Cambridge, Mass.: Harvard University Press; London: William Heinemann, 1929–1962.

Philo of Byzantium. *Liber Philonis de ingeniis spiritualibus.* Ed. with German translation by Wilhelm Schmidt. *Heronis Alexandrini Opera,* vol. 1. Lepzig: Teubner, 1899, pp. 459–489.

See Vaux, Carra de.

Philoponus, John. *Ioannis Philoponi in Aristotelis Physicorum libros quinque posteriores commentaria.* Ed. by H. Vitelli, *Commentaria in Aristotelem Graeca,* vol. 17. Berlin: G. Reimer, 1888.

Ioannes Grammatici, cognomento Philoponi, in Aristotelis Physicorum libros quattuor explanatio. Io. Baptista Rasario Novariensis, interprete. Venice, 1569.

Johannes Philoponos Grammatikos von Alexandrien (6 Jh. n. Chr.), Christliche Naturwissenschaft im Ausklang der Antike, Vorläufer der modernen Physik, Wissenschaft und Bibel, Ausgewählte Schriften, übersetzt, eingeleitet und kommentiert von Walter Böhm. Munich/Paderborn/Vienna: Verlag Ferdinand Schöningh, 1967.

See Wolff, Michael.

Pico della Mirandola, Gian Francesco. *Examen vanitatis doctrinae gentium et veritatis Christianae disciplinae.* Mirandola, 1520. Reprinted in facsimile in *Opera Omnia (1557–1573) [di] Giovanni Pico della Mirandola [e] Gian Francesco Pico.* Con un introduzione di Cesare Vasoli. 2 vols. Hildesheim: Georg Olms Verlag, 1969.

Pitigianus, Franciscus. See Scotus, John Duns.

Plato. *Plato's Cosmology. The "Timaeus" of Plato.* Trans. with a running commentary by Francis MacDonald Cornford. New York: Liberal Arts Press, 1957. First printed 1937.

Plotinus. *Plotini Opera.* Tomus II: Enneades IV–V. Ed. by Paul Henry and Hans-Rudolf Schwyzer. Paris: Desclée de Brouwer; Brussels: L'Edition Universelle, 1959.

The Enneads. Trans. by Stephen MacKenna. 4th ed. revised by B. S. Page. With a foreword by Prof. E. R. Dodds and an introduction by Prof. Paul Henry, S. J., London: Faber and Faber, 1969.

Porta, Giovanni Battista della. *Pneumaticorum libri tres.* Naples, 1601.

Pseudo-Siger of Brabant. See Siger de Brabant.

Quidort. See Johannes Quidort Parisiensis.

Raphson, Joseph. *Analysis aequationum universalis, seu ad aequationes algebraicas resolvendas methodus generalis et expedita. Ex nova infinitarum serierum methodo deducta ac demonstrata.* Editio secunda . . . cui etiam annexum est *De spatio reali seu ente infinito conamen mathematico-metaphysicum.* London, 1702.

Richard of Middleton. *Clarissimi Theologi Magistri Ricardi de Media Villa . . . Super quatuor libros Sententiarum Petri Lombardi quaestiones subtilissimae.* 4 vols. Brixiae [Brescia],

1591. Reprinted under the title *Richardus de Mediavilla*, Frankfurt: Minerva, 1963.

Ripa, Jean de. See Jean de Ripa.

Risner, Frederick. See Alhazen.

Schott, Gaspar. P. *Gasparis Schotti Regiscuriani, e Societate Jesu . . . Mechanica hydraulico-pneumatica . . . Opus bipartitum, cujus Pars I. Mechanicae Hydraulico-pneumaticae Theoriam continet. Pars II. Ejusdem Praxin exhibet, . . . Accessit Experimentum novum Magdeburgicum quo vacuum alij stabilire, alij evertere conatur.* Würzburg [Herbipolis]: Henricus Pigrin, 1657.

Scott, Walter (ed. and trans.). *Hermetica, the Ancient Greek and Latin Writings which Contain Religious or Philosophic Teachings Ascribed to Hermes Trismegistus.* 4 vols. Oxford: Clarendon Press, 1924–1936.

Scotus, John Duns. *God and Creatures: The Quodlibetal Questions.* Trans. with an introduction, notes, and glossary by Felix Alluntis, O. F. M., and Allan B. Wolter, O. F. M. Princeton, N.J.: Princeton University Press, 1975.

——— *Opera Omnia.* Vol. 2: *R. P. F. Ioannis Duns Scoti, Doctoris Subtilis Ordinis Minorum, in VIII libros Physicorum Aristotelis Quaestiones, cum annotationibus R. P. F. Francisci Pitigiani Arretini. . . .* ; Vol. 5, part 2: *. . . Quaestiones in Lib. I Sententiarum*; Vol. 6, part 1: *. . . Quaestiones in Lib. II Sententiarum.* Luke Wadding ed., Lyon, 1639. Reprinted in facsimile Hildesheim: Georg Olms Verlag, 1968.

——— *Doctoris subtilis et Mariani Ioannis Duns Scoti Ordinis Fratrum Minorum Opera Omnia.* Ed. by Carl Balić. Vol. 17: *Lectura in librum primum Sententiarum a distinctione octava ad quadragesimam quintam.* Rome: Vatican City, 1966.

Sextus Empiricus. *Sextus Empiricus* with an English translation by the Rev. R. G. Bury. 4 vols. Loeb Classical Library. London: William Heinemann; Cambridge, Mass.: Harvard University Press, 1933–1949.

Siger de Brabant. *Questions sur la Physique d'Aristote (texte inédit).* Ed. by Philippe Delhaye. *Les Philosophes Belges, Textes et Etudes,* vol. 15. Louvain: Editions de l'Institut Supérieur de Philosophie, 1941. Falsely ascribed to Siger.

Simplicius. *Simplicii philosophi acutissimi commentaria in quatuor libros De celo Aristotelis. Guillermo Morbeto interprete.* Venice, 1540.

——— *Simplicii philosophi perspicacissimi, clarissima commentaria in octo libros Arist. de Physico Auditu.* Venice: apud Hieronymum Scottum, 1566.

——— *Simplicii in Aristotelis Physicorum libros quattuor priores commentaria.* Ed. by H. Diels. *Commentaria in Aristotelem Graeca,* vol. 9. Berlin: Berlin Academy, 1882.

Smith, T. V., and Marjorie Grene. *From Descartes to Kant. Readings in the Philosophy of the Renaissance and Enlightenment.* Chicago: University of Chicago Press, 1940.

Soto, Domingo, O. P. *Super octo libros Physicorum Aristotelis commentaria. Secunda aeditio nuperrime ab authore recognita, multisque in locis aucta et a mendis quam maxime fieri potuit repurgata.* Salamanca, 1555.

Spinoza, Benedictus de. *The Chief Works of Benedict de Spinoza: On the Improvement of the Understanding, The Ethics, Correspondence.* Trans. from the Latin with an introduction by R. H. M. Elwes. New York: Dover Publications, 1951; reprinted from the edition of 1883.

Suarez, Francisco. *Disputationes Metaphysicae.* 2 vols. Paris, 1866. Reprinted in facsimile Hildesheim: Georg Olms Verlag, 1965. First printed 1597.

Sunczel, Friedrich. *Collecta et exercitata Friderici Sunczel Mosellani liberalium studiorum Magistri in octo libros Phisicorum Arestotelis [sic] in almo studio Ingolstadiensi.* Hagenau, 1499.

Telesio, Bernardino. *Bernardini Telesii De rerum natura.* 3 vols. Ed. by Vincenzo Spampanato. Modena/Genoa/Rome: A. F. Formíggini, 1910–1923.

Themon Judaeus. *In quatuor libros Meteororum.* Paris, 1518. For the full title of the edition, see Albert of Saxony.

Thomas de Vio, Cardinalis Caietanus. *Thomas de Vio Cardinalis Caietanus (1469–1534), Scripta philosophica. Commentaria in Porphyrii Isagogen ad Praedicamenta Aristotelis.* Ed. by P. Isnardus M. Marega, O. P. Rome: apud Institutum 'Angelicum,' 1934.

Thorndike, Lynn. *The "Sphere" of Sacrobosco and Its Commentators.* Chicago: University of Chicago Press, 1949.

Toletus, Franciscus. *D. Francisci Toleti Societatis Iesu Commentaria una cum Quaestionibus in octo libros Aristotelis De physica auscultatione, nunc secundo in lucem edita.* Venice: apud Iuntas, 1580.

Vasquez, Gabriel. R. P. *Gabrielis Vazquez Societatis Iesu Disputationes Metaphysicae desumptae ex variis locis suorum operum.* Antwerp: apud Ioannem Keerbergium, 1618.

Vaux, Carra de (Baron). "Le Livre des appareils pneumatiques et des machines hydrauliques par Philon de Byzance." *Notices et Extraits des Manuscrits de la Bibliothèque Nationale et Autres Bibliothèques.* Paris: L'Académie des Inscriptions et Belles-Lettres, 1903.

William of Auvergne. *Guilelmi Alverni . . . Opera Omnia.* Paris, 1674. Reprinted in facsimile. Frankfurt: Minerva, 1963.

William of Auxerre. *Summa aurea in quattuor libros Sententiarum a doctore Magistro Guillermo Altissiodorensi edita, quam nuper a mendis quamplurimis doctissimus sacre theologie professor Magister Guillermus de Quercu diligenti admodum castigatione emendavit ac tabulam huic pernecessarium edidit. Impressa est Parisijs maxima Philippi Pigoucheti cura. Impensis vero Nicolai Vaultier et Durandi Gerlier alme universitatis Parisiensis librariorum iuratorum.* Paris, n. d. Reprinted in facsimile Frankfurt: Minerva, 1964.

Zabarella, Giacomo. *Jacobi Zabarellae Patavini, De rebus naturalibus libri XXX.* Frankfurt, 1607; first published 1590. Reprinted in facsimile Frankfurt: Minerva, 1966.

Zimmermann, Albert (ed.). *Ein Kommentar zur Physik des Aristoteles aus der Pariser Artistenfakultät um 1273.* Berlin: Walter de Gruyter, 1968.

Secondary sources

Armstrong, A. H. (ed.). *The Cambridge History of Later Greek and Early Medieval Philosophy.* Cambridge: Cambridge University Press, 1970.

Blanchet, Léon. *Campanella.* Paris, 1920. Reprinted in facsimile New York: Burt Franklin, n. d.

Bloch, Olivier René. *La Philosophie de Gassendi, Nominalisme, Matérialisme et Métaphysique.* The Hague: Martinus Nijhoff, 1971.

Bonansea, Bernardino M. *Tommaso Campanella, Renaissance Pioneer of Modern Thought.* Washington, D.C.: Catholic University of America Press, 1969.

Burtt, Edwin Arthur. *The Metaphysical Foundations of Modern Physical Science,* rev. ed. London: Routledge & Kegan Paul, 1932. First ed. 1924.

Čapek, Milič. "Was Gassendi a Predecessor of Newton?" *Actes du Dixième Congrès International d'Histoire des Sciences,* vol. 2. Paris: Hermann, 1964, pp. 705–709.

Castañeda, Hector-Neri. "Leibniz's Meditation on April 15, 1676 About Existence, Dreams, and Space." *Leibniz à Paris (1672–1676).* Symposion de la G. W. Leibniz-Gesellschaft (Hannover) et du Centre National de la Recherche Scientifique (Paris) à Chantilly (France) du 14 au 18 Novembre 1976. Vol. 2: *La Philosophie de Leibniz* (Wiesbaden: Franz Steiner Verlag, 1978), pp. 91–129.

Clagett, Marshall. *The Science of Mechanics in the Middle Ages.* Madison, Wis.: University of Wisconsin Press, 1959.

"Adelard of Bath." *Dictionary of Scientific Biography,* vol. 1. New York: Scribner, 1970, pp. 61–64.

Cohen, I. Bernard. "Newton, Isaac." *Dictionary of Scientific Biography,* vol. 10. New York: Scribner, 1974, pp. 42–101.

Coleman, Janet. "Jean de Ripa O. F. M. and the Oxford Calculators." *Mediaeval Studies* 37 (1975), 130–189.

Copleston, Frederick, S. J. *A History of Philosophy.* Westminster, Md.: Newman Press, 1957 (vol. 2), 1953 (vol. 3).

Cornford, F. M. "The Invention of Space." *Essays in Honour of Gilbert Murray.* London: Allen & Unwin, 1936, pp. 215–235. Reprinted in Čapek (ed.), *Concepts of Space and Time,* pp. 3–16.

Cranz, F. Edward. "Alexander Aphrodisiensis." *Catalogus translationum et commentariorum: Medieval and Renaissance Latin Translations and Commentaries,* vol. 1. Ed. by

P. O. Kristeller. Washington, D.C.: Catholic University of America Press, 1960, pp. 77–135.

Crombie, A. C. *Robert Grosseteste and the Origins of Experimental Science 1100–1700.* Oxford: Clarendon Press, 1953.

Crombie, A. C., and North, J. D. "Roger Bacon." *Dictionary of Scientific Biography*, vol. 1. New York: Scribner, 1970, pp. 377–385.

Dick, Steven J. "Plurality of Worlds and Natural Philosophy: An Historical Study of the Origins of Belief in Other Worlds and Extraterrestrial Life." Ph.D. dissertation, Indiana University, 1977.

"The Origins of the Extraterrestial Life Debate and its Relation to the Scientific Revolution." *Journal of the History of Ideas 41* (1980), 3–27.

Donahue, William Halsted. "The Dissolution of the Celestial Spheres 1595–1650." Ph.D. dissertation, University of Cambridge, 1972.

Drachmann, A. G. *Ktesibios, Philon and Heron, A Study in Ancient Pneumatics. Acta Historica Scientiarum Naturalium et Medicinalium*, vol. 4. Copenhagen: Ejnar Munksgaard, 1948.

Duhem, Pierre. *Etudes sur Léonard de Vinci, ceux qu'il a lus et ceux qui l'ont lu.* 3 vols. Paris: Hermann, 1906–1913; reprinted Paris: F. de Nobele, 1955.

Le Système du monde. Histoire des doctrines cosmologiques de Platon à Copernic. 10 vols: Paris: Hermann, 1913–1959.

"Roger Bacon et l'horreur du vide." *Roger Bacon Essays.* Contributed by various writers on the occasion of the commemoration of the seventh centenary of his birth. Collected and ed. by A. G. Little. Oxford: Clarendon Press, 1914, pp. 241–284.

Einstein, Albert. *Relativity, The Special and the General Theory, A Popular Exposition*, 15th ed. New York: Crown Publishers, 1961.

Feldman, Seymour. "Platonic Themes in Gersonides' Cosmology." *Salo Wittmayer Baron Jubilee Volume.* Jerusalem: American Academy for Jewish Research, 1975, pp. 383–405.

Ferguson, James P. *The Philosophy of Dr. Samuel Clarke and Its Critics.* New York/Washington/Hollywood: Vantage Press, 1974.

Fierz, Markus. "Ueber den Ursprung und die Bedeutung der Lehre Isaac Newtons vom absoluten Raum." *Gesnerus 11* (1954), fasc. 3/4, 62–120.

Fritz, Kurt von. "Philo of Byzantium." *Dictionary of Scientific Biography*, vol. 10. New York: Scribner, 1974, pp. 586–591.

Fuerst, Rev. Adrian, O. S. B. *An Historical Study of the Doctrine of the Omnipresence of God in Selected Writings between 1220–1270.* Ph.D. dissertation. The Catholic University of America, *Studies in Sacred Theology* (second series), no. 62. Washington, D. C.: Catholic University of America Press, 1951.

Funkenstein, Amos. "The Dialectical Preparation for Scientific Revolutions: On the Role of Hypothetical Reasoning in the Emergence of Copernican Astronomy and Galilean Mechanics." In Robert Westman (ed.), *The Copernican Achievement.* Contributions of the UCLA Center for Medieval and Renaissance Studies, 7. Berkeley: University of California Press, 1975, pp. 165–203.

"Descartes, Eternal Truths, and the Divine Omnipotence." *Studies in History and Philosophy of Science 6* (1975), 185–199.

Furley, David. "Epicurus." *Dictionary of Scientific Biography*, vol. 4. New York: Scribner, 1971, pp. 381–382.

"Lucretius." *Dictionary of Scientific Biography*, vol. 8. New York: Scribner, 1973, pp. 536–539.

Garnett, Christopher Browne, Jr. *The Kantian Philosophy of Space.* New York: Columbia University Press, 1939.

Gibson, James, *Locke's Theory of Knowledge and Its Historical Relations.* Cambridge: Cambridge University Press, 1917.

Gilbert, Neal W. "Telesio, Bernardino." *Dictionary of Scientific Biography*, vol. 13. New York: Scribner, 1976, pp. 277–280.

Gilson, Etienne. *History of Christian Philosophy in the Middle Ages.* London: Sheed and Ward, 1955.

Grant, Edward. "Jean Buridan: A Fourteenth Century Cartesian." *Archives internationales d'histoire des sciences 16* (1963), 251–255.

"Motion in the Void and the Principle of Inertia in the Middle Ages." *Isis 55* (1964), 265–292.

"Aristotle, Philoponus, Avempace, and Galileo's Pisan Dynamics." *Centaurus 11* (1965), 79–95.

"Bradwardine and Galileo: Equality of Velocities in the Void." *Archive for History of Exact Sciences 2*, no. 4 (1965), 344–364.

"On the Origin of the Medieval Version of Equality of Fall for Unequal Bodies in the Void: A Critique of Duhem's Explanation." *Actes du XIe Congrès International d'Histoire des Sciences, Varsovie-Cracovie 24–31 Août 1965*, vol. 3. Warsaw/Cracow, 1967, pp. 19–23.

"The Arguments of Nicholas of Autrecourt for the Existence of Interparticulate Vacua." *Actes du XIIe Congrès International d'Histoire des Sciences*, tome IIIA, Paris, 1968. Paris: Albert Blanchard, 1971, pp. 65–68.

"Medieval and Seventeenth-Century Conceptions of an Infinite Void Space beyond the Cosmos." *Isis 60* (1969), 39–60.

"Henricus Aristippus, William of Moerbeke and Two Alleged Mediaeval Translations of Hero's Pneumatica." *Speculum 46* (1971), 656–669.

"Medieval Explanations and Interpretations of the Dictum that 'Nature Abhors a Vacuum.'" *Traditio 29* (1973), 327–355.

"Place and Space in Medieval Physical Thought." *Motion and Time, Space and Matter, Interrelations in the History of Philosophy and Science*. Ed. by Peter K. Machamer and Robert G. Turnbull. Columbus, Ohio: Ohio State University Press, 1976, pp. 137–167.

"The Concept of *Ubi* in Medieval and Renaissance Discussions of Place." *Science, Medicine, and the University: 1200–1550, Essays in Honor of Pearl Kibre*, part 1. Special ed. Nancy G. Siraisi and Luke Demaitre. *Manuscripta 20*, no. 2 (1976), 71–80.

"The Principle of the Impenetrability of Bodies in the History of Concepts of Separate Space from the Middle Ages to the Seventeenth Century." *Isis 69* (1978), 551–571.

"Scientific Thought in Fourteenth-Century Paris: Jean Buridan and Nicole Oresme." *Machaut's World: Science and Art in the Fourteenth Century*. Ed. by Madeleine Pelner Cosman and Bruce Chandler. *Annals of the New York Academy of Sciences*, vol. 314. New York: New York Academy of Sciences, 1978, pp. 105–124.

"The Condemnation of 1277, God's Absolute Power, and Physical Thought in the Late Middle Ages." *Viator 10* (1979), 211–244.

"The Condemnation of 1277." *Cambridge History of Later Medieval Philosophy*. Ed. by Norman Kretzmann, Anthony Kenny, and Jan Pinborg. Cambridge: Cambridge University Press, in press.

"The Medieval Doctrine of Place: Some Fundamental Problems and Solutions." *Filosofia e scienze nella tarda scolastica: Studi in memoria di Anneliese Maier*. Ed. Alfonso Maierù and Agostino Paravicini Bagliani. Rome: Edizioni di Storia e Letteratura; in press.

Guthrie, W. K. C. *A History of Greek Philosophy*, vol. 2. Cambridge: Cambridge University Press, 1965.

Hahm, David E. *The Origins of Stoic Cosmology*. Columbus, Ohio: Ohio State University Press, 1977.

Harries, Karsten. "The Infinite Sphere: Comments on the History of a Metaphor." *Journal of the History of Philosophy 13* (1975), 5–15.

Heath, P. L. "Nothing." *The Encyclopedia of Philosophy*, vol. 5. New York: Macmillan and Free Press, 1967, pp. 524–525.

Henry, John. "Francesco Patrizi da Cherso's Concept of Space and its Later Influence." *Annals of Science 36* (1979), 549–573.

Hisette, Roland. *Enquête sur les 219 articles condamnés à Paris le 7 mars 1277. Philosophes Médiévaux*, vol. 22. Louvain: Publications Universitaires; Paris: Vander-Oyez, 1977.

Hurter, H., S. J. *Nomenclator Literarius Theologiae Catholicae, Theologos Exhibens Aetate, Natione, Disciplinis Distinctos*, 5 vols. Innsbruck: Universitäts-Verlag Wagner, 1906–1926. Reprinted in facsimile New York: Burt Franklin, 1962.

Jammer, Max. *Concepts of Space, The History of Theories of Space in Physics*, 2d ed. Cambridge, Mass.: Harvard University Press, 1969.

Jeans, Sir James. *The Universe Around Us*. New York: Macmillan, 1929.

Johnson, Francis R. *Astronomical Thought in Renaissance England, A Study of the English Scientific Writings from 1500 to 1645*. New York: Octagon Books, 1968. First published by Johns Hopkins University Press, 1937.

Kargon, Robert. "Charleton, Walter." *Dictionary of Scientific Biography*, vol. 3. New York: Scribner, 1971, pp. 208–210.

Kauffeldt, Alfons. "Otto von Guericke on the Problem of Space." *Actes du XIᵉ Congrès International d'Histoire des Sciences*, vol. 3. Aug. 24–31, 1965. Wroclaw/Warsaw/Cracow: Ossolineum Maison d'Edition de l'Académie Polonaise des Sciences, 1968, pp. 364–368.

Otto von Guericke Philosophisches über den leeren Raum. Berlin: Akademie-Verlag, 1968.

Keller A. G. "Schott, Gaspar." *Dictionary of Scientific Biography*, vol. 12. New York: Scribner, 1975, pp. 210–211.

Kenny, André. "Vacuum Theory and Technique in Greek Science." *Transactions, Newcomen Society* 37 (1964–1965), 47–56.

Kirk, G. S., and Raven, J. E. *The Presocratic Philosophers, A Critical History with a Selection of Texts*. Cambridge: Cambridge University Press, 1957.

Koyré, Alexandre. *Études Galiléennes*, 3 vols. Actualités Scientifiques et Industrielles, Nos. 852–854. Paris: Hermann, 1939.

"Le vide et l'espace infini au xivᵉ siecle." *Archives d'histoire doctrinale et littéraire du moyen âge 24* (1949), 45–91.

From the Closed World to the Infinite Universe. Baltimore: Johns Hopkins Press, 1957.

Newtonian Studies. Cambridge, Mass.: Harvard University Press, 1965.

Galileo Studies. Trans. by John Mepham. Sussex, England: Harvester Press, 1978.

Koyré, Alexandre, and Cohen, I. Bernard. "The Case of the Missing *Tanquam*: Leibniz, Newton & Clarke." *Isis 52* (1961), 555–566.

"Newton & the Leibniz-Clarke Correspondence with Notes on Newton, Conti, & Des Maizeaux." *Archives Internationales d'Histoire des Sciences 15* (1962), 63–126.

Krafft, Fritz. "Guericke (Gericke), Otto von." *Dictionary of Scientific Biography*, vol. 5. New York: Scribner, 1972, pp. 574–576.

Kristeller, Paul Oskar. *Eight Philosophers of the Italian Renaissance*. Stanford, Calif.: Stanford University Press, 1964.

Lawn, Brian. *The Salernitan Questions. An Introduction to the History of Medieval and Renaissance Problem Literature*. Oxford: Clarendon Press, 1963.

Leff, Gordon. *Paris and Oxford Universities in the Thirteenth and Fourteenth Centuries*. New York: Wiley, 1968.

Leyden, W. von. *Seventeenth-Century Metaphysics, An Examination of Some Main Concepts and Theories*. London: Gerald Duckworth, 1968.

Lohr, Charles H., S. J. "Medieval Latin Aristotle Commentaries." *Traditio 23* (1967), 24 (1968), 26 (1970), 27 (1971), 28 (1972), 29 (1973), 30 (1974).

"Renaissance Latin Aristotle Commentaries: Authors A–B." *Studies in the Renaissance 21* (1974), 228–289; "Authors C." *Renaissance Quarterly 28* (1975), 689–741; "Authors D–F." *Renaissance Quarterly 29* (1976), 714–745; "Authors G-K." *Renaissance Quarterly 30* (1977), 681–741; "Authors L-M." *Renaissance Quarterly 31* (1978), 532–603.

Lovejoy, Arthur O. *The Great Chain of Being. A Study of the History of an Idea*. New York: Harper, 1960; first published 1936.

McGuire, J. E. "Existence, Actuality and Necessity: Newton on Space and Time." *Annals of Science 35* (1978), 463–508.

"Newton on Place, Time, and God: An Unpublished Source." *The British Journal for the History of Science 11* (1978), 114–129.

Mahnke, Dietrich. *Unendliche Sphäre und Allmittelpunkt. Beiträge zur Genealogie der mathematischen Mystik. Deutsche Vierteljahrsschrift für Literaturwissenschaft und Geistesgeschichte*, herausgegeben von Paul Kluckhohn und Erich Rothacker. Buchreihe. 23 band. Halle/Saale: Max Niemeyer Verlag, 1937.

Maier, Anneliese. *Die Vorläufer Galileis im 14. Jahrhundert. Studien zur Naturphilosophie der Spätscholastik*, vol. 1. Rome: Edizioni di Storia e Letteratura, 1949.

Zwei Grundprobleme der scholastischen Naturphilosophie, 2d ed. Rome: Edizioni di Storia e Letteratura, 1951.

An der Grenze von Scholastik und Naturwissenschaft. Die Struktur der materiellen Substanz; Das Problem der Gravitation; Die Mathematik der Formlatituden. Studien zur Naturphilosophie der Spätscholastik, 2d ed., vol. 3. Rome: Edizioni di Storia e Letteratura, 1952.

Metaphysische Hintergründe der spätscholastischen Naturphilosophie. Studien zur Naturphilosophie der Spätscholastik, vol. 4. Rome: Edizioni di Storia e Letteratura, 1955.

Ausgehendes Mittelalter: gesammelte Aufsätze zur Geistesgeschichte des 14. Jahrhunderts. 3 vols. Rome: Edizioni di Storia e Letteratura, 1964–1977.

Mandonnet, Pierre F., O. P. *Siger de Brabant et l'Averroïsme Latin au XIIIme siècle.* Deuxième édition revue et augmentée: Ire Partie: Edition Critique. Louvain: Institut Supérieur de Philosophie de l'Université, 1911. IIme Partie: Textes inédits. Louvain: Institut Supérieur de Philosophie de l'Université, 1908.

Manuel, Frank E. *The Religion of Isaac Newton.* Fremantle Lectures, 1973. Oxford: Clarendon Press, 1974.

Maurer, Armand. *Medieval Philosophy.* New York: Random House, 1962.

Michel, Paul Henri. *The Cosmology of Giordano Bruno.* Trans. by Dr. R. E. W. Maddison. Paris: Hermann; London: Methuen; Ithaca, N.Y.: Cornell University Press, 1973. Original French text published 1962.

Moody, Ernest A. "Ockham and Aegidius of Rome." *Franciscan Studies* 9 (1949), 417–442.

"Galileo and Avempace: The Dynamics of the Leaning Tower Experiment." *Journal of the History of Ideas* 12 (1951), 163–193, 375–422.

Morry, W. F. "Omnipresence." *New Catholic Encyclopedia.* 17 vols. New York: McGraw-Hill, 1967–1979. vol. 10 (1967), p. 689.

Murdoch, John E. "*Mathesis in Philosophiam Scholasticam Introducta*, The Rise and Development of the Application of Mathematics in Fourteenth Century Philosophy and Theology." *Arts Libéraux et Philosophie au Moyen Age. Actes du Quatrième Congrès International de Philosophie Médiévale*, Montréal 27 août – 2 septembre 1967. Montréal: Institut d'Etudes Médiévales; Paris: Librairie Philosophique J. Vrin, 1969, pp. 215–254.

Murdoch, John, and Sylla, Edith. "Burley, Walter." *Dictionary of Scientific Biography*, vol. 2. New York: Scribner, 1970, pp. 608–612.

Pancheri, Lillian Unger. "The Atomism of Pierre Gassendi: Cosmology for the New Physics." Ph.D. dissertation, Tulane University, 1972.

Pines, Salomon (Shlomo). *Beiträge zur Islamischen Atomenlehre.* Inaugural dissertation. Berlin: A. Heine, 1936.

"La dynamique d'Ibn Bājja." In *Mélanges Alexandre Koyré publiés à l'occasion de son soixante-dixième anniversaire.* 2 vols. *Histoire de la pensée*, vols. 12, 13. Paris: Hermann, 1964. Vol. 1: *L'aventure de la science*, pp. 442–468.

"Philosophy, Mathematics, and the Concepts of Space in the Middle Ages." In Y. Elkana (ed.), *The Interaction Between Science and Philosophy.* Atlantic Highlands, N.J.: Humanities Press, 1974, pp. 75–90.

"Ibn Bājja (Avempace)." *Dictionary of Scientific Biography*, vol. 1. New York: Scribner, 1970, pp. 408–410.

Poulet, Georges. *The Metamorphoses of the Circle.* Trans. from the French by Carley Dawson and Elliott Coleman in collaboration with the author. Baltimore: Johns Hopkins Press, 1966.

Randall, John Herman, Jr. *The Career of Philosophy from the Middle Ages to the Enlightenment.* New York: Columbia University Press, 1962.

Rescher, Nicholas. *Leibniz, An Introduction to His Philosophy.* Totowa, N.J.: Rowman and Littlefield, 1979.

Rochot, Bernard. *Les Travaux de Gassendi sur Epicure et sur l'atomisme 1619–1658.* Paris: Librairie Philosophique J. Vrin, 1944.

"Gassendi (Gassend), Pierre." *Dictionary of Scientific Biography*, vol. 5. New York: Scribner, 1972, pp. 284–290.

Rome, Beatrice K. *The Philosophy of Malebranche. A Study of his Integration of Faith, Reason, and Experimental Observation.* Chicago: Henry Regnery, 1963.

Sabra, A. I. "Ibn al-Haytham." *Dictionary of Scientific Biography*, vol. 6. New York: Scribner, 1972, pp. 189–210.

Sambursky, S. *The Physical World of Late Antiquity.* New York: Basic Books, 1962.

"John Philoponus." *Dictionary of Scientic Biography*, vol. 7. New York: Scribner, 1973.

"Place and Space in Late Neoplatonism." *Studies in History and Philosophy of Science 8* (1977), 173–187.

Sarton, George. *Introduction to the History of Science.* 3 vols. in 5 parts. Baltimore: Williams & Wilkins, 1927–1948.

Saunders, Jason Lewis. *Justus Lipsius, The Philosophy of Renaissance Stoicism.* New York: Liberal Arts Press, 1955.

Schmitt, Charles B. *Gianfrancesco Pico della Mirandola (1469–1533) and His Critique of Aristotle.* The Hague: Martinus Nijhoff, 1967.

"Experimental Evidence For and Against a Void: the Sixteenth-Century Arguments." *Isis 58* (1967), 352–366.

"Towards a Reassessment of Renaissance Aristotelianism." *History of Science 11* (1973), 159–193.

"Campanella, Tommaso." *Dictionary of Scientific Biography*, vol. 15, supplement 1. New York: Scribner, 1978, pp. 68–70.

Scott T. K., Jr. "John Buridan on the Objects of Demonstrative Science." *Speculum 40* (1965), 654–673.

Shapiro, Herman. "Walter Burley and Text 71." *Traditio 16* (1960), 395–404.

Smith, A. Mark. "Galileo's Theory of Indivisibles: Revolution or Compromise?" *Journal of the History of Ideas 37*, no. 4 (1976), 571–588.

Sommervogel, Carlos, S. J. *Bibliothèque de la Compagnie de Jésus.* Première Partie: Bibliographie par les Pères Augustin et Aloys de Backer; Second Partie: Histoire par le Père Auguste Carayon. Nouvelle édition par Carlos Sommervogel, S. J. 12 vols. Brussels: Oscar Schepens; Paris: Alphonse Picard, 1890–1911.

Steenberghen, Fernand Van. *Siger de Brabant d'après ses oeuvres inédites*, vol. 2. *Les Philosophes Belges*, vol. 13. Louvain: Editions de l'Institut Supérieur de Philosophie, 1942.

Sweeney, Leo J., S. J. "Divine Infinity: 1150–1250." *The Modern Schoolman 35* (1957–1958), 38–51.

"John Damascene and Divine Infinity." *The New Scholasticism 35* (1961), 76–106.

Sweeney, Leo J., S. J., and Ermatinger, Charles J. "Divine Infinity According to Richard Fishacre." *The Modern Schoolman 35* (1957–1958), 191–211.

Thomas, P.-Félix. *La Philosophie de Gassendi.* Paris: Ancienne Librairie Germer Baillière, Félix Alcan Editeur, 1889.

Thomson, S. Harrison. "Unnoticed *Questiones* of Walter Burley on the Physics." *Mitteilungen des Instituts für österreichische Geschichtsforschung 62* (1954), 390–405.

Thorndike, Lynn. *A History of Magic and Experimental Science.* 8 vols. New York: Columbia University Press, 1923–1958.

Tigerstedt, E. N. *The Decline and Fall of the Neoplatonic Interpretation of Plato. An Outline and Some Observations. Commentationes Humanarum Litterarum*, vol. 52. Helsinki: Societas Scientiarum Fennica, 1974.

Tuveson, Ernest. "Space, Deity, and the 'Natural Sublime'." *Modern Language Quarterly 12* (1951), 20–38.

Villiers, Cosmas de (Cosme de Villiers de Saint-Etienne). *Bibliotheca Carmelitana, notis criticis et dissertationibus illustrata.* 2 vols. Aurelianis: M. Couret de Villeneuve, 1752.

Waard, C. de. *L'Expérience barométrique, ses antécédents et ses explications. Etude historique.* Thouars: Imprimerie Nouvelle, 1936.

Weinberg, Julius R. *Nicholas of Autrecourt. A Study in 14th-century Thought.* Princeton, N.J.: Princeton University Press, 1948.

Weisheipl, James A., O. P. *Friar Thomas d'Acquino, His Life, Thought, and Work.* Garden City, N.Y.: Doubleday, 1974.

"Motion in a Void: Aquinas and Averroes." *St. Thomas Aquinas 1274–1974, Commemo-*

rative Studies. Foreword by Etienne Gilson. Toronto: Pontifical Institute of Mediaeval Studies, 1974, pp. 467–488.

Westfall, Richard S. *Force in Newton's Physics. The Science of Dynamics in the Seventeenth Century.* London: MacDonald; New York: American Elsevier, 1971.

"The Foundation of Newton's Philosophy of Nature." *British Journal for the History of Science 1* (1962), 171–182.

Weyl, Hermann. *Space-Time-Matter.* Trans. from the German by Henry L. Brose. First American printing of the 4th ed. 1922. New York: Dover Publications, n. d.

Whitmore, P. J. S. *The Order of Minims in Seventeenth-Century France.* The Hague: Martinus Nijhoff, 1967.

Wilson, Curtis. *William Heytesbury, Medieval Logic and the Rise of Mathematical Physics.* Madison, Wis.: University of Wisconsin Press, 1956.

Wippel, John F. "The Condemnations of 1270 and 1277 at Paris." *The Journal of Medieval and Renaissance Studies 7* (1977), 169–201.

Wippel, John F., and Wolter, Allan, O. F. M. (eds.). *Medieval Philosophy from St. Augustine to Nicholas of Cusa.* New York: Free Press; London: Collier Macmillan Publishers, 1969.

Wolff, Michael. *Fallgesetz und Massebegriff. Zwei wissenschaftshistorische Untersuchungen zur Kosmologie des Johannes Philoponus. Quellen und Studien zur Philosophie,* vol. 2. Ed. by Gunther Patzig, Erhard Scheibe, and Wolfgang Wieland. Berlin: Walter de Gruyter, 1971.

Wolfson, Harry A. *The Philosophy of Spinoza.* 2 vols. Cambridge, Mass.: Harvard University Press, 1934.

The Philosophy of the Kalam. Cambridge, Mass.: Harvard University Press, 1976.

Worthen, Thomas D. "Pneumatic Action in the Klepsydra and Empedocles' Account of Breathing." *Isis 61* (1970), 520–530.

Yates, Frances A. *Giordano Bruno and the Hermetic Tradition.* New York: Vintage Books, 1969. First published by the University of Chicago Press, 1964.

Zimmermann, Albert. *Verzeichnis ungedruckter Kommentare zur Metaphysik and Physik des Aristotles aus der Zeit von etwa 1250–1350.* Leiden/Cologne: Brill, 1971.

Index

Page numbers cited subsequent to a semicolon following a final textual subentry signify relatively minor mentions of the major entry.

Aaron, R. I., 406n314
absolute power, see God
Abu'l Barakāt al-Baghdadi, 291n83, 332n18
accident: space is not an, 158, 166, 204, 209
Adam, Charles, 273n50, 399n235
Adami, Tobias, 381n72
Adams, Francis, 310n77
Addison (Joseph), 412n383
Adelard of Bath, 67, 68, 69, 83, 84, 312nn96, 97
Aegidius Romanus (Giles of Rome): and fall of bodies in void, 287n54, 290n73, 299nn135, 136; and instantaneous motion, 24, 36–7; on vacuum as privation, 268–9n13; rejected corpus quantum, 39–40; 32, 41–2, 278n4, 283n35, 288n61, 289n67, 290nn71, 72, 305n26, 307n56, 363n90
aether, see ether
Agrippa, (Henry) Cornelius, 231, 374n2, 402n264
Alan of Lille, 346n108, 347n108, 351n134
Albert of Saxony: and the distantia terminorum, 29, 280n17; and internal resistance, 46; and natural place, 297n126; imagines world placed in millet seed, 274 n55; on equality of fall in a void, 58, 298 nn129, 134; on infinite velocity in the void, 47–8; on mixed bodies, 51, 53–5, 66, 295 n114, 296n123; rejected separate space, 18; void as privative concept, 11; 63, 64, 65, 123, 267n1, 269nn17, 18, 278n4, 280n17, 288n59, 290n72, 297n123, 299n137, 305 n24, 306n39, 323n20, 333n23, 334n29, 335 n34, 340n48, 341n53, 363n90
Albertus Magnus: and Avicenna, 272n36; and separation of surfaces, 316n118; on the place of the world, 327n42; on vacuum as body, 283n30; rejected distantia terminorum, 285n49; rejected pre-creation void,

325n35; 30, 278n4, 283n26, 305n26, 306 nn30, 31, 363n90
Alexander of Aphrodisias: and prevention of vacuum, 68–9; Problemata falsely ascribed to, 301n1; 78, 80, 302n10, 309n74; see also Pseudo-Alexander of Aphrodisias
Alexander of Hales, 346n108, 368n125
Alexander, H. G., 410n373, 412n383, 412–13nn386–93, 413nn395–6, 413nn398–400, 414nn402–4, 414–15nn408–12, 416nn419, 425
Alhazen, see Haytham, Ibn al-
Aliquié, Ferdinand, 399n235
Alluntis, Felix, 339n44
Ambrose, St., 60, 161, 308n67, 328n44
Amicus, Bartholomeus, 165–71, 362n87–370 n130; and Albert of Saxony, 295n115; and Augustine, 328–9n53, 348n116; and Barrow, 406n309; and Gassendi, 211, 213, 214, 394n200; and Hobbes, 401n256; and imaginary space, 260, 270n28, 327n40, 334 n26, 345n98, 350n124; and Patrizi, 386 n135; and the imagination, 364n101, 368 n125; and violent motion in a void, 291 n78; distinguished imaginary space from vacuum, 169–71, 366n111, 396n211; experiments on vacuum, 309n75, 312nn92, 93, 320n150; on space and quantity, 364 nn97, 98; space and body coexist, 408n343; 60, 153, 160, 175, 177, 214, 278n4, 290 n72, 334n29, 344n76, 387n150, 398n234
angel(s), 131
annihilation of matter: as methodological device, 324n30, 390–1n169, 408n343; within lunar orb, 12, 333n22, 336n35, 337n37; 339n44, 411n381; see also God
anonymous author: denied distance in void space. 338n40; on contact of bodies to avoid vacuum, 337n37; on mixed bodies, 294

439

anonymous author (*cont.*)
n110, 296–7n123; on successive motion in
a vacuum, 280n12, 281n17
Anselm of Canterbury, 162, 360n71, 408n349
antiperistasis, 308n62
Appollodorus, 376n21
Aquinas, St. Thomas: and condemned articles, 324n26; and *corpus quantum*, 38–9;
and creation, 129, 341–2n59; and eternity
of world, 324n31; and God's presence, 146;
and imaginary space, 118–19; and *Liber
de causis*, 305n22; and Newton, 254, 264;
and Suarez, 157; God does not act at a
distance, 146; on *distantia terminorum*, 28,
281n18; on motion in a vacuum, 278n4;
on pre-creation void, 325n35; 29, 116, 145,
161, 282n19, 285n49, 286n51, 288n63,
292n94, 299n137, 300n149, 323n20, 346
n108, 363n90
Archytas of Tarentum: and Gassendi, 389
n168; unlimited extension beyond world,
106; 107, 108, 119, 386n131
Aristotle, 5–8; and Amicus, 165–6; and Bruno,
184, 378n41; and clepsydra, 312n98; and
distantia terminorum argument, 29–30, 280
n16, 282n22; and Euclid, 16; and al-
Ghazali's argument, 322n12; and mixed
bodies, 44–5, 49; and More, 225; and motion in a void, 6–7, 24, 57, 59; and Stoics,
107; attacked by Autrecourt, 74–7; definition of void, 9–10, 141–1, 169; denied
direct contact between surfaces, 92; existence beyond the world, 323n18, 329n57;
on condensation and rarefaction, 72–4; on
force of air, 42, 290n75; on Hesiod, 267n2;
on imagination, 117, 331n8; on instantaneous motion, 24, 277n1, 282n24; on interstitial vacua, 71, 72, 306n33; on Leucippus
and Democritus, 265n1; on place, 57, 111,
141–2, 267n4; on void, 5–8, 10, 68, 110,
259, 266n18, 269n14, 321n1; on void as
privation, 267n24; rejected plurality of
worlds, 331n11; rejection of extracosmic
void, 13, 103, 105, 260, 262, 271n31, 329
n57; the *generans* in natural motion, 288
n65; *De anima*, 92, 282n24, 318n135; *De
caelo*, 9, 68, 103, 105, 110, 184, 268n6,
291n83, 292nn86, 89, 321nn2–4, 322n12,
325n32, 329n57, 331n10, 345n101, 347
n108; *Categories (Predicaments)*, 9, 73, 267
n4, 268n4, 307n49; *De generatione et corruptione*, 292n87, 293n105, 307n47; *Metaphysics*, 267n24, 267n5, 269n13; *Meteorologica*, 86; *Physics*, 4, 9, 29, 40, 70, 71,
103, 111, 165, 192, 199, 266nn15, 17–19,
268n7, 280n16, 282n22, 288n57, 290n75,
292n90, 298n134, 306nn33–5, 307nn45, 46,

312n98, 317n130, 321nn1, 4, 323n18, 330
n58, 404n298; *Problemata (Problems)*, 304
n19, 312n98; *De respiratione*, 312n98; *De
sensu*, 30, 282n24; 3, 4, 14, 18, 19, 20, 22,
23, 27, 33, 34, 35, 43, 46, 47, 50, 61, 62,
65, 66, 70, 80, 93, 96, 115, 116, 118, 137,
157, 161, 187, 195, 201, 203, 204, 216,
229, 231, 261, 284n36
Armstrong, A. H., 328n44
Arnobius of Sicca, 113, 231
Arnold, Richard E., 306n26
Arriaga (Roderigo de), 179
Asclepius, 114, 115, 142, 329n54, 330n58,
359n57
Athanasius, 161
atomism, Greek (atomists): and Burley, 283–
4n36; defended by Autrecourt, 76–7; designated void as "nonbeing," 3; on full and
empty, 265n1; 4, 22, 71, 111, 141, 188, 204,
207, 210, 216, 234
atoms, 207
attribute, 413n394
Augustine, St.: and De Ripa, 344n82; and
extracosmic space, 114, 172–3, 178–9, 354
n17; and whole in every part doctrine, 222,
350n127, 356n37; God is in Himself, 113–
14, 132, 146; his definition of place, 21,
344n83; metaphor comparing creator and
world, 140–1, 172, 179; *City of God*, 150,
163, 172, 354n17; *Confessions*, 140–1, 160;
Contra Maximum, 163; *De diversis quaestionibus*, 276n70; 115, 130, 142, 161, 163,
368n125
Augustinian dictum that God is in Himself,
149, 150, 160, 163, 176, 219, 353n11
Aureoli, Peter, 286n51, 324n27
Avempace (Ibn Bajja): and Aristotle's description of motion, 279n8; and Campanella, 292n94; and "limitation of the
agent," 65; and Philoponus, 280n13; material medium and finite motion, 26–7; 28,
29, 30, 40, 45, 46, 62, 300n149, 301n152
Averroes: and Blasius of Parma, 315n116;
and pre-creation void, 110, 325n33; and
the imagination, 117–18; criticized by Autrecourt, 75; on Aristotle, 9, 267n4; on
Avempace, 26–7, 29; on candle experiment, 78–80; on dense and rare, 307n51;
on motive forces, 40–1; on vacuum, 14,
68–9, 71, 266n16, 271n36, 333n23; Commentary on *De caelo*, 68, 78, 302n8, 9,
308n68, 310n77, 325nn33, 34, 331nn10,
11; Commentary on the *Metaphysics*, 9;
Commentary on the *Physics*, 272n36, 279
nn7, 10, 11, 280n13, 306n36, 325n34;
*Tahafut al-Tahafut (The Incoherence of
the Incoherence; Destructio Destruction-*

um), 322n12, 325n33, 327n40; 19, 35, 39–40, 52, 74, 80, 137, 275n62, 284n36, 289 n69, 300n149
Avertissement au lecteur, 252, 254
Avicenna, 30, 272n36, 283n26

Bacon (Francis), 207
Bacon, Roger: and separation of surfaces, 87–8, 89, 316n117, 318n135; and universal nature (or agent), 69, 82, 85, 304nn17, 21; and William of Ware, 271n33; equated vacuum with prime matter, 272n37; no distances in vacua, 123–4; on clepsydra, 84–5, 314n106; on *distantia terminorum*, 27–8, 31, 282n22; on extracosmic void, 13, 106, 321n5–322n7, 336n35; on motions in a void, 24; on sealed vessels, 82; on speed of light, 30; the place of the world, 327n42; vacuum not a cause, 30; *Liber primus Communium naturalium*, 303n11; *De multiplicatione specierum*, 283n29; *Opus Majus*, 336n35; *Physics* (4 bks), 306 n29; *Physics* (8 bks), 305nn23, 26, 306 n28, 307n56; 29, 35, 39, 92, 93, 303n11, 305nn23, 26, 307n56, 311n84, 330n1, 332nn14, 16, 336n35, 337n38
Baconthorpe, Johannes, 129, 341n54
Baeumker, Clemens, 346n108
Bailey, Cyril, 315n114
Barbay, Petrus, 328–9n53, 354n17
Bardenhewer, Otto, 305n22
Barefoot Brothers of the College of Alcala de Henarez, *see* Complutenses
Barrow, Isaac, 236–8; and Bonaventure, 406n308; and Fonseca, 405n305; and Gassendi, 393n187; and Locke, 238; and More, 405nn301, 304; did not geometrize space, 233; God and space, 237, 393n191; no coexistence of space and body, 406n309; *Mathematical Works of Isaac Barrow*, 405n302–406n307, 406nn310–13; 192, 207, 213, 404n297
Barry, Frederick, 320n151
Bartholemew the Englishman, 346n108
Basil, St., 161
Bassolis, John de, 129
Baur, Ludwig, 326n35
bellows, 82–3, 96
Bellutus, Bonaventura, 25, 60, 278n4, 290n72
Benedetti, Giovanni, 65, 300n149
Bentley, Richard, 379n58
Berkeley, George: attacked Newton, 411n380; on absolute space, 247, 416n425; on problems of external space, 411–12n382; 264, 407n337, 411n381
Bernard of Clairvaux, 368n125
Biblical citations: Acts *17*, *28*, 396n216,

407n329; II Chronicles *2*, *6* and *6*, *18*, 407n329; Job *11*, verses *8–9*, 373n162; Kings *3*, *8* (Vulgate; *1*, *8*, *27* in King James Version), 149, 359n68, 373n162, 407n329; Psalm *8*, verse *1*, 369n126
bishop of Paris, *see* Tempier, Etienne
Blanchet, Léon, 381n76
Blasius of Parma: on separation of surfaces, 82–92, 317n124–318n133; 87, 96, 99, 315n116, 319n142
Bloch, Olivier René: on Pascal and Gassendi, 392–3n185; significance of Gassendi, 388–9n163; 389n164, 391nn170, 174, 178, 393n186, 394n191
body (bodies): and extension, 408n346; and space do not coexist, 406n309; and spirit, 225, 342–3n66; and void, 31, 34; expands to fill space, 197; mixed, 44–5, 57–60, 62–4, 279n4, 292n87, 294nn111, 112, 295 n117; on coexistence of two, 6, 32–3
Boehner, Philotheus, 339n44
Boethius, 113, 360n70
Böhm, Walter, 275n61
Bona Spes, Franciscus: and Augustine, 329 n53, 348n116; and Clarke, 412n386; and Descartes, 399n235; and God as infinite sphere, 347n108; and Guericke, 216; and Kircher, 396n222; on imaginary and real space, 178–81, 216–17, 372n153–373n161, 373n163–374n164; 153
Bonansea, Bernardino M., 380nn68, 70, 381 n76, 382n81, 384nn104, 109, 112, 115, 117
Bonaventure, St.: and Barrow, 406n308; and *ubi repletive*, 368n125; 160, 179, 346n108, 363n90
Bonetus, Nicholas: and Galileo, 63; defined resistance to motion in void, 35–6; impressed force in a vacuum, 42–4; 38, 129, 286n51, 291n78
Book of the XXIV Philosophers, 138–9, 145, 346n108
Bourke, Vernon, 348n116
Bradwardine, Thomas, 135–43; and actual infinite, 341n53, 349n122; and *Book of the XXIV Philosophers*, 346n108; and Bruno, 329n55; and Clarke, 248; and *De Ripa*, 134, 142; and Galileo, 63; and Gassendi, 389n165; and God as infinite sphere, 138–40, 348n112; and Henry of Ghent, 339n41; and Koyré, 374n4; and Major, 353n6; and More, 228; and Newton, 243; influence of, 142–3, 144; on equality of fall in void, 58–60, 301n156; on extracosmic infinite void, 135–44, 262, 344n90; on God, 260, 346n106, 353n11, 356n33; on pre-creation void, 111, 326n39; vacuum conceived in two ways, 10; 51, 120, 148,

Bradwardine, Thomas (*cont.*)
154, 156, 162, 163, 164, 165, 169, 218, 268n9, 298n128, 298n134, 301n152, 324 n27, 326nn38, 39, 327n40, 343n76, 352 n137, 355n23, 403n284
Brickman, Benjamin, 320n152, 384n118–387 n140, 387nn144–8, 388n152, 388n154–6, 388nn158–60
Bridges, J. H., 283n30
Brose, Henry L., 273n46
Bruers, Antonio, 381nn72, 77, 80
Bruno, Giordano: and Aristotle, 378n41, 379 n50; and Guericke, 395n210; and Hermetic literature, 329n55; and pantheism, 379 n57; and Raphson, 403n287; on God and space, 190–2; on infinite space, 186–90; on penetrability of space, 188; rejected Aristotelian and Stoic cosmologies, 184–5, 348n115; space neither substance nor accident, 377n38; *Camoeracensis Acrotismus*, 377n38, 378–9n50, 379n51; *Concerning the Cause Principle, and One*, 379n55; *De immenso*, 186, 188, 190, 377nn29–32, 34–8, 378n48; *On the Infinite Universe* (*De l'infinito universo*), 184, 188, 190, 375 nn13–14, 375n16–376n25; *De minimo*, 379 n52; 19, 21, 138, 140, 198, 204, 206, 210, 234, 347n108, 380n62
Brush, Craig B., 389n165–394n198
Burgundio of Pisa, 276n67, 350n127
Buridan, John: and Albert of Saxony, 275n58; and Aristotle's *Predicaments*, 267–8n4; and Guericke, 311–12n91; and God's power, 127, 343n72; and impetus theory, 43–4, 291n84; and imagination, 119–20; and the theologians, 341n52; arguments against vacuum, 278n3, 340n46; definition of succession, 277–8n1; experiments against vacuum, 71–2, 82, 84, 310n81; negative propositions on vacuum, 269n21; no distance measurements in void, 123, 336n35, 339n44; on actual infinite, 127–9, 340n47–341n52, 343n72; on extracosmic void, 121, 127–9, 270n23, 334n27–335n32; on instantaneous motion, 45–6, 36–7, 287n55; on internal space, 122; on mixed bodies, 51, 294–5n112, 295nn114, 115, 117; on natural place, 297n126; on predominant element theory, 293n106; on separate space, 18; void as containing surface, 12, 270 n24; *Questions on De caelo*, 270n23, 295 n115, 296n117; *Questions on the Ethics*, 269n21; *Questions on the Physics*, 267n1, 275n59, 295nn115, 117, 305n23; 15, 16, 24, 42, 156, 205, 298n129, 314n109, 323 n20, 324n27, 344n76, 355n23, 376n18
Burke, Robert B., 336n35

Burley, Walter: and Bradwardine, 298n128; and *distantia terminorum* argument, 282 n22, 288n61; and Galileo, 63; and Greek atomists, 283–4n36; and *Lives* of Diogenes, 376nn21, 24; animals in a vacuum, 290 n72, 299n136; definitions of vacuum, 268 n8; on the clepsydra, 84, 85–6, 304n20, 314 n107–315n111; on mixed bodies, 50, 54, 292n87, 294n109, 295n114; on motion in a vacuum, 47, 58, 278–9n4, 299n136; on resistance in motion, 289–90n70; on separation of surfaces, 88, 93–5, 318n136–320n148; on treatment of questions on vacuum, 269n17; on universal nature, 84, 85–6, 304n17, 21, 305n15, 314n107–315n111; on yielding of void, 32–5; Shapiro on, 285–6n50; *Commentary on the Physics*, 290n72, 305n26, 306n39; *Questions on the Physics*, 303n15, 304n20; 25, 52, 55, 59, 96, 97, 201, 205, 278n4, 299 n137, 306n31, 324n27, 341n53, 344n76
burning candle experiment, 77–80
Burtt, E. A.: on Descartes, 400n254; on Malebranche, 398n235; on mutual influence of More and Barrow, 405n301; 393n187, 393n191, 400nn246, 249, 252, 410nn258, 259, 402n265, 405n305
Bury, Rev. R. G., 276n67, 376n21
Butler, Joseph, 416n425
Buytaert, Eligius M., O.F.M., 348n116, 350n127
Byrne, Paul M., 324n31

Cabbalists, 182, 231, 402n264
Cajori, Florian, 408n357, 415n418
Calo Calonymos, 325n33
Calonymos ben Calonymos, 325n33
Cambridge University, 241
Campanella, Tommaso, 194–9, 380n68–384 n117; and Gassendi, 389n165; and Patrizi, 196, 382n87, 383n94; and Telesio, 380 n68; and the *limitatio agentis*, 292n94; denied extracosmic space, 198; denied nature's abhorrence of a vacuum, 198; on finitude of space, 381–2n81; on plurality of worlds, 381n72; space has sense and feeling, 196–7; *Del senso delle cose*, 197, 292n94, 381nn72, 77; *Epilogo magno*, 382 n81; *Universalis philosophiae*, 194, 380 n69–384n117 *passim*; 19, 207, 210, 228, 234, 404n294
candle experiment, 77–80; 96, 309nn74, 75
Canonicus, Johannes, *see* Johannes Canonicus
Čapek, Milič, 160, 277n74, 322n10, 323n19, 391nn171, 176, 393nn187–9
Capreolus, Johannes, 179, 358n60, 363n90
Carra de Vaux, B., 302n3, 308n69

Castañeda, Hector-Neri, 413n395
Chandler, Bruce, 341n52
Charleton, Walter: and annihilation of matter, 389n168; and uncreated space independent of God, 393n191; influence on Newton, 241; linked time and space, 392n182; on natural impossibilities, 309n169; 207, 208, 212, 215, 388n162, 391n171, 392 n182, 394n191
Chatelain, Emil, 274n56, 323n21, 324nn25, 28, 326n37, 339n45
Church (Holy) Fathers, 172–3, 182, 229, 231, 393n191, 394n200
Cicero, 375n7
Clagett, Marshall: "Adelard of Bath," 302n2; *Science of Mechanics in the Middle Ages*, 287n52, 290n76, 291n78, 291n83, 291 n84, 298nn131–3, 353n3
Clarke, John, 416–17n425
Clarke, Joseph, 416–17n425
Clarke, Samuel: and "attribute," 410n373, 413n394; and Bona Spes, 412n386; and Bradwardine, 248; and Gassendi, 393n187; and God's operation, 416n422; and motion of world, 344n76; and Newton, 248, 410n373, 415n414; and Oresme, 249; correspondence with Leibniz, 412n386–415n414 *passim*; critique of Leibniz on space, 248–52, 414–15n412; on spiritual beings, 415n413; reaction to, in the eighteenth century, 416–17n425; *A Demonstration of the Being and Attributes of God*, 413n394; *The Works of Samuel Clarke*, 413n394; 241, 261, 262
Clement of Alexandria, 161
Cleomedes the Stoic: and Aristotle's definition of void, 322n13; and Stoic conception of universe, 375n8; on extracosmic void, 107; 183, 201, 213, 231, 233, 324n29, 388n153, 389n168
clepsydra, 83–6, 312–15; Adelard on, 67; Burley on, 304n20; Duhem on, 312n95; 96, 97, 98, 123
Coffa, J. Alberto, 273n43
Cohen, I. Bernard: on the missing *tanquam*, 409n371; "The Case of the Missing *Tanquam*," 409n371; *Isaac Newton's Papers*, 409n370; *Isaac Newton's Philosophiae Naturalis Principia Mathematica*, 411n375; "Newton & the Leibniz-Clarke Correspondence," 407n336, 411n375, 412nn383–5, 415nn414, 415, 416n421; "Newton, Isaac," DSB, 407n338
Cohen, Morris R., 266n18, 291n83, 299n138, 320n149
Coimbra Jesuits, *see* Conimbricenses
Coleman, Elliott, 347n108

Coleman, Janet, 342n63, 343n71
Colliber, Samuel, 416n425
Colson, F. H., 327–8n44
Combes, André, 341n57–344n89 *passim*; 339n42, 346n106
Complutenses, 60, 279n4, 316nn117, 119
Comptonus (Thomas Compton-Carleton, S.J.), 179
Condemnation of *1277*, 108–10; *article 34*, 109; *article 49*, 109, 343–4n76, 361n73, 413n397; *article 139*, 126, 339n45; *article 140*, 126, 339n45; *article 141*, 126, 274n56, 339n45; *article 147*, 108; *article 201*, 326n37; 104, 110–11, 116, 121, 127, 140, 336n35, 340n48
condensation and rarefaction: and history of vacuum, 305n24; and microvacua, 71–4; explained by Autrecourt, 74–7; in plenum, 95–6; of empty space, 33; 96, 141, 313n103
Conimbricenses (Coimbra Jesuits), 160–3, 358n58–362n84 *passim*; and Aristotle, 330 n57; and Augustine, 328–9n53, 348n116; and Gassendi, 211, 213, 394n200; and God as infinite or intelligible sphere, 347 n108; and Guericke, 221, 397n230; and Hobbes, 401n256; and imaginary space, 260; and separation of surfaces, 316nn117, 119; 60, 120, 153, 157, 163, 165, 169, 177, 215, 306n30, 311n90, 363n90
continuum, 206
Copernicus, 139
Copleston, Frederick, 360n71
Cornford, F. M., 322n10, 323n18
Coronel, Luis, 81–2
corpus quantum, 38–40
Cosman, Madeleine Pelner, 341n52
Cranz, F. Edward, 302n10
Created Something, 215
Crescas, Hasdai: and extracosmic vacuum, 271n33, 321n5; and Pico della Mirandola, 379n59; and possibility of other worlds, 277n71; distinguished material from immaterial dimensions, 22; on divisibility of space, 404n298; 23, 119, 277n72, 322n20
Crew, Henry, 300n146
Crombie, A. C., 336n35, 342n62
Crosby, H. Lamar, Jr., 297n128, 298nn130, 132, 133
Cudworth, Ralph: denied infinity of space, 398n233; denied infinity of world, 402 n273; on God and space, 230; 402n272, 416n420
cupping glass, 303n10, 309n77, 310n77
Cusa, Nicholas of, *see* Nicholas of Cusa

Cyprian, St., 113, 231
Czwalina, Arthur, 323n13

Dales, Richard C., 330n4
Dawson, Carley, 347n108
Delhaye, Philippe, 267n1, 280n14
Delorme, Ferdinand M., 270n30, 278n2
Democritus, 141, 185, 210, 271n35
Denifle, Heinrich, 274n56, 323n21, 324nn25, 28, 326n37, 339n45
Denis, St. (Pseudo-Dionysius), 161
Denomy, Alexander J., 330n3
density, 33–4
Derham, William, 379n58
Des Maizeaux, 252
Descartes, René: and imaginary space, 399 n235; and infinite space, 341n49, 398n233; and internal space, 15, 16–17; and Locke, 239; and More, 223, 225, 229, 399n238, 402n262; and Newton, 243, 408n349; on the collapse of a vessel, 335n33, 336n34; 122, 156, 175, 218, 226, 231, 269n21, 274n51
Dick, Steven J., 376nn26, 27, 379n58, 399n236
Diels, H., 275n61, 387n142
Digby, Kenelm, 207, 389n165
dimension(s): distinction between material (corporeal) and immaterial (incorporeal), 19–23, 203, 261; interpenetration of, 5–6, 16–17
Diogenes Laertius, 201, 210, 213, 375n7, 376nn21, 24
distantia terminorum (*incompossibilitas terminorum*), 27–38; and *corpus quantum*, 39–40; and impenetrable vacuum, 33; as kinematic argument, 27–8, 29; as resistance, 28–9; criticisms of, 31–8; described, 259; origins of, 29–30; 34, 45, 47, 49, 51, 62, 76
Dod, Bernard G., 330n2
Dods, Marcus, 328n52
Donahue, William H., 392n182
Drabkin, Israel E.: on Aristotle and velocity in the void, 266n18; on Benedetti and Philoponus, 300n149; on Galileo and Benedetti, 299n144; 291n83, 299nn138, 140, 320n149
Drachmann, A. G., 309n70, 310n78
Drake, Stillman, 299n140, 300n146, 320n156
Dressler, Hermigild, 271n34
Du Bois, Jacques, 218
Duhem, Pierre: on actual infinite, 340n47; on the separation of surfaces, 316n120; *Etudes sur Léonard de Vinci*, 337n39, 340 n47; *Le Système du monde*, 271n33, 298n 134, 301–2n1, 304n19, 308n70, 309n74,

312n95, 315n113, 323n22, 339n43, 341 n54, 374n5, 375n9, 387n142, 388n153
Dullaert, Johannes, 297n123, 298n129, 352n3
Dumbleton, John, 70, 304n19, 315n113
Dümmler, Ernest, 271n34
Duns Scotus, John: and internal space, 15–16; and Leibniz, 264; and Marsilius of Inghen, 288n59; and Suarez, 157; God can act at a distance, 146, 254; on yielding of void space, 31–2; positive distance in void, 339n44; rejected extracosmic void, 146, 147; rejected pre-creation void, 352 n143; 145, 160, 179, 270n22, 299n137, 300n149, 324n27, 341n53, 363n90, 371 n141
Durandus de S. Porciano, 160, 358n60, 363n90

Edelston, J., 407n337
Einstein, Albert, 273n43, 418n427
element(s): in mixed body, 50, 52, 56–7
Elie, Hubert, 340n47, 353nn4–6, 355n24
Elkana, Y., 331n9
Elwes, R. H. M., 402n270
Emge, Walter, 391n171
Empedocles, 30, 282n24
Engels, Friedrich, 395n203
Epicurus, 108, 182, 185, 210, 375n7, 376n18
Ermatinger, Charles J., 350nn126, 127
eternity: analogy between, and immensity, 360n70; defined by Boethius, 360n70
ether (aether): its relation to void (Bruno), 188–90; Newton on, 247; 204, 210, 220
Euclid: nature of geometric space, 16, 232, 233, 273nn46, 49; 108
Eudemus, 322n10
Eugubinus, 161
experiments: for and against vacuum, 70–4, 77–100

Faber, Phillip: and Aquinas, 352n139; and Augustine, 328–9n53; on capacity and vacuum, 354n19; on extracosmic existence, 329–30n57, 334n29, 369n125; on God and space, 361n73, 366n110; on imaginary space, 355n29, 365n108; 356n39, 358n62, 398n235
Fabri, Honoré, 367–8n122
Fallico, Arthur B., 381nn72, 77, 80, 382n82, 383nn97–102, 384nn103–6, 114
Favaro, Antonio, 299n140
Feldman, Seymour, 326n39
Ferguson, James P.: on Clarke's argument against Leibniz, 414n401; on post-Newtonian disputes on space, 416–17n425; 410 n373, 412n383
Festugière, A. J., 329nn54, 56

Ficino, Marsilio, 161, 183, 231, 329n54, 347n108
Fierz, Markus, 393n187
Fiorentino, F., 375n18
fire, 77–80
Fishacre, Richard, 143–4, 350n127
Fludd (Robert), 297, 347n108
Fonseca, Pedro, 157–9, 357n47–358n56 *passim*; and Barrow, 405n305; and Philoponus, 357n52; Aristotle on space, 357 n48; on imaginary space, 350n124; on pure negation, 159; space neither substance nor accident, 205, 357n53; space not a true being, 327n40; 153, 162, 163, 165, 169, 177, 179, 214, 363n90, 364n93
forma mixti uniformis, 50
Fortin, Ernest, 323n23
Franciscus de Marchia, 286n51, 291n78
Franciscus de Mayronnes (Mayronis), 120, 129, 286n51
Fraser, Alexander Campbell, 411nn380, 381
Fridugis the Deacon, 271n34
Fromondus, 179
Fuerst, Adrian, O.S.B., 350n127
fuga vacui, 67
Funkenstein, Amos, 324nn24, 30
Furley, David, 375n7

Gaietanus de Thienis, 344n76
Galen, 157
Galileo Galilei, 60–6; and self-expending impetus, 300–1n150; on falling bodies, 299–300n145, 300n146; on motion in a void, 61–2; on the separation of surfaces, 98–9, 320n156–321n159; *De motu*, 299n140–301n151; 44, 87, 175, 259
Garnett, Christopher Browne, Jr., 410n373
Gassendi, Pierre, 206–14; adopted Stoic universe, 207–8; and Barrow, 237; and Descartes, 391n174; and Guericke, 217, 220, 397n228; and Hobbes, 401n256; and More, 227; and Newton, 241–2, 244, 388n162; and Patrizi, 391n179; and properties of space, 209–10, 232; and scholastic tradition, 213–15; distinguished corporeal from incorporeal dimensions, 22–3, 261; on the continued existence of space, 408n343; on the methodology of imaginary annihilation, 390–1n169; on the relation between the *Syntagma* and *Animadversiones*, 390n168; on space and God, 210, 211–12, 222, 262; probably a Copernican, 391n170; significance of, 388–9 n163; space three-dimensional, 221; whole in every part doctrine assumed, 210; *Animadversiones*, 389nn165, 167; *Syntagma philosophicum*, 389n165–394n198 *pas-*

sim; 23, 108, 110, 192, 218, 219, 226, 228, 231, 234, 237, 255, 320n155, 344n76, 398n234, 403nn279, 287
Gaye, R. K., 282n22
General Scholium (Newton), 241, 246, 247, 253, 407n337, 415n418
Gentile, Giovanni, 375n10
Gerard of Bologna, 129, 341n54
Gerard of Cremona, 70
Gerard of Odo, 286n51
Ghazali, al-, 322n12, 325n33, 327n40
Gibb, Jocelyn, 406n314
Gibson, James, 406n317
Gilbert, Neal W., 380n60
Gilbert, William, 207
Gilbertus Porretanus, 116
Gilby, Thomas, O.P., 331n14
Gilson, Etienne, 351n134
Ginsberg, Morris, 398n235
Glanvil, Joseph, 399n238, 400n248
God: and creation of actual infinites (including space), 127–33, 152, 200, 341 n53; and creation of other worlds, 150, 351n130; and extracosmic void, 141–2; and His absolute power, 103, 108–9, 116, 121, 127, 128, 130, 131, 134, 137, 184, 326n37, 330n3, 339–40nn45, 46, 345n96; and His relationship to space, 141–2, 154, 176 (Maignan), 191 (Bruno), 218–20 (Guericke), 223, 243–4 (Newton), 250, 253, 255; and imaginary space, 179, 366 n10; and motion of world, 137, 249, 361 n73; and real space, 179, 180; and space in Leibniz-Clarke correspondence, 248–52; and three-dimensional space, 212–13; annihilates all or part of material world, 12, 23, 44, 47, 52, 55, 109, 124, 150–1, 170, 208–9, 213, 237, 242, 270n23, 335 n33, 337n37, 339n44; as infinite or intelligible sphere, 129, 138–40, 145, 151, 161, 179, 403n279; as place of the world, 113; departed space, 262, 263, 264; extension of, 141, 142–3, 154, 177, 210, 239–40, 245, 261, 402–3n276; His extracosmic existence, 163; immensity of, and space, 129, 133, 140–1, 163, 165, 173, 234, 249, 346n106, 360n70, 367n116, 406–7n329; immutability of, 135–42; infinitely exceeds infinite space, 134; is He present where He acts?, 146, 153–4, 157, 253–4, 371n141, 402n275; is in Himself, 113–14, 211, 344n82, 351n136, 353n11, 365n108, 366n110, 368n125, 371n143, 398n235; is wholly in every part, 143, 144, 191, 210, 212–13, 222, 356n33; Leibniz on space and, 413nn391, 396; location of, 103, 104, 112; nondimensional, 142, 144,

God (*cont.*)
178, 180, 221, 222, 223; on His eternity, 360n70; space and omnipresence of, 131–2, 132–3, 135–6, 136–8, 143, 144, 156–7, 160, 211, 222, 237, 239, 254, 346n106; space as sensorium of, 246, 409n371; space and other properties and attributes of, 142, 227, 232, 414–15n412; space independent of, 190, 211–12, 411–12n382; unlimited and uncircumscribed, 131, 134, 135; 148, 150, 155, 164–5, 175, 184, 228
Godfrey Fontaines, 333n22
Gollancz, Hermann, 302n2
Googe, Barnabe, 378n39
Gorlaeus, David (Van Goorle), 392n182
Gossen, Hans, 395n209
Gouhier, H., 398n235
Grant, Edward: "The Arguments of Nicholas of Autrecourt," 308n61; "Aristotle, Philoponus, Avempace," 279n5, 292n94, 299 n138, 301n154; "Bradwardine and Galileo," 279n7, 288n56, 295n116, 298n129, 298–9n134, 301n156; "The Concept of *Ubi*," 343n67, 350n127; "The Condemnation of *1277*" (*Viator*), 323n21, 324n24, 339n45, 340n46, 388n151, 391n169, 413 n397; "The Condemnation of *1277*" (*Cambridge Medieval Philosophy*), 323 n21; "Henricus Aristippus, William of Moerbeke," 302n3, 311n82; "Jean Buridan: A Fourteenth Century Cartesian," 334n29; "The Medieval Doctrine of Place," 327n41; "Medieval Explanations and Interpretations of the Dictum that 'Nature Abhors a Vacuum,' " 302n1, 304 n19, 315nn112, 115; "Medieval and Seventeenth-Century Conceptions of an Infinite Void Space," 322n12, 323n20, 323nn13, 16, 332n17, 334nn25, 26, 344n91, 348 n117; "Motion in the Void," 266nn14, 17, 279n7, 289n70, 291nn81, 82, 292n85; *Nicole Oresme and the Kinematics of Circular Motion*, 298n130; *Nicole Oresme: De proportionibus proportionum*, 298 n130, 301n153; "On the Origin of the Medieval Version of Equality of Fall," 298–9n134; "Place and Space in Medieval Physical Thought," 274n55, 334n27, 335n32, 337n36, 385n125; "The Principle of the Impenetrability of Bodies," 272n38, 283n32, 386n133; "Scientific Thought in Fourteenth-Century Paris," 341n52; *A Source Book in Medieval Science*, 279nn9–11, 280n13, 280n17, 282 n24, 283nn28, 29, 288n65, 289n70, 292 nn93, 95, 293nn100, 103, 105, 295n115, 296n118, 297n125, 298n129, 301nn152–

4, 306nn26, 27, 307n52, 310nn81, 82, 311nn85, 89, 312n91, 313nn99, 104, 321 n160, 326nn36–8, 334n26, 344n90–346 n104 *passim*, 348n117–349n120, 358n59, 394n201, 395n209–397n230 *passim*
Greenberg, Sidney, 346n105, 348n115, 379n55
Greenwood, J. C., 309n70
Gregory, David, 245–6, 415n417
Gregory Nazianzenus, 161
Gregory of Rimini, 129
Grene, Marjorie, 402n269
Grosseteste, Robert: on infinites, 118, 342 n62; on place, 339n43; 116, 268n10, 330n4
Guericke, Otto von, 215–21, 394n201–397 n232 *passim*; and Barrow, 237; and Buridan, 311–12n91; and Coimbra Jesuits, 397 n230; and Gassendi, 393n187, 397n228; and Leibniz, 412n386; and Malebranche, 398n235; and Raphson, 396n218, 403 nn280, 287; critique of Kircher, 219; description of nothing, 3, 216; identified space with God's immensity, 396n223; meanings of imaginary space, 120–1; nondimensionality of universal space, 220; scholastic influence on, 215, 262; two spaces distinguished, 217; 14, 173, 192, 222, 231, 374n1, 405n305
Guillaume de Conches, 207
Guthrie, W.K.C., 302n8, 313n98

Hahm, David E., 323nn13, 14, 17
Haldane, Elizabeth S., 274n51–4, 335n33
Hall, A. R. and M. B., 405n300, 407n338, 408 nn340–56, 414n408, 415n418
Hardie, R. P., 282n22
Harries, Karsten, 347nn109–11
Hartel, William, 328n46
Hartsoeker (Nicolaas), 411n375
Haytham, Ibn al- (Alhazen), 30, 331n9
Heath, P. L., 269n15
Heath, Sir Thomas L., 273n47
heaviness and lightness: and location of body, 56; Galileo's attitude toward, 63–4, 301 n151; in mixed bodies, 49–53; 78
heliocentric system, 183
Henry, John, 387nn140–2, 388n161, 391n179
Henry of Ghent: and imagination, 119; measurement of distances, 124–5; 120, 126
Hermes Trismegistus: and *Asclepius*, 161; and God as infinite sphere, 151, 179, 347n112; on void beyond world, 114–15; pseudo-Hermetic *Book of the XXIV Philosophers*, 138–9, 145; 182, 183, 201
Hero of Alexandria, 97, 98, 182, 183, 308–9 n70
Heron, Germain, O.F.M., 348n113

Hesiod, 5, 267n2, 375n9
Hilary, St., 161
Hipparchus, 291n83
Hiscock, W. G., 409n369
Hisette, Roland, 324n22
Hobbes, Thomas: agreed with More and
Newton, 408n344; on real and imaginary
space, 226, 401nn255–7; on the methodol-
ogy of imaginary annihilation, 390n169; *De
corpore*, 401nn255, 257; *Thomas White's
"De mundo dialogi tres,"* 401n255; 110,
207, 231, 411n382
Hoffman, Ernest, 348n113
Holenmerism (Holenmerians), 223, 224, 225,
253, 400n239
Holkot, Robert, 129, 350–1n130
horror vacui, 67
Hume, David, 416n425, 417n425
Hurtado (Diego Hurtado de Mendoza), 179,
363n90
Hyman, Arthur, 323n21

Iamblichus, 231
Ideler, Julius Ludwig, 301n1
imaginary: meaning of term, 117–21
imaginary space, *see* space
immensity of God, 129, 133, 153, 156–7,
346n106
impenetrability: and yielding, 396n218; of
space, 370n130
impetus, 42, 43–4, 63, 300n150
incompositas terminorum, 38, 280n14, 288
n57; *see also distantia terminorum*
*incompossibilitas terminorum, see distantia
terminorum*
inertia, 266n16
infinite(s): actual, 127–33, 152, 200, 340n47,
367n116, 385nn122, 123; categorematic
and syncategorematic, 48; compared to
finite extension, 143–4; derivation of infinite space, 403n285; extension and God,
143–4; on simultaneous existence of two,
397n225; ordinary and super, 129; potential infinite space, 397–8n233; space, 104,
156; space, concept unclear, 239; *see also*
God; space
instants, 90–92, 287n54
interpenetration of dimensions, 5–6, 201, 231

Jackson, John, 416n425
Jacquot, Jean, 401n255
Jammer, Max: on Campanella, 380n69, 382
n81; on Gassendi, 393n187; on Leibniz,
413n395, 414n401; *Concepts of Space*, 322
n10, 323n20, 328n50, 380nn61, 63, 399
n237, 403n287, 408n357, 409nn364, 365,
418n427

Javelli, Chrysostomus, 305n24, 311n85
Jean de Ripa, 129–34, 341n57–344n89 *passim*; and actual infinite, 164, 341n50, 385
n123; and Bradwardine, 142; and measurement in void, 125–6; God as infinite sphere,
347nn108, 112, 348n112; on positive and
imaginary places, 343n75; 116, 124, 140,
144, 145, 148, 156, 163, 165, 169, 262,
324n27, 345n101, 363n90
Jeans, Sir James, 119
Jesuits, 152
Joachim, Harold H., 307n48
Johannes Canonicus (Johannes Marbres): and
Bonetus, 286n51; and candle experiment,
79–80; and Philoponus, 299n139; on impressed forces, 42, 291n78; relationship of
speed and heaviness in void, 292n94; violent motion in void, 42, 290–1n77; 36,
38, 299n137, 311n85
Johannes Quidort, 303n14, 304n16
John Damascene, St.: and whole in every part
doctrine, 222, 350n127; 348n116
John of Jandun: and Vasquez, 270n28, 358
n57; experiments against vacuum, 72, 83;
interpenetration of void and body, 17–18,
32; on Aquinas and his followers, 288n64;
on separation of surfaces, 89, 316nn117,
119, 318n135; on vacuum and privation,
10; superfluousness of void, 275n60; 11,
13, 25, 92, 93, 268n12, 278n4, 283n34, 299
n137, 305nn23, 26, 306nn27, 31, 38, 39,
307nn40–4, 311n85, 362n86
John Major, 148–52, 353n4–355n25 *passim*;
and actual infinite, 358n123; and Amicus,
367n116; God as infinite sphere, 347–8
n112; God beyond heaven, 151; misinterpreted Augustine, 150, 328–9n53; 161,
163–4, 169, 179, 347n108, 352n3, 406n308
John of Saint Thomas (John Poinsot), 60,
331n14
Johnson, Francis R., 378n39
Jones, Harold Whitmore, 401n255

Kalam, 271n35
Kargon, Robert, 389n165
Kauffeldt, Alfons: scholastic influences on
Guericke, 397n232; 394–5n203, 395n209,
396nn216, 223, 397n228
Keill, John, 379n58, 403n287, 404n299, 405
n305
Keller, A. G., 395n203
Kendzierski, Lottie H., 324n31
Kennedy, Leonard A., 306n26
Kenny, André, 309–10nn77, 78
Kenny, Anthony, 323n21
Kepler, Johannes, 207
Kircher, Athanasius: criticized by Guericke,

Kircher, Athanasius (*cont.*)
219; imaginary space is God, 396n222; 175, 215, 218, 395n203
Kirk, G. S., 284n38
Kirkby, John, 404n297, 405n302
Klibansky, Raymond, 348n113
Knust, Hermann, 376nn21, 24
Koyré, Alexandre: a curious claim, 398n234; and scholastic tradition, 374n4; ignored Gassendi and Patrizi, 388–9n163; interpretation of Bradwardine, 349n124; interpretation of Raphson, 403n286; omitted Campanella, 380n69; omitted Gassendi, 393 n187; on Bruno's void, 378n45; on Euclidean space, 232–3; on Leibniz, 250; on limitation of God's power, 343n72; on Malebranche, 398–9n235; on More, 399 n237; "The Case of the Missing *Tanquam*," 409n371; *From the Closed World*, 335n33, 336n34, 341n49, 348n115, 375n13, 376 n23, 378nn42, 43, 391n174, 393n187, 398 n233, 400n247–402n264 *passim*, 403n277–404n292 *passim*, 407nn335–7, 409n369, 412n383, 416n421, 417n426; *Isaac Newton's Philosophiae Naturalis Principia Mathematica*, 411n375; "Newton & the Leibniz-Clarke Correspondence," 407n336, 411n375, 413n394, 412nn383–5, 415 nn414, 415, 416n421; *Newtonian Studies*, 343n72, 404n293, 412n383; "Le vide et l'espace infini," 334n26, 337n39, 344n90–346n104 *passim*, 348n117; 384n111
Krafft, Fritz, 394n202, 395nn209, 210, 397 n225
Kren, Claudia, 349n123
Kretzmann, Norman, 323n21
Kristeller, P. O. 303n10, 375n10, 379nn54, 55

Latham, R. E., 315n114
Law, Edmund, 416–17n425
Leff, Gordon, 323n22
Leibniz, Gottfried, 248–52; and distinguishable spatial parts, 414–15n412; and Duns Scotus, 264; and missing *tanquam*, 409–10 n371; and Newton, 261–2, 407n337, 409 n371, 410n373; denied whole in every part doctrine, 415–16n419; his principle of sufficient reason, 249; on Clarke's relationship to Newton, 412n383; on God and space, 250–1, 254, 264, 413n396, 416n422; on imaginary space, 412n386; on *ubi*, 343 n67; 231, 241, 247, 411n375
Leibniz-Clarke correspondence, 248–52, 412 n383–415n415 *passim*; 241, 245, 247, 409 n371, 410n373
Lerner, Ralph, 323n23

Lessius, Leonard, 215, 219, 231, 334n26, 396 n222
Leucippus, 141, 271n35
Levi ben Gerson (Gersonides), 326–7n39
Leyden, W. von, 382n81, 416n422
Liber de causis (pseudo-Aristotle), 70, 305 n22
Liber de ingeniis spiritualibus, 68, 77
Liber sex principiorum, 116
light: Patrizi on, 203, 387n142; speed of, 29–30, 46–7; 204, 210
lightness, *see* heaviness and lightness
limitatio agentis, 46, 292nn94, 95
Lindberg, David C., 283nn28, 29
Lipsius, Justus, 376n21
Little, A. G., 302n1
Locke, John, 238–40, 406n314–407n334 *passim*; and Archytas of Tarentum, 322n10; and Berkeley, 411n382; and endlessness of space, 332n19; and Gassendi, 393n187; and More, 407n334; and scholastics, 374 n1; and three interpretations of space, 238; corporeal and incorporeal extension distinguished, 23; on idea of vacuum, 277n76; on space as substance or accident, 240; vacuum and the annihilation of matter, 277 n77; *Essay Concerning Human Understanding*, 406n319–407n333; 22, 110, 192, 207, 231, 236
Lohr, Charles H., S.J.: "Medieval Latin Aristotle Commentaries," 267n4, 286n51, 303 n12; "Renaissance Latin Aristotle Commentaries," 355n29, 357n47; 372–3n153
Lokert, George, 352n3
loquentes, 110, 118
Luce, A. A., 411n382
Lucretius: on the separation of two surfaces, 86–7; 4, 97, 183, 185, 188, 190, 201, 210, 213, 282n23, 315nn114, 115, 316n116, 322n10, 376nn18, 24

McCormack, Thomas J., 418n427
McCracken, George E., 328n47
McDermott, Timothy, O.P., 352n138
McGuire, J. E.: on Newton and the whole in every part doctrine, 416n420; "Existence, Actuality and Necessity: Newton on Space and Time," 388n162, 408n349, 415n416, 416n420; "Newton on Place, Time, and God," 407n338, 409n368, 410n371
Mach, Ernst, 418n427
Machamer, Peter K., 267n1, 334n27
MacKenna, Stephen, 350n127
MacKinnon, F. I., 374n2, 398n233, 399n238–400n248 *passim*, 402n271
Maddison, R. E. W., 375n11
Mahdi, Muhsin, 323n23

Mahnke, Dietrich, 346–7n108
Maier, Anneliese: *An der Grenze von Scholastik und Naturwissenschaft*, 279n7, 282n21, 287n54, 287–8n56, 292n87, 293n105, 295 nn114, 115, 296n117, *Ausgehendes Mittelalter*, 286n51; *Metaphysische Hintergründe*, 341nn54, 55; *Die Vorläufer Galileis*, 307n51, 340n47, 341nn54, 56; *Zwei Grundprobleme*, 283n26, 286n51, 290n76, 291nn77, 78; 129
Maignan, Emanuel, 174–8, 370n131–372n151 *passim*; and Augustine's metaphor, 348 n116; on imaginary space, 175, 260; 153
Maimonides, Moses, 77, 308n65
Major, John, *see* John Major
Malebranche, Nicolas: on God and space, 398–9n235; on God's immensity and the whole in every part doctrine, 222; 231, 398 n234
Mandonnet, Pierre F., 323nn21, 22
Manuel, Frank, 409nn366, 371
Manuscripts: Basel Universitäts-bibliothek *F.V.12*, 303n15; Bibliothèque Nationale, fonds latin, *6752*, 280n12, 281n17, 294 n111, 297n123, 337n37; Bibliothèque Nationale, fonds latin, *16146*, 304n19; Bibliothèque Nationale, fonds latin, *16621*, 304n19; British Museum (Library), Sloane MS *2030*, 302n3; Cambridge, Gonville and Caius *512*, 303n15; Madrid, Escorial, Real Biblioteca, Manuscritos, *F.1.11*, 301n1, 310n77; Vatican, Biblioteca Vaticana, Latin MS *3013*, 291n77; Vatican, Biblioteca Vaticana, MS Reg. *747*, 302n1
Marbres, Johannes, *see* Johannes Canonicus
Marcus, R., 328n44
Marega, P. Isnardus M., O.P., 352n3
Marsilius of Inghen: and experiments against vacuum, 309n74, 311n85; on clepsydra, 84; on experience with vacuum, 296n119, 333n23; on internal resistances, 292n95; on measurement of distances in void, 122–3, 336n35; on mixed bodies, 51, 294nn110, 112; on motion in a vacuum, 37–8, 293 nn102, 104; rejects two types of vacua, 270n24; 47, 280n14, 298n129, 305n23, 307n52, 343n76
Mastrius, Bartholomeus, 25, 60, 278n4, 290 n72
matter: annihilation of, within world, 170, 247, 333n22; Campanella on, 197, 384 n106; prime, and extension, 272n40; prime, and vacuum, 272n37; *see also* God
Maurach, Gregor, 395n209
Maurer, A., 307n54
Maxwell, James Clerk, 418n427
measurement: in vacuum, 122–26; intensive

and extensive, 58; paradoxes of, 125–6
Menger, Karl, 418n427
Menut, Albert D., 330n3, 332n17, 413n397
Mersenne, Marin, 175
Michael Scot, 80–1, 302n8
Michel, Paul-Henri, 375nn11, 12, 15, 377 n39, 378n49, 379nn52–4, 56–8
Millward, Arthur E., 306n26
Minio-Paluello, L., 330n2
mixed body (bodies): and equal velocities in the void, 57–60; and the generation of strange results, 66; and the homogeneity of the vacuum, 55–7; and natural places, 53–5; imperfect, 50, 51, 292n87, 293n107, 294nn110, 112, 295n113; and internal resistance, 49–55; their natural motion in a vacuum, 49–55; velocity of, in plenum and vacuum, 54
mixta perfecta, 50
mixtum imperfectum, 50
Moerbeke, William of, 106, 108, 291n83, 302n8, 305n22, 310n77, 313n98
Molland, George, 333n21
Moncho, J. R., 276n67
Mongeus, Philip, 363n90
Montaigu, Collège de, 149, 352–3n3
Moody, Ernest A.: "Galileo and Avempace," 279n9, 282n21, 301n154; *Iohannis Buridani Quaestiones . . . De caelo et mundo*, 293n106, 295n115, 297n126, 334n27; "Ockham and Aegidius of Rome," 280n14, 281n18
More, Henry, 223–8, 399n236–402n265 *passim*; and Barrow, 237, 238, 405nn301, 304; and Berkeley, 411n382; and Descartes, 229; and Hobbes, 401n260, 408n344; and indivisibility of space, 234; and Leibniz, 252; and Locke, 238, 407n334; and Malebranche, 398–9n235; and Newton, 244, 245, 253, 254, 408n344; and the Nullibists, 223, 399–400n238; and Raphson, 403n286; and space as substance or accident, 227, 240; and substance-accident metaphysics, 227; the Cabbalists and space, 374n2; charted a new course, 261; direct influence of, 230–2; matter and distance, 401n260; on an extended God and extended space, 227–8, 262; opposed Spinoza, 402n271; rejection of whole in every part doctrine, 223–5, 415n419; space cannot be "disimagined," 226, 417n425; *Antidote Against Atheism*, 226, 400n248, 401n258, 259; *A Collection of Several Philosophical Writings of Dr More*, 400nn246, 248, 401n258; *Democritus Platonissans*, 399n236; *Enchiridion metaphysicum*, 227, 228, 399 n238–402n265 *passim*; *The Immortality of*

More, Henry (*cont.*)
the Soul, 400nn246, 249, 250; *Psychodia Platonica*, 399n236; 192, 207, 236, 394 n199, 398n234
Morry, M. F., 346n106
motion, 24–66; absolute, 413n397; and impenetrable void, 234; in void, *ex parte mobilis*, 35, 44–55, 57–60; inertial, in void, 43–4; infinitely great velocities in vacuum, 47–8; instantaneous, 7, 24–5, 36, 47, 76, 92, 93, 99, 277n1, 293n102; natural, 6–7, 40, 44–55, 57–60; rectilinear, and interstitial vacua, 75–6; rectilinear, of world, 84, 131, 136, 249; uniform or accelerated, in void, 266n18; violent, 7, 41–4
Motte, Andrew, 408n357, 415n418
Muller, Jean-Pierre, O.S.B., 303n12
Müller, Martin, 302n2
Murdoch, John E., 279nn7, 9, 289n70, 298 n128, 342n63, 344n90
Mutakallimun, 110, 118, 325n34

Napoli, Giovanni di, *Metafisica* (Campanella), 380n69–384n110 *passim*
natura abhorret vacuum, 67
natura communis, 81
natura universalis, 83
natural place, *see* place
Neckham, Alexander, 312n95, 346n108
negatio pura, 159
negation(s): Amicus's description of, 368 n123; pure, 13, 159, 165, 270n28; vacuum as, 10–11
Nemesius, bishop of Emesa, 21, 22, 108, 276n67
Newton, Isaac, 240–55, 407n335–417n425; and Aquinas, 264; and Barrow, 238; and Clarke, 248, 410n373, 415n414; and Descartes, 243–4; and Gassendi, 388–9 n163, 393n187; and geometrization of space, 233; and indivisibility of space, 234; and Leibniz, 247–52, 411n375; and More, 225, 244, 254, 261, 408n344; and scholastics, 263; and use of "attribute," 413 n394; Berkeley's attack on, 411n380; his absolute space attacked, 416–17n425; infinite extension intelligible, 243; on divisibility or indivisibility of space, 410n373; on relations of God and space, 243–4, 246, 247, 252, 253, 261, 262, 410n373; space neither substance nor accident, 242; spatial ideas of, in *Principia*, 244; *De gravitatione*, 233, 241, 244, 245, 405n300, 414n408, 415n418; *Opticks*, 245, 246, 247, 248, 409n370–410n373; *Philosophiae Naturalis Principia Mathematica* (*Mathematical Principles of Natural Philosophy*),

241, 244, 245, 246, 408n357–409n365, 411n375, 414n401, 414n408, 415n418; *Questiones quaedam philosophiae*, 407 nn338, 339; 110, 164, 207, 229, 231, 236, 379n58; *see also* General Scholium; Leibniz
Niceron, Jean-François, 175
Nicholas Alfanus, 276n67
Nicholas of Autrecourt: and al-Ghazali, 325 n33; defense of interstitial vacua, 74–7; nature abhors a vacuum, 84; on the clepsydra, 313n104; 100, 305n26, 306nn27, 31, 313n103
Nicholas of Cusa: and Koyré, 374n4; and metaphor of God as infinite sphere, 139–40, 347n108; universe an unlimited plenum, 348n114
Nidditch, Peter H., 277n7, 332n19, 374n1, 406n319–407n333
Nifo, Agostino, 325n33
Nock, A. D., 329nn54, 56
North, J. D., 336n35
nothing: as conceived by Guericke, 215–16; conceived as something, 3; defined by Kircher, 396n222; has same attributes as space (Berkeley), 412n382; two senses of, 12, 417n425 (Joseph Clarke)
Nullibists, 223, 225, 253

Ockham, William (of); on relationship of vacuum and nothing, 11–12; 269n19, 274 nn56, 57, 282n21, 324n27, 339n44
O'Donnell, J. Reginald, C.S.B., 305n26, 307 n55, 308nn57–60, 63, 64, 313n104
Olivi, Peter John, 282n21
O'Neill, Peter D., 323n23
Or Adonai (*The Light of the Lord*), 22
Oresme, Nicole: and Clarke, 249; and rectilinear motion of world, 343n76, 413n397; God and extracosmic void, 262, 349n123; God and other worlds, 345n101; God is wholly in every part, 350n127; on existence of extracosmic void, 119; on shape of world and the existence of vacua, 330 n3; 120, 140, 145, 147, 148, 162, 163, 164, 165, 169, 184, 323n20, 348n112
Ottaviano, Carmelo, 382n81
Oviedus (Franciscus de), 179, 373n158

Page, B. S., 350n127
Palingenius, Marcellus (Pietro Manzoli), 377–8n39
Pancheri, Lillian Unger, 394n198
Parmenides, 11
Parsons, Coleman O., 399n238, 400n248
particular natures, 69–70
parts: analogy between parts of atom and

space, 405n300; distinguishable, 235, 250–2, 254, 404n299; in an indivisible space, 234–5; in space, in the Leibniz-Clarke correspondence, 414–15n412

Pascal, Blaise: space and time are neither body nor substance, 392–3n185; 100, 140, 207, 320n151

Patrizi, Francesco, 199–206; and Aristotle's concept of space, 385n121; and Barrow, 237; and Campanella, 196, 380n69, 382 n87, 383n94; and Gassendi, 389n165, 391 n179; and geometrization of space, 233; and Guericke, 217, 220; and More, 227; and Newton, 242, 244, 245; and Philoponus, 261; and Raphson, 231–2; and Stoic universe, 199–206, 385n125, 385–6 n127; assumed possibility of artificial vacua, 97–8; distinguished three parts of universe, 387n142; ignored by Koyré, 388–9n163; on God and space, 200; space is neither substance nor accident, 204, 240, 388n152; *Nova de Universis Philosophia,* 384n118–388n160 *passim;* 19, 21, 192, 194, 198, 207, 209, 210, 213, 214, 218, 221, 226, 228, 231, 234, 255, 320n155, 398n234, 403n287

Pattin, Adriaan, 305n22

Paul, St., 407n329

Paul of Perugia, 129, 341n54

Paul Scriptor, 352n143

Paul of Venice, 307n56

Pelzer, A., 333n22

Pereyra, Martin: God and imaginary space, 366n110, 371n143; on imaginary time and imaginary space, 372n148; 356n39

Peter of Abano: and avoidance of vacuum, 68; and *Problemata,* 301n1, 304n19, 310 n77; on Avicenna, 304n19; 309n74, 314 n109

Peter of Auvergne, 267n4

Peter Damian, 324n23

Peter Lombard: and definition of place, 344 n83; and God's presence in things, 353 n8; and Jean de Ripa, 344n82; and whole in every part doctrine, 222, 350n127; God is in Himself, 132; importance of his *Sentences* for discussions of God and space, 115; places of spiritual and corporeal substances, 129–30; 21, 145, 150, 153, 324n23, 330n59

Philo of Byzantium: and burning candle experiment, 77–8; and the pneumatic tradition, 308n70; denied existence of vacua, 67–8; *Liber de ingeniis spiritualibus,* 310n78, 310–1n82, 312n98; 69, 79, 80, 302n3

Philo Judaeus (Philo of Alexandria), 112–13, 328n51

Philoponus, John: and Averroes' report on, 271–2n36; on Bruno, 186, 378n41; and Crescas, 22; and Fonseca, 158, 357n52; and Gassendi, 210; and internal space, 15, 273n44; and Patrizi, 201; and specific gravity, 301n154; denied extracosmic void, 276n64; distinguished material from immaterial dimensions, 229, 261; extension common to terrestrial and celestial bodies, 272n39; on motion in void, 25–6, 60–1, 280n13, 292n94; rejected interpretation of dimensions, 19–21; transmitted Stoic conception of universe, 375n8; void never empty of matter, 192; *De aeternitate mundi,* 15; *Commentary on Aristotle's Physics,* 275n61, 276n63, 299n138, 375n8; *De opificio mundi,* 272n39; 62, 65, 157, 166–7, 182, 183, 189, 201, 203, 220, 233, 300n149, 402n266

pia philosophia, 183

Pico della Mirandola, Giovanni Francesco: and Crescas, 332n20, 379n59; and Philoponus, 275–6n63; 19, 192

Pinborg, Jan, 323n21

Pines, Shlomo (Salomon), 273n48, 279n6, 308n65, 323n20, 331n9, 332n18

Pirotta, Angeli-M., 281n19

Pitigianus, Franciscus, 205, 357n53, 364 n93, 387n150, 392n180

place(s): Aristotle on, 5, 9, 111, 125, 142, 174, 376n19; as defined by Augustine and Peter Lombard, 344n83; differences between positive and imaginary, 132; identity of, 414–15n412; infinity of coincidental, 20–1; natural, in plenum, 297 n126; natural, in void, 51–3, 56–7, 64–5; of spiritual and corporeal substances, 129–30; paradoxes of, 339n43

Plato, 161, 182, 183, 201, 229, 231, 326n39, 331n10, 347n108, 385n121

Platonists, 268n8

plenum: as envisioned by Bruno, 188; compared to vacuum with respect to falling bodies, 45–6; condensation and rarefaction in, 95–6; continuity of, 69, 197; 86

Plotinus: and infinity as perfection, 350n125; and prime matter, 272n40; and whole in every part doctrine, 222, 350n127, 400 n239, 416n420; cited by Zabarella, 272 n41; 183, 201, 229

Plutarch, 182, 183, 201, 265n1, 375n8

pneuma, 112

Poggio, Gian Francesco, 4

Poinsot, John, *see* John of St. Thomas

Poliziano, Angelo, 303n10

Porta, Giovanni Battista della, 303n10
potentia Dei absoluta, 108
Poulet, Georges, 347n108
pre-creation void, *see* vacuum
prime matter, 72, 272nn37, 40
principle of plenitude, 184, 188
principle of sufficient reason, 413n396
privation: and pure negation, 159, 270n28; and vacuum, 10–11, 12–13; Aristotle's definition of, 267n24
Problemata, 303n10
Proclus, 182, 183, 305n22, 375n9, 387n142
Pseudo-Aegidius Romanus, 304n21, 315n112
Pseudo-Alexander of Aphrodisias: *Problemata* of, 301n1, 302–3n10, 310n77
Pseudo-Grosseteste: *Summa philosophiae* of, 325n35, 362n86, 332n16
Pseudo-Siger of Brabant: and Averroes, 275 n62; and imaginary places, 118–19; and resistance, 29, 288–9n66; and separation of surfaces, 316nn117, 119, 121; and vacuum, 29, 304n19; on the place of the world, 327n42; 19, 30, 87, 267n1, 280 n14, 282n20
Pythagoreans, 106, 141, 283n36

quantum, 15–16
Quidort, Johannes, *see* Johannes Quidort
Quinn, John M., 287n54

Rand, E.K., 328n48, 360n70
Randall, John Herman, Jr., 383n96
Raphson, Joseph: and Berkeley, 411n382; and Cyprian and Arnobius, 113; and geometrization of space, 233; and God as infinite sphere, 403n279; and Guericke, 396n218, 403n280; and impenetrability of space, 187–8, 235, 378n42, 396n218; and indivisibility of space, 234; God is an infinitely extended being, 232; influenced by More, 230–2; properties assigned to space, 231–2; *De spatio reali*, 403n277–404n291 *passim*; 192, 236, 252, 328n47, 380n62, 402n264
rarefaction and condensation, 33, 70
rarity, 33–4
Rasario, Johannes, 275n61, 276n64, 375n8
ratio of ratios, 58
Raven, J. E., 284n38
Razi, al- (Rhazes), 119, 323n20, 332n18
reed(s) (*canna*), 78, 80–2
repugnantia terminorum, 29, 280n14
Rescher, Nicholas, 413n395
resistance: absence of, in void space, 7, 159, 209; and *distantia terminorum* argument, 288n62; and successive, finite motions, 259; Bradwardine's concept of, 58–9; *ex parte medii*, 29, 38, 259; *ex parte mobilis*, 38–55, 58–60, 259; in material media, 25–7, 65; in vacuum, 28–9, 35, 38–40, 41–55, 289n66; internal, 46, 47, 49–55, 59, 60, 64; is property of body, 201; primary function of, 28; privative, 34, 35; produces finite motion, 41; "pure mathematical," 286n50; space as negation of, 365n107
rest, 76
Richard of Middleton: and God as infinite or intelligible sphere, 346n108, 347nn108, 112; and rectilinear motion of world, 343 n76; divorced God's immensity from void space, 145–6; no distances in vacua, 123–4; on God's extracosmic existence, 145–6; rejected extracosmic space, 146, 147; 145, 323n20, 324n27, 358n62, 363n90
Richter, Vladimir, 303n15
Ripa, Jean de, *see* Jean de Ripa
Risner, Frederick, 283n28
Robinet, André, 398n235, 412nn383, 384, 413n394
Rochot, Bernard, 387n150, 390n168
Rome, Beatrice K., 398n235
Ross, G. R. T., 274nn51–4, 335n33
Ruello, Francis, 339n42

Sabra, A. I., 283n27
sacred doctors, 394n200
Sacrobosco, John of, 80, 116, 330n3
Salvio, Alfonso de, 300n146
Sambursky, S., 272n39, 275n61, 327n43, 328n44, 374n5
Saunders, Jason Lewis, 376n21
Savage, John J., 308n67
Savile, Henry, 148, 344n90
Sceptics, 366n109
Schimank, H., 395n209
Schmidt, Wilhelm, 302n3, 320n155
Schmitt, Charles B.: "Campanella, Tommaso," 380n69, 389n165; "Experimental Evidence For and Against a Void," 305 n24, 311nn85, 90, 312n92, 320nn150–2, 380n63; *Gianfrancesco Pico della Mirandola*, 272n36, 276n63, 332n20; "Towards a Reassessment of Renaissance Aristotelianism," 356–7n46, 370n134
Schofield, Robert E., 409n370
scholastics: and Leibniz, 412n386; conceived void as body, 234; contribution to spatial doctrines, 262–3; influenced Gassendi, 213–14, 221; influenced Guericke, 397n232; Raphson on, 232; views ignored, 182; 231, 262, 374nn1, 3, 4, 387n150
Schott, Gaspar, 395n203
Scotists, 153
Scott, T. K., Jr., 269n21

Scott, Walter, 329n56, 359n69
secundum imaginationem, 390n169
Sentences (of Peter Lombard), 21, 115
Sextus Empiricus: and Stoic distinction, 376
n21; on relations of body and space, 276
n67, 366n109, 375n9; 201, 375n7
Shapiro, Herman: and interpretation of
Burley, 285–6n50; Renaissance Philosophy,
381nn72, 77, 80, 382n82, 383nn97–102,
384nn103–6, 114
Siger of Brabant, 267n4; see Pseudo-Siger
of Brabant
Simplicius: and the clepsydra, 313n98; and
his Physics Commentary, 275n61, 322n11;
and impetus, 291n83; and Philoponus,
276n67; and prime matter, 272n40; preserved Archytas' argument, 106–7; 60,
108, 157, 182, 183, 201, 213, 231, 310
n77, 322n10, 375n9
Singer, Dorothea Waley, 348n115, 375n13–
376n25 passim
siphon(s), 80–2, 96, 98, 310–11n82
Smith, T. V., 402n269
Sommervogel, Carlos, S.J., 357n47, 358n58,
362n87
Soto, Domingo, 179, 311n90
space: absolute motion in, 249, 413n397; and
body, 167, 203, 204, 386–7n139, 387
n142; and ether, 188–90; and geometrization of, 206, 232–4, 242–3, 349n124; and
matter, 188–90, 197–8, 203, 384n106;
and mind, 196, 383n96, 388n158; and
nothing, 215–16, 362n86, 396n222, 412
n382; and time, 177, 179, 372n148, 392
n182, 407n329; as attribute or property
of a subject, 227, 248, 413nn391, 394, 396;
as dimensional entity, 133–4, 221, 227,
242; as intermediate entity, 195, 206; as
pure potentiality, 236–8; can receive
bodies, 159, 162, 167, 168, 171, 173, 201;
coexists with body, 168, 169, 209, 375n9;
different kinds of, distinguished, 201, 217,
219–20, 238; divinization of, 142–3, 234;
does not coexist with body, 33, 406n309;
does not need a space, 8, 158; external,
17–23; filled with light, 203; finitude of,
195, 198, 381–2n81; first systematic description of, 200; has sense and feeling,
196–7; imaginary (analogy between, and
time), 179, 372n148, (and God and negations), 171–2, (and God's immensity),
154, 161, 163–5, 249, 367n116, 396n223,
406–7n329, (and real space), 175, 216–
17, 226, (and vacuum), 169–71, 173,
366n11, 395–6n211, (as God-created,
independent, extrascosmic void), 121–
34, (as God-filled, dependent, extracos-

mic void), 112, 135–47, 248–55, (as
negation), 159, 164–5, 167–9, 171, 172,
178, 180, 214, 363n90, (attack on, by
Bona Spes), 179, (coexists with body),
168, (equated with nothing), 216, 395
n210, (has no positive nature), 137, 158,
162, 166–7, 168, 172, 214, (meanings
of), 117–21, (non-dimensionality of), 173,
397n226, (ontological status of), 260,
(penetrates world), 173–4, (properties
of), 132, (three major opinions on), 363
n90, (mentioned), 112, 113, 117, 223,
248, 255, 263, 387n142; impenetrability
of, 187, 234, 378n42, 403n287; in eighteenth century, 263, 416–17n425; indivisibility of, 228, 234, 410n373; infinity of,
127–31, 156, 164, 239, 243, 245, 385n122,
397–8n233, 391n174; internal, 14–17,
122, 156, 272–3n41; lacks dimensions,
123, 143, 173, 220, 236; neither substance
nor accident, 158, 166, 187, 193–4, 199,
204–5, 206, 209–10, 211, 237, 240, 242,
243–4, 261, 357–8n53, 392n182; not a
quantity, 158, 162, 167, 174–5; objections
to acceptance of, 18–19; on imagining it
away, 226, 417n425; parts of, 235, 250–
2, 404n299, 414–15n412; penetrability of,
235, 403n287; penetrates body, 201;
properties of, 186–7, 190–1, 193, 195–6,
201, 202, 205–6, 209, 217–18, 228, 231–
2, 235, 237, 242, 243, 244, 365n108, 414–
15n412; real, 175, 216–17, 226, 365n104,
370n130; and resistance of, 159, 168, 173,
199; separate, (always occupied by body),
19–21, (superfluousness of), 6, 18, 275
n60; uncreated and independent, 155,
171, 190, 191, 223, 237, 253, 261, 365
n108, 397n225; yielding of, 202, 204, 217–
18, 396n218; see also God; resistance;
vacuum
Spampanato, Vincenzo, 379n60
spatium imaginarium, 117
species, perfection of, 342n63
specific weight(s): concept of, used by
Galileo, 63–4, 300n148; 58, 65, 279n5
sphere(s): infinite, as metaphor of God, 129,
138–40, 161; 48
Spiers, I. H. B. and A. G. H., 320n151
Spinoza, Benedict: God extended and material, 229; read Crescas, 379n59; Ethics,
402n270, 403n284; 164, 231, 234, 344
n91, 394n199, 398n234, 402n267, 402
n269, 411n382, 413n394
spirit(s), 143–4, 225; see also "whole in
every part doctrine"
Spirit of Nature, 225
spiritus mundi, 387n143

Steele, Robert, 270n13, 278n2, 303n11
Stewart, H. F., 328n48, 360n70
Stocks, J. L., 321n2
Stoic(s), 106–8; and Aristotle's definition of vacuum, 107, 271n33; and doctrine of re-creation, 385n127; and void beyond world, 107, 385n125, 389n168; argument for extracosmic infinite void, 243, 263, 375n18, 382n82, 395n210 (Guericke), 406n326 (Locke); conception of universe, 185 (rejected by Bruno), 192, 201, 203 (altered by Patrizi), 207 (Gassendi), 236 (Barrow), 231 (Raphson), 239; (Locke), 241 (Newton), distinguished world from universe, 376n21; Simplicius on, 322n12; 106, 108, 111, 112, 119, 141, 182, 204, 207, 210, 216, 231, 234, 261
Stone, Edmund, 406n307
Strato of Lampsacus, 275n61
Strauss, Leo, 308n65
Suarez, Francisco, 153–7, 355n26–356n45 passim; and Amicus, 364n94, 365n104; and Boethius's definition of eternity, 360 n70; and Faber, 369n125; and God as infinite or intelligible sphere, 347n108; and God's immensity, 361n78; and internal space, 15, 156; assumed whole in every part doctrine, 356n33; Newton aligned with, 254; on God's "extension," 154–5; on infinity of space, 156–7; 110, 120, 122, 163, 164, 169, 179, 349n124, 363n90
substance(s): and accident metaphysics applied to space, 227; and intensive degrees of perfection, 130–1, 132; God the only, 229; no clear idea of, 240; places of corporeal and spiritual, 129–30; space as, 158; space is not a, 166, 204, 209, 248
succession, definition of, 277–8n1
Sunczel, Friedrich, 80, 81, 305n23
surfaces, separation of, 86–95, 315n114–320n148; Galileo's discussion of, 98–9; part after part, 87, 88, 89; 83, 96
Sweeney, Leo J., S.J., 348n116, 350nn125, 126
Sylla, Edith, 298n128

Tannery, Paul, 273n50, 399n235
Telesio, Bernardino, 192–4, 379n60–380n68; and Guericke, 220; and impenetrability of space, 380n62; and Philoponus, 192–3; artificial vacuum possible, 97, 98, 380n63; influenced Campanella, 381n74; 19, 198, 199, 205, 207, 389n165
Tempier, Etienne (bishop Stephen of Paris),

108, 111, 136, 137, 326n39, 345n98
Tertullian, 368n125
Themistius, 78, 79, 80, 302n10, 309n74
Themon Judaeus, 315n112
Theodore of Gaza, 303n10
theologians, 341n52, 366n110
Thomas Anglicus, 286n51
Thomas of Strasbourg, 363n90
Thomas de Vio, Cardinal Cajetan (Caietanus), 148–9
Thomas, Félix, 394n199
Thomson, S. Harrison, 303n15
Thorndike, Lynn, 310n80, 337n37
Tigerstedt, E. N., 230
time: analogy with space, 177, 179, 372n148, 392n182; God's eternity and infinite extent, 239; neither substance nor accident, 209, 210; space and duration linked, 407n329
Toletus, Franciscus: and bellows, 312nn92, 93; and internal space, 15, 273nn42–4; and water frozen in sealed vessel, 311n85; 16, 120, 122, 156, 275n58
Tractatus de inani et vacuo, 81
Trincavelli, Victor, 275n61
Turnbull, Robert G., 267n1, 334n27
Turnèbe, Adrian, 98
Turner, John, 230, 402n275
Tuve, Rosemond, 378n39
Tuveson, Ernest, 399n236, 402n274, 402–3n276, 411n382

ubi circumscriptivum, 130, 342n66, 343n67
ubi definitivum, 126, 130, 132, 177, 342 n66, 343n67
ubi repletive, 368n125
Uncreated Something, 215, 219, 397n226
universal nature (or celestial agent), 69–70: different interpretations of, 304–5n21; invoked by Burley, 85–6, 94; status of, 85–6; 83, 95, 97

vacuum (void): and ether, 188–90; and mathematical dimensions, 10, 268n10; and nothing, 11–12, 215–16, 396n222; and prime matter, 268n8, 272n37; and privation, 10, 12–13, 159, 268–9n13, 372n151; and relationship to body, 31, 34, 193, 370n129, 379n51; and spirit, 370n129; and spiritus mundi, 387n143; and yielding to body, 31–5, 188, 202, 217–18; animals in, 41, 290nn72, 73, 299n136; artificial, 97–100, 197, 383n101; as containing surface, 12–13, 270n24; as imaginary space, 13; definitions of, 5, 105–6, 142, 169,

174, 268n8, 370n129; different types of, distinguished, 10, 14, 32–3, 271–2n36, 306 n33; distinguished from imaginary space, 169–71, 173, 175, 366n111; equality of velocities in, 7, 57–60, 62–3, 293n102; experiments against existence of, 77–95; homogeneity of, 6; impenetrability of, 31, 32–3; impressed forces in, 42–4, 290 n77–291n78, 291n81; interstitial, 74–7, 100, 201, 203, 208, 306n33, 320–1n159, 370n132; lack of experience with, 54, 333 n23; motion in, 24–66, (impossible), 6–7, (instantaneous), 7, 24–38, 45–9, (uniform or accelerated), 266n18; nature's abhorrence of a, 67–100, 260; non-homogeneity of, 55–7; of air, 55, 64; of fire, 55, 64; order of questions in an analysis of, 269n17; pre-creation, 110–12, 137, 141, 325n35–327n40; what it is, 9–14; see also, God; resistance; space

vacuum aeris, 55
vacuum diffusum, 71
vacuum ignis, 55
vacuum imbibitum, 71
vacuum immixtum, 71
vacuum infusum, 71
vacuum interceptum, 71, 74
vacuum negativum, 332n14
vacuum privativum, 331n14
vacuum separatum: nature's abhorrence of, 77–100; 74, 321n159
Valla, Giorgio, 303n10
Van den Bergh, Simon, 322n12, 325n33
Van Helmont (Johannes Baptista), 231
Van Steenberghen, Fernand, 323n22
Vasquez, Gabriel: and concept of pure negation, 13, 159; and God as infinite sphere, 347n112; correctly interpreted Augustine, 328–9n53; denied existence to extracosmic imaginary space, 358n62; made unjust accusation, 374n165; on dictum that "God is in Himself," 363n90, 366 n110, 368–9n125; on privation, 13, 270 n27; on rectilinear motion of world, 361n73; 148, 179, 334n29, 363n90
velocity (velocities): and specific weights, 300n148; equality of, in void, 57–60, 62–3, 64, 66; finite or infinite, in void, 266 n18; in void, 37–8; infinitely great, in void, 47–8; uniform or accelerated, in void, 266n18
Verbeke, G., 276n67
Victorinus, Marius, 161
Vignaux, Paul, 339n42, 343n69
Villiers, Cosmas de, 372n153

virtus imaginativa, 118
virtus regitiva universi, 304n17
Vitelli, H., 275n61, 276n64
void, *see* vacuum
Vollert, Cyril, 324n31, 342n59

Waard, C. de, 282n19, 304n19, 310n79, 315 n114, 387n143
Walsh, James J., 323n21
water: actions of, to prevent vacuum, 81–2; frozen in sealed vessel, 97
Watts, Isaac, 416–17n425
Watts, V. E., 360n70
Weinberg, Julius R., 307n53
Weisheipl, James A., O.P., 282n19, 285 n49, 288n63, 289n70
Westfall, Richard, 388n162, 407n338, 407–8 n339, 409n369, 411nn376–9
Westman, Robert, 324n30
Weyl, Hermann, 273n46
Whewell, William, 404n297, 405n302–406 n313
Whiston, William, 379n58
Whitaker, G. H., 328n44
Whitmore, P. J. S., 370nn131, 133, 135
whole in every part doctrine: Bruno's version of, 191; equivalent to *ubi definitivum*, 177; Gassendi and, 210, 393n188; Leibniz rejected it, 415–16n419; Newton and, 416n420; origin of, 350n127; repudiated by More, 223–5; 143, 212–13, 222, 228, 356n33
Wicksteed, P. H., 323n18
Wiesen, David S., 354n17
William of Auvergne, 284n38, 301n1
William of Auxerre (Guillelmus Altissiodorensis), 161, 351n136, 363n90
William of Ware, 271n33, 282n21
Wilson, Curtis, 317n130
Wippel, John F., 271n34, 323n22, 324n23
Wolff, Michael, 272n39
Wolfson, Harry A.: on Crescas's explanation of divisibility of infinite space, 404n298; *Crescas' Critique of Aristotle*, 272n40, 277n71, 321n5, 328n45, 332n20, 379n59, 404n298; *Philosophy of the Kalam*, 271 n35, 325n34; *Philosophy of Spinoza*, 277 n71, 332n20, 402nn267–9
Wolter, Allan B., O.F.M., 271n34, 324n23, 339n44
world(s): distinguished from universe, 376 n21; eternity of, 110, 111, 137; five kinds of, distinguished by Campanella, 194; has feelings, 383n99; imaginary annihilation of, 390–1n169; infinity of, denied,

world (*cont.*)
402n273; moved rectilinearly by God, 131, 136, 137, 249, 343–4n76, 361n73; on material existence prior to creation of, 405–6n307; place of, 111; seeks to preserve its continuity, 69–70, 197; unity of, depends on space, 199; 383n98; *see also* annihilation of matter; God
Worthen, Thomas D., 313n98
Wright, Sir Thomas, 312n95
Wulf, M. de, 333n22

Yates, Frances, 329nn54, 55, 108
yielding, 201, 217–18, 396n218

Zabarella, Giacomo, 272n40, 272–3n41
Zedler, Beatrice, 325n33
Zeno of Elea, 8, 96, 377n37
Zeno the Stoic, 376n21
Ziegler, H., 323n13, 388n153
Zimara, Marc Antonio, 307n51
Zimmermann, Albert, 267n4, 304n15
Zycha, Joseph, 328n51